CRITICAL VALUES OF STUDENT'S t DISTRIBUTION

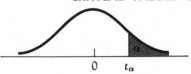

$0 \quad t_\alpha$

The entries in this table are the critical values for Student's t for an area of α in the right-hand tail. Critical values for the left-hand tail are found by symmetry.

Degrees of freedom	Amount of α in one-tail				
	.1	.05	.025	.01	.005
1	3.078	6.314	12.706	31.821	63.657
2	1.886	2.920	4.303	6.965	9.925
3	1.638	2.353	3.182	4.541	5.841
4	1.533	2.132	2.776	3.747	4.604
5	1.476	2.015	2.571	3.365	4.032
6	1.440	1.943	2.447	3.143	3.707
7	1.415	1.895	2.365	2.998	3.499
8	1.397	1.860	2.306	2.896	3.355
9	1.383	1.833	2.262	2.821	3.250
10	1.372	1.812	2.228	2.764	3.169
11	1.363	1.796	2.201	2.718	3.106
12	1.356	1.782	2.179	2.681	3.055
13	1.350	1.771	2.160	2.650	3.012
14	1.345	1.761	2.145	2.624	2.977
15	1.341	1.753	2.131	2.602	2.947
16	1.337	1.746	2.120	2.583	2.921
17	1.333	1.740	2.110	2.567	2.898
18	1.330	1.734	2.101	2.552	2.878
19	1.328	1.729	2.093	2.539	2.861
20	1.325	1.725	2.086	2.528	2.845
21	1.323	1.721	2.080	2.518	2.831
22	1.321	1.717	2.074	2.508	2.819
23	1.319	1.714	2.069	2.500	2.807
24	1.318	1.711	2.064	2.492	2.797
25	1.316	1.708	2.060	2.485	2.787
26	1.315	1.706	2.056	2.479	2.779
27	1.314	1.703	2.052	2.473	2.771
28	1.313	1.701	2.048	2.467	2.763
29	1.311	1.699	2.045	2.462	2.756
30	1.310	1.697	2.042	2.457	2.750
40	1.303	1.684	2.021	2.423	2.704
60	1.296	1.671	2.000	2.390	2.660
120	1.290	1.661	1.984	2.358	2.626
∞	1.282	1.645	1.960	2.326	2.576

Abridged from Table III of Fisher and Yates, *Statistical Tables for Biological, Agricultural, and Medical Research*, published by Longman Group Ltd., London (previously published by Oliver and Boyd Ltd., Edinburgh), and by permission of the authors and publishers.

INTRODUCTORY STATISTICS

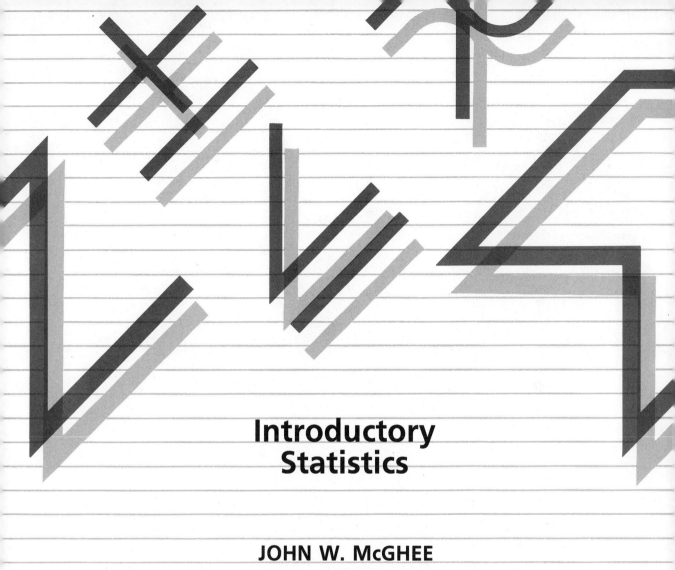

Introductory Statistics

JOHN W. McGHEE

Professor of Mathematics
California State University
Northridge

West Publishing Company
St. Paul   New York   Los Angeles   San Francisco

Copyediting and Technical Art Coordination: Editing, Design & Production, Inc.
Cover and Chapter-Opening Art: Stephen Scott Green
Composition: The Clarinda Company
Indexer: Linda L. Thompson

COPYRIGHT © 1985 By WEST PUBLISHING CO.
                       50 West Kellogg Boulevard
                       P. O. Box 43526
                       St. Paul, Minnesota 55164

Library of Congress Cataloging in Publication Data
McGhee, John.
    Introductory statistics.
    Includes index.
    1. Statistics.   I. Title.
QA276.M384   1985   519.5   84–17255
ISBN 0–314–85277–8

Contents

Preface

Introductory Statistics is intended for the one semester or one year foundations course in elementary statistics. As prequisites it assumes a year of high school algebra and reasonable arithmetic skills.

The development stresses concept understanding and the relationship of these concepts to statistical procedures. Explanations are made to the fullest degree consistent with the anticipated background of the student. All definitions, theorems, and procedures are set out in boxes for ease of reference. The reader should not have to look between the lines to find a definition or an important result.

All important concepts, theorems, and definitions are illustrated with detailed examples. Several exercise sets are located throughout each chapter to coincide with units of instruction. Each set is divided into two parts, A and B; the first part contains basic mastery problems, the second contains more advanced problems, some of which use data from published research. At the end of each chapter is a vocabulary and technique mastery checklist and a chapter test.

To enhance understanding of the use and application of statistics, **Research Studies** and **Case Studies** are used to identify real life applications. **Computer Printouts** are used to show how the analysis of data is performed. The **Library Projects** provide experience with locating and reading published statistical research.

The research and case studies located throughout the chapters provide insight into the breadth of application of statistics, and illustrate the use of statistical procedures as well as the reporting of results.

The computer printouts located at the end of each chapter illustrate how procedures are carried out with the aid of a computer. These were produced using the Minitab Statistical Package and include the commands and input data, in addition to the resulting analysis. These printouts are easily interpreted since Minitab commands and labels are practically self explanatory.

The library projects, which begin in chapter four with assignments on the use of abstracts and continue with assignments on research reports, develop skills for future course work and research. These reports span nearly all disciplines and involve statistical procedures from the chapters in which they are located.

The topics outline follows traditional lines. Occasionally there are slight departures. Random samples are discussed in the first chapter since they play such an important role in subsequent discussions. Density histograms are used to relate areas under curves to probabilities. p values are emphasized throughout all hypothesis testing. The chi square distribution is introduced early so that students in a one semester program will have an opportunity to become familiar with this useful distribution. Linear correlation and regression are extended to the multilinear case for those who wish to pursue these topics. Likewise the analysis of variance includes two way ANOVA based on the factorial design with multiple cell entries. Finally the chapter on non parametric statistics contains enough variety to provide substantially more than an introduction to the subject.

A **student study guide** is available for use with this text. It contains chapter outlines with over 200 additional examples, detailed solutions to the odd numbered exercises, solutions to all exercises of the chapter tests, and calculator usage illustrations.

This textbook evolved out of notes and material developed over 15 years of teaching statistic courses. In assembling it I have tried to identify all possible sources and have obtained permissions to reprint and adapt material where I felt it to be necessary. Over 120 permissions requests were honored by the authors and publishers. I wish to express my gratitude to these individuals and companies for their cooperation. The **Bibliography** details these sources along with many texts, from which I have taught or used as resources over the years. I have included these texts since inevitably they have shaped my attitude on the development of topics used in this book.

In expressing my gratitude, I would be remiss not to single out Thomas F. Ryan, Brian L. Joiner, and Barbara P. Ryan of the Statistics Department of Pennsylvania State University. They generously consented to the use of the Minitab printouts found in the text. This software system is available to educational institutions for a nominal charge and is in the libraries of many university and college computer systems. I have used it successfully with my statistics course for several years and have introduced several collegues to it. It should be noted that Professors Ryan, Joiner, and Ryan are the authors of the *Minitab Student Handbook* which is available from Duxbury Press, Boston, Ma.

During development, the manuscript of this textbook underwent two major reviews. Many of the reviewers comments and suggestions have been included in the text and helped shape its final form. Special thanks go to these reviewers: C. P. Barton, Steven F. Austin State University, Texas; Phillip M. Beckman, Black Hawk College, Illinois; Ben B. Bockstege, Broward Community College, Florida; Col. William S. Carter, U.S.A.F. (Ret.), Bob Goosey, Eastern Michigan State University; Dan Kemp, South Dakota State University; Eric Lubot, Bergen Community College, New Jersey; Stanley M. Lukawecki, Clemson University, South Carolina; David R. Lund, University of Wisconsin, Eau Claire; Maurice L. Monahan, South Dakota State University; Howard Reiter, University of North Carolina; L. Thomas Shiflett, Southwest Missouri State University; Bill Stines, North Carolina State University.

The typing of the manuscript as well as the student and instructors guides was done by Laurie Beerman of the University of California at Los Angeles. She successfully mastered my handwriting, helped me meet my deadlines, and produced beautiful manuscripts. Many thanks.

Finally my appreciation to my wife Sue and daughters Anne and Lynne for their support and encouragement.

INTRODUCTORY STATISTICS

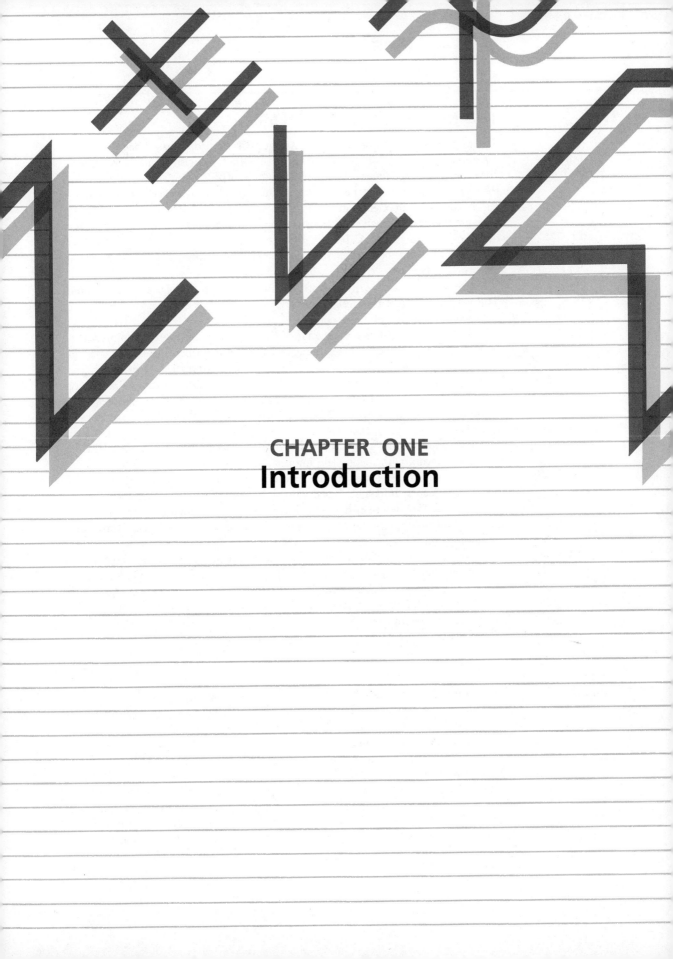

CHAPTER ONE
Introduction

1.1   THE NATURE OF STATISTICS

Statistics is the branch of mathematics that deals with data-based decision making. Generally there is some population about which we wish to make a decision; uncertainty arises when we have only a sample of this population. In addition, the data we gather may be "noisy" because of chance errors that inevitably occur when measurements are made or when there are variables for which we have not controlled. Such conditions are commonplace for anyone who must make decisions and predictions based on data. Statistics provides the researcher with the tools for both designing experiments and interpreting the resulting data.

For our purposes, we divide statistics into two areas: *descriptive* and *inferential*. The field also covers the design and implementation of experiments, but these are beyond the scope of a first course.

Descriptive statistics are summaries that describe the more prominent features of a mass of data. These summaries include descriptive measures such as means and percentiles, as well as tables, charts, and graphs, all of which the reader has no doubt encountered. Statistical inferences are the generalizations or conclusions we can form from the data under study. Most statistical problems are inferential in nature, and it is with the inferential role that modern statistics is identified.

Probability is important in the inferential process. Inferences are based on samples: we work with incomplete information, possessing only a partial picture. Our view is further clouded by the presence of chance variations and errors in our measurements. Inferences based on such information will be probable at most. Probability theory describes how samples are related to a population and provides the basis for assigning a numerical degree of certainty or confidence to any conclusions we form.

Statistical inference is the main topic of this book. In particular, we develop the theory and techniques for treating three principal classes of inference problems: (1) estimation, (2) hypothesis testing, and (3) correlation and regression. Rather than attempting at this time to describe these terms in detail, we will illustrate each by an example.

Example 1.1   *Estimation.* In planning a budget for a major city, the budget director needs an estimate of the change in the assessed value of all taxable property. Sometimes such an estimate is needed long before the survey of all taxable property is available from the assessor's office. To estimate this change, the assessor will select a sample of parcels from those having undergone resale during the year and determine their new values. Each area of the city and each class of property (residential, commercial, etc.) is sampled. By comparing these new values with the previous ones, the assessor is able to predict the change in assessed valuation for each area and property classification and ultimately for the whole city. Since a poor estimate of valuation may lead to wrong decisions, the budget director will also need a measure of the confidence that can be placed in the estimate. ■

Example 1.2

Hypothesis Testing. Consumer protection agencies must all deal with false representation. Suppose an agency has received several complaints leading it to doubt the 2000-hour average service life claimed for a certain brand of calculator battery. The agency cannot immediately contest the manufacturer's claim in court; such unsubstantiated accusations can prove quite costly. What is needed is a controlled test of the hypothesis that the average battery life is less than 2000 hours. Suppose now that a sample of 400 batteries is selected and tested. If the sample shows, say, an average life of 1340 hours, then even the manufacturer should agree that the agency has an indisputable case. Suppose, however, the average is 1945 hours. Can the agency then convince the manufacturer or the court that this is grounds for rejecting the manufacturer's claim? After all, only 400 batteries have been examined. With what certainty can the 2000-hour average service life be rejected? ■

Example 1.3

Correlation and Regression. Medical schools use information such as scores on the Medical College Admissions Test (MCAT), grade point averages, and letters of recommendation to rank their applicants. They also rank the members of their graduating classes. Is the initial ranking worthwhile? Possibly so, if it is an indicator of the quality of the corresponding graduates. Thus a question of interest is whether there is a correlation or dependence between these two rankings. There are two problems: (1) How high is the correlation between the two rankings, if indeed there is one? (2) Is there a formula that relates these two variables, and if so, how successful will it be when applied to an entering class of, say, 200 students? Regression analysis is concerned with the second problem, namely establishing formulas showing the relationship between variables. ■

As these examples indicate, the essence of the statistical process is to make inferences from a sample about a variable or a population. When we actually perform such an analysis, we find descriptive techniques to be very valuable. Descriptive statistics is an important topic and will not be neglected.

Statistical studies fall into two classes: *observational* and *controlled* studies. In a controlled study some "treatment" is generally being investigated—a new medication, a new tobacco mix, or a new advertising campaign. In any event, the experiment is controlled in the sense that we select the individuals who will receive the treatment and try to isolate or eliminate the effects of extraneous variables. Individuals may be selected who do not receive the treatment and are used for comparative purposes. These are called *controls*.

Example 1.4

A Research Study

Does Alcohol Facilitate Learning?

Researchers have repeatedly demonstrated that alcohol administered before a learning experience impairs memory retention. The possibility that using alcohol after training might actually improve retention was investigated by Dr. R. L. Alkana of the University of Southern California's School of Pharmacy and Dr. Elizabeth Parker of the National Institute on Alcohol and Alcoholism. Various factors, such as the amount of alcohol, the time it was given, and the measurement of the ability to recall, all dicatated a controlled study. The learning experience involved using a mild electric shock to teach mice to avoid a darkened compartment of a trough-shaped box. Immediately after the experience the mice were injected with either alcohol or a harmless saline solution. Since the pain of the injection could itself be a stimulus, a second sample of mice were also given the injections but did not receive the electric shock while in the darkened compartment. Thus there were two sets of controls: those of the first sample that did not receive the alcohol and those of the second that did not receive the shock treatment. When tested after one week, the memory retention of the mice that had received the alcohol was more than double that of their counterparts that had received the saline injections. As might be expected, those receiving only moderate amounts of alcohol outperformed those that were inebriated. Further, the injection process was eliminated as a factor when it was found that the alcohol did not enhance the retention of the mice in the second sample that did not receive the shock treatment. Thus the improved performance was directly related to the alcohol. Source: *Journal of Psychopharmacology*, Vol. 66, 1979. ■

Observation studies address existing conditions in a population. We may be interested in the percentage of the population who favor a political candidate or the distribution of yearly gross incomes. Or it may be that there is a treatment, such as exposure to radiation or to a pesticide, to which we cannot subject individuals but which we can study because certain individuals have undergone it because of their life style, occupation, ignorance, or whatever.

Example 1.5

A Research Study

What Factors Are the Greatest Predictors of Mortality?

In 1962, the American Cancer Society concluded a study of more than one million people that sought to determine what symptoms and habits were strong predictors of death. Obviously this was an observational study, since we can not ask people to adopt potentially life-threatening habits so their effect on the mortality rate could be studied. Again, there were many variables that needed to be controlled: age, sex, occupation, and the like. One of the many conclusions that resulted when the subjects were grouped according to age was that the amount of time an older person sleeps is a crucial factor, along with such factors as diabetes, cigarette smoking, and hypertension. Those who slept fewer than 4 hours or more than 10 hours a day had a much higher likelihood of dying, almost twice as great as older people with normal sleep habits. Recently Dr. William Dement of Stanford University explained this finding by relating both severe insomnia and excessive sleeping to a disease, sleep apnea, in which the individual momentarily stops breathing, perhaps as many as 680 times a night and experiences nighttime hypertension. The accompanying unusual rise in blood pressure is apparently the life-threatening factor first suggested by the American Cancer Society study. ■

In the remainder of this chapter we develop in detail some of the more important terms that have been introduced in this discussion.

1.2   VARIABLES, POPULATIONS, AND SAMPLES

For the moment let us think of a population as a collection of objects, such as people, families, automobiles, or whatever. A sample is a subcollection or part of the population. Populations are studied because they have some property or characteristic that varies among different members of the population. Such a characteristic is thus a *variable*. Examples of variables include the time effectiveness of a medication on adults, the incomes of families, and the yearly maintenance costs of automobiles. A variable identifies a property of interest and is the basis on which values are associated with members of the population. These values, of course, simply describe the degree to which the property is possessed. For our purposes, a variable is defined as follows.

DEFINITION 1.1.

A *variable* is a rule that associates a numerical value with each member of a set (the population). The set of values that are assigned is called the *range* of the variable.

When we speak of the *observations* or *data* of a statistical investigation we are simply referring to the values that have been observed for a variable.

Since we often refer to variables and their values in general discussions, some symbols are needed. Commonly, a capital letter, say X, is used as the name of a variable; its lower-case counterpart, x, is then used to indicate a value of the variable. If we wish to enumerate several values of X, as in recording observations, then subscripts are added to x. Thus x_1 denotes the first value of X, x_2 the second, x_{75} the 75th, and so forth.

Example 1.6.

A statistics course has five sections. Let X associate the length of the waiting list with a section and let x_i be the length of the list for the ith section. Thus if $x_1 = 8$, $x_2 = 5$, $x_3 = 12$, $x_4 = 12$, and $x_5 = 0$, then section 1 has a waiting list of eight students, section 2 has five, sections 3 and 4 both have twelve, and there are none on the list of section 5. ■

We commonly speak of a population as a set of objects in which we wish to determine the behavior of some variable. In this sense the population and the variable are inseparable. A population may, however, be associated with several variables, and this can produce some difficulty in mathematical exposition. We avoid this problem by defining populations in terms of the values of the different variables.

DEFINITION 1.2.

A *population* is the collection of all values of the variable under study.

The same convention extends to samples.

> **DEFINITION 1.3.**
> A *sample* is any subcollection of a population.

Thus in the mathematical sense, populations and samples are composed of numbers. This approach offers a great deal of flexibility in making statements. For example, rather than talking about the average of the measurements made on a sample, we can simply refer to the average of the sample. In practice, there is no confusion between the mathematical and common usages of these terms; they complement one another. If we should need to refer to a particular individual in a study, the term *experimental subject* is used.

1.3   DISCRETE AND CONTINUOUS VARIABLES

In subsequent developments we encounter two important classes of variables: *discrete* and *continuous* variables. The distinction is mainly whether observed values of the variable are exact, as in counting, or only approximations, as in measuring. As examples of discrete variables we have the number of children in families, the number of shares traded each day on the New York Stock Exchange, the number of vacant units in the different buildings of an apartment complex, and the daily attendance at a theater. Each of these variables has a range that can be listed or enumerated.

> **DEFINITION 1.4.**
> A variable X is *discrete* if its range can be enumerated as a sequence x_1, x_2, x_3, . . . (finite or infinite).

When plotted on the number line, the range of a discrete variable appears as a set of isolated points. Such is not the case with the weights of adults, the diameters of bolts, and the running times of computer programs. Each has a range that is a continuous interval.

> **DEFINITION 1.5.**
> A variable X is *continuous* if all numbers intermediate to any two numbers in its range are also in the range.

We should emphasize that the *possible* range of values is important with a continuous variable, irrespective of experimental limitations that may exist on measuring these values. Even though tread mileages of automobile tires are measured only to the nearest 100 miles or computer runs are measured only to the nearest thousandth of a second, these variables are still continuous. These are limitations imposed by our ability to measure accurately.

Whether a variable is discrete or continuous is important because different models are required for predicting the occurrence of its values. For a discrete variable, such as the number of registered voters in a precinct, we can state exactly how many precincts had 200 voters, how many had 201, and so forth. Contrast this with, say, the number of miles per gallon for different automobiles. An observation of 16.3 miles per gallon indicates only that the automobiles tested produced between 16.25 and 16.35 miles per gallon. Thus we cannot say how many automobiles produced exactly 16.3 miles per gallon, or any other number for that matter. For a continuous variable we must confine our attention to predicting the occurrence of values on an interval.

While both discrete and continuous variables are important, we shall see that ultimately continuous variables have a greater role in inferential statistics.

EXERCISE SET 1.1

Part A

1. Are the following variables discrete or continuous? Explain.
   a. The number of correct answers on a true-false test.
   b. The duration of the effectiveness of a pain medication.
   c. The number of commercials aired daily by a television station.
   d. The weights of Sunday newspapers.
2. Give both discrete and continuous variables that might be studied in connection with:
   a. Computer programs written by students.
   b. Student use of a chemical supply room.
3. The placement center of a university is interested in how successful the publicity has been for a new interview preparation program it has developed. The center contacts 200 graduating seniors and questions them about their knowledge of the program.
   a. What is the population of interest?
   b. What is the sample?
   c. What type of inference is involved in this study?
4. What type of inference is suggested by each of the following?
   a. A statistics teacher wishes to know whether homework grades and examination grades are related.
   b. A television network wishes to find out what percent of the viewing audience watches its six o'clock news program.
   c. A medical benefit analyst for an insurance company needs to predict the average daily cost of a private hospital room.
   d. An insurance company wishes to know if the accident rate of college graduates is below that of adults in general.
5. In a study of whether certain vitamins may prevent cancer, the employees of a large European transit system have been monitored for several years. Vitamin levels are determined from workups of blood samples taken as part of their annual physicals, and the employees are not aware of the study. Is this an observational or controlled study? Explain.

Part B

6. The learning resource center of a school questioned eight students about their grade point averages and the number of hours they worked on part-time jobs. The results are as follows:

Student number	Hours worked	Grade point Average
1	12	2.8
2	8	2.9
3	20	2.3
4	15	2.7
5	0	3.6
6	30	1.9
7	40	2.4
8	11	3.4

Describe an inferential problem that is suggested by this information.

7. Below are the grades of a student on the four hourly examinations in a statistics course. Let x_i be the grade on the ith examination.

Examination Number	Grade
1	50
2	60
3	90
4	80

   a. What is x_3?
   b. Find $x_1 + x_2 + x_3 + x_4$.
   c. Find $(x_1 + x_2 + x_3 + x_4)/4$.
   d. What does the quantity in part c represent?
   e. The instructor says she will disregard the lowest examination grade. Using subscripted variables, give the formula for computing the new average.

8. A metropolitan newspaper surveyed 1340 people by telephone between the hours of 10:00 and 11:00 PM on a Sunday night. Give at least four populations of which they have a sample.

9. A psychologist believes that coronary-prone adults tend to have higher aspirations for themselves than adults in general. To determine whether this is the case, 100 coronary-prone adults are placed in highly competitive problem-solving situations where their achievements can be compared with performance projections. These results are then compared with established achievement/projection norms for the same problem-solving situations.
   a. What is the population of interest?
   b. What is the sample?
   c. What is the hypothesis?

10. Refer to problem 6. Let x_i and y_i be the number of hours worked and the grade point average, respectively, of the ith student.
   a. What quantity is represented by $(x_1 + x_2 + x_3 + \ldots + x_8)/8$?

b. What quantity is represented by $(y_1 + y_2 + y_3 + \ldots + y_8)/8$?

c. If the points (x_1, y_1), (x_2, y_2), $\ldots$ (x_8, y_8) are plotted, do they show a tendency to fall on or near a straight line?

1.4 RANDOM SAMPLES

Rarely are we able to examine an entire population. The most obvious reason for this is that most are undergoing continuous change and thus exist only hypothetically. Other factors such as cost, time, and accessibility also come into consideration. Often the only practical way to study a population is through a sample.

Ideally we want to select a sample that will be a miniature copy of the population—a difficult feat since the population is generally not available for comparison. The problem, of course, is that not all samples of any particular size will be representative of the population; any method we devise for selecting a sample will have occasional failures.

Short of examining the entire population, there is no way to guarantee absolutely that a sample is a close facsimile of the population. An important question, then, is how often properties of a sample deviate significantly from those of the population. This is the subject matter of Chapter 7, and the answer requires the use of random samples.

DEFINITION 1.6.

A *random sample* of size n is a sample selected by a process whereby all samples of size n have an equal chance of being chosen.

Although we expect properties of the sample to better approximate those of the population as the sample size increases, completely unrepresentative results are very infrequent, even when the sample sizes are relatively small. When difficulties with sampling do occur, it is usually because the sampling process is not random. In effect, certain individuals have been excluded, and the resulting sample fails to represent the population in question.

Example 1.7

A Research Study

The Literary Digest Poll

The *Literary Digest*, a prominent magazine in the early part of this century, accumulated the enviable record of correctly predicting every national election from 1916 until 1936. In that year, on the basis of more than 2.4 million responses to a mailed questionnaire, it predicted an overwhelming victory for the Republican candidate, Alf Landon. Yet Franklin Delano Roosevelt won in a landslide, garnering more than 62% of the popular vote. What went wrong? The main source of trouble was the mailing list. It was compiled from telephone directories and auto registrations and thus consisted principally of upper and middle income families who could afford such items during the Depression of the 1930s. The large Democratic registration contained in the lower income brackets was excluded from the sampling process. In effect, only one stratum of the population was sampled. A second difficulty lay in the mechanics of the process. Sampling by mail means that responses are received only from those who are deeply committed or cooperative enough to voluntarily participate. In this poll more than 10 million questionnaires were mailed out and fewer than 25% were returned. It is doubtful whether the 2.4 million responses were even representative of the families on the mailing list. ■

It is an easy matter to give other examples of sampling plans that are not random or even close to being so. Suppose a student placement center at a university needs information on the work habits of students and arranges to interview students at random as they arrive for their 8:00 AM classes. Those who do not have such classes are excluded from the sample, and in fact most such samples will contain a disproportionate number of students who have part-time jobs and thus arrange an early schedule for themselves.

Or consider an attempt to describe the physical fitness of young adults by studing a random sample of fitness evaluations performed on military enlistees. There is little hope of obtaining a miniature cross-section of young adults in this manner, since many groups (for example, college students) will be excluded or at least underrepresented in any sample that is obtained.

The next section briefly introduces techniques for obtaining random samples.

1.5 · RANDOM NUMBERS

Most random selection processes make use of a table of random numbers. The use of such a table is so simple and contributes so much to one's understanding of sampling that we choose to introduce the topic at this time.

First we consider the concept of a random digit. A table of random digits consists of row after row and page after page of the digits 0, 1, 2, . . ., 9. These numbers have been produced by a selection process in which each digit had the same chance of being selected at each step of the process. One can think of making repeated drawings with replacement from a hat containing 10 poker chips, numbered 0 through 9. The digits of such a table are often arranged in groups of five for readability.

Each of the 10 digits occurs in a random digit table with about the same relative frequency, $1/10$. If these digits are grouped by pairs, the 100 possible two-digit numbers 00, 01, 02, . . ., 99 have about the same relative frequency, $1/100$. Similarly, each of the three-digit numbers 000, 001, 002, . . ., 999 has a relative frequency close to $1/1000$, and so forth. Tables with millions of random digits are available. Additionally, most computers include a random number generator.

For ease of reference in the following discussion we reproduce in Table 1.1 a few lines of a table of random digits. A more extensive list is found in Table I of the appendix.

In selecting a random sample from a finite population, we first assign a number to each member of the population. Then, using an appropriately grouped table of random digits, we select the population members. To ensure randomness, the starting place in the table should be selected in advance. Ordinarily we move

TABLE 1.1 FIVE ROWS OF A RANDOM DIGIT TABLE						
75242	15945	40866	58879	30841	75365	10994
15532	87708	49025	85490	95310	96768	80844
33980	53563	98785	61270	73753	74549	57306
23455	24827	19650	85421	19278	09982	12354
93878	89405	02360	54271	26242	93735	20752

from left to right across the rows; however, any fixed pattern can be employed. One should not try to select "randomly" from a table of random digits; this defeats the purpose of the table.

Example 1.8

A list contains 10 questions from which a random sample of four is to be selected. Label the questions 0, 1, 2, . . ., 9. Pick a starting point in the table of random numbers, say row 3 of Table 1.1 above. In succession we read off: 3, 3, 9, 8, 0. We ignore the second 3 of the list and use the questions numbered 3, 9, 8, and 0. ■

Notice in this example that the selection process is free of any personal bias we might have about "favorite" numbers, location of questions near the beginning or end of a list, and so forth.

Example 1.9

A low-cost housing complex has 21 new units and a list of 93 qualified applicants numbered 1, 2, 3, . . ., 93. Use random numbers to select 21 of the applicants.

Imagine the digits of the table of random numbers to be grouped by pairs and ignore the numbers 00, 94, 95, 96, 97, 98, and 99. The remaining 93 numbers 01, 02, 03, . . ., 93 each has the same chance of occurring, $\frac{1}{93}$. Starting in row 1, column 2 of our abbreviated table, and moving from left to right, we obtain in succession:

15, 94, 54, 08, 66, 58, 87, 93, 08, 41, 75, 36, 51, 09, 94, 15, 53, 28
77, 08, 49, 02, 58, 54, 90, 95, 31, 09, 67, 68, 80, 84, 43, 39, 80, . . .

From this list we have removed all the numbers above 93 and any duplicates and retaining the first 21 numbers that remain. Thus, arranged in order, we select the applicants numbered 2, 8, 9, 15, 28, 31, 36, 39, 41, 43, 49, 51, 53, 54, 58, 66, 67, 68, 75, 77, 80. ■

Example 1.10

Suppose a sheet of glass is assumed to be acceptable if it is distortion-free over 90% of its surface. Since it is not practical to examine the entire sheet of glass, the quality control technician will superimpose a 1 cm by 1 cm grid on it and examine the glass within a sample of these squares. Suppose the glass is 20 cm by

FIGURE 1.1
A SAMPLE OF 16 SQUARES FROM A 20 CM BY 30 CM GRID

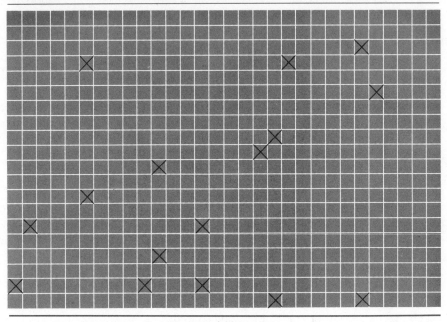

30 cm, as shown in Figure 1.1, and a sample of $n = 16$ squares is needed. Label the squares of the grid successively as 000, 001, 002, . . ., 599 (moving across each row and then to the beginning of the next). Imagine the digits of Table 1.1 to be grouped by threes, ignore all numbers above 599, and choose a starting place, say row 1, column 1. We find:

752	421	594	540	866	588	793	084	175	365	109
941	553	287	708	490	258	549	095	310	967	688
084	433									

The squares selected are indicated on the grid in Figure 1.1. ∎

The technique of the last example may be employed in selecting samples to study the bacterial growth in a culture, the density of weeds in a field of wheat, the diameters of trees in a grove, and so forth.

In sampling a population with several strata, we generally require that the proportion of each stratum in the sample should be the same as in the population. We then speak of a stratified sample.

Example 1.11.   A factory has 400 workers: 300 men and 100 women. The management wants to sample their opinions of the suitability of a new fringe benefit package it is considering. A sample of $n = 20$ employees is to be selected and questioned in depth about the various features of the package. Since 25% of the work force are women, 25% of the sample, or 5, should be women and 15 should be men. Thus management selects a random sample of size $n_1 = 5$ from the women and one of size $n_2 = 15$ from the men, using a technique similar to that of Example 1.8. ■

A random sample is not a panacea for all the problems of a statistical study. We often find that a second sample or possibly several must be used to control for other variables. One of the most insidious of these factors is the *placebo effect*. In many investigations, experimental subjects give a positive response that is not caused by the treatment under study. Sometimes this arises from subject's desire to be helpful or from a sick person's wish to recover. Both psychological and physiological factors, which are not at all understood, cause some people to respond just from receiving a treatment. To combat the placebo effect, one sample should receive the treatment while another receives a placebo—that is, a treatment with no specific action. The subjects are then "blind" to the nature of the treatment they receive. Another factor is the enthusiasm of the researcher for the study. In spite of good intentions, this may cause the researcher to interpret borderline responses in a way that biases the results. In the evaluation phase of a study the investigator's feelings may also become apparent to the subject. Thus the evaluator should also be blind to the nature of the treatment the subject received. In a *double blind* experiment, both the subjects and the evaluators are blind to the nature of the treatment.

Example 1.12

A Research
Study

Angina Pectoris and the Placebo Effect

In the June 1979 issue of the *New England Journal of Medicine*, Herbert Benson, M.D., and David McCallie, Jr., present a detailed study of the placebo effect in the treatment of angina pectoris (chest pains). Several treatments that have been proposed and discarded over the last 40 years are examined. Included are details of some researchers' attempts to control for both the placebo effect and their own enthusiasm, long before the need for double blind studies was fully recognized. The recurrent pattern when a new treatment is introduced appears to be a 70% to 90% cure rate. This was inferred by summarizing the results of non-blind, single trial, single treatment reports of practicing physicians. When double blind experiments are carried out, the effective cure rate drops to between 30% and 40%. This may be attributed to the placebo effect. Indeed, it seems that something on the order of a 35% response is a universal constant for the placebo effect, at least in angina pectoris treatments. ■

EXERCISE SET 1.2

Part A

1. A sample of 25 families on a particular street is needed. The interviewer simply walks down the street ringing doorbells until 25 families are found who are willing to be surveyed. Is this a random sample? Explain.

2. A student newspaper is interested in the amount of money that full-time students spend on textbooks each semester. What is wrong with the following sampling plans?
   a. Print a survey form in the paper that can be left off at the editorial office when completed.
   b. Instruct a reporter to survey as many students as he can in his classes.
   c. Interview a sample of students as they arrive on campus for their 8:00 AM class.

3. The occurrence of the word *the* is to be determined on 10 pages of a book whose pages are numbered 1 to 400. Use the table of random numbers in the Appendix to select the pages. (The answer we give is based on starting in row 26, column 1 of the first page.)

4. A machine shop has 22 workers, numbered 1, 2, 3, . . ., 22. A random sample of seven of them are needed to evaluate a health insurance plan that is being considered. Assuming Table I of the Appendix is used, what employees are selected if we begin on the first page in row 2 and column 1?

5. A new drug is to be tested for patients who have received open heart surgery. What would be necessary if this were to be a double blind experiment?

6. One hundred chronic headache sufferers are told of a marvelous new medication and asked to participate in a test of its effectiveness. If all agree to participate and the drug is in fact worthless, how many would be expected to report relief from their headaches?

Part B

7. In connection with the placebo effect, interpret the following statement: "Your best chance for a cure with a new treatment is when the treatment is relatively new on the scene."

8. If all else has failed, do you see anything wrong with a doctor trying a placebo?

9. A sample of the opinions of the 80 families in an apartment complex is needed regarding the location of a proposed playground. There are eight floors with 10 units to a floor. If the families of each floor constitute a stratum, explain how to select a random sample of 24 of these families in which all strata will be equally represented.

10. Discuss some factors that should be controlled in a good teacher evaluation study.

11. Select a random sample of five names from a list of 30 that are numbered from 1 through 30, as follows: Group the random digits by pairs to obtain 100 two-digit numbers as usual. Since 100 ÷ 30 is a little more than 3, we can use nearly all of these numbers if we assign three pairs to each of the 30 names. If we assign 00, 01, 02 to the first name, 03, 04, and 05 to the second, and so forth, with 87, 88, 89 finally being assigned to the 30th name, then only the numbers from 90 to 99 need be ignored. Which names are selected if we begin in Table I in the first row and first column?

12. Proceeding as in problem 11, describe an efficient method of selecting a random sample of 25 names from a list of
    a. 200 names, using three-digit random numbers.
    b. 400 names, using four-digit random numbers.

13. Give several reasons why the opinion polls conducted by radio talk shows do not produce a random sample of public opinion.

COMPUTER PRINTOUT

Computer printouts of interesting procedures are included throughout the text. These were produced by a general purpose statistical program called Minitab that was developed by the Statistics Department at the Pennsylvania State University. As the examples will show, Minitab is versatile and extremely easy to use.

We do not try to teach Minitab in this book. A few words about it will, however, make the printouts easier to understand. Minitab acts on data that we can think of being arranged on a worksheet with columns C1, C2, C3, ... In most cases a column will contain data from a sample or information derived from a sample. There are commands we can use to put information into these columns, cause procedures to be performed on the data of these columns, store results in other columns, and print out results.

Commands are entered after the prompt symbol MTB> in the University of Toledo interactive version of Minitab that we use in this text. Minitab is continuously being updated and expanded with new versions being released every year or so. As a result, output formats and commands will vary somewhat depending on the version available on the computer being used. The reader wishing to use this statistical program is directed to the *Minitab Student Handbook*, by T. A. Ryan, B. L. Joiner, and B. F. Ryan, published by Duxbury Press.

In Figure 1.2 we show the printout and the commands used to obtain a stratified sample of six of the 16 administrative and nine of the 24 clerical employees of a regional insurance office, whom we assume to be numbered 1 through 16 and 17 through 40, respectively. The sequence of commands needs no explanation. Note, however, that in practice there would be no need to print out columns C1 and C2, since the employee numbers we need are in columns C3 and C4.

FIGURE 1.2   MINITAB PRINTOUT FOR GENERATING A STRATIFIED SAMPLE

```
MTB >NOPRINT
MTB >GENERATE THE INTEGERS FROM 1 TO 16 INTO COLUMN
C1
MTB >GENERATE THE INTEGERS FROM 17 TO 40 INTO COLUMN
C2
MTB >SAMPLE 6 MEMBERS FROM C1, PUT INTO C3
MTB >SAMPLE 9 MEMBERS FROM C2, PUT INTO C4
MTB >PRINT C1,C2,C3,C4
```

COLUMN	C1	C2	C3	C4
COUNT	16	24	6	9
ROW				
1	1.	17.	9.	27.
2	2.	18.	10.	34.
3	3.	19.	15.	31.
4	4.	20.	8.	23.
5	5.	21.	6.	35.
6	6.	22.	1.	40.

Continued.

```
FIGURE 1.2   (CONTINUED)
           7              7.         23.              29.
           8              8.         24.              38.
           9              9.         25.              24.
          10             10.         26.
          11             11.         27.
          12             12.         28.
          13             13.         29.
          14             14.         30.
          15             15.         31.
          16             16.         32.
          17                         33.
          18                         34.
          19                         35.
          20                         36.
          21                         37.
          22                         38.
          23                         39.
          24                         40.

        MTB >STOP
```

CHAPTER CHECKLIST

Vocabulary

Descriptive statistics
Inferential statistics
Hypothesis testing
Correlation
Regression
Population
Variable
Continuous variable
Discrete variable

Subscripts
Sample
Random sample
Random digits
Stratified sample
Placebo effect
Blinds
Double blind experiment

Techniques

Using a table of random digits to select a random sample
Using a table of random digits to select a stratified sample

Notation

X Variable name
x A specific value of the variable X
x_i The ith value of the variable X (a subscripted value)

CHAPTER TEST

1. Are the following variables discrete or continuous?
   a. The temperature of the drinking water from a fountain.
   b. The number of units completed by randomly selected graduating seniors.

2. Give two continuous and two discrete variables that might be studied in connection with school classrooms.
3. Make up an interesting question about coffee consumption that might be the basis for a study involving
   a. Estimation.
   b. Hypothesis testing.
   c. Correlation and possibly regression.
4. A military supplier claims that a new hand-held launch system for a small rocket is so foolproof that any soldier can use it after a brief introduction. To demonstrate this, a salesman drives around a Marine Corps base and picks personnel at random from various job classifications. After 5 minutes of instruction, the Marines then try their hand at hitting a target 400 yards away. Of the 56 personnel involved in the test, 53 scored a direct hit on the first try.
   a. What is the population of interest?
   b. What is the sample?
   c. Is the sample a random sample of all military personnel?
   d. What type of inference is involved?
5. The number of units being taken by students is reported as follows: $x_1 = 14$, $x_2 = 16$, $x_3 = 15$, $x_4 = 12$, $x_5 = 15$, $x_6 = 18$.
   a. Find $(x_1 + x_2 + x_3 + x_4 + x_5 + x_6)/6$.
   b. Find $(x_1 - 15) + (x_2 - 15) + (x_3 - 15) + (x_4 - 15) + (x_5 - 15) + (x_6 - 15)$.
6. A classroom has seats for 42 students. Because of an error, 49 students have been enrolled and the instructor must drop seven at random. Assuming the students are numbered 1, 2, 3, . . . , 49, and beginning in Table I on page 2 in row 51 and column 1, which students are dropped?
7. Describe the placebo effect.
8. Describe how a double blind experiment might be designed to test a new medication for hypertension.
9. The houses on one block of an east-west street are numbered as shown. A random sample of five of these houses is needed, with three from the north side and the remaining two from the south side. Describe a process for obtaining this sample.

North side									
1	3	5	7	9	11	13	15	17	19
0	2	4	6	8	10	12	14	16	
South side									

CHAPTER TWO
Descriptive Statistics: Tables and Graphs

Populations are studied by means of random samples. Unfortunately, when a sample is first obtained, it is often an unorganized accumulation of observations that reveals little information. This chapter is concerned with efficient ways of grouping data into tables called *distributions* and then displaying them graphically.

The reader should not be unfamiliar with the notion of a distribution. Instructors commonly describe the results of an examination by giving the number of grades between 90 and 100, 80 and 90, 70 and 80, and so forth; this is a distribution of grades. When similar observations are grouped into classes, the distribution describes the count or frequency of each class. Properly executed, this type of summary proves in many ways to be more useful than the original data.

The graph of a distribution is often more informative than the distribution itself. The adage that a picture is worth a thousand words is indeed true here. Two types of graphs are discussed: the *histogram* and the *frequency polygon*. A histogram is a type of bar graph; a frequency polygon is a line graph. While both of these are important, the histogram is the workhorse of the statistician. It is both a useful way of summarizing data and an important analytical tool. Frequency polygons are of greatest value when one is comparing two distributions.

HISTOGRAMS AND FREQUENCY DISTRIBUTIONS FOR CONTINUOUS VARIABLES

The raw data of a sample often appear as a confused mass of numbers. The first step in analyzing the sample is to tally the number of times each value of the variable was observed and to summarize the results in a table.

DEFINITION 2.1

The *frequency f* of x in a sample is the number of times the value x occurs.

In place of frequencies, proportions or relative frequencies are often used.

DEFINITION 2.2

In a sample with n observations, the *relative frequency rf* of x is the frequency f of x divided by the sample size n, that is

$$rf = \frac{f}{n}$$

In Table 2.1 we have the observations of a study of the operating altitude of a new model of a low-level meteorological balloon. These altitudes were obtained by testing 20 balloons with the same load under similar conditions.

TABLE 2.1   ALTITUDES ACHIEVED BY METEOROLOGICAL BALLOONS

Balloon	Altitude (feet)	Balloon	Altitude (feet)
1	23,700	11	24,600
2	25,200	12	22,900
3	24,200	13	23,100
4	22,400	14	24,200
5	22,500	15	23,700
6	23,600	16	23,400
7	23,900	17	24,300
8	23,700	18	24,000
9	23,200	19	23,000
10	22,800	20	23,800

Both the frequencies and relative frequencies of the altitudes are summarized in Table 2.2. From this we can readily see that $x = 23,700$ had the highest

TABLE 2.2 FREQUENCY TABLE FOR ALTITUDES

x	f	Relative frequency
22400	1	1/20 = .05 = 5%
22500	1	1/20 = .05 = 5%
22800	1	1/20 = .05 = 5%
22900	1	1/20 = .05 = 5%
23000	1	1/20 = .05 = 5%
23100	1	1/20 = .05 = 5%
23200	1	1/20 = .05 = 5%
23400	1	1/20 = .05 = 5%
23600	1	1/20 = .05 = 5%
23700	3	3/20 = .15 = 15%
23800	1	1/20 = .05 = 5%
23900	1	1/20 = .05 = 5%
24000	1	1/20 = .05 = 5%
24200	2	2/20 = .10 = 10%
24300	1	1/20 = .05 = 5%
24600	1	1/20 = .05 = 5%
25200	1	1/20 = .05 = 5%

frequency and that all observations fell between 22,400 and 25,200. This arrangement is clearly superior to that of Table 2.1.

When searching for patterns and trends in a study involving measured data, it is not realistic to focus too sharply on the precise values obtained. For example, while balloon 20 of the altitude study reached 23,800 feet, we would not expect this height to be duplicated if the balloon were to be retested. This time it might reach 24,000 feet or perhaps only 23,600 feet. The same can be said of a balloon 15, which reached 23,700 feet. Both these balloons performed about the same.

Such considerations suggest that observations of a continuous variable should be grouped into classes according to the interval on which they fall; a frequency can then be determined for each class. Experience suggests that somewhere between 5 and 20 intervals should be used. In most cases they should be of equal width.

Consider again the altitude study and suppose we decide to group the data using seven intervals, each of 500 feet in width, with the first starting at 22,050 feet. The resulting intervals are then 22,050–22,550, 22,550–23,050, . . ., 25,050–25,550. These rather strange end points were chosen so that all observations would fall *within* the intervals. Otherwise, we would have followed the convention that any observation falling on the boundary of two intervals will be tallied in the interval immediately to its left.

Having decided on the intervals, the next step is to tally the observations (Table 2.3). From this we can read off, for example, that seven or 35% of the altitudes were between 23,550 and 24,050 feet. We describe this information by saying that the class associated with the interval 23,550–24,050 had a frequency of 7. To avoid such cumbersome constructions, we shall in the future freely interchange the terms *class frequency* and *interval frequency*.

TABLE 2.3   FREQUENCIES ARE OBTAINED BY TALLYING THE OBSERVATIONS

Interval	Tallies	f	Relative frequency
22050–22550	/ /	2	2/20 = .10(10%)
22550–23050	/ / /	3	3/20 = .15(15%)
23050–23550	/ / /	3	3/20 = .15(15%)
23550–24050	〰〰 / /	7	7/20 = .35(35%)
24050–24550	/ / /	3	3/20 = .15(15%)
24550–25050	/	1	1/20 = .05(5%)
25050–25550	/	1	1/20 = .05(5%)

FIGURE 2.1   HISTOGRAM OF BALLOON ALTITUDES

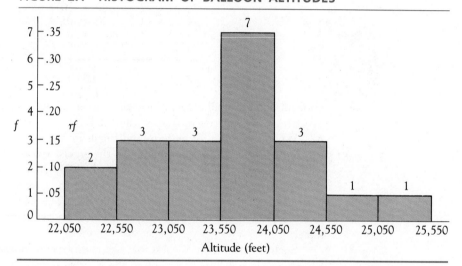

The information in columns 1 and 4 of Table 2.3 forms a relative frequency distribution for the sample, which is presented separately in Table 2.4. Similarly, a table formed by using columns 1 and 3 is called a frequency distribution. Either of these may be termed a distribution for the sample.

The simplest means of picturing a distribution is a histogram, consisting of a bar for each interval whose height is equal to its frequency or relative frequency of the interval.

The histogram for the distribution of balloon altitudes (Table 2.4) is given in Figure 2.1. Both the frequency and the relative frequency scales are shown. Notice how much additional information concerning the distribution of altitudes is now apparent. In particular, we sense a clustering and possibly symmetry about some central value near 23,550 feet. This would be very important if we felt that the sample was representative of the performance of all balloons of this model. Often only the first and last vertical lines of a histogram are retained, so that it assumes the appearance of a silhouette of a city skyline (Figure 2.2).

TABLE 2.4
DISTRIBUTION
OF ALTITUDES

Interval	Relative frequency
22050–22550	.10(10%)
22550–23050	.15(15%)
23050–23550	.15(15%)
23550–24050	.35(35%)
24050–24550	.15(15%)
24550–25050	.05(5%)
25050–25550	.05(5%)

FIGURE 2.2   BALLOON ALTITUDES

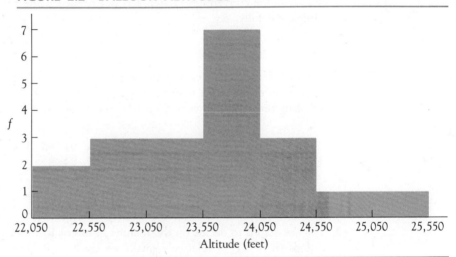

Histograms and distribution tables are popular ways of summarizing data in studies. The histogram is, of course, easily constructed from the distribution. A sequence of steps is given below for classifying a set of data and arriving at a distribution.

PROCEDURE FOR CONSTRUCTING A RELATIVE FREQUENCY DISTRIBUTION FOR A SAMPLE

1. Estimate the approximate number k of intervals that will be used.
2. Determine an interval width as follows:
   a. As an approximation to the width, compute

$$w = \frac{\text{largest observation} - \text{smallest observation}}{k}$$

      b. Adjust this number w by rounding or other means to obtain some relatively simple number for the width.

3. Find the interval end points as follows:

      a. Select some number less than the smallest observation as the first end point.

      b. Obtain the remaining end points by adding multiples of the interval width to this first end point.

4. Obtain the class frequencies by tallying the observations according to the intervals on which they fall. The following convention may be used: Should an observation fall on an end point, tally it in the interval immediately to the left of the point.

5. Compute the relative frequencies and summarize the results in a table.

Note that the convention in step 4 is not needed if the interval end points are selected carefully.

Example 2.1

A study of the delays experienced by a sample of 51 applicants for a vehicle registration change at a busy department of motor vehicles office contained the following times (in minutes). Construct a distribution and its histogram.

```
79 159 89   68 132  97 112  91 124 91 80 104  88 123  88  86
82  77 73  103 100 125 130  86  57 84 89  82 107 102 120  92
141 70 81  114  81 108 107 120  65 73 74  78  95 102  84 104
86  86 80
```

The first step is to decide on the number of intervals. Since the number of observations ($n = 51$) is relatively small, we try $k = 8$. The approximate interval width is next found by using the largest (159) and smallest (57) of the observations:

$$w = \frac{159 - 57}{8} = 12.75$$

Since it is desirable to have a reasonably simple number for the interval width, this is rounded up to 15.

    If we now select $x = 50$ as the left end point of the first interval, the intervals are then found to be 50–65, 65–80, 80–95, . . . , 155–170. With this choice of intervals, some observations fall at the end point of two intervals. The convention is to tally any such point in the interval immediately to its left. Thus, for example, the time $x = 80$ is counted in the interval 65–80.

    The actual tallying and construction of the histogram are done simultaneously by tallying with X's above the appropriate intervals as we proceed through the list of times (Figure 2.3). The distribution of Table 2.5 is then read off from this.

FIGURE 2.3 HISTOGRAM OF DELAY TIMES IN PROCESSING VEHICLE REGISTRATIONS

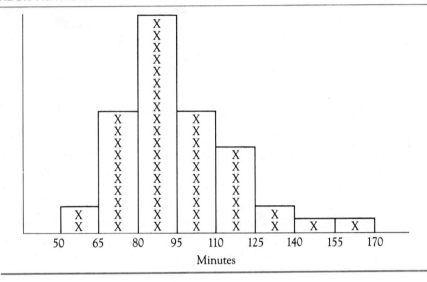

TABLE 2.5 DISTRIBUTION OF DELAY TIMES

Interval	Relative frequency
50–65	2/51 = .039
65–80	10/51 = .196
80–95	18/51 = .353
95–110	10/51 = .196
110–125	7/51 = .137
125–140	2/51 = .039
140–155	1/51 = .020
155–170	1/51 = .020

Obviously some distributions may be more informative than others. As the next example illustrates, one must be willing to experiment.

Example 2.2

A large mail order house has received a number of complaints about the amount of time needed to assemble a 10-speed bicycle it markets. Forty-three adults having no prior experience with such bicycles are selected and timed. The time needed to complete the assembly to the nearest 10 minutes is given in Table 2.6.

TABLE 2.6 DISTRIBUTION OF ASSEMBLY TIMES

Time	f	Time	f
130	2	280	1
150	2	300	3
160	2	310	1
170	1	320	2
180	4	330	1
190	3	340	2
200	2	350	4
210	1	360	2
230	1	370	4
250	2	380	3

To obtain a broad perspective of the distribution, intervals 40 minutes wide beginning at 115 are used first. This distribution and the accompanying histogram are given in Table 2.7 and Figure 2.4.

TABLE 2.7 DISTRIBUTION OF ASSEMBLY TIME (40 minute intervals)

Interval	f
115–155	4
155–195	10
195–235	4
235–275	2
275–315	5
315–355	9
355–395	9

FIGURE 2.4 HISTOGRAM OF ASSEMBLY TIMES

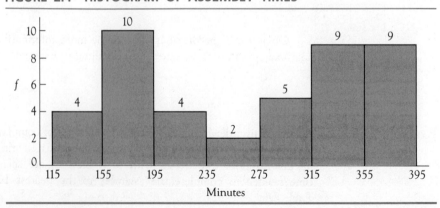

Recognizing now that the distribution has two clusterings, we next try a smaller interval width of 20 minutes, again starting at 115. The results are shown in Table 2.8 and Figure 2.5.

TABLE 2.8   DISTRIBUTION OF ASSEMBLY TIME (20 minute intervals)

Interval	f
115–135	2
135–155	2
155–175	3
175–195	7
195–215	3
215–235	1
235–255	2
255–275	0
275–295	1
295–315	4
315–335	3
335–355	6
355–375	6
375–395	3

FIGURE 2.5   HISTOGRAM OF ASSEMBLY TIMES

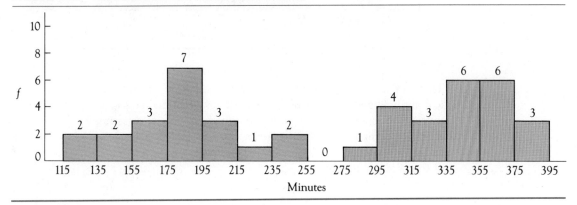

Finally, using an interval width of 30, we also consider the effect of the starting point by using both 115 and 125 as initial boundaries. The resulting histograms are given in Figures 2.6 and 2.7.

It would be very difficult to say that one of these distributions is more informative than the others. Each has contributed to our understanding of what is happening.

FIGURE 2.6   HISTOGRAM OF ASSEMBLY TIMES

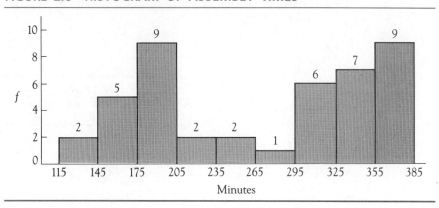

FIGURE 2.7   HISTOGRAM OF ASSEMBLY TIMES

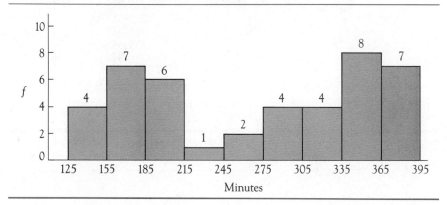

There are apparently two types of assemblers—those with perhaps some mechanical aptitude and those of questionable talents. Or perhaps there are those who follow instructions and those who consult them only as a last resort. Or perhaps there is a combination of such factors. In any event, there are certainly grounds to question whether the bicycle is easily assembled, at least by a sizable proportion of the population.   ■

2.2   HISTOGRAMS AND FREQUENCY DISTRIBUTIONS FOR DISCRETE VARIABLES

Our discussion so far has dealt only with continuous variables. If the variable is discrete, the basis for deciding whether or not to group the data should be whether or not small differences between individual observations are important. For example, one would probably group for such variables as the number of shares traded daily on a stock exchange, the number of engineers who graduate annually, and the number of airline flights daily from a metropolitan airport. If we do not group, a histogram can still be constructed by using the observations as the centers of the intervals.

Example 2.3

An insurance company requires each applicant to report the number of traffic citations during the previous three years. The distribution of the number of reported citations and the corresponding histogram are given in Table 2.9 and Figure 2.8.

TABLE 2.9
DISTRIBUTION OF CITATIONS

Number of citations	Relative frequency
0	.43 (43%)
1	.37 (37%)
2	.13 (13%)
3	.04 (4%)
4	.02 (2%)
5	.01 (1%)

FIGURE 2.8   HISTOGRAM OF THE DISTRIBUTION OF CITATIONS

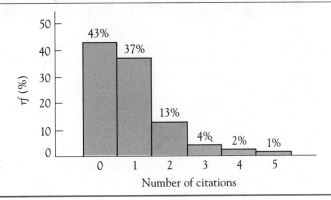

Example 2.4

A Research Study

Can a job be made so interesting that it becomes unsafe? Jobs that are too simple and repetitive are often shunned by workers. In recent years, job designers have tried to combat this tendency by designing jobs with high degrees of variety and autonomy and a product or component the worker can identify with. Whether this approach increases the risk of an accident is detailed in a study from Finland by J. T. Saari and J. Lahtela in the October, 1978 issue of *Industrial Engineering.* Nine light metal factories in the Helsinki area, each having between 100 and 500 workers, were involved in the study. Five of the factories were classified as having high accident rates; the remaining four had low rates (as compared to the national average for this industry). One of every six workers was selected from each factory, interviewed, and observed on the job. There were 131 subjects from the high-accident-rate factories and 115 from the lower rate factories. Two histograms showing accident frequency plotted against the number of tasks in the job in which the accident occurred are shown in Figure 2.9. Both have been combined on one graph for ease of comparison.

FIGURE 2.9   NUMBER OF TASKS BELONGING TO A JOB IN COMPANIES WITH HIGH AND LOW ACCIDENT RATES

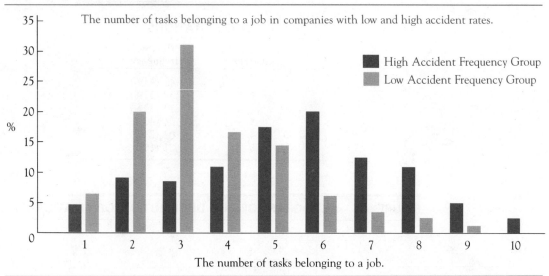

Note that the number of tasks belonging to a job is significantly higher for the high accident group. Similar results were obtained when high-accident-rate jobs of the companies were compared with jobs having an average rate. The study also included factors such as the cycle time for the tasks, the degree of autonomy, and the extent to which the jobs were preplanned. The conclusion is that "accidents occurred more often than could be expected in tasks that are often used for job enrichment when the work structure is changed to diminish the feeling of monotony and to improve job satisfaction among workers." The authors then go on to recommend that in the enrichment of a job, the additional tasks must be well planned and the workers well trained in performing them.   ■

2.3   STEM-AND-LEAF DIAGRAMS

When observations are summarized in a report by means of a histogram or distribution table, the individual observations are lost to the reader. As with any type of rounding, some information is lost. Stem-and-leaf diagrams, which we now discuss, were invented by Professor John Tukey of Princeton University to avoid this problem. These diagrams are increasing in popularity, both as an efficient way of displaying data and as an aid in preparing histograms.

In a stem-and-leaf diagram, each observation is represented in terms of a leaf and a stem or simply as a leaf on a stem. The leaf is generally the last or rightmost digit of the observation; the stem is made up of the remaining first digits. Thus for $x = 154$, the leaf is 4 and the stem is 15. The use of a decimal point is optional if its location is clear from the nature of the data. Thus an infant's weight of $x = 7.8$ pounds has 8 as the leaf and 7. or simply 7 as the stem. In a stem-and-leaf diagram, the stems are listed to the left of a vertical line, which is often called the *trunk*. The leaves then extend horizontally to the right of the stems to which they belong.

Example 2.5

Low-lead gasoline was obtained from a sample of 24 distributors throughout the western United States and tested. The resulting octane measurements are given below and a stem-and-leaf diagram is shown in Figure 2.10. ■

| 84.5 | 87.6 | 87.3 | 87.8 | 88.0 | 86.5 | 86.2 | 84.9 | 84.8 | 85.7 | 85.4 | 85.6 |
| 86.7 | 86.9 | 86.2 | 86.7 | 86.9 | 87.2 | 87.4 | 88.2 | 86.6 | 86.5 | 86.8 | 85.2 |

FIGURE 2.10 STEPS IN THE CONSTRUCTION OF A STEM-AND-LEAF DIAGRAM OF OCTANE DATA

84		84	5	84	5	84	598
85		85		85		85	7462
86		86		86		86	5279279658
87		87		87	6	87	63824
88		88		88		88	02
Trunk with stem		First leaf attached		First two leaves attached		Complete stem-and-leaf diagram	

Note that if the leaves of each stem of a stem-and-leaf diagram are blocked in, we obtain a histogram at a right angle to the orientation of the figure (Figure 2.11).

FIGURE 2.11 HISTOGRAM FROM A STEM-AND-LEAF DIAGRAM

84	598
85	7462
86	5279279658
87	63824
88	02

■

Example 2.6

A stem and leaf diagram for a set of examination scores is given in Figure 2.12. Find the scores. ■

FIGURE 2.12 STEM-AND-LEAF DIAGRAM FOR EXAMINATION SCORES

```
0 |
1 |
2 | 8
3 |
4 | 692
5 | 89301
6 | 624793879
7 | 50854
8 | 458
9 | 39
```

The lowest score is 28. The scores in the 40s are 46, 49, and 42, those in the 50s are 58, 59, 53, 50, and 51. The complete list of scores is given below:

28 46 49 42 58 59 53 50 51 66 62 64 67 69 63 68
67 69 75 70 78 75 74 84 85 88 93 99

Note that a stem-and-leaf diagram also provides a partial ordering of the observations by size. ■

If most of the variability occurs in the last digit of the observations, the number of stems can be doubled by using the same stem twice, the first with the digits 0 through 4, second time with the digits 5 through 9.

Example 2.7

The octane data from Example 2.5 are reproduced below. Find a stem-and-leaf diagram with twice as many stems as previously used (Figure 2.13).

FIGURE 2.13 DOUBLED STEM-AND-LEAF DIAGRAM OF OCTANE MEASUREMENTS

```
84 |
84 | 598
85 | 42
85 | 76
86 | 22
86 | 57979658
87 | 324
87 | 68
88 | 02
88 |
```

84.5 87.6 87.3 87.8 88.0 86.5 86.2 84.9 84.8 85.7 85.4 85.6
86.7 86.9 86.2 86.7 86.9 87.2 87.4 88.2 86.6 86.5 86.8 85.2
                                                                    ■

When preparing a histogram or distribution table one can first construct a stem-and-leaf diagram, where the stems correspond to the intervals that will be used with the histogram.

Example 2.8   A stem-and-leaf diagram for interest rates on a sample of short-term construction loans for residential housing is given in Figure 2.14. Interpret the diagram, give the distribution, and draw the histogram.

FIGURE 2.14   STEM-AND-LEAF DIAGRAM OF LOAN RATES

15.0–15.4	23
15.5–15.9	5889
16.0–16.4	00430
16.5–16.9	6588978
17.0–17.4	2402
17.5–17.9	55

The loan rates from the first stem are 15.2 and 15.3. From the second stem the rates are 15.5, 15.8, 15.8, and 15.9. The complete list is given below:

15.2 15.3 15.5 15.8 15.8 15.9 16.0 16.0 16.4 16.3 16.0
16.6 16.5 16.8 16.8 16.9 16.7 16.8 17.2 17.4 17.0 17.2
17.5 17.5

The distribution is easily found from the stem-and-leaf diagram (Table 2.10). The corresponding histogram is given in Figure 2.15.

TABLE 2.10   DISTRIBUTION OF LOAN RATES

Interval	Frequency	Relative frequency
$15.0 \leq x < 15.5$	2	.0833
$15.5 \leq x < 16.0$	4	.1667
$16.0 \leq x < 16.5$	5	.2083
$16.5 \leq x < 17.0$	7	.2917
$17.0 \leq x < 17.5$	4	.1667
$17.5 \leq x < 18.0$	2	.0833

FIGURE 2.15   HISTOGRAM OF LOAN RATES

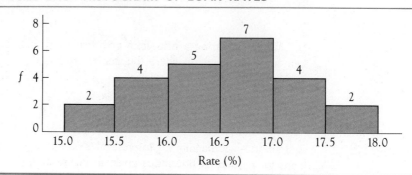

Rate (%)

**EXERCISE
SET 2.1**

Part A

1. A histogram of the distribution of salaries of nonadministrative personnel of a company is given in the figure below.
   a. What percent of the salaries are between $14,500 and $15,500?
   b. What percent of the salaries exceed $17,500?
   c. Construct the distribution table that corresponds to this histogram.
   d. If there were 400 employees in the study, how many had salaries between $16,500 and $17,500?

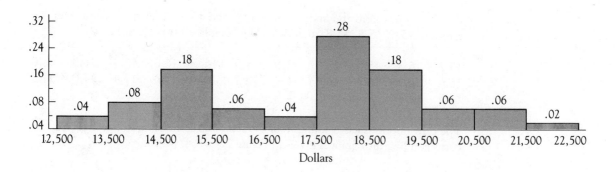

Dollars

2. A distribution of a sample of vehicle speeds at a freeway location is given in the table below. Construct a corresponding histogram.

Speed (mph)	f
46.5–50.5	8
50.5–54.5	27
54.5–58.5	49
58.5–62.5	148
62.5–66.5	84
66.5–70.5	78
70.5–74.5	16

3. The depths (in feet) at which ground water was found below the surface at 25
   test sites located throughout a valley in the High Sierras are given below.
   a. Group the data using six intervals, with the left end point of the first interval
      at 2.55; obtain a distribution.
   b. Construct the histogram of the distribution of part a.

   8.3, 7.4, 2.9, 5.4, 6.8, 3.4, 3.0, 3.9, 4.7, 4.9, 2.6, 3.3, 4.2,
   6.4, 5.9, 3.1, 4.4, 4.0, 3.8, 5.8, 4.3, 4.7, 4.3, 4.6, 3.7

4. State agencies regulate the content of different grades of ground beef. A sample
   of a particular grade from different stores of a supermarket chain produced the
   following percentages of fat:

   25.0, 24.3, 24.7, 24.2, 25.1, 24.2, 24.5, 24.6, 24.5, 24.4, 24.1, 24.8, 24.4,
   24.5, 24.9, 24.2, 24.3, 24.6, 24.7, 24.0, 24.5, 24.7, 24.9, 24.4, 24.5

   Using intervals of width .2% and starting with 23.95, find the distribution and
   sketch its histogram.

5. A random sample of the earnings per share of corporations reporting a first
   quarter 1983 operating profit is given below. Group the data to obtain a distribution and construct the histogram.

   .12, .31, .04, .28, .03, .28, .27, 1.58, .95, .10, .02, .28, .35, 1.62,
   .45, .05, .18, .08, .12, 1.90, .68, .52, .01, .48, 1.91, .59, 1.21,
   1.82, .48, .35, .51, .53, .10, .44, .13, .47, .05, .18, .96, 1.32, .51,
   .10, .04, .09, 1.23, .95, .07, .25, .95, .02, .24, 1.08, 1.31, 1.67,
   .16, .69, 1.84, 1.80

6. A distribution of the IQ scores of students at an elementary school is given.
   Assuming the reported scores are the midpoints of the class intervals, construct
   the histogram.

x	f	x	f
72	4	108	35
76	8	112	26
80	14	116	16
84	25	120	10
88	40	124	4
92	60	128	3
96	82	132	2
100	68	136	1
104	50	140	2

7. The following is a list of the prices observed in a metropolitan newspaper for a
   certain make and model of auto:

   2395, 2000, 2350, 3295, 2895, 3150, 3200, 2195, 2495, 2850, 2500,
   2950, 2995, 2600, 3050, 2975, 2750, 2100, 2450, 2800, 2695, 2200,
   2550, 2625, 1995, 2600

   Group the data and draw the histogram of the resulting distribution.

8. A die is thrown repeatedly and produces the following results:

2, 6, 4, 3, 5, 4, 5, 3, 1, 2, 4, 6, 1, 4, 4, 1, 5, 6, 4, 4, 3, 2,
4, 4, 4, 4, 2, 1, 3, 2, 5, 1, 4, 3, 3, 5, 4, 6, 6, 1, 5, 2, 4, 3,
4, 2, 4, 6, 6, 3, 3, 4, 1, 2, 2, 2, 5, 4, 1, 5, 2, 1, 3, 6, 2, 5,
2, 4, 3, 3, 5, 2, 4, 4, 3, 2, 2, 6, 6, 2, 3, 5, 1, 1, 2, 2, 1, 1.

    a. Tabulate the frequencies and construct the histogram.
    b. Does the die appear to be fair?

9. Thirty weak transmissions of 100 binary characters each (0s and 1s) were made to a receiver in a test of equipment designed to receive communications from deep space. The number of receiver errors in each of these transmissions is given below. (An error is made if a 0 is changed to a 1 or vice versa.) Construct the histogram.

0, 3, 1, 1, 2, 1, 1, 2, 3, 5, 0, 0, 1, 1, 2,
1, 3, 1, 2, 1, 3, 4, 2, 4, 1, 0, 2, 4, 3, 1

10. Six assistants in a psychology experiment are strategically placed along a campus walkway so that students using it will have the opportunity to make eye contact with the assistants. The number of assistants with whom eye contact was made by each of 30 students using the walkway are given below. Construct a histogram for these results.

0, 0, 5, 5, 1, 2, 3, 3, 2, 1, 6, 6, 5, 6, 4,
4, 1, 6, 0, 6, 2, 5, 4, 1, 6, 5, 1, 4, 5, 6

11. Forty-two retail stores were surveyed as to the percentage markup that they apply to all goods to compensate for inventory shrinkage resulting from shoplifting and employee theft. The results are given below. Prepare a stem-and-leaf diagram of the data as in example 2.8 using the intervals 8.0-8.1, 8.2-8.3, and so forth.

8.6, 8.5, 9.4, 9.4, 9.5, 8.5, 8.8, 8.3, 8.8, 8.9, 9.2, 9.3, 9.5, 9.4
8.5, 8.6, 8.7, 8.9, 8.5, 9.5, 9.4, 9.3, 9.5, 9.0, 9.0, 8.7, 8.9, 8.5
8.5, 8.6, 9.4, 9.5, 9.0, 8.5, 8.4, 8.6, 9.5, 9.5, 9.3, 9.1, 9.1, 8.6

12. A sample of small eastern cities were surveyed as to their minimum charge for a speeding ticket (first-time violator). The results (in dollars) are given below. Use a stem-and-leaf diagram to produce a histogram with the intervals $30 \leq x < 35$, $35 \leq x < 40$, ...

30, 50, 60, 35, 50, 60, 35, 50, 62, 63, 50, 40, 50, 42, 65, 32, 52, 52,
42, 45, 45, 54, 37, 30, 48, 55, 55, 58, 58, 48, 48, 40, 50, 52, 65

Part B

13. The relative brightness of each of the 25 star systems nearest to the earth are given below. Find a distribution for these data and construct a histogram.

Star	Magnitude
α Centauri C	11.05
α Centauri AB	−0.29
Barnard's	9.54
W 359	13.53
+36°2147	7.50
Sirius AB	−1.46
L 726-8 AB	12.50
Ross 154	10.60
Ross 248	12.29
ε Eridani	3.73
L 789-6	12.18
Ross 128	11.10
61 Cygnus AB	4.80
ε Ind	4.68
α 2398 AB	8.47
Procyon AB	0.37
+43°44 AB	8.00
G51-15	14.81
-36°15693	7.36
+5°1668	9.82
ε Ceti	3.50
L 725-32	11.60
-39°14192	6.67
Krug 60 AB	9.60
Kapteyn's	8.81

Source: "Look Whats Moving into Our Neighborhood," H. Cohen and J. Oliver, Astronomy Magazine, April 1979.

14. Twenty cartons of fragile ceramic castings were shipped on each of two air freight carriers. On delivery at their destination the cartons were opened and inspected. The number of damaged items per carton were as follows:

   17, 20, 1, 22, 18, 5, 19, 14, 9, 18, 10, 6, 19, 20, 14, 15, 9, 16, 18, 22, 23, 2, 7, 8, 5, 13, 12, 5, 7, 6, 3, 8, 8, 15, 18, 8, 10, 11, 12, 18

   a. Find the frequency distribution using the groupings: 1–4 inclusively, 5–8 inclusively, 9–12 inclusively, and so on.
   b. Draw the histogram for the distribution of part a, using the intervals 0.5–4.5, and so on.
   c. Suppose the first 20 observations are from carrier A and the remaining ones from carrier B. Using the same intervals as in parts a and b, obtain the frequency distributions for the samples from each carrier and draw the corresponding histograms.
   d. How is the appearance of the histogram in part b explained by those obtained in part c?

15. The distribution below is for retail refrigerator prices in a sample of 150 stores in 11 cities. Three makes of refrigerators were surveyed but no significant difference was found in their prices (within a given store or locale). Sketch the histogram of the distribution obtained by using intervals of length 20 beginning at 379.5.

RETAIL REFRIGERATOR PRICES IN ELEVEN CITIES

Price	Frequency	Price	Frequency	Price	Frequency
$380	3	$420	7	$450	13
385	1	424	1	458	2
386	1	425	7	459	7
387	1	426	1	460	8
389	4	428	3	465	1
390	3	429	12	467	1
391	1	430	14	469	6
395	4	432	1	470	13
399	9	435	4	475	1
400	10	437	1	479	8
405	1	438	3	480	4
409	13	439	6	485	1
410	12	440	11	489	1
414	2	444	3	490	4
415	2	445	1	495	2
417	1	446	1	499	1
418	2	448	2	500	1
419	22	449	5	510	1

Mean $434.04; range: $380 to $510; median: 430; *standard deviation:
±$28.07. All prices have been rounded to the nearest dollar.
Source: "Price Variation for Refrigerators Among Retail Types and
Cities," A.F. Jung, Journal of Consumer Affairs, vol. 13, no. 1,
Summer, 1979.

16. Below is a tabulation of aircraft flow at O'Hare Airport in Chicago. The first
table is for a typical day in November (1979); the second for a day when one
runway was closed between 1700 and 2100 hours because of a snow storm. The
columns show the time, number of arrivals, landing, average number of aircraft

ORIGINAL TRAFFIC

TIME	ARR	LAND	AVHLD	AVDEL	PKHLD	PKDEL
0800	6	6	0	0	0	0
0900	5	5	0	0	2	1
1000	19	19	0	1	3	2
1100	20	20	0	0	2	1
1200	5	5	0	0	2	1
1300	60	52	6	3	9	8
1400	57	61	4	4	13	11
1500	51	53	5	7	17	15
1600	75	65	9	6	16	15
1700	58	65	8	11	27	24
1800	54	59	1	3	10	8
1900	68	64	6	7	17	15
2000	76	66	11	7	16	15
2100	51	63	7	11	19	17
2200	74	66	9	7	16	14
2300	69	63	11	6	18	16
0000	68	66	14	10	21	19
0100	64	66	18	18	29	26
0200	51	66	7	12	22	19
0300	33	34	1	1	5	4
0400	18	18	0	0	3	2
0500	19	19	0	0	2	1
0600	8	7	0	0	1	1
0700	13	14	0	0	1	1
0800						

TRAFFIC WITH SNOW STORM						
TIME	ARR	LAND	AVHLD	AVDEL	PKHLD	PKDEL
0800	6	6	0	0	0	0
0900	5	5	0	0	2	1
1000	19	19	0	1	3	2
1100	20	20	0	0	2	1
1200	5	5	0	0	2	1
1300	60	52	6	3	9	8
1400	57	61	4	4	13	11
1500	51	53	5	7	17	15
1600	75	65	9	7	16	18
1700	58	45	22	27	28	37
1800	54	45	32	40	36	47
1900	68	45	52	63	60	79
2000	76	45	80	75	89	82
2100	51	66	79	78	93	84
2200	74	66	80	72	87	79
2300	69	66	79	68	86	78
0000	68	66	82	72	89	81
0100	64	66	86	79	97	88
0200	51	66	75	74	90	81
0300	33	66	44	48	73	66
0400	18	54	13	23	37	33
0500	19	19	0	0	2	1
0600	8	7	0	0	1	1
0700	13	14	0	0	1	1
0800						

Source: "ATC Flow Management," Interavia, vol. 4, 1980

held during that hour, average delay in minutes, peak number of aircraft held, and peak delay in minutes.

  a. Draw the histograms for the landing times on these two days.

  b. How is the runway closure reflected in the second histogram?

17. The distribution of the number of catalogued earthquakes in California since 1927 (the year of the first seismological laboratory in California) is given.

  a. Does this indicate that earthquakes are a recent phenomenon in California?

  b. Does there appear to be an increase in earthquake activity since 1974?

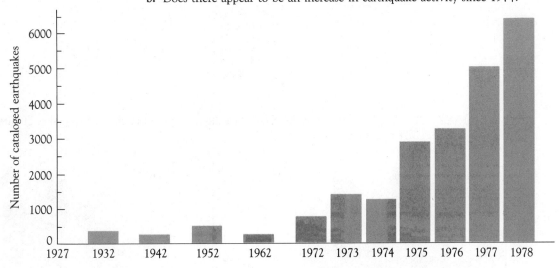

    c. New scanning techniques and new stations were introduced in 1974. How is this reflected in the histogram?

Source: "The Southern California Seismic Network," Earthquake Information Bulletin, U.S. Geological Survey, vol. 11, no. 6, Nov.-Dec., 1979

18. What should the histogram look like for the first 1,000,000 digits of a random digit table?

19. A cumulative frequency distribution for a sample describes the number of observations at or below each of the right end points of the intervals. The cumulative frequency distribution of the oxides of nitrogen emissions of 40 automobiles is given below.

    a. How many automobiles had emissions at or below 4.05 parts per million (ppm)?

    b. How many automobiles had emissions at or below 5.05 ppm?

    c. How many automobiles had emissions exceeding 5.05 ppm?

    d. Find the (frequency) distribution implied by this cumulative distribution.

Interval	Cumulative frequency
$x \le 1.05$	0
$x \le 2.05$	8
$x \le 3.05$	18
$x \le 4.05$	30
$x \le 5.05$	36
$x \le 6.05$	38
$x \le 7.05$	40

20. A distribution of the reading scores of sixth grade children is given. Find the corresponding cumulative (relative) frequency distribution. (See Problem 19.)

Scores	Relative frequency
20.5–30.5	.12
30.5–40.5	.22
40.5–50.5	.46
50.5–60.5	.14
60.5–70.5	.04
70.5–80.5	.02

21. The distributions of the miles per gallon (mpg) of samples of two makes of automobiles are given.

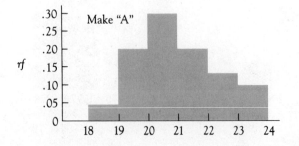 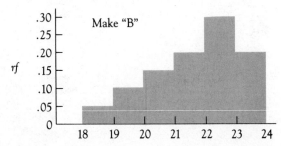

      a. What percentage of the sample of make A autos tested in excess of 21 mpg? Of make B?

      b. Which make of auto produced the better gas mileages?

      c. Give the distributions in table form that correspond to these histograms.

      d. Suppose 60 autos of make A and 140 of make B were tested. Give the distribution for the combined sample of 200 autos using the intervals 18–19, 19–20, . . . as above. Compute both the frequencies and the relative frequencies.

      e. Draw the histogram for the distribution of part *d*.

22. A sample of weights (in pounds) of male students produced the following double stem-and-leaf diagram. Construct a histogram using the intervals $130 \leq X < 140$, $140 \leq X < 150$, and so forth.

13	21
13	55896
14	0040203020
14	5567957
15	021314
15	56769
16	04023214213
16	56559868
17	042
17	895
18	42
18	5

2.4   THE AVERAGE DENSITY SCALE

In working with grouped data it is often useful to regard the observations as being uniformly distributed over the interval on which they lie. Loosely speaking, this means that the observations are evenly spaced across the interval under consideration. Thus if .2 = 20% of the weights of a sample of adult males are on the interval from 150 to 160 pounds, then the subinterval 150–151 would contain .02 = 2% of all observations, as would 151–152, 152–153, and so forth. Each unit subinterval from 150 to 160 would contain the fraction .20/10 = .02 = 2% of all the observations of the study. We describe this state of affairs by saying that the average (unit) density for the interval 150–160 is .02 or 2%.

DEFINITION 2.3

The *average density* d of an interval of length ℓ with relative frequency *rf* is given by

$$d = \frac{rf}{\ell}$$

Example 2.9

Following the airing of a controversial documentary, a television station received several telephone complaints. The distribution of the number of hours after the show at which calls were received is given in Table 2.11.

	Relative	
Interval	frequency	d
0–1	.40	.40 (40%)
1–2	.25	.25 (25%)
2–4	.15	.075 (7.5%)
4–8	.20	.05 (5%)

TABLE 2.11   DISTRIBUTION OF COMPLAINT TIMES

For the interval 0–1, the relative frequency rf is .40, the length ℓ is 1 (hour) and the density is

$$d = \frac{.40}{1} = .40 = 40\% \text{ (per hour)}$$

For the interval 1–2, $rf = .25$, $\ell = 1$ and

$$d = \frac{.25}{1} = .25 = 25\% \text{ (per hour)}$$

For the interval 2–4, $rf = .15$, $\ell = 2$ and

$$d = \frac{.15}{2} = .075 = 7.5\% \text{ (per hour)}$$

For the interval 4–8, $rf = .20$, $\ell = 4$ and

$$d = \frac{.20}{4} = .05 = 5\% \text{ (per hour)}$$

The interval from 2 to 4 hours had a higher rate than the interval from 4 to 8, even though its relative frequency was smaller. ■

Histograms using average densities rather than relative frequencies provide a means of controlling variations resulting from large samples and intervals of uneven lengths. If large samples are used, actual frequencies become large and unmanageable. Relative frequencies control this. However, intervals of different lengths with the same relative frequencies may produce different relative frequencies or densities when a standard interval of length one is used. When densities are used, the resulting histogram is what would have been obtained if the distribution had been uniform over each interval and we had chosen to regroup using smaller intervals of length 1. In Figure 2.16 we have both the relative frequency and density histograms for the distribution of complaint times in the last example.

It is an easy matter to recover the relative frequencies from a histogram using densities: simply multiply the length of an interval by its density. Thus $rf = \ell d$. This is obtained from the definition

$$d = \frac{rf}{\ell}$$

by solving for rf. Now consider the rectangle in Figure 2.17, which we have extracted from a histogram. The base is of length ℓ, the height is d, and the area

FIGURE 2.16 HISTOGRAM OF COMPLAINT TIMES

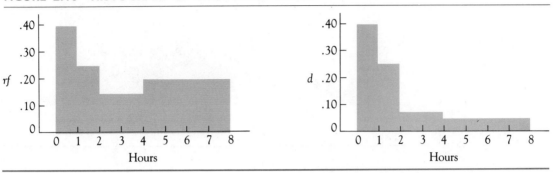

FIGURE 2.17 $rf = \ell \times d;$ **Area** $= \ell \times d$

is base × height = $\ell \times d$. Thus the relative frequency of an interval is numerically equal to the area of that part of the histogram lying above that interval. We therefore interpret areas as relative frequencies. Further, since the relative frequencies must sum to 1, this means that the total area bounded by any of these density histograms is 1. We illustrate this concept with an example.

Example 2.10

A pesticide is studied in different applications. The distribution of the duration of its effectiveness (in days) is given by the histogram of Figure 2.18. The relative frequency of the interval 0–20 is the area of the first rectangle, whose height is .01 and whose base is 20. Thus $rf = .01 \times 20 = .2 = 20\%$. For the interval 20–30, the length is 10, the height is .02 and

$$rf = .02 \times 10 = .2 = 20\%.$$

For the interval 30–40 we find

$$rf = 10 \times .03 = .3 = 30\%,$$

For the interval 40–50, we find again that $rf = 10 \times (.02) = .2 = 20\%$, and finally for the interval 50–60

$$rf = 10 \times .01 = .10 = 10\%$$

You should be able to look at the areas of the rectangles and decide without computation which intervals have the greater relative frequencies.

FIGURE 2.18   DISTRIBUTION OF EFFECTIVENESS TIMES

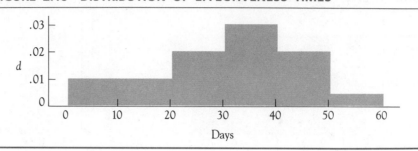

We shall in time consider theoretical distributions. Their graphs can be thought of as generalizations of histograms where densities have been employed. It is not surprising that the total area bounded by such graphs is 1 and that we can interpret areas under these curves as relative frequencies (or probabilities), as we have done with the histograms.

2.5   DISTRIBUTION CURVES FOR A CONTINUOUS VARIABLE

What is the form of the histograms we meet in a study involving a continuous variable? If we focus on a particular population we find that there is not a great deal of difference in most of the samples, as long as the sample sizes are not too small. True, there will be small differences, but these tend to disappear from the histograms as the sample sizes become larger and larger, provided we use smaller and smaller intervals for grouping the observations and make use of the density scale. The reason for this tendency is that there is a distribution curve for the variable in the population that most of the histograms will approximate. We discover the shape of this distribution curve by examining the shapes of the histograms of larger and larger samples from the population (Figure 2.19).

FIGURE 2.19   HISTOGRAMS APPROXIMATING A FREQUENCY CURVE

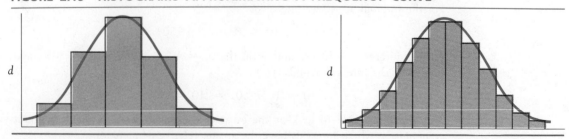

While there are many distribution curves, we find in practice that the following four forms shown in Figure 2.20 are most commonly encountered.

FIGURE 2.20   FOUR COMMONLY OCCURRING DISTRIBUTION CURVES

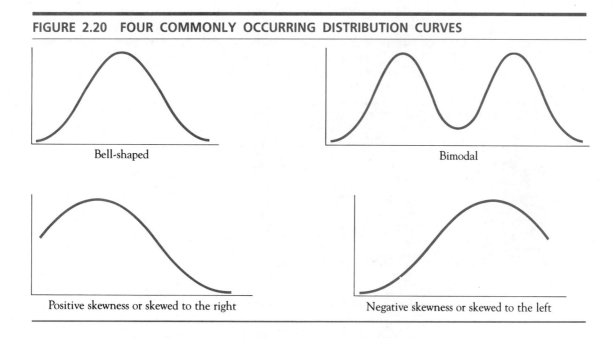

Bell-shaped

Bimodal

Positive skewness or skewed to the right

Negative skewness or skewed to the left

The bell-shaped curve is representative of a very important distribution called the *normal distribution*. Continuous variables having such a distribution are said to be normally distributed and will be studied in Chapter 6. Many variables, both discrete and continuous, show a strong symmetrical clustering about a central value—as exhibited by the bell-shaped curve—and may be approximated by the normal distribution. The distribution of heights of 25-year-old males in Figure 2.21 is typical.

FIGURE 2.21   DISTRIBUTION OF MALE HEIGHTS IN INCHES

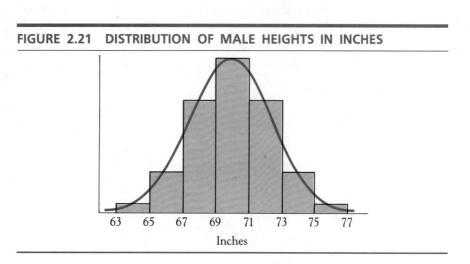

Inches

Bimodal distributions result from two clusterings, in essence two numbers that compete as central values. This was observed in the distribution of assembly times in Example 2.2. Similar results are obtained for the weights of adults, since the males and females will cluster about two different values.

Skewed distributions result from a strong asymmetric clustering centered near the beginning or end of the range of the variable. The histogram of infant mortalities during the first few weeks after birth (Figure 2.22) exhibits this characteristic: the frequency of deaths is highest in the first week after birth.

FIGURE 2.22   INFANT MORTALITY DISTRIBUTION CURVE

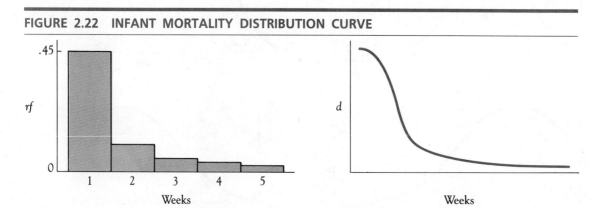

The vertical scale used for a distribution curve is the density scale, since it is used for the histograms that produced the curve through successive approximations. As a result, the total area under the curve is 1 and the relative frequency of an interval is the area under the curve above that interval.

Example 2.11

The distribution curve for the coverage (square feet) of a one-gallon can of a certain brand of paint is given in Figure 2.23. Approximately what percentage of all cans of this paint will cover between 395 and 405 square feet?

We need to approximate the area under this curve above the interval 395–405. Think of an approximating histogram and draw the rectangle corresponding to

FIGURE 2.23   APPROXIMATING THE AREA UNDER A CURVE BY USING A RECTANGLE

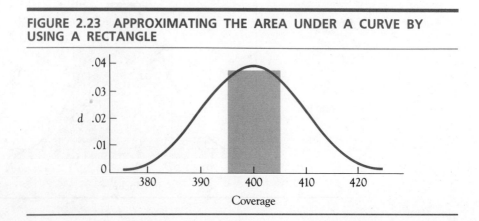

this interval. Its height should be somewhere between .035 and .040, so we use .0375. The interval 395–405 is of length 10. Thus

$$rf = \text{area} = 10 \times .0375 = .375 = 37.5\% \text{ (approximately)}$$

Thus about 37.5% of all cans of this paint should cover between 395 and 405 square feet of area. ■

In later chapters we introduce in more detail several distributions that are useful in making inferences. For these it will be possible to provide a table of areas and relative frequencies (or probabilities).

In the literature we often find a smooth curve used in place of a histogram in reporting the distribution of a sample. Frequencies rather than densities may also be used. Cne should simply interpret these *frequency curves* as approximations of the shape and form of the corresponding histogram.

Example 2.12

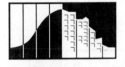

A Research Study

How fast do breast tumors grow? The growth rate of cancer cells is important both in designing regular screening schedules and for subsequent treatment once a tumor has been identified. A report by D. V. Fournier and coworkers in the *American Cancer Society Journal* of April 1980 details a study of the doubling times for breast tumors. The participants in the study resulted from a screening of some 22,000 women at the University of Heidelberg in the period from 1960 to 1977.

FIGURE 2.24   FREQUENCY DISTRIBUTION OF TUMOR DIAMETER, TUMOR VOLUME, AGE, AND DOUBLING TIME AT TIME OF DIAGNOSIS

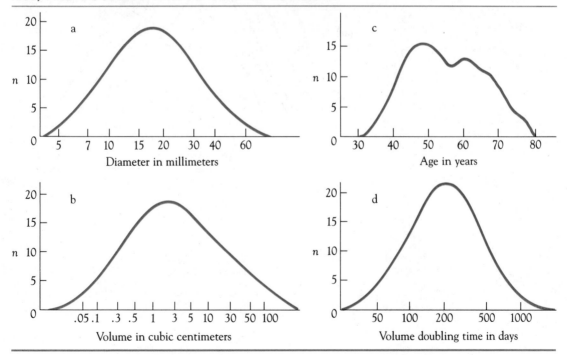

to 1977. Of the 972 breast cancers discovered, it was possible to track 147 over an extended period. These included some women who refused treatment, as well as others for whom there were delays in the final diagnosis and treatment. Individuals with abnormally fast or slow growing tumors were excluded. Between 2 and 11 mammographies were used on each subject; based on these tests, the rate of change in the volume of the tumors was computed. One hundred of the 147 cases had been completely analyzed at the time of the report and were the basis for the findings. Ninety-five percent of the times needed for the tumors to double in size were between 65 and 627 days, with the average being 202 days. In Figure 2.24 we have the frequency curves for the distribution of tumor diameters, volumes, age at detection, and volume doubling times. Figure 2.25 shows the predicted distribution of ages based on different tumor diameters. Thus women with a 20-millimeter tumor had an average age of 50 years, while for an 0.01-millimeter size the average was 35 years. The authors emphasize that these distributions are simply extrapolations and are certainly not confirmed by the data under study. However, should the distributions prove true, they would show that the growth time for a 20-millimeter tumor is about 20 years, that most breast cancers of the type studied begin in the age range from 30 to 40 years, and that there is hope for early detection by reliable screening methods.

FIGURE 2.25 FREQUENCY DISTRIBUTION OF ESTIMATED AGES AT EXTRAPOLATION ON DIFFERENT TUMOR DIAMETERS

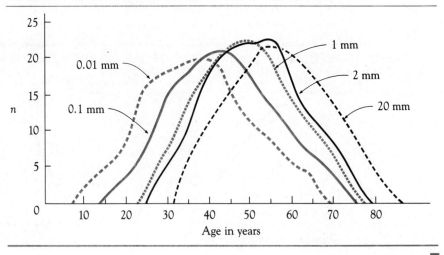

EXERCISE SET 2.2

Part A

1. For the distribution below:

Interval	Density
0–30	.004
30–40	.030
40–50	.040
50–60	.010
60–80	.004

    a. Which of the intervals has the highest relative frequency?

    b. Estimate the percent of the observations that are between 42 and 43.

    c. Estimate the percent of the observations that are between 52 and 56.

    d. Find the distribution in terms of relative frequencies and construct the histograms using both the density and relative frequency scales.

2. For the distribution below, give the histograms, using both the density scale and the relative frequency scale.

Interval	Density
40–45	.05
45–50	.06
50–55	.04
55–60	.05

3. A distribution of examination scores is given below. Compute the densities for each interval and draw the histogram, using the density scale.

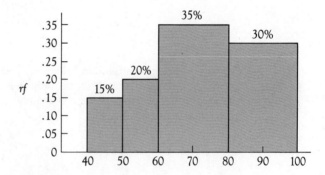

4. Give examples of variables and a population that should show:

    a. A skewed distribution.

    b. A bimodal distribution.

    c. A bell-shaped distribution.

    d. A symmetric distribution that is not bell-shaped.

5. Predict the form of the following distributions and sketch the distribution curves.

    a. Ages of drivers involved in accidents.

    b. Starting salaries of engineering graduates in 1985.

    c. Ages of women at the birth of their first child.

    d. Auto sales in the United States since 1948.

    e. Hours of service of electric motors before a first failure.

6. Suppose the variable X takes on values between 0 and 2 and has the distribution given below.

    a. What percent of the observations are between .75 and 1.5?

    b. Find the relative frequency of the interval from .2 to .5.

7. A symmetric distribution is given. What is the area under the curve to the left of $x = 80$?

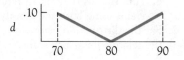

Part B

8. A symmetric distribution for a variable Z is given. The area of the shaded region is .40, as indicated.
   a. What is the area under the curve to the right of $z = 1.65$?
   b. What is the area under the curve to the left of $z = 1.65$?
   c. What is the area under the curve between -1.65 and $+1.65$?

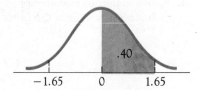

9. For the symmetric distribution below, the area of the shaded region is .95.
   a. What is the area of each of the unshaded regions under the tails of the curve?
   b. Approximately what percent of the observations that produced this distribution exceed 140?

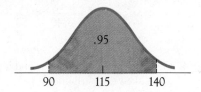

10. After studying the outcomes of many spins of a wheel of fortune at a casino, you believe the distribution of outcomes is as shown below. Which numbers appear to be good bets?

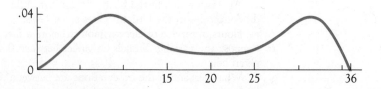

11. Three hundred people are asked to rate a perfume fragrance on a scale of 0 to 100. These data are grouped and the histogram below is obtained (density scale).
    a. How many ratings were between 30.5 and 50.5?
    b. How many ratings were greater than 50.5?

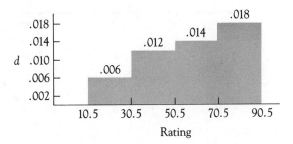

2.6   THE FREQUENCY POLYGON

In comparing the distributions of two samples we often superimpose the graph of one on the other, as we did in the accident-task study of Example 2.4. With histograms it is often difficult to distinguish which rectangle belongs to which distribution. Frequency polygons avoid some of this difficulty. Such graphs are constructed by assigning the interval frequencies to the midpoints of the respective intervals, plotting these points, and joining them with line segments.

Example 2.13   Two instructors, A and B, teach the same course using the same text and cover the same material. At the end of the term their students take a standardized test, producing the distributions of Table 2.12. Their frequency polygons are shown in Figure 2.26.

TABLE 2.12   RELATIVE FREQUENCIES

Interval	Midpoint	Instructor A	Instructor B
29.5–39.5	34.5	.10	.02
39.5–49.5	44.5	.08	.13
49.5–59.5	54.5	.14	.31
59.5–69.5	64.5	.28	.34
69.5–79.5	74.5	.24	.12
79.5–89.5	84.5	.13	.06
89.5–99.5	94.5	.03	.02

Now, which class performed better on the test? Don't let the high frequencies of Instructor B's class confuse you; these occur for the low grades. Instructor A's class excelled here for the most part. Now look at the high grades. Instructor A's class had the higher frequencies, excelling here also. Overall, Instructor A's class outperformed Instructor B's. However, this does not mean that Instructor A is a better teacher than Instructor B. Many other factors should be considered: Did the backgrounds of students in both classes match? Did either instructor teach to the test? What are the students' attitudes toward the subject after completing the course?

FIGURE 2.26   COMPARISON OF GRADE DISTRIBUTIONS

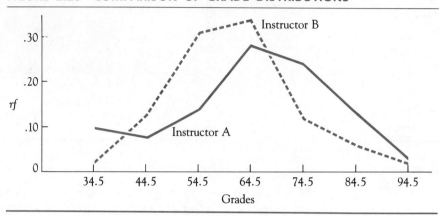

It should be noted that if the density scale is used, then the frequency polygons of large samples will approximate the distribution curve of the population variable in the same manner as the histograms.

EXERCISE SET 2.3

Part A

1. Give the frequency polygon for the distribution below.

Interval	rf
155–158	.12
158–161	.26
161–164	.34
164–167	.22
167–170	.06

2. Draw the frequency polygon corresponding to the histogram below.

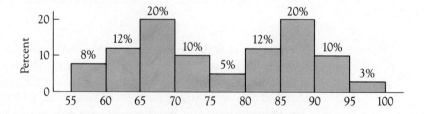

3. The distributions of the miles per gallon of samples of autos from two weight classifications are given below. Draw the frequency polygon for each (using the same set of axes) and compare the performances of these two classifications.

Miles per gallon	Under 2500 pounds rf	Over 2500 pounds rf
11.5–14.5	.025	.050
14.5–17.5	.065	.165
17.5–20.5	.145	.315
20.5–23.5	.200	.250
23.5–26.5	.245	.105
26.5–29.5	.185	.085
29.5–32.5	.125	.030
32.5–35.5	.010	.000

4. A study of the smoking habits of lung cancer patients involving the average number of cigarettes smoked per day and the type of cigarette produced the following two distributions:

Number of cigarettes	SMOKERS f (Filter)	f (Nonfilter)
1–10	5	10
11–20	12	15
21–30	18	33
31–40	38	53
41–60	29	40

  a. Draw the (relative) frequency polygons for both distributions and compare.
  b. Why should relative frequencies be used if one is going to make a comparison?
5. When the same examination was given to two sections of a statistics course, the following grades were obtained:

Section 1

Grade x	51	54	55	58	63	65	66	70	71	72	73	76	77	79	83	88	92	98
Frequency f	1	1	2	1	2	1	2	1	2	3	1	2	3	2	1	3	1	2

Section 2

Grade x	53	57	61	64	67	68	72	74	75	76	78	79	83	85	86	91	94	97
Frequency f	1	1	1	2	1	1	4	3	2	1	1	1	2	3	4	3	4	1

Using five intervals of length 10 starting at 50.5, group the data and obtain a relative frequency distribution for the grades of each section. Then draw the frequency polygons and compare the performances of the two sections.

Part B

6. Two graphs are combined in the figure below. The first shows the ages of people for whom non-B viral hepatitis is reported in the population. The second charts data from a Louisiana study of the ages of household members who contracted hepatitis as a result of having their children in daycare centers that had outbreaks of hepatitis. Note that different scales are used.
  a. What age intervals were used for each report?
  b. If there were 22,000,000 people between 20 and 25 years of age, how many cases of hepatitis were reported for this age group?

AGE-SPECIFIC RATE OF CLINICAL HEPATITIS

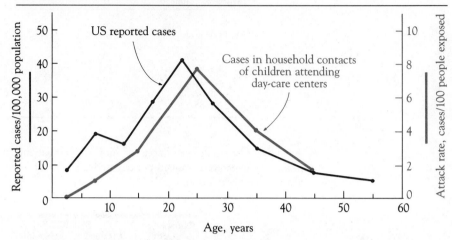

Data are for household contacts of children attending affected daycare centers and for hepatitis A cases reported to the Center for Disease Control.

Source: "Hepatitis in Day Care Centers," Storch et al., *Journal of the American Medical Association*, Oct. 5, 1979.

7. Studies of PCB levels in water sources in 1972, 1973, and 1974 produced the distribution below. Assuming that no readings exceeding 4 parts per million were found, draw the frequency polygons for these three distributions. Use the same scale for all three and compare for low levels of contamination.

Number of tests	Parts per million detected	rf 1972	rf 1973	rf 1974
1972: 4102	0	.260	.245	.091
1973: 1277	<1	.155	.402	.506
1974: 1047	1–2	.507	.296	.354
	>2	.079	.055	.049

Note: Relative frequencies do not total 1 because of rounding.

Source: *Review of PCB Levels in the Environment*, U.S. Environmental Protection Agency, Jan. 1976.

8. A survival rate is the fraction of the original sample who survive to a given time. Below we have two survival rate graphs for acute myocardial infarction (heart attack) patients. The solid orange line is for those receiving anticoagulant therapy, while the black line shows subjects who did not receive the therapy. The first group included 572 patients, the second 361.

   a. After 5 years, how many of each group were still alive? (Estimate.)

   b. After 10 years, what percent of each group was still alive? (Estimate.)

   c. Is the therapy effective?

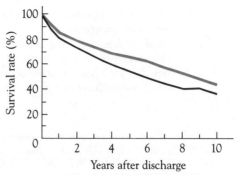

Years after discharge

Source: Additional Data Favoring Use of Anticoagulant Therapy in Myocardial Infarction, M. Szklo et al., *Journal of the American Medical Association*, vol. 242, no. 12, Sept. 1980

9. A study of the relationship between respiratory symptoms and smoking habits according to the sex of the individual was carried out with 899 adults (39.55% males) in a Paris industrial center. All graphs appearing in this study are reproduced below.

   a. What conclusion can you draw about the ages of the smokers and nonsmokers in this study? Use Figure A.

A   AGE DISTRIBUTION OF MALES AND FEMALES BY SMOKING STATUS

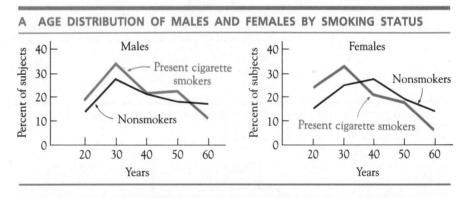

B   FREQUENCY OF SUBJECTS WITH CHRONIC RESPIRATORY SYMPTOMS ACCORDING TO SMOKING STATUS AND SEX

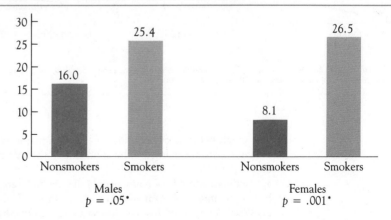

b. Judging from Figures B, C, and D, which group has the greater prevalence of respiratory symptoms?

c. Based on this study, the authors concluded that when women smoked, "the relative risk of having respiratory symptoms was 3.3 times higher than when they did not smoke, whereas in males that risk was 1.6 times higher." How were these risks computed?

C FREQUENCY OF SUBJECTS WITH CHRONIC RESPIRATORY SYMPTOMS ACCORDING TO SMOKING CATEGORIES AND SEX

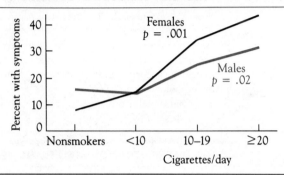

D PREVALENCE OF DYSPNOEA AND WHEEZING ACCORDING TO SMOKING CATEGORIES AND SEX

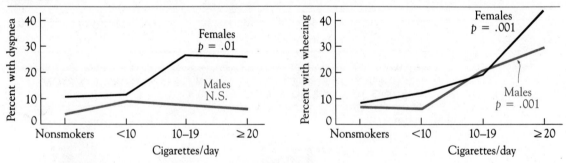

(Source: "Smoking and Chronic Respiratory Symptoms: Prevalence in Male and Female Smokers," R. Liard et al., *American Journal of Public Health*, vol. 70, no. 3, March 1980.)

10. The graph below shows the loss of life from major earthquakes since 1900. Individual events are marked. The solid line is the annual average (taken five years at a time).

a. Approximately how many people died in the San Francisco quake?

b. Approximately how many people died in the Tangshan quake?

c. What time period had the lowest frequency of earthquakes?

d. Approximately how many people died in 1960 as a result of earthquakes?

LOSS OF LIFE CAUSED BY MAJOR EARTHQUAKES SINCE 1900

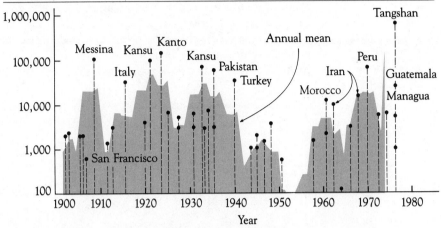

Individual events are marked. The solid line is the annual average (taken five years at a time).
Source: "The Size of Earthquakes," H. Kanamori, *Earthquake Information Bulletin*, Jan.–Feb. 1980.

11. The graph of a cumulative distribution (see problem 17, set 2.1) shows the frequency of observations less than or equal to *x*. Such a graph is called a cumulative frequency histogram. A prime is a whole number with exactly two positive factors, namely 1 and itself. Thus the first eight primes are 2, 3, 5, 7, 11, 13, 17, and 19. The cumulative frequency histogram for the primes less than 100 is given below on the left. On the right is the approximate cumulative frequency curve for the primes less than 50,000.
   a. From the histogram, how many primes are less than 50?
   b. From the frequency curve, estimate the number of primes less than 50,000.
   c. Estimate the number of primes between 30,000 and 40,000 and compare with the number of primes less than 10,000.

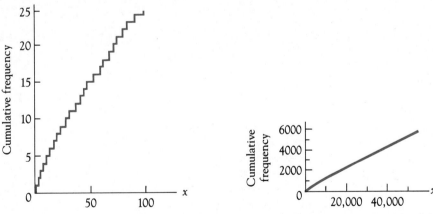

Source: "The First 50 Million Primes," Don Zagier, *Mathematical Intelligencer*, Springer-Verlag NYC, vol. 0, 1977

COMPUTER PRINTOUT

Minitab printouts were discussed briefly at the end of Chapter 1. It is an easy matter to produce histograms using this versatile program. As we illustrate below (Figure 2.27), either the computer can generate the histogram or the intervals can be specified by giving the midpoint of the first and their common length. The data supplied are a set of examination scores.

FIGURE 2.27 PRINTOUT OF DISTRIBUTIONS GENERATED BY MINITAB

```
MTB >SET THE FOLLOWING DATA INTO C1
DATA>89 78 87 65 74 90 63 76 90 98 99 50 81 82 96
DATA>73 69 73 81 73 72 54 48 957 4
DATA>END
MTB >HISTOGRAM OF C1

     MIDDLE OF      NUMBER OF
     INTERVAL       OBSERVATIONS
        50,             2       **
        55,             1       *
        60,             0
        65,             2       **
        70,             2       **
        75,             6       ******
        80,             4       ****
        85,             1       *
        90,             3       ***
        95,             2       **
       100,             2       **

MTB >HISTOGRAM OF C1, FIRST MIDPOINT 45, INTERVAL
WIDTH 10

     MIDDLE OF      NUMBER OF
     INTERVAL       OBSERVATIONS
        45,             1       *
        55,             2       **
        65,             3       ***
        75,             8       ********
        85,             5       *****
        95,             6       ******

MTB >STOP
```

CHAPTER CHECKLIST

Vocabulary

Frequency
Relative frequency
Class interval
Class frequency
Interval frequency
Frequency distribution
Relative frequency distribution
Stem-and-leaf diagram

Average density
Distribution curve
Bell-shaped distribution
Bimodal distribution
Skewed right distribution
Skewed left distribution
Normal distribution
Frequency polygon

Techniques

Constructing a frequency table

Constructing a distribution (table) for a sample

Constructing and interpreting the histogram of a distribution

Constructing a stem-and-leaf diagram

Interpreting stem-and-leaf diagrams

Finding the average density of each interval of a distribution

Finding the relative frequency of a class interval from the corresponding average density

Finding the relative frequency of a class interval from the area under the density curve above the given interval

Giving examples of distributions of the following types: bell-shaped, bimodal, skewed right, skewed left

Sketching the shape of the density curve for the four types of distributions above

Constructing the frequency polygon of a distribution

Comparing two distributions by using their frequency polygons

Notation

f Frequency

rf Relative frequency

d Average density of an interval of a distribution

CHAPTER TEST

1. The effective relief times (in minutes) of an oral medication for a certain allergy condition was studied on 25 adults. The following results were obtained.

   182, 211, 187, 188, 203, 200, 195, 185, 205, 206, 205, 215, 191, 203, 199
   218, 202, 197, 214, 195, 193, 202, 208, 196, 200

   Construct a distribution and its histogram, using seven intervals with the left end point of the first at 181.5.

2. Electronic assemblers working in two different work environments are studied to determine how long it takes them to assemble a circuit board. The distributions are given below. Draw the frequency polygon of each and compare the performances of the two groups.

Environment A		Environment B	
Time (minutes)	Relative frequency	Time (minutes)	Relative frequency
10.0–10.5	.12	10.0–10.5	.15
10.5–11.0	.14	10.5–11.0	.28
11.0–11.5	.21	11.0–11.5	.24
11.5–12.0	.26	11.5–12.0	.14
12.0–12.5	.17	12.0–12.5	.12
12.5–13.0	.10	12.5–13.0	.07

3. Give an example of a population whose distribution should be
   a. Bell-shaped.
   b. Bimodal.
   c. Skewed right.
   d. Skewed left.
4. Students are surveyed as to their number of close friends (individuals to whom they would not be afraid to confide a deep secret). The frequency distribution of replies is given below.
   a. How many students were surveyed?
   b. Compute the relative frequencies.
   c. What percent of the students had 2 or fewer close friends?
   d. Construct the histogram of the distribution.

Number of friends x	f
0	2
1	3
2	5
3	5
4	4
5	1

5. A distribution of the dollar value of auto accident claims is given below.
   a. Construct the histogram of this distribution.
   b. Find the distribution if intervals of length 600 are used with the first interval beginning at 100.
   c. Using the average density scale, construct the histogram of the distribution of part b.

Interval	Relative frequency
100–400	.06
400–700	.08
700–1000	.04
1000–1300	.04
1300–1600	.08
1600–1900	.10
1900–2200	.12
2200–2500	.14
2500–2800	.09
2800–3100	.07
3100–3400	.06
3400–3700	.04
3700–4000	.03
4000–4300	.05

6. The (average) density of an interval of a distribution is .008. What is the relative frequency of the interval if its length is:
   a. 20.
   b. 15.
7. When the average density scale is used, what is the total area bounded by the histogram and the x-axis?
8. A sample of the boarding records of a regularly scheduled Los Angeles–New York flight showed the following number of no-shows.

   30, 24, 20, 17, 10, 7, 5, 2, 4, 6, 32, 37, 34, 33, 35, 15, 19, 24, 20, 22, 22, 18, 14, 23, 23, 12, 18, 15, 16, 26, 28, 25, 26, 25, 27, 13, 14, 28, 24, 22

   a. Prepare a stem-and-leaf diagram of these data.
   b. By using each of the stems 0, 1, 2, and 3 twice, prepare a double stem-and-leaf diagram.
9. The average dollar expenditure per family member of a sample of 48 families visiting a large East Coast amusement park are given below.
   a. Prepare a stem-and-leaf diagram and draw the corresponding histogram.
   b. Construct a double stem-and-leaf diagram, using each of the stems 2, 3, 4, 5, 6, and 7 twice. Then draw the corresponding histogram.
   76, 72, 65, 60, 55, 50, 45, 43, 40, 42, 35, 35, 30, 30, 30, 25, 20, 20, 22, 70, 67, 68, 36, 37, 25, 28, 26, 22, 62, 56, 58, 63, 59, 55, 60, 45, 50, 50, 52, 47, 54, 52, 53, 45, 34, 32, 32, 30
10. Using the data of the stem and leaf diagram below, construct a (relative) frequency distribution table based on the intervals $5.00 \le x < 5.20$, $5.20 \le x < 5.40$, and so forth.

5.2	34
5.3	
5.4	573
5.5	220435
5.6	35670123
5.7	2135460987
5.8	55649023458
5.9	2313458745231551
6.0	42305
6.1	315

CHAPTER THREE
Descriptive Statistics: Numerical Measures

Graphs and distribution tables are important aids in summarizing and presenting the data of a study. In most cases, however, it is virtually impossible to base sound decisions on these rather general summaries. What is needed are a few concise numerical measures that describe characteristic features of the data.

One of the more important features of a collection of data is its tendency to cluster about a center. Depending upon one's particular interests, there are many choices as to how this center should be defined. Consider, for example, the grades on an examination. The *median* or 50th percentile locates a center based on numerical ranks. The class average or *mean* provides a center based on the actual numerical value of the grades. The most frequently occurring grade, the *mode*, (literally, the fashion), can be regarded as a center since it is the most common. While there are many other measures of central tendency, it is these three, the median, the mean, and the mode, that are the most useful for our purposes.

A second important feature of a collection of data is its scattering or variation. The center, whatever is used, provides no information on this. The reader can easily visualize many sets of examinations where the average grade is 70. Some will have grades scattered all the way from 0 to

100, others from only 55 to 90, and so forth. Two important measures of this scattering, the *variance* and the *standard deviation,* are discussed.

The final two topics in this chapter are *z scores* and *percentiles.* The *z* scores locate individual observations relative to the mean, using the standard deviation as the unit of measure. Percentiles describe locations of individual observations relative to the sample as a whole.

Before concluding these introductory remarks, we point out that many of the topics developed in this chapter also have important roles in inferential statistics. As a result they will find extensive use throughout the remainder of the text.

3.1   MEASURES OF CENTRAL TENDENCY

In elementary statistics the most frequently used measures of the center of a collection of observations are the mode, the median, and the mean.

The mode is that observation that dominates a sample by virtue of its frequency of occurrence.

DEFINITION 3.1

The *mode* of a collection of observations is that value that occurs most frequently.

Example 3.1   For the jean waist size data shown in Table 3.1, the mode is $x = 7$

TABLE 3.1   JEAN WAIST SIZES OF 573 SALES TO YOUNG WOMEN

Waist size x	f
5	118
7	233
9	116
11	62
13	44

It should be noted that there may be several values that have the same "highest" frequency. We may then speak of the observations being multimodal (e.g., bimodal, trimodal, etc.). When working with grouped data, the term *modal class* is occasionally used for that class having the greatest frequency.

The mode is used primarily to identify active categories and preferred positions that occur with clothing sizes, makes of automobiles, political views (in opinion sampling), and so forth. It is particularly useful with large samples of data, such as occur with census studies and industry wide production surveys. Since the mode may not exist for a sample and also does not lend itself to an algebraic treatment, it finds little use in inferential statistics.

When the observations in a sample are listed according to size, one's interest naturally focuses on the middle of the list. This is the median, denoted by $\tilde{x}$. If the number of observations is odd, there is a middle observation in the list and this is the median. If the number is even, there is no middle observation. We then take the median to be the average of the two most central observations.

DEFINITION 3.2

Let the observations of a sample be arranged in order of increasing size (duplications included) and labeled $x_1, x_2, \ldots, x_n$. If n is odd, then $n = 2k - 1$ and the *median* is

$$\tilde{x} = x_k \text{ (the middle observation)}$$

If n is even, then $n = 2k$ and the median is

$$\tilde{x} = \frac{x_k + x_{k+1}}{2}$$

Example 3.2

Find the median of the following examination scores:

$$80, 56, 34, 67, 55, 91, 82, 47, 75, 31, 90$$

First we arrange the scores according to size:

$$34, 34, 47, 55, 56, 67, 75, 80, 82, 90, 91$$

Since the number of scores is odd, $2k - 1 = 11$ and $k = 6$. The median is the sixth or middle observation of the ranked list. Thus $\tilde{x} = x_6 = 67$. ∎

Example 3.3

The grade point averages of the pledge classes of 10 campus sororities are given below. Find the median.

$$2.1, 2.4, 2.5, 2.5, 2.6, 2.7, 2.8, 2.9, 3.2, 3.4$$

Note that the observations are already ordered according to size. Since the number of observations (10) is even, $2k = 10$ and $k = 5$. Thus the median is midway between the fifth and sixth observation:

$$\tilde{x} = \frac{x_5 + x_6}{2} = \frac{2.6 + 2.7}{2} = 2.65 \quad ∎$$

The median of a sample, unlike the mode, always exists. it has the useful property that it tends to be insensitive to the occurrence of extreme values in large data sets, such as occur with nationwide economic and public health surveys. It is easily computed and is often included with statistical summaries. The full importance of the median will, however, not become apparent until nonparametric tests based on ranks are considered in Chapter 13.

Neither the mode nor the median fully take into account the sizes of the observed values. The mode focuses on that value that occurs most frequently and ignores all else. The median is oblivious to all values except the one which is at the center of the rankings. The average or mean, which we discuss next, uses all of the observed values and as a result has several properties that recommend it for use as the center.

DEFINITION 3.3

The *mean* of the observations $x_1, x_2, \ldots, x_n$ is defined by

$$\bar{x} = \frac{x_1 + x_2 + \ldots x_n}{n} = \frac{\sum_{i=1}^{n} x_i}{n}$$

Example 3.4.

The commissions charged by four full-service stock brokers on a sale of 100 shares of stock are as follows:

$$43.60, \ 48.20, \ 53.40, \ 44.20 \ \text{(dollars)}$$

The mean or average commission charge is

$$\bar{x} = \frac{43.60 + 48.20 + 53.40 + 44.20}{4} = \frac{189.40}{4} = 47.35 \ \text{(dollars)}$$

A word about notation. For a variable X, we use $\bar{x}$ to denote its mean in a sample. For the mean of a population we use the greek letter μ (mu), equivalent to our m. To emphasize the role of the variable, we shall also write μ_X, which is read simply as "the mean of X." The Greek letter Σ (sigma) corresponds to our letter S and is used to indicate a summation. Thus

$$\sum_{i=1}^{n} x_i$$

denotes the sum of x_i values starting with x_1 ($i = 1$) and ending with x_n ($i = n$). Dividing this sum by n yields the mean:

$$\bar{x} = \frac{\sum_{i=1}^{n} x_i}{n} \quad \blacksquare$$

Example 3.5.

If $x_1 = 20$, $x_2 = 12$, $x_3 = 16$, $x_4 = 20$, $x_5 = 24$, and $x_6 = 10$ then

$$\sum_{i=1}^{6} x_i = x_1 + x_2 + x_3 + x_4 + x_5 + x_6$$

$$= 20 + 12 + 16 + 20 + 24 + 10 = 102$$

and

$$\bar{x} = \frac{\sum_{i=1}^{6} x_i}{6} = \frac{102}{6} = 17 \quad \blacksquare$$

If it is clear that the summation is to extend over all of the observed values of x, then we conserve on notation and write either Σx_i or Σx in place of

$$\sum_{i=1}^{n} x_i$$

If the observations $x_1, x_2, \ldots, x_k$ have associated frequencies $f_1, f_2, \ldots f_k$ then the mean is computed by

$$\bar{x} = \frac{x_1 f_1 + x_2 f_2 + \ldots + x_k f_k}{f_1 + f_2 + \ldots + f_k} = \frac{\sum_{i=1}^{k} x_i f_i}{n}$$

where

$$n = \sum_{i=1}^{k} f_i$$

is the sum of the frequencies. $\blacksquare$

Example 3.6

The distribution of word lengths in a message is given in Table 3.2.

TABLE 3.2 DISTRIBUTION
OF WORD LENGTHS

x	f
2	16
3	5
4	38
5	20
6	18
7	6
8	4
9	3

The mean word length is

$$\bar{x} = \frac{2{\cdot}16 + 3{\cdot}5 + 4{\cdot}38 + 5{\cdot}20 + 6{\cdot}18 + 7{\cdot}6 + 8{\cdot}4 + 9{\cdot}3}{16 + 5 + 38 + 20 + 18 + 6 + 4 + 3}$$

$$= \frac{508}{110} = 4.6 \text{ (rounded)}$$

Note that this is exactly what would have resulted from computing the mean of a list containing sixteen 2s, five 3s, thirty-eight 4s, and so forth. $\blacksquare$

When the observations in the sample have been grouped, we treat all observations of any particular class as if they are uniformly distributed over the class interval. In computing the mean, this is equivalent to assuming that all observations of a class are located at the midpoint of the corresponding interval.

DEFINITION 3.4

If $x_1, x_2, \ldots, x_k$ are the midpoints of the class intervals of a sample of grouped data with associated frequencies $f_1, f_2, \ldots, f_k$, then the mean of the sample is defined by

$$\bar{x} = \frac{x_1 f_1 + x_2 f_2 + \ldots x_k f_k}{f_1 + f_2 + \ldots f_k} = \frac{\sum\limits_{i=1}^{k} x_i f_i}{n}$$

where

$$n = \sum_{i=1}^{k} f_i$$

We often refer to this mean as the mean of the (sample) distribution.

Example 3.7

A study of the upper division grade point averages of 150 graduate school applicants produced the distribution of Table 3.3. Find the mean.

TABLE 3.3 DISTRIBUTION OF GRADE POINT AVERAGES OF GRADUATE SCHOOL APPLICANTS

Grade point average interval	Midpoint	Frequency
2.8–3.0	2.9	10
3.0–3.2	3.1	24
3.2–3.4	3.3	53
3.4–3.6	3.5	34
3.6–3.8	3.7	17
3.8–4.0	3.9	12

$$\bar{x} = \frac{2.9(10) + 3.1(24) + 3.3(53) + 3.5(34) + 3.7(17) + 3.9(12)}{10 + 24 + 53 + 34 + 17 + 12}$$

$$= \frac{29 + 74.4 + 174.9 + 119 + 62.9 + 46.8}{150} = \frac{507}{150} = 3.38$$

Naturally the mean of a set of grouped data usually will not agree with that of the ungrouped data. Generally, however, they are quite close.

3.2 PROPERTIES OF THE MEAN

We now wish to demonstrate two important properties of the mean as a center. Both are related to the distance or *deviation* of the individual observations from the mean.

DEFINITION 3.5

The *deviation* of the observation x_i from $\bar{x}$ is the number $d_i = x_i - \bar{x}$.

The deviation d_i is the directed distance from $\bar{x}$ to x_i. The magnitude of the deviation (the absolute value $|x_i - \bar{x}|$) describes how far x_i is located from $\bar{x}$; the sign indicates on which side of $\bar{x}$ it is located.

Example 3.8

The weights of five coins were reported in ounces as follows: 1.523, 1.531, 1.528, 1.521, 1.522. The mean is

$$\bar{x} = \frac{1.523 + 1.531 + 1.528 + 1.521 + 1.522}{5} = 1.525$$

The deviations from the mean are

$$d_1 = 1.523 - 1.525 = -.002 \text{ (thus 1.523 is 0.002 ounces below } \bar{x})$$
$$d_2 = 1.531 - 1.525 = +.006 \text{ (thus 1.531 is 0.006 ounces above } \bar{x})$$
$$d_3 = 1.528 - 1.525 = +.003 \text{ (thus 1.528 is 0.003 ounces above } \bar{x})$$
$$d_4 = 1.521 - 1.525 = -.004 \text{ (thus 1.521 is 0.004 ounces below } \bar{x})$$
$$d_5 = 1.522 - 1.525 = -.003 \text{ (thus 1.522 is 0.003 ounces below } \bar{x})$$

The locations of the observations relative to the mean are given in Figure 3.1.

FIGURE 3.1 COIN WEIGHTS AND THEIR MEAN

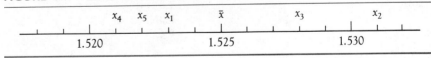

The sum of the deviations is

$$\sum_{i=1}^{5} d_i = (-.002) + .006 + .003 + (-.004) + (-.003) = 0 \quad \blacksquare$$

In this last example the sum of the individual deviations from the mean turned out to be 0. This is not an accident.

THEOREM 3.1

The sum of the deviations of the observations $x_1, x_2, \ldots, x_n$ from their mean $\bar{x}$ is 0.

Thus the mean is a measure of the center, in the sense that the sum of the deviations from it is 0. It is an interesting fact that this same choice minimizes the sum of the squares of the deviations.

THEOREM 3.2

Let $\bar{x}$ be the mean of $x_1, x_2, \ldots, x_n$. The quantity

$$\sum_{i=1}^{n} d_i^2 = (x_1 - x)^2 + (x_2 - x)^2 + \ldots + (x_n - x)^2$$

is a minimum when $x = \bar{x}$.

We omit the proof and illustrate the results with an example.

Example 3.9

Find the value of x that minimizes

$$y = (3 - x)^2 + (5 - x)^2 + (7 - x)^2$$

Theorem 3.2 states that $\bar{x} = (3 + 5 + 7)/3 = 5$ produces the minimum. The graph of y versus x is given in Figure 3.2 along with a Table 3.4, which shows values of y for selected values of x. Note that y increases as x moves away from 5.

TABLE 3.4 THE MINIMUM VALUE $y = 8$ OCCURS AT $x = 5$

x	y
2	35
3	20
4	11
5	8
6	11
7	20
8	35

FIGURE 3.2   GRAPH OF $y = (3 - x)^2 + (5 - x)^2 + (7 - x)^2$

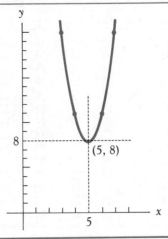

The results of this section are just two of many that ultimately form the foundation of an extensive theory relating sample means to population means. This theory, which we take up in Chapter 7, leads to many significant applications in both estimation and hypothesis testing. As a result, the mean is the most popular and possibly the most important measure of the center of a sample or a population.

Example 3.10

A Research Study

Does the death penalty cause an increase in the number of homicides? In social science, there is a "brutalization" hypothesis, which states that "the use of the death penalty as a punishment diminishes people's respect for life and thus increases the incidence of homicide." This hypothesis is the subject of a study entitled "The Brutalization Effect," by David King (*Social Forces*, Vol. 57, 2, 1978).

The investigation is centered on newspaper stories of executions and statistics of reported homicides in South Carolina from 1950 to 1963. The premise is that if the brutalization hypothesis holds, then the month in which the story of an execution appears (a story month) should have a greater incidence of homicide than would be normally expected.

The expected number of homicides for a story month was computed as an average obtained from the same month of the preceding or succeeding years not having an execution story. If the hypothesis holds, then there should be a large positive difference between the observed and expected number of homicides in a story month; while there should be some variation, this difference should carry over to the mean of the differences.

Table 3.5 gives homicides by the month and year for the period of the study. Table 3.6 gives the differences between the observed and the predicted number of homicides for the story month and the month immediately following. The mean of $-.6$ indicates that on the average there were *fewer* homicides than expected, offering some support to a deterrent hypothesis. Results for the follow-up months offer some support to the notion of a delayed effect, but the author's conclusion is that "at best . . . newspaper publicity about executions bears little relationship to the incidence of homicide."

TABLE 3.5   HOMICIDES BY MONTH AND YEAR: SOUTH CAROLINA, 1950–63

Year	Jan	Feb	Mar	April	May	June	July	August	Sept	Oct	Nov	Dec
1950	20	20	16	14	17	23	10	19	21	31	12	20
1951	12	14	11	35	25	22	12	12	32	21	15	21*
1952	20	15*	21	13	14*	20*	14	13	23	18	25	28
1953	12	22	21	18	14	14	12	20	22	15	20	18
1954	18	12	22*	20	22*	13	17	18	20*	18	11	16
1955	19	18	18	13*	22	20	22	21	21	13*	19	14*
1956	12*	20*	17	21	17	15	19	23*	18*	22	20	23
1957	12	17	22	18	17	23*	19*	23	14	18	14	19
1958	14	14	17	9	14	15	17	17	24	13	16	19
1959	25	6	17	20	21	15	16	24	25	15	24	22
1960	14	20	18	26	15	24	30	22	27	28	19	23
1961	17	29	21	15	11*	19*	16	18	23	13	15	18
1962	25	28	33	14	19*	14	16	22	21	26	28	22
1963	14	11	16	15	17	18	24	33	21	16	20	21

*Indicates a month in which an execution story appeared.
Source: Compiled from *South Carolina Vital Statistics*

TABLE 3.6   DIFFERENCE BETWEEN OBSERVED AND EXPECTED NUMBERS OF HOMICIDES FOR THE MONTH OF AND THE MONTH IMMEDIATELY FOLLOWING THE APPEARANCE OF AN EXECUTION STORY

Month and year of story	Month of story	One month after story
December 1951	−3.0	+8.0
February 1952	−3.0	+5.0
May 1952	−5.5	—†
*June 1952	+2.0	+2.0
March 1954	+2.5	+0.5
May 1954	+4.0	−4.0
September 1954	+1.5	−0.5
April 1955	−7.5	+6.5
October 1955	−7.0	+4.5
December 1955	+5.5	—†
*January 1956	−3.5	—†
*February 1956	+2.5	−3.0
August 1956	+1.0	—†
*September 1956	+0.5	+4.0
June 1957	+8.0	—†
*July 1957	+1.0	+4.0
May 1961	−4.3	—†
*June 1961	0.0	—†
*July 1961	−7.0	+4.0
May 1962	+3.0	−7.0
Mean	−0.6	+1.8
Standard	4.5	4.2

*Story month preceded by a story month.
†A story month which follows a story month.

It is possible to raise many questions regarding this experiment. One can for example, question attempts to predict the number of homicides in a given month. Certainly there are seasonal adjustments and linear trends, but the exact number of homicides fluctuates erratically from month to month. One can also question whether the people responsible for the homicides knew about the executions. And is there not the possibility of both a brutalization and a deterrent effect that cancel out one another? But **how** could one resolve such issues with the data at hand? The questions are so numerous and complex that one doubts that they can be addressed in a better manner than has been done here. ■

3.3 THE STANDARD DEVIATION

The mean by itself provides little information about a distribution. In Figure 3.3, we show three samples, each including eight observations. They are quite different and yet each has the same mean, $\bar{x} = 7$.

FIGURE 3.3 THREE SAMPLES WITH THE SAME MEAN

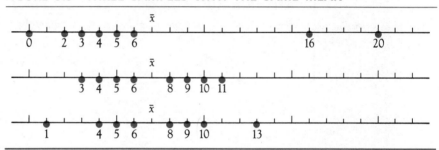

We need a measure or measures of how the observations are scattered about the mean. The individual deviations $d_i = x_i - \bar{x}$, while providing such measures, are too numerous. The average or mean of the deviations

$$\frac{(x_1 - \bar{x}) + (x_2 - \bar{x}) + \ldots + (x_n - \bar{x})}{n}$$

is useless since it is always zero (Theorem 3.1). The average of the absolute deviations

$$\frac{|x_1 - \bar{x}| + |x_2 - \bar{x}| + \ldots + |x_n - \bar{x}|}{n}$$

while certainly a possibility, is difficult to use in mathematical developments because of the absolute values. A more tractable expression is obtained by averaging the squares of the deviations and then taking their square root to return to the original dimensions of the data:

$$s' = \sqrt{\frac{(x_1 - \bar{x})^2 + (x_2 - \bar{x})^2 + \ldots + (x_n - \bar{x})^2}{n}}.$$

This quantity is defined as the *standard deviation* when the x values comprise the entire population.

Generally, however, we are trying to predict the standard deviation of a population from that of a sample. It then turns out that to obtain an unbiased estimate, a divisor of $(n - 1)$ rather than n should be used.

DEFINITION 3.6

The *standard deviation* of the sample $x_1, x_2, \ldots, x_n$ is defined by

$$s = \sqrt{\frac{(x_1 - \bar{x})^2 + (x_2 - \bar{x})^2 + \ldots + (x_n - \bar{x})^2}{n - 1}} = \sqrt{\frac{\sum_{i=1}^{n}(x_i - \bar{x})^2}{n - 1}}$$

The square of the standard deviation, s^2, is called the (sample) variance.

DEFINITION 3.7

The *variance* of the sample $x_1, x_2, \ldots, x_n$ is defined by

$$s^2 = \frac{(x_1 - \bar{x})^2 + (x_2 - \bar{x})^2 + \ldots + (x_n - \bar{x})^2}{n - 1} = \frac{\sum_{i=1}^{n}(x_i - \bar{x})^2}{n - 1}$$

At present, we will simply regard the variance as an intermediate quantity used in computing the standard deviation.

Example 3.11

The number of semester hours of science and mathematics reported by five candidates for elementary teaching credentials are as follows: 7, 12, 10, 20, 6. Find the mean and standard deviation.

$$\bar{x} = \frac{7 + 12 + 10 + 20 + 6}{5} = 11 \text{ hours}$$

$$s^2 = \frac{(7 - 11)^2 + (12 - 11)^2 + (10 - 11)^2 + (20 - 11)^2 + (6 - 11)^2}{5 - 1}$$

$$= \frac{16 + 1 + 1 + 81 + 25}{4} = \frac{124}{4} = 31 \text{ (hours)}^2$$

$$s = \sqrt{31} = 5.6 \text{ hours (rounded)} \quad \blacksquare$$

In journals and reports, one often sees *SD* and *Var* being used for the sample standard deviation and variance. We have used s and s^2, as is common in most textbooks. The corresponding small greek letter sigma σ and its square σ^2 are used for the standard deviation and variance of the population.

For grouped data, the standard deviation is computed by using the interval midpoints, as with the mean.

DEFINITION 3.8

If $x_1, x_2, \ldots, x_k$ are the interval midpoints of a sample of grouped data with corresponding frequencies $f_1, f_2, \ldots, f_k$, then the sample standard deviation is defined by:

$$s = \sqrt{\frac{(x_1 - \bar{x})^2 f_1 + (x_2 - \bar{x})^2 f_2 + \ldots + (x_k - \bar{x})^2 f_k}{n - 1}} = \sqrt{\frac{\sum_{i=1}^{k} (x_i - \bar{x})^2 f_i}{n - 1}}$$

where

$$n = \sum_{i=1}^{k} f_i$$

Example 3.12

Find the mean and standard deviation of the distribution represented by the histogram of Figure 3.4.

FIGURE 3.4 HISTOGRAM FOR EXAMPLE 3.12

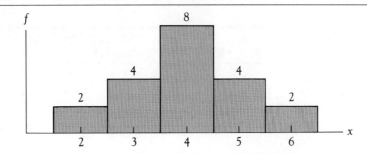

TABLE 3.7 DATA AND INTERMEDIATE CALCULATIONS FOR THE MEAN AND VARIANCE

	x	f	xf	$(x - \bar{x})$	$(x - \bar{x})^2$	$(x - \bar{x})^2 f$
	2	2	4	−2	4	8
	3	4	12	−1	1	4
	4	8	32	0	0	0
	5	4	20	1	1	4
	6	2	12	2	4	8
Sum		20	80			24

The interval midpoints are treated as the observations. These, their frequencies, and related products needed in the computations are given in Table 3.7. Note that columns 4, 5, and 6 are not completed until the mean is found by using the sum from column 3.

Using columns 2 and 3,

$$\bar{x} = \frac{\Sigma\, x_i f_i}{\Sigma\, f_i} = \frac{80}{20} = 4$$

This is also clear from the symmetry of the distribution about $x = 4$. From columns 2 and 6,

$$s^2 = \frac{\Sigma\, (x_i - \bar{x})^2 f_i}{\Sigma\, f_i - 1} = \frac{24}{20 - 1} = \frac{24}{19} = 1.26 \text{ (rounded)}$$

and $s = \sqrt{1.26} = 1.1$ (rounded).   ∎

We proceed in exactly the same way if we have a frequency table for a discrete variable where the observations have not been grouped.

Example 3.13

The distribution of daily requests for assistance by motorists on a freeway network is given in Table 3.8. After computing the mean from columns 2 and 3, the remainder of the table was completed.

TABLE 3.8 DATA AND INTERMEDIATE CALCULATIONS FOR THE MEAN AND VARIANCE

x (Requests)	f	xf	$(x - \bar{x})$	$(x - \bar{x})^2$	$(x - \bar{x})^2 f$
1	1	1	-3.3	10.89	10.89
2	2	4	-2.3	5.29	10.58
3	4	12	-1.3	1.69	6.76
4	5	20	-0.3	.09	.45
5	8	40	0.7	.49	3.92
6	4	24	1.7	2.89	11.56
7	1	7	2.7	7.29	7.29
Sum	25	108			51.45

$$\bar{x} = \frac{108}{25} = 4.3 \text{ (rounded)}$$

$$s^2 = \frac{51.45}{24} = 2.14 \text{ (rounded)}$$

$$s = \sqrt{2.14} = 1.5 \text{ (rounded)}   ∎$$

3.4

AN ALTERNATE FORMULA FOR THE STANDARD DEVIATION

If we use the expansion $(x_i - \bar{x})^2 = x_i^2 - 2x_i\bar{x} + \bar{x}^2$ and replacing $\bar{x}$ by $\Sigma x_i/n$ in the formula

$$s^2 = \frac{\Sigma(x_i - \bar{x})^2}{n - 1}$$

we obtain an alternate formula for computing the standard deviation:

$$s = \sqrt{\frac{\Sigma x_i^2 - \frac{(\Sigma x_i)^2}{n}}{n - 1}}$$

where n is the number of observations. If the observations have assigned frequencies, the corresponding result is

$$s = \sqrt{\frac{\Sigma x_i^2 f_i - \frac{(\Sigma x_i f_i)^2}{n}}{n - 1}}$$

where $n = \Sigma f_i$ (the sum of the frequencies).

Example 3.14

Consider again the distribution of motorists' requests for assistance in the last example. When the alternate formula for the standard deviation is used, the calculations are arranged as in Table 3.9.

TABLE 3.9 DATA AND INTERMEDIATE CALCULATIONS FOR THE MEAN AND VARIANCE

x	f	xf	x^2f
1	1	1	1
2	2	4	8
3	4	12	36
4	5	20	80
5	8	40	200
6	4	24	144
7	1	7	49

$\Sigma f_i = 25$ $\Sigma x_i f_i = 108$ $\Sigma x_i^2 f_i = 518$

Using column 3, $\bar{x} = \Sigma x_i f_i/n = 108/25 = 4.3$ (rounded). Using both columns 3 and 4,

$$s^2 = \frac{\Sigma x_i^2 f_i - \dfrac{(\Sigma x_i f_i)^2}{n}}{n - 1}$$

$$= \frac{518 - \dfrac{(108)^2}{25}}{24} = \frac{518 - 466.56}{24} = 2.14 \text{ (rounded)}$$

Thus $s = \sqrt{2.14} = 1.5$ (rounded). ■

The alternate formula is easy to use and does not make use of the mean $\bar{x}$, which is generally rounded. Further, it is readily adaptable to hand-held calculators which are able to accumulate sums. Its one weakness is that for large samples, the sums $\Sigma x_i^2 f_i$ and $\Sigma x_i f_i$ can become so large and unwieldy that one experiences a loss of significant digits in the final result.

Some rounding is inevitable in most computations. In the future when stating a result, we shall not bother to indicate that it is rounded. Thus if we write $s = 1.5$, as in the last example, it should be understood that this is the value of s correctly rounded to one decimal place. The following procedure for rounding off is suggested:

PROCEDURE FOR ROUNDING

1. Round the results of the final computations to one more decimal place than is present in the data.
2. Avoid rounding intermediate results. If you must do so, always retain at least twice as many places as will be used in the final result.

In the final analysis, problems of rounding and significant digits are best handled by retaining calculated values in the memory of the calculator for recall as needed in subsequent calculations.

EXERCISE SET 3.1

Part A

1. Find the median and mode of the following:
   a. 5.6, 2.4, 3.8, 7.2, 1.0, 4.6, 2.4.
   b. 8, 1, 2, 4, 3, 3, 7, 10.
2. Let $x_1 = 8$, $x_2 = 3$, $x_3 = 5$, $x_4 = 0$, $x_5 = 2$, $x_6 = 3$, $x_7 = 3$, and $x_8 = 8$.
   a. Find

   $$\sum_{i=1}^{8} x_i$$

   b. Find

   $$\sum_{i=1}^{8} x_i^2$$

   c. Find $\bar{x}$.

   d. Find the individual deviations d_i and find $\displaystyle\sum_{i=1}^{8} d_i$.

3. In a test of the ability of people to judge speed, an auto is driven past 10 experimental subjects at 35 miles per hour. Their estimates of the speed are as follows: 30, 42, 30, 46, 35, 40, 40, 32, 30, 25. Find the mode, median, and mean.

4. The weight (in milligrams) of tar in a sample of five cigarettes of a certain brand were reported as: 2.2, 2.4, 2.7, 2.4, and 2.3.
   a. Find the mean $\bar{x}$.
   b. Find the deviations from the mean and show that their sum is 0.
   c. Find the variance and standard deviation.

5. A data processing firm sampled 75 small businesses to find the number of days their computer systems were down during the previous three months. The distribution of responses is given below. Find the mean and standard deviation.

Days of down time	Frequency
0	16
1	10
2	14
3	20
4	8
5	7

6. The earnings (in dollars) of 20 first-year college students during summer "vacation" are as follows:

800, 2000, 1100, 1700, 1700, 1650, 1450, 2000, 1750, 1550,
1450, 1100, 1650, 2000, 1650, 1450, 1750, 1650, 1450, 1650

   a. Prepare a frequency table and extend it to contain columns for xf, $x - \bar{x}$, $(x - \bar{x})^2$, and $(x - \bar{x})^2 f$. (See Table 3.7.)
   b. Find the mean.
   c. Find the variance and standard deviation.

7. A distribution of the miles per gallon of 30 light trucks is given below. Find the mean and standard deviation. Use the alternate formula for the standard deviation and prepare a table similar to Table 3.9.

Interval	f
15.5–16.5	1
16.5–17.5	4
17.5–18.5	8
18.5–19.5	12
19.5–20.5	3
20.5–21.5	0
21.5–22.5	1
22.5–23.5	1

8. Twenty-five people visually estimate the length of a cord (in inches) from a distance of 10 feet. The distribution of estimates is given. Arrange the computations as in Table 3.8 and compute the mean and standard deviation.

x	f
13	3
14	3
15	4
16	8
17	5
18	2

9. The distribution of errors when 60 entering college students took a spelling test on 40 commonly misspelled words is given.

x	f
5	5
6	4
7	7
8	10
9	13
10	12
11	6
12	3

    a. Find the mean and standard deviation. Use the alternate formula for the standard deviation.

    b. What percent of the sample is within one standard deviation of the mean? Within two standard deviations?

10. Six coins are thrown 64 times to produce the following distribution of heads per throw. Arrange the computations as in Table 3.8 and compute the mean and standard deviation.

x	f
0	1
1	5
2	14
3	21
4	16
5	5
6	2

Part B

11. Give two situations where μ and σ rather than $\bar{x}$ and s would be used in reporting the mean and standard deviation.

12. In 1975, the employees of a company had a mean age of 34 years and a standard deviation of seven years. In six years what will be:

    a. The mean age of the same employees?

    b. The standard deviation of these employees' ages. Hint: Has there been any change in the variation of the ages?

13. The mean and standard deviation of a sample are 16 and 3 respectively.

    a. If the mean $\bar{x} = 16$ is subtracted from each observation, what is the mean of the resulting set of numbers?

   b. What would be the standard deviation. Hint: Has there been any change in the variation of the data?

14. What do you know about a collection of observations whose standard deviation is 0?

15. The distribution of ages in a collection of artifacts was reported in 1975 to have a mean of 1260 and a standard deviation of seventy years. What will be their mean and standard deviation in the year 2020?

16. A 100 point maximum test has been extensively tested and found to have a mean of 70 and a standard deviation of 10. A change of scale is being considered. What will be the mean if:

   a. Each test grade is increased by 430 points?

   b. Each test grade is multiplied by 4?

17. The following pair of distributions each have a mean of 20. Which has the largest standard deviation? No computations need be made.

   a. 10, 10, 10, 18, 20, 22, 30, 30, 30 or 10, 10, 18, 18, 20, 22, 22, 30, 30

   b. 16, 17, 18, 19, 20, 21, 22, 23, 24, or 16, 17, 20, 20, 20, 20, 20, 23, 24

18. The F.D.A. recommended amount of copper in a diet is 2 mg/day. A study of 15 hospitals and institutions produced the following distribution of copper in their daily diets.

Find the mean and standard deviation of this distribution.

Amount of copper (mg)	f
0–.5	5
.5–1.0	3
1.0–1.5	3
1.5–2.0	2
2.0–3.5	2

Source: Journal of the American Medical Association, 1979, vol. 241, no. 18. Copper and Zinc Deficiencies, L.M. Klevay et. al.

19. The assets of 50 small companies had a mean of $27.3 million. What was the sum of their assets?

20. The Test of Consumer Competencies is a standardized test involving 55 questions that measures knowledge of consumer affairs. When the test was administered to 4309 prospective education majors, the following results were obtained.

ACHIEVEMENT BY MAJOR OF PROSPECTIVE TEACHER

Major	N	Mean score	Standard deviation
Social studies, history, or geography	264	35.18	5.87
Science	170	35.09	5.91
Home economics	139	34.99	6.84
Business	171	34.77	5.95
Industrial arts or trades and industry	111	34.76	6.15
Mathematics	116	34.38	5.88
Sociology and psychology	84	34.35	6.28
English	224	33.87	5.63
Agriculture	13	33.77	8.11

Continued.

ACHIEVEMENT BY MAJOR OF PROSPECTIVE TEACHER—CONT'D			
Speech	77	33.09	5.95
Foreign languages	93	32.52	5.81
Elementary education	1,464	31.91	5.79
Special education	340	31.68	5.79
Art and music	268	31.21	6.48
Physical education and health	425	31.02	5.62
Errors in recording major	94	—	—
Omitted reporting major	256	—	—
TOTAL	4,309	32.67	6.06

Source: "Consumer Knowledge of Prospective Teachers, "E. T. Garman, *The Journal of Consumer Affairs*, vol. 13, no. 1, Summer 1979.

a. For which major did the scores show the greatest variation?

b. How could we find the mean of all the students who took this test if it was not given? (Ignoring those which involved errors.)

c. Is this mean of 32.67 very representative of the typical major involved in the test? Explain.

21. Three sets of numbers are plotted below on the number line. The first set is symmetric about $x = 5$. The second has a slight positive skewness as a result of changing one observation, and the third has a slight negative skewness as the result of a similar change.

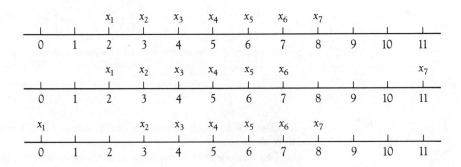

a. Find the median of each distribution and mark it on the number lines.

b. Find the mean of each distribution and mark it on the number lines.

c. Repeat the above using the sample 2, 2, 3, 4, 6, 7, 8, 8. Produce the two skewed distributions by first changing one 8 to a 10 and then a 2 to a 1.

d. Make a prediction of the relationship of the median to the mean for distributions that are (i) symmetric, (ii) positively skewed, (iii) negatively skewed.

22. a. For what value of x is $y = (2 - x)^2 + (6 - x)^2$ a minimum?

   b. For what value of x is $y = [(2 - x)^2 + (6 - x)^2]/2$ a minimum?

23. For the distribution below, find

$$\sum_{i=1}^{5} (x_i - \bar{x})f_i.$$

Note that there is an easy way to do this.

x	f
1	1
3	2
5	4
6	3
10	1

3.5   THE IMPORTANCE OF THE STANDARD DEVIATION

The standard deviation was introduced as a measure of the scattering of the data about the mean. To see how it functions in this respect, first note that if it is zero, then *all* the deviations $(x_i - \bar{x})$ must be zero and thus all observations are clustered at one point. If all the observations are clustered close to the mean, then the sum $\Sigma(x_i - \bar{x})^2$ is small, which results in a standard deviation near zero. As the data become more spread out, the standard deviation will increase in size.

While there are formulas that use the standard deviation to make predictions as to the scattering of the data of a sample about its mean, they must of necessity be quite conservative since they cover every conceivable type of distribution. For distributions having a bell-shaped appearance, the following empirical results more nearly reflect what is often encountered.

EMPIRICAL PREDICTIONS

The interval from $\bar{x} - s$ to $\bar{x} + s$ will contain about 68% of the data.
The interval from $\bar{x} - 2s$ to $\bar{x} + 2s$ will contain about 95% of the data.
The interval from $\bar{x} - 3s$ to $\bar{x} + 3s$ will contain about 99.7% of the data.

Alternatively, we can say that about 68% of the observations are within one standard deviation of the mean, about 95% within two standard deviations, and virtually all are within three standard deviations. See Figure 3.5.

FIGURE 3.5   THE EMPIRICAL PREDICTIONS

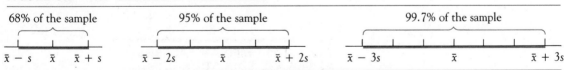

As with any result derived through observation, these results should not be expected to be exact. They furnish good approximations in many important distributions. As we shall see later, these predictions are exact for a normal distribution and accordingly produce excellent results when a histogram has the characteristic bell shape.

Example 3.15   For the observations: 0, 5, 6, 7, 8, 8, 8, 9, 9, 10 we find

$$\bar{x} = \frac{0 + 5 + 6 + 7 + 8 + 8 + 8 + 9 + 9 + 10}{10} = 7$$

$$s = \sqrt{\frac{(0 - 7)^2 + (5 - 7)^2 + (6 - 7)^2 + (7 - 7)^2 + (8 - 7)^2 \cdot 3 + (9 - 7)^2 \cdot 2 + (10 - 7)^2}{9}}$$

$$= 2.9$$

A comparison of the predicted and observed distributions for these data is given in Table 3.10.

TABLE 3.10

z	$\bar{x} - zs$	$\bar{x} + zs$	Number Observed	Percent Observed	Empirical Prediction, %
1.	4.1	9.9	8	80	68
2.	1.2	12.8	9	90	95
3.	−1.7	15.7	10	100	99.7

The next example is somewhat more realistic.

Example 3.16   The median salaries of nonmedical administrators at private universities and colleges for the 1980–1981 school year are given in Table 3.11. Their mean and standard deviation in dollars are

$$\bar{x} = 25,152.20 \qquad s = 8662.60.$$

The observations that are within one standard deviation of the mean lie on the interval from 16,489.60 to 33,814.80. By tallying we find that 74.1% (20 of 27) of the salaries fall here. Similarly, the interval from $\bar{x} - 2s$ to $\bar{x} + 2s$ extends from 7827.00 to 42,477.40 and contains 92.6% of the salaries (25 of 27). Finally, we find that 100% of the salaries are within three standard deviations of the mean. These results compare favorably with the empirical predictions of 68%, 95%, and 99.7%.

TABLE 3.11   MEDIAN SALARIES AT PRIVATE EDUCATIONAL INSTITUTIONS

Position	Salary
President (chief exec. officer)	45,000
Chief academic officer	32,610*
Registrar	19,450*
Director of admissions	23,220*

Continued.

TABLE 3.11 MEDIAN SALARIES AT PRIVATE EDUCATIONAL INSTITUTIONS—CONT'D

Head librarian	21,497*
Director, computer center	23,625*
Chief business officer	31,049*
Purchasing agent	18,000*
Director, personnel services	22,500*
Director, physical plant	21,900*
Director, food services	20,300*
Comptroller	24,500*
Director, student housing	15,750
Manager, book store	13,300
Staff legal counsel	35,000
Chief development officer	30,250*
Chief public relations officer	19,800*
Director, information office	19,500*
Chief student life officer	26,000*
Director, student union	16,500*
Director, student placement	17,808*
Director, student financial aid	18,050*
Director, student counseling	20,245*
Director, athletics	24,600*
Director/dean, arts & sciences	38,000
Director/dean, engineering	46,636
Director/dean, graduate programs	34,020

*Starred salaries are within one standard deviation of the mean.

Sources: Administrative Compensation Surveys, 1967–1968, 1975–1976, and 1980–1981, College and University Personnel Assn. (CUPA)

The above empirical predictions will be used repeatedly throughout the text.

Example 3.17

A military procurement study for enlisted men's uniforms showed that for soldiers who have completed basic training, the distribution of waist sizes has a mean of 31 and a standard deviation of 1.4 inches. Since such measurements ordinarily have a distribution that is approximately normal, they concluded that about 68% of the waist sizes needed will be between $31 - 1.4$ and $31 + 1.4$, that is between 29.6 and 32.4 inches. Similarly, about 95% will be between $31 - 2(1.4)$ and $31 + 2(1.4)$, that is, between 28.2 and 33.8 inches. Why, with this type of information, most service uniforms seemed to be size 34 or larger is a mystery to many a serviceman. ■

The fact that about 99.7% of observations are within three standard deviations of their mean provides a basis for roughly estimating the standard deviation of a distribution. The interval $\bar{x} - 3s$ to $\bar{x} + 3s$ is of length $6s$ and should contain most of the observations.

Example 3.18

A study of yearly food expenditures by middle-income families showed food budgets ranging between \$4150 and \$6880. Estimate the standard deviation of the food expenditure data.

$$6s \approx 6880 - 4150 = 2730$$
$$s \approx \frac{2730}{6} = 455 \quad \blacksquare$$

3.6　STANDARD UNITS

The last section showed the importance of knowing how many standard deviations an observation is located above or below the mean. This number is called a standard score or a z score; when an observation has been so expressed, it is said to have been converted to standard units. The computation of the standard score is discussed below.

DEFINITION 3.9

If a sample has a mean $\bar{x}$ and a standard deviation s, then the *standard score z* associated with the observation x is defined by

$$z = \frac{x - \bar{x}}{s}$$

To illustrate the origin of this formula, consider a test on which the mean is $\bar{x} = 68$ and the standard deviation is $s = 8$. A grade of, say, $x = 84$ is $84 - 68 = 16$ points above the mean. Since one standard deviation is 8 points, this grade is $z = 16/8 = 2$ standard deviations above the mean. Tracing these computations backward we see

$$z = 2 = \frac{16}{8} = \frac{84 - 68}{8} = \frac{x - \bar{x}}{s}$$

Now consider the grade $x = 56$. The difference $56 - 68 = -12$ shows that this grade is 12 points below the mean and $z = (56 - 68)/8 = -12/8 = -1.5$ shows that it is 1.5 standard deviation units below the mean. Observations below the mean have negative z scores; those above the mean have positive z scores. In Figure 3.6, we show on the number line several grades and their corresponding z scores.

FIGURE 3.6　GRADES AND CORRESPONDING Z SCORES

z	-3		-2	-1.5	-1		0		1		2		3
x	44		52	56	60		68		76		84		92

$$\bar{x}$$

If the sample consists of x_1, x_2, . . ., x_n, then successively replacing x by these values in the above formula produces a corresponding set of z scores z_1, z_2, . . ., z_n where

$$z_i = \frac{x_i - \bar{x}}{s}$$

We emphasize that the score z_i is the distance of the observation x_i from the mean *measured* in standard deviation units.

Example 3.19

A sampling of auto speeds at a test point showed $\bar{x} = 50$ and $s = 10$ miles per hour. The standardized scores corresponding to $x_1 = 60$, $x_2 = 30$, and $x_3 = 50$ are found from

$$z_i = \frac{x_i - 50}{10}$$

Thus $z_1 = (60 - 50)/10 = 1$ and $x_1 = 60$ is one standard deviation above the mean; $z_2 = (30 - 50)/10 = -2$ and thus $x_2 = 30$ is two standard deviations below the mean. Similarly $z_3 = (50 - 50)/50 = 0$ and $x_3 = 50$ is exactly at the mean. The speed that is 2.5 standard deviations above the mean is (clearly) 75. If it were not obvious, we could set $z = 2.5$ and solve

$$2.5 = \frac{x - 50}{10}$$

Both x and the z scores can be interpreted as speeds observed on two speedometers with different scales. See Figure 3.7.

FIGURE 3.7 AUTO SPEEDS AND CORRESPONDING Z SCORES

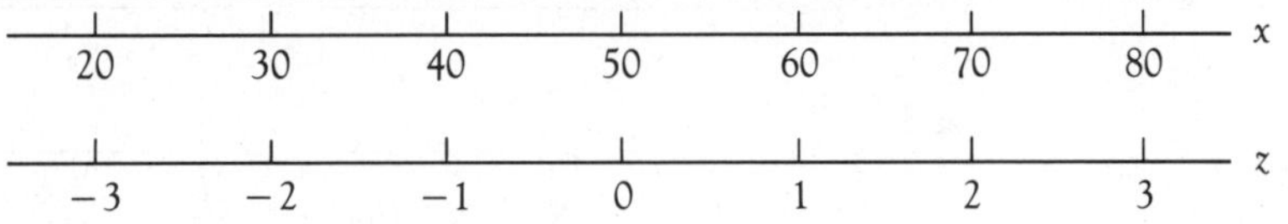

Note that the standard scores identify exactly the same outcomes as the original observations; the only difference is the scale employed. Figure 3.8 shows graphically the steps in producing the standard scores. The transformation $y = x - \bar{x}$ centralizes the data about the mean. The transformation $z = y/s = (x - \bar{x})/s$ changes the unit of measure to s. The final z scores have a mean of 0 and a standard deviation of 1.

FIGURE 3.8   HISTOGRAMS REPRESENTING STEPS IN TRANSFORMING OBSERVATIONS TO STANDARD SCORES

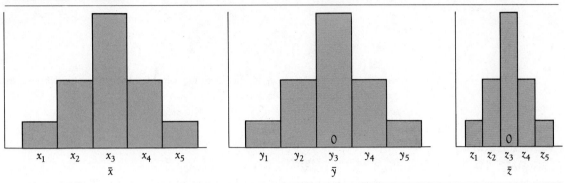

We summarize these results in the following theorem.

THEOREM 3.3

Let the sample $x_1, x_2, \ldots, x_n$ have a mean $\bar{x}$ and a standard deviation s. Then the standardized scores $z_1, z_2, \ldots, z_n$ produced from $z = (x - \bar{x})/s$ have mean $\bar{z} = 0$ and standardized deviation $s_z = 1$.

Example 3.20

Five students estimated the weight, in ounces, of a book as $x_1 = 33$, $x_2 = 37$, $x_3 = 35$, $x_4 = 33$, and $x_5 = 37$. We quickly find that $\bar{x} = 35$ and $s = 2$. Using $z_i = (x_i - 35)/2$ we obtain the standard scores: $z_1 = -1$, $z_2 = 1$, $z_3 = 0$, $z_4 = -1$, and $z_5 = 1$. Then

$$\bar{z} = \frac{z_1 + z_2 + z_3 + z_4 + z_5}{5} = \frac{-1 + 1 + 0 + (-1) + 1}{5} = 0$$

and

$$s_z = \sqrt{\frac{\sum_{i=1}^{5} (z_i - \bar{z})^2}{5 - 1}} = \sqrt{\frac{1 + 1 + 0 + 1 + 1}{4}} = 1$$

Thus the standard scores have a mean of 0 and a standard deviation of 1. ■

When the standard scores of a sample are interpreted in terms of the empirical predictions of the previous section, we have a powerful means of judging whether the deviations from the mean are significant. At present, we can only preview what will later be described in more quantitative terms, once sampling theory has been developed.

Example 3.21

When a certain breed of rat is implanted with cancer cells, it is found that after three weeks, the resulting tumor masses have a mean of 2.56 and a standard deviation of .35 grams. Six experimental rats are implanted with the cancer cells and are also treated at regular intervals with a possible anticancer drug. The resulting weights of the tumors after three weeks are found to be: 1.84, 1.76, 1.53, 1.69 and 1.47 grams.

Now, if the drug is ineffective (nonactive), then these rats can be regarded as members of the original population insofar as the tumor growth is concerned. The standard scores are computed from

$$z = \frac{x - 2.56}{.35}$$

and are found to be: $z_1 = -2.06$, $z_2 = -2.29$, $z_3 = -2.94$, $z_4 = -2.49$, and $z_5 = -3.11$. All observations are at least two standard deviations below the mean. For a normal distribution, which these weight increases should have, only about 2.5% of the population should show this behavior (see Figure 3.9). The fact that one of the rats produced a weight change more than two standard deviations below the mean is of little importance, since there must be observations both above and below the mean. The fact that all five would show this behavior is extremely unlikely, unless the treated rats belong to a different population. In other words, the drug has had some effect in controlling the tumor growth.

FIGURE 3.9   z SCORES FOR TUMOR GROWTHS

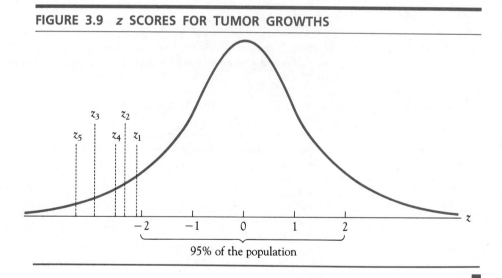

In computing z scores with reference to a population, we replace $\bar{x}$ and s by their population counterparts μ and σ in the formula relating z to x.

DEFINITION 3.10

An observation x from a population with mean μ and standard deviation σ has an associated z score given by

$$z = \frac{x - \mu}{\sigma}$$

Example 3.22

A student takes admission tests at two universities. At the first $\mu = 74$ and $\sigma = 11$ and her grade is $x_1 = 88$. At the second, $\mu = 63$, $\sigma = 9$, and her grade is $x_2 = 85$. The corresponding z scores are

$$z_1 = \frac{x_1 - 74}{11} = \frac{88 - 74}{11} = \frac{14}{11} = 1.27$$

$$z_2 = \frac{x_2 - 63}{9} = \frac{85 - 63}{9} = \frac{22}{9} = 2.44$$

Her grade at the second university (85) is 2.44 standard deviations above the mean, while that at the first (88) is only 1.27. Relatively speaking, her performance at the second university is far better than at the first. ■

EXERCISE SET 3.2

Part A

1. If $\bar{x} = 112$ and $s = 10$, find the observation that is:
   a. Two standard deviations above the mean.
   b. Three standard deviations below the mean.
   c. 1.4 standard deviations above the mean.
   d. One-half standard deviation below the mean.
2. If $\bar{x} = 508$ and $s = 43$ find the end points of the following intervals:
   a. $\bar{x} \pm 2s$.
   b. $\bar{x} \pm 1.4s$.
3. If $\mu = 18$ and $\sigma = 6$, what is the location of all observations that are:
   a. Within 1.5 standard deviations of the mean.
   b. Within two standard deviations of the mean.
   c. More than three standard deviations above the mean.
   d. Between one and two standard deviations below the mean.
4. Let $\bar{x} = 165$ and $s = 18$.
   a. Find the z score corresponding to (i) $x = 174$, (ii) $x = 130$.
   b. Find the observation x corresponding to (i) $z = 2.5$, (ii) $z = -1.3$.
5. Let $\bar{x} = 175$ and $s = 16$.
   a. Find z if $x = 200$.
   b. Find z if $x = 163$.
   c. Find z if $x = 175$.
   d. Find x if $z = 2.3$.
   e. Find x if $z = -3.6$.
   f. Find x if $z = 1.5$.
6. Suppose the pulse rates of students in your aerobics class have a mean of 76 and a standard deviation of 6.
   a. If your rate is $x = 90$, how many standard deviations above the mean is this?
   b. Find the z score corresponding to a rate of $x = 100$.

    c. Find the z score corresponding to a rate of 65.

    d. The instructor claims that anyone whose rate is more than 2.2 standard deviations above the mean is "out of shape." To what pulse rate does this correspond?

    e. Find the pulse rate corresponding to $z = -2.8$.

7. A somewhat bell-shaped distribution of 300 observations has a mean of 46 and a standard deviation of 8. Estimate the number of observations that are:

    a. Between 38 and 54.

    b. Between 30 and 62.

    c. Greater than 62.

    d. Between 54 and 62.

    e. Below 30.

8. Liquid morphine produces a sedative effect to which the body rapidly adjusts. After that adjustment period, it can be given on a regular basis to terminally ill patients to reduce pain and the patient will remain lucid and alert. The hospice ward of a metropolitan hospital reported that adjustment periods are normally distributed with a mean of 48 and a standard deviation of 3 hours. In a sample of 180 patient records, approximately how many should show an adjustment period

    a. Between 45 and 51 hours?

    b. Above 54 hours?

    c. Below 39 hours?

9. An instructor reported that on a test administered to 60 students, the mean was 65 and the standard deviation was 10. Estimate:

    a. The number of grades between 45 and 85.

    b. The number of grades above 85.

    c. What assumption is implicit in these estimates?

10. In a survey of starting salaries of marketing majors, the highest and lowest salaries reported were $21,400 and $12,600. Estimate the standard deviation.

11. Give rough estimates of the mean and standard deviations of each of the following distributions.

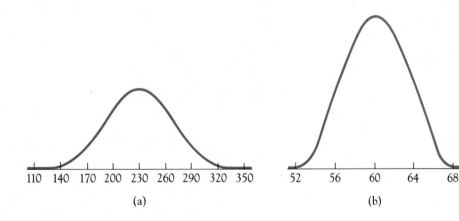

(a)   (b)

12. Verify Theorem 3.3 for the sample $x_1 = 2$, $x_2 = 6$, $x_3 = 10$, and $x_4 = 10$.

Part B

13. A certain medication has a mean effective time of 190 minutes with a standard deviation of 20 minutes. Five test subjects tried an "improved version" of the medication and produced the following effective times (in minutes): 200, 210, 184, 190, and 170.

a. Compute the standard z scores using $\mu = 190$ and $\sigma = 20$.

b. Is there any evidence of an improvement in the effective time of this medication?

14. The distribution of the horsepower output of a certain diesel engine has a mean of 160 and a standard deviation of 3. Equipped with a soot removal antipollution device, six test engines showed the following horsepowers: 150, 155, 154, 153, 154 and 150. Compute the standard scores and discuss whether the device has altered the performance of the engine. You may assume a bell-shaped distribution.

15. The nickel concentration in lung tissue from nine deceased victims of Legionnaires' Disease and in nine controls is given in the accompanying table.

NICKEL LEVELS

| Patient | Nickel in tissue (μg/100 g, dry weight) | |
	Lung	*Kidney*
Legionnaire Cases		
L1	65	3
L2	24	—
L3	52	12
L4	86	26
L5	120	17
L6	82	—
L7	399	—
L8	87	—
L9	139	—
L10	—	10
L11	—	20
Controls		
C1	12	5
C2	10	13
C3	31	12
C4	6	9
C5	5	15
C6	5	21
C7	29	—
C8	9	—
C9	12	—
Other cases		
N1(BSP)	12	9
N2(Congress)	21	5
N3(Congress)	17	11

In lung and kidney tissues of the Legionnaire cases, controls, the "Broad Street Pneumonia" (BSP) and Eucharistic Congress cases.

Source: Legionnaires' Disease, J. R. Chen et al., *Science*, May 20, 1977, vol. 196, no. 4292.

a. Compute the mean and standard deviation for the nickel in the control group.

b. Using the results from the control group as being representative of the general population, compute the average Z score for the nickel concentrations in the Legionnaire victims.

CHARACTERISTICS OF THE SAMPLE

Variable	Mean	Standard deviation	Minimum	Maximum
Organization age (years)	45.76	37.34	2	134
Organization size (total workers)	303.17	653.12	14	3292
Avarage worker tenure	4.48	2.79	1	12
Centralization	47.72	10.04	28	69
Formalization	41.70	6.70	28	56
Productivity (clients/week/ worker)	27.60	17.46	5	82
Efficiency (clients/week/$10,000)	35.65	45.85	7	228

Source: "Productivity Efficiency . . .," C. A. Glisson and P. Y. Martin, *Academy of Management Journal,* 1980, vol. 23, no. 1.

16. A study of production and efficiency in human service organizations contained the above summary of the characteristics of the sample studied.
   a. Which of the variables did not have all observations within three standard deviations of their sample mean?
   b. For those that do not, how many standard deviations are needed?

17. The distribution of job interviews undergone by graduating seniors before receiving an acceptable offer is given.

x	f
1	21
2	25
3	24
4	11
5	7
6	5
7	5
8	2

   a. Draw the histogram.
   b. Why would you not expect the empirical predictions to be particularly good for this distribution?
   c. Compute the mean and standard deviation.
   d. What percent of the observations are within one standard deviation of $\bar{x}$?

18. The unemployment rates for the period from May 1982 through May 1983 are given below. Compute the mean and standard deviation and determine what percent of the observations are within one standard deviation of the mean.

Month	year	Unemployment rate (%)
May	1982	9.5
June	1982	9.5
July	1982	9.9
August	1982	9.9
September	1982	10.1
October	1982	10.4
November	1982	10.8
December	1982	10.8

Continued.

Month year	Unemployment rate (%)
January 1983	10.3
February 1983	10.3
March 1983	10.2
April 1983	10.2
May 1983	10.1

Source: The Unemployment Rates for the Period from 1948–1978, Agricultural Economic Research, Vol. 32, No. 1, Jan. 1980.

19. Tchebycheff's theorem states that within h standard deviations of the mean will be found at least the fraction $1 - 1/h^2$ of the observations of the sample. For example, within $h = 2$ standard deviations will be found at least $1 - 1/2^2 = 1 - 1/4 = 3/4 = 75\%$ of the sample. What prediction does the theorem make for each of the following?
    a. Within three standard deviations of the mean.
    b. Within 1½ standard deviations of the mean.
    c. Within one standard deviation of the mean.

20. Outliers are observations so far removed from the mean as to be suspect. Chauvenet's criterion for identifying (and possibly rejecting) an outlier states that an observation from a sample of size n is an outlier if deviations of its size or larger (from the mean) occur with a relative frequency of less than $1/2n$. For example, with a sample of size $n = 10$, $1/2n = 1/20 = .05$. Thus observations that are at least two standard deviations from the mean are outliers for samples of size $n = 10$ from a bell shaped distribution.
    a. For a sample of size $n = 10$ with mean 189 and standard deviation 18, is $x = 250$ an outlier? (Assume a bell shaped distribution.)
    b. Show for a sample of size $n = 25$ with $\bar{x} = 42$ and $s = 8$ that $x = 76$ is an outlier. (Assume a bell shaped distribution.)

21. A symmetric distribution has a mean of 162 and a standard deviation of 14. Using intervals of one standard deviation and starting three standard deviations below the mean, sketch the histogram, assuming the relative frequencies can be found by using the empirical rule.

3.7   PERCENTILES AND PERCENTILE RANKS

Standard z scores describe the location of observations relative to the mean. Percentiles locate observations relative to the sample as a whole. If you have ever taken a standardized test such as the SAT or the GREA, then your result was no doubt accompanied by a statement indicating the percentile of your score. If, say, it was at the 88th percentile, then 88% of the scores were less than yours and 12% were greater.

> **DEFINITION 3.11**
>
> The kth *percentile* of a distribution of grouped observations is a number P_k such that at most k% of the observations are less than P_k and at most $(100 - k)\%$ are greater.

Thus if the 88th percentile of the test scores was 640, we would write $P_{88} = 640$.

Now consider the histogram of Figure 3.10. This represents the distribution of reported miles by the participants in a charity walkathon. The relative frequency of each interval has been marked on the corresponding rectangle. We can easily read off that the 15th percentile is 5 miles.

FIGURE 3.10   FREQUENCY DISTRIBUTION OF MILES WALKED

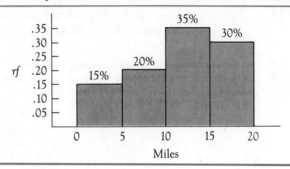

Thus $P_{15} = 5$. Similarly we see that $P_{35} = 10$, since 15% + 20% = 35% of the observations are less than 10.

If we assume that the observations are uniformly distributed over the intervals, then we can find approximations to any percentile by a process called *interpolation*. To illustrate, suppose we want the 88th percentile of the walkathon data. Clearly it is on the interval 15–20. Now the left end point, 15, is the 70th percentile, so we need to move to the right of 15 a distance sufficient to obtain an additional 18% of the observations. Since the interval 15–20 contains 30% of the observations, we need to move 18/30 of the distance across this interval. Since this interval is of length 5, we move to the right of the end point 15 by the amount (18/30) × 5 = 3 (miles). Thus

$$P_{88} = 15 + (18/30) \times 5 = 15 + 3 = 18$$

Approximately 88% of the participants walked less than 18 miles and 12% walked more than 18.

In many studies only the 25th, 50th, and 75th percentiles are reported. The 50th percentile locates the center of the observations in the same sense as the median and is so named. The 25th and 75th percentiles are called the upper and lower quartiles.

DEFINITION 3.12

The *lower or first quartile* is the 25th percentile and is denoted Q_1. The *upper or third quartile* is the 75th percentile and is denoted Q_3. The median is the 50th percentile and is commonly denoted by $\tilde{x}$ when X is the variable under investigation.

Note that 50% of the observations are located between the lower and upper quartiles and are centered about the median.

Example 3.23

A report on the reading scores of a citywide test stated that $Q_1 = 43$, $\bar{x} = 62$, and $Q_3 = 79$. This means that 25% of the students scored below 43, 50% scored below 62, 75% below 79, and 50% of all scores were between 43 and 79. ∎

Example 3.24

A study of the weights of passenger carry-on luggage by an airline produced the histogram of Figure 3.11. Estimate the lower quartile, median, and upper quartile.

FIGURE 3.11   DISTRIBUTION OF CARRY-ON LUGGAGE WEIGHTS

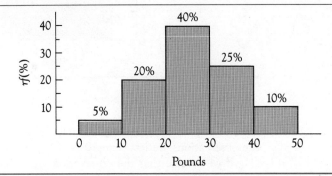

We easily read off that $Q_1 = 20$ (pounds). The median is on the interval 20–30, which has a length of 10 and contains 40% of the data. Since the left end point, 20, is the 25th percentile we need to move to the right of this point a distance sufficient to obtain an additional 25% of the data. This distance is 25/40 of the distance 10. Thus

$$\bar{x} = P_{50} = 20 + (25/40) \times 10 = 26.25$$

Similarly,

$$Q_3 = P_{75} = 30 + (10/25) \times 10 = 34 \quad ∎$$

When one is reporting on raw data where no grouping has been used, *percentile ranks* are assigned to the individual scores. To illustrate how this is done, consider the brief list of examination grades:

$$68, \ 76, \ 77, \ 80, \ 80, \ 80, \ 85, \ 88, \ 90, \ 94$$

Consider the score 77. We would like to assign it a ranking similar to a percentile. There are 2 of the 10 or 20% of the scores below 77. Similarly there are 3 of 10 or 30% of the scores below the next highest score, 80. We opt for the average and assign 77 a percentile rank of 25 and write $PR(77) = 25$. Next consider the score 80. Three tenths or 30% of the scores are less than 80 and

6/10 = 60% are less than the next highest score, 85. We assign 80 the percentile rank that is halfway between these, namely 45. Thus $PR(80) = 45$. In the following definition we incorporate a formula for finding the percentile rank of any observation.

DEFINITION 3.12

Let a study contain n observations and let these be arranged in order of increasing size. Suppose k of these observations are below an observation x whose frequency is f. Then the *percentile rank* of x, denoted $PR(x)$ is defined as

$$PR(x) = \frac{k + \frac{1}{2}f}{n} \times 100$$

Example 3.25

Of the 510 students in the graduating class of an engineering school, 243 had starting salaries below $25,000, 15 were at $25,000, and 252 were above $25,000. Find the percentile rank of the $25,000 salary.

The salary $x = 25,000$ has the frequency $f = 15$ and there are $k = 243$ salaries below this. Thus

$$PR(25,000) = \frac{243 + \frac{1}{2}(15)}{510} \times 100 = 49 \text{ (rounded)} \quad \blacksquare$$

Example 3.26

A Research Study

Do overweight children have high blood pressure? Hypertension in adults is often related to obesity. Whether this is true for children is the subject of a report, "The Relationship Between Elevated Blood Pressure and Obesity in Black Children," by Barbara Gentry Lands, et al. in the February 1980 issue of the *American Journal of Public Health*. Black children were chosen since "it is known that hypertension is more prevalent and severe among blacks than among whites."

This study was funded by the U.S. Department of Public Health and involved 1692 black elementary school children from a large school district in North Carolina. The sample was obtained by randomly selecting 82 of the 177 available classrooms and including every black child who was present at the time the data were gathered. The data included age, sex, weight, and blood pressure.

Using the 1976 Growth Charts developed by the National Center for Health Statistics, normative values of weight for height were derived. Using these, the children were then grouped into five weight classes as follows:

- Normal: between the 25th and 75th percentiles
- Underweight: between the 10th and 25th percentiles
- Lean: below the 10th percentile
- Overweight: between the 75th and 90th percentiles
- Obese: above the 90th percentile.

Defining elevated blood pressure (EBP) proved somewhat of a problem, since most previous studies had involved white children. Also, there are reports which show that blood pressure distributions and norms do not remain stable over extended periods. The definition finally adopted was based solely on the investigators' own data and established EBP as "any systolic or diastolic reading at or above the 90th percentile in the distribution of these readings by sex and age."

Tables 3.12 and 3.13 give the mean systolic and diastolic EBP readings by age and by sex. Seventy-seven of the children appear in both tables. In reading Table 3.12, we see that 195 children (102 boys and 93 girls) had elevated systolic readings. For a specific age group, say 7 years, there were 15 boys whose pressures ranged between 106 and 130 and whose mean was 112.0. Similarly, there were 14 girls in this age group and their mean was 114.1. Combined, the 29 blood pressures had a mean of 113.0.

TABLE·3.12   ELEVATED SYSTOLIC PRESSURES FOR STUDY POPULATION BY AGE AND SEX IN mmHg[a]

Age	Boys Range	Boys Mean	Girls Range	Girls Mean	Total Number	Total Mean
5	106–120 (12)	111.0	100–132 (17)	105.3	29	107.7
6	102–120 (21)	108.3	102–120 (11)	106.4	32	107.7
7	106–130 (15)	112.0	106–124 (14)	114.1	29	113.0
8	112–120 (8)	117.0	110–120 (11)	113.1	19	114.7
9	112–130 (12)	117.6	114–148 (14)	120.1	26	119.0
10	118–146 (13)	127.4	118–140 (11)	124.9	24	124.3
11	118–135 (21)	122.7	122–140 (15)	127.2	36	124.6
TOTAL	(102)	116.3	(93)	115.8	195	116.1

[a]Number of children shown in parentheses

TABLE 3.13   ELEVATED DIASTOLIC PRESSURES FOR STUDY POPULATION BY AGE AND SEX IN mmHg[a]

Boys Range	Boys Mean	Girls Range	Girls Mean	Total Number	Total Mean
68–80 (15)	72.4	72–80 (10)	75.5	25	73.6
72–80 (15)	75.3	72–80 (14)	76.1	29	75.7
78–84 (14)	79.4	80–94 (17)	82.1	31	80.9
80–90 (14)	81.3	82–90 (9)	84.2	23	82.4
80–90 (16)	84.7	80–94 (17)	83.4	33	84.1
82–94 (13)	86.3	80–94 (15)	84.7	28	85.4
84–110 (16)	89.6	84–100 (14)	88.9	30	89.2
(103)	81.3	(96)	82.4	199	81.8

[a]Number of children shown in parentheses

In Figure 3.12, the distribution of these EBPs is given according to weight categories and compared with the distribution of weights by these categories. Comparing, say, the elevated systolic pressures with the total population we see that a much greater likelihood exists for an EBP in the overweight and obese groups.

For example, the normal weight group contained about 58% of all the children but only 40% of those with systolic EBPs. On the other hand, the obese category, which only contained about 12% of the sample, had almost 34% of the elevated systolic readings.

The report concludes that "the relationship between elevated blood pressure and obesity consistently observed in black adults exists in young children as well." And it recommends the regular screening of young children for weight and blood pressure.

FIGURE 3.12   PERCENT DISTRIBUTION OF BLACK CHILDREN BY WEIGHT CATEGORIES, TOTAL POPULATION AND ELEVATED SYSTOLIC AND DIASTOLIC PRESSURES

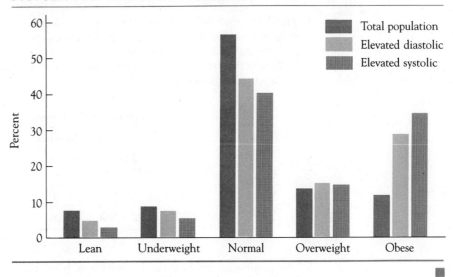

EXERCISE SET 3.3

Part A

1. The distribution of radio signals (AM) that can be received in a certain area is given.
   a. Estimate the lower quartile.
   b. Estimate the median.
   c. Estimate the upper quartile.

x (kHz)	Relative frequency
55–65	.25
65–80	.32
80–100	.14
100–130	.22
130–160	.07

2. The distribution of hourly starting salaries of a sample of nursing aides is given below. Using interpolation as needed, estimate:
   a. The 90th percentile.
   b. The median.
   c. The lower and upper quartiles.

Salary ($/hour)	Relative frequency
5.00–5.25	.16
5.25–5.50	.29
5.50–5.75	.45
5.75–6.00	.10

3. The distribution below resulted from a study of second trust deed home mortgage rates in a metropolitan area during the second quarter of 1983
   a. Estimate the median loan rate.
   b. Ninety percent of the loans were at or below what rate?
   c. Estimate the lower and upper quartiles.

Rate (%)	Relative frequency
13.4–14.9	.10
14.9–16.4	.15
16.4–17.9	.40
17.9–19.4	.20
19.4–20.9	.10
20.9–22.4	.05

4. The distribution below gives the number of hours per week spent in the library by 50 students. Estimate the median number of hours spent in the library.

x (Hours)	f
0–1	12
1–2	16
2–3	8
3–4	10
>4	4

5. Of the 1000 people who took an examination, 480 scored below you and 109 had the same score. Find the percentile rank for your score. (Note: $f = 109 + 1 = 110$)

6. Twenty-five pages are selected at random from the galley proofs for a manuscript and checked for typing errors. The number of errors per page are as follows:

   0, 1, 2, 2, 3, 1, 5, 6, 0, 0, 1, 4, 5, 3, 3, 2, 1, 0, 5, 2, 3, 4, 3, 6, 0

   Find the percentile rank for:
   a. $x = 4$.
   b. $x = 2$.

    c. $x = 0$.
    d. $x = 6$.
7. Twenty people are questioned about the number of books they read during the previous month. The distribution of responses is below. Find the percentile rank of each response.

x (Number read)	f
0	7
1	3
2	5
3	2
5	3

Part B

8. True or false. If the median of a set of examination scores is 100, then all scores are 100 (assuming a 100-point examination).
9. In a study of the blood pressure of students, your diastolic pressure was reported as being at the 90th percentile. Does this necessarily mean that you have high blood pressure?
10. The distribution of the depth/diameter d/D ratios of 1933 lunar craters based on Orbiter IV photography is given below. Note that if a crater was a hemisphere then its diameter would be twice its depth and the depth/diameter ratio would be $\frac{1}{2} = .5$. A ratio of, say, .2 indicates that the depth is $.2 = 2/10$ of the diameter. Estimate the median depth/diameter ratio.

CLASSIFICATION OF 1933 LUNAR CRATERS FROM ORBITER IV PHOTOGRAPHY, WITH RESPECT TO THE *d/D* RATIOS

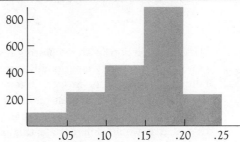

Source: Impact Spacecraft Imagery and Comparative Morphology of Craters, M. Moutsoulas and S. Petri, *The Moon and the Planets*, Vol. 21, No. 3, 1979.

11. Describe a situation where a low percentile ranking would be desirable.
12. The distribution of the ages of a sample of graduating college students is given below. Estimate:
    a. The lower quartile.
    b. The upper quartile.
    c. The 90th percentile.

Interval	Relative frequency
19.5–21.5	.08
21.5–23.5	.25
23.5–25.5	.24
25.5–27.5	.18
27.5–29.5	.16
29.5–31.5	.09

13. The interquartile range of a distribution is defined as the difference $Q_3 - Q_1$. Find the interquartile range of the distribution in problem 12 and interpret this number on the number line.

14. A study of the sales records of 1400 new car purchases contained the following information: $Q_1 = \$6800$, $\bar{x} = \$8900$, $Q_3 = \$11,500$, and $P_{90} = \$16,000$.
    a. How many of the purchase prices were between $6800 and $8900?
    b. How many of the purchase prices were between $6800 and $11,500?
    c. How many of the purchase prices exceeded $16,000?
    d. How many of the purchase prices were between $11,500 and $16,000?

3.8   A WORD TO THE WISE

In the preceding sections, we have enthusiastically praised the virtues of descriptive statistics. We pause at this time to note, or better to warn, that these summaries smooth over the individual observations for the sake of a description of the sample as a whole. One must be constantly on guard that these summaries are not used in place of the individual observations or to hide important features, either accidentally or intentionally. Occasionally, it pays to examine the original data. We reprint here an experience by an educational evaluator with the Wechsler Intelligence Scale for Children–Revised (WISC-R) test.

Example 3.27

A Research Study

"Upon completion of an exhausting research effort involving the WISC-R scores of a group of severely emotionally disturbed (SED) children, the realization crystallized. Numbers can, in fact, distort truth. The research in this case involved studying the WISC-R subtest profile of a sample group of SED children. The test protocols of 113 subjects were examined by computer to determine if a profile existed that would differentiate this particular psychodiagnostic group from the standardization sample for the WISC-R. Analysis of the data naturally involved compilation of all 113 protocols. It was at this early point that the number game began to cover important interpretive clues for the sake of statistical significance.

"The subtest profile analysis proved less than overwhelming in terms of the statistical results obtained. Though the SED group did score significantly lower on all subtests than the standardization sample, no definite profile existed that could be used diagnostically to define this group.

"Analysis of overall intellectual functioning proved similarly unimpressive. Mean IQ levels for the group as a whole were significantly depressed. The Verbal Scale mean IQ was 83.07, the Performance Scale IQ ws 83.81 and the Full Scale score was 81.96. Further probing of a statistical nature yielded little information of clinical usefulness.

Then the 'number crunching' began to lose its hypnotism. Eyeball analysis began to surface. Where statistical significance did not exist, interpretive clues did. By looking at *individual* scores this researcher discovered many important facts about the SED group.

"Though no verbal vs. performance discrepancy existed for the 113 children considered as a whole, 29 children showed a discrepancy of 5 to 9 points, 19 children showed a discrepancy of 10 to 14 points, 12 children showed a discrepancy of 20 or more points. Certainly these important figures were stolen by the computer quest for "average score."

"It was also discovered through inspection of individual protocols that 24 children displayed a subtest scatter of three to five points, 77 children displayed a six- to nine-point scatter, and 12 children displayed a scatter of 10 or more points. Again, the 'lumping together' drive of man and computer had cheated away an important aspect of the WISC-R analysis.

"Lastly, it was discovered through inspection that only 11 children fell into the 80 to 85 IQ range. Thus, the 'average' IQ range for the group was only representative of 10% of the 113 children in this sample.

"The point to be made is a simple one. Use computers. Seek statistical significance. Obtain and utilize means, standard deviations, analysis of variance or multiple regressions. But, don't overlook the power and significance of human inspection of research data. Also, don't let the search for statistical significance on the part of the entire sample lie, cheat, or steal away the knowledge that could be obtained from appraisal of individual performance." Source: J. G. Evans, "Numbers Sometimes Lie, Cheat, and Steal," *Psychology*, Fall 1979 vol. 16, no. 3. ■

COMPUTER PRINTOUT

The analysis with Minitab of the evaluations (scale 0–10) of an instructor by 14 students is given in Figure 3.13. Additionally the standard z scores were generated and placed in column C2 by the command:

$$\text{LET C2} = (\text{C1} - \text{AVER(C1))/STAN(C1)}$$

The ordered versions of the original scores and the z scores were generated and placed in columns C3 and C4. The mean and standard deviation of the z scores should be 0 and 1, respectively. Apart from a small rounding error this is seen to be true.

FIGURE 3.13 MINITAB PRINTOUT

```
MTB >SET THE FOLLOWING SCORES INTO COLUMN C1
DATA>8,3,3,6,4,5,5,5,4,9,10,8,7,2
DATA>END
MTB>COUNT C1
    COUNT    =        14.000
MTB >MAXIMUM C1
    MAXIMUM  =        10.000
```

Continued.

FIGURE 3.13   MINITAB PRINTOUT—CONT.

```
MTB >MINIMUM C1
   MINIMUM  =       2.0000
MTB >AVERAGE C1
   AVERAGE  =       5.6429
MTB >STANDARD DEVIATION C1
   ST.DEV.  =       2.4371
MTB >MEDIAN C1
   MEDIAN  =        5.0000
MTB >LET C2 = (C1 - AVER(C1))/STAN(C1)
MTB >ORDER C1 AND C2, PUT INTO C3 AND C4
MTB >PRINT C1,C2,C3,C4
COLUMN      C1              C2          C3              C4
COUNT       14              14          14              14
ROW
  1          8.         0.96718         2.        -1.49474
  2          3.        -1.03442         3.        -1.08442
  3          3.        -1.08442         3.        -1.08442
  4          6.         0.14654         4.        -0.67410
  5          4.        -0.67410         4.        -0.67410
  6          5.        -0.26378         5.        -0.26378
  7          5.        -0.26378         5.        -0.26378
  8          5.        -0.26378         5.        -0.26378
  9          4.        -0.67410         6.         0.14654
 10          9.         1.37750         7.         0.55686
 11         10.         1.78782         8.         0.96718
 12          8.         0.96718         8.         0.96718
 13          7.         0.55686         9.         1.37750
 14          2.         1.49474        10.         1.78782

MTB >AVERAGE OF C2
   AVERAGE : 0.000000040
MTB >STANDARD DEVIATION C2
   ST.DEV. :       1.0000
MTB >STOP
```

CHAPTER CHECKLIST

Vocabulary

Center	Standard deviation
Mode	Empirical rule
Median	z score
Mean	Percentile
Variation	Lower quartile
Deviation	Upper quartile
Variance	Percentile rank

Techniques

Finding the mode

Computing the median

Computing the mean, variance, and standard deviation of a sample of ungrouped data, using both formulas for the variance.

Computing the mean, variance, and standard deviation for a distribution
(grouped data), using both formulas for the variance.
Using the empirical rule
Computing and interpreting z scores
Interpolating to find percentiles of a distribution
Computing percentile ranks of individual observations

Symbols

$\bar{x}$ The mean of a sample
$\tilde{x}$ The median of a sample
s^2 The variance of a sample
s The standard deviation of a sample
μ The mean of a population
σ^2 The variance of a population
σ The standard deviation of a population
P_k The kth percentile
$PR(x)$ The percentile rank of x

**CHAPTER
TEST**

1. Find the median and mode of the following observations:
   a. 2, 3, 5, 2, 2, 4, 2, 5, 5.
   b. 0, 3, 0, 3, 0, 3.
2. For the sample below, find
   a. The mean.
   b. The standard deviation.
   c. The number of observations that are within 1.5 standard deviations of the mean.

   2, 3, 5, 5, 6, 3, 6, 5, 6, 9, 2, 5, 3, 5, 6, 3, 5, 6, 6, 9
3. In a study of the monthly electrical usage of 1400 homes, the resulting distribution had a bell-shaped distribution with a mean of 1250 and a standard deviation of 200 kilowatt hours (kwh).
   a. Estimate the number of homes in the study that had a monthly usage between 1050 and 1450 kwh.
   b. Estimate the number of homes in the study that had a usage exceeding 1650 kwh.
4. Suppose $\mu = 140$ and $\sigma = 14$. Find x if:
   a. x is located 2.5 standard deviations above the mean.
   b. x is located 1.4 standard deviations below the mean.
   c. The z score associated with x is $z = 1.8$.
   d. The z score associated with x is $z = -2.6$.
5. Forty-five pregnant rats were repeatedly subjected to controlled dosages of alcohol. The distribution of the number of live births per litter is given at the right. Find the mean and standard deviation. Arrange the computations as in Table 3.9 and use the alternate formula for the standard deviation.

Number of live births, x	Frequency, f
0	2
1	7
2	8
3	14
4	10
5	3
6	1

6. A study of the checkout times of 100 customers at a supermarket resulted in the distribution below. Find the mean and standard deviation.

x (minutes)	f
.5–1.5	15
1.5–2.5	20
2.5–3.5	15
3.5–4.5	20
4.5–5.5	30

7. Write out the expression corresponding to each of the following:

a. $\displaystyle\sum_{i=1}^{4} x_i^2$

b. $\displaystyle\left(\sum_{i=1}^{4} x_i\right)^2$

c. $\displaystyle\sum_{i=1}^{3} x_i f_i$

d. $\displaystyle\sum_{i=1}^{3} x_i^2 f_i - \left(\sum_{i=1}^{3} x_i f_i\right)^2$

8. A study of the yearly cost of nonrisk auto insurance for residents of a major metropolitan area stated that the insurance costs of the 2300 families surveyed varied between $390 and $798. Estimate the standard deviation of the cost data.

9. The grades on an examination have a mean of 66 and a standard deviation of 10. Find the mean and standard deviation of the grades that would result from:
   a. Subtracting 66 from each of the original grades.
   b. Subtracting 66 from each of the original grades and dividing by 10.

10. One hundred joggers were surveyed as to the number of miles which they jog or run weekly. The distribution is given at the right. Find

a. The lower quartile.
b. The upper quartile.
c. The median.
d. The 90th percentile.

Miles run	f
0–2	3
2–4	12
4–6	30
6–8	20
8–10	14
10–12	6
12–14	5
14–16	10

11. Thirty-eight graduate students were surveyed as to the number of units in which they were enrolled. The distribution is given below. Find the percentile rank of each response.

x	f
6	10
9	14
10	8
12	6

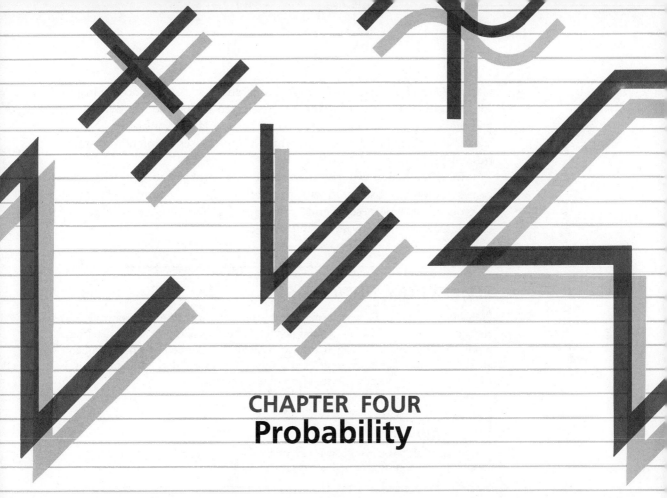

CHAPTER FOUR
Probability

The preceding two chapters have dealt with techniques for describing the data in a sample. Our goal now is to examine the processes for using samples to draw inferences about populations. Basic to all such processes are measures of the reliability or certainty of the inferences that are made. These measures have their origins in probability theory.

Probability theory originated with calculations in games of chance. History has it that the 17th century nobleman Antoine Gombaud, Chevalier de Mere, approached the mathematician Blaise Pascal with two problems relating to the odds and wagering stakes in a game of dice. Pascal subsequently conveyed the problems to another well-known mathematician, Pierre de Fermat, and out of their correspondence was born the subject of probability.

While probability theory is often identified with games of chance, it is in fact applicable to any type of experiment that can be repeated and for which there is an element of chance regarding the outcome. Selecting a sample from a known population and computing, say, its mean is just such an experiment. And just as there are probability distributions that cover games of dice, cards, and roulette, so there are distributions that pertain to the sampling process.

Probability theory provides a means of reasoning from a known population to randomly selected samples. Inferential statistics proceeds in just the opposite direction, using randomly selected samples to test hypotheses and make estimates about populations that are largely unknown. Nevertheless it is probability theory on which the processes of inferential statistics are based.

In this chapter we consider elementary probability concepts: outcomes and events associated with probability experiments, the assignment of probabilities to these events, and rules for obtaining probabilities of complex events from those of simpler events.

4.1   ELEMENTARY NOTIONS

Probability theory deals with experiments where the outcomes are controlled by chance. Generally, there are several possible outcomes and we cannot determine a priori which will occur when the experiment is performed. Such experiments are called *probability experiments*. Obviously, in a probability experiment, it is important that we specify the set of possible outcomes.

DEFINITION 4.1

The set of all possible outcomes of a probability experiment is called a *sample space* for the experiment and is represented by S.

Since there are many ways to interpret an experimental result, we stipulate that no result shall yield more than one outcome of the sample space.

Example 4.1

1. A coin is tossed. If we agree not to consider the possibility of the coin landing on its edge, then a sample space is $S = \{heads, tails\}$ or more simply $S = \{H, T\}$.
2. Consider the experiment of randomly selecting a family and counting the number of children. If no family has more than, say, 15 children, then a sample space of possible outcomes is

$$S = \{0, 1, 2, 3, . . ., 15\}$$

Another sample space is comprised of pairs of the form

(number of boys, number of girls)

A pair such as (3, 2) indicates a family with five children, consisting of three boys and two girls.
3. A sample of 1250 voters is to be polled to determine what percent favor lowering the drinking age. This is a probability experiment, since we cannot predict in advance what will occur, except that the result will fall between 0% and 100%. ∎

Our immediate goal is twofold:
1. To assign numbers, called *probabilities,* to the individual outcomes of the sample space that predict, insofar as possible, their relative frequencies when the experiment is performed many times.
2. To assign probabilities to sets of outcomes. Again, these are to be predictors of relative frequencies.

DEFINITION 4.2

A set of outcomes with an assigned probability is called an *event.*

Events are denoted by capital letters such as E, F, and G; outcomes by subscripted lower-case letters such as e_1, e_2, e_3, ... The probability of an event E is denoted $P(E)$.

As an illustration, suppose a student is selected at random and asked to choose a soft drink from a display of three colas numbered 1, 2, and 3 and two noncolas numbered 4 and 5. The numbers 1, 2, 3, 4, and 5 represent the possible outcomes and would be denoted by e_1, e_2, e_3, e_4, and e_5. The set of outcomes $E = \{1, 2, 3\}$ represents the event "cola drink is selected." An event is simply a collection of outcomes that may be of particular interest. We say that an event has occurred if any of the outcomes that constitute it occurs. Thus if the student selects any one of the drinks numbered 1, 2, or 3, then the event E occurs.

As a second illustration, if we are playing roulette and the wheel stops on 12, then not only has the outcome 12 occurred, but also the event "red number," since 12 is one of the red numbers. Also, the event "even number" has occurred. Whether these events are of any interest depends on the type of bet you have made.

We shall try to avoid the use of symbols such as E and F for events and simply work with specific descriptions. However, in some instances, particularly in the statements of theorems and formulas, their use is necessary.

Now let us look at the matter of interpreting probabilities. Our view is simply that they represent estimates of relative frequencies over the long run. For example, the probability of a coin coming up heads is

$$P(\text{head}) = \frac{1}{2} = .5 = 50\%.$$

This means that in many throws of a coin, we can expect a head to occur in about ½ or 50% of all tosses. It tells us very little, if anything, about what we will see in, say, four tosses. But in 10,000 tosses, the coin should show about 5000 heads. We should not be surprised if only 4950 or possibly 5025 occurred, but a result like 7800 or 1400 would certainly test our belief that the probability of a head is ½ for the particular coin being used.

Similarly, suppose we know that the probability of dying from a particular influenza virus is $p = .047$. This will be more meaningful if we convert it to 4.7%. This still will not tell you very much if you have this virus, but if we examine 10,000 cases, we would expect around 470 deaths.

Obviously, it is one thing to interpret probabilities and another matter to decide how they should be assigned. The simplest approach is to list all the outcomes in the most elementary fashion possible. If all outcomes are equally likely, they have the same probability. For this to happen, there must be no condition that favors one outcome over another. Then, if the sample space has n outcomes, each should occur with about the same frequency and have probability $1/n$. If an event contains k of these outcomes, then it has probability k/n.

Example 4.2

Consider the toss of a single die with faces numbered 1, 2, 3, 4, 5, and 6. Each of these faces is equally likely to turn up (assuming an honest die) and would have probability 1/6. ∎

Example 4.3

The distribution of 33 grades on an examination is given in Table 4.1. Consider the experiment of selecting a paper at random. Each is equally likely and has probability 1/33 of being selected. Consider the event "score = 63." There are four such papers. Thus $P(\text{score} = 63) = 4/33$. Similarly,

$$P(\text{score above } 80) = \frac{3}{33} = \frac{1}{11}$$

$$P(\text{score between } 63 \text{ and } 70) = \frac{12}{33} = \frac{4}{11} \quad ∎$$

TABLE 4.1
DISTRIBUTION OF GRADES

Grade	Frequency
54	2
60	4
63	4
66	7
67	5
70	3
77	5
84	2
85	1

In many experiments there will be several stages, treatments, or perhaps objects to deal with. Often it is then possible to write the outcomes as ordered arrangements: pairs, triples, and so on.

Example 4.4

Two coins are tossed. Record the outcomes as pairs of the form (first coin result, second coin result). The sample space is then the set {(H, H), (H, T), (T, H), (T, T)}. There is no reason to expect any one of these pairs to be favored over any other; thus they are equally likely. We assign each a probability of 1/4. The event "coins show different results" is represented by the set of outcomes $E =$ $\{(H, T), (T, H)\}$ and should occur about $2/4 = 1/2$ of the time. Thus $P(E) =$ $1/2$. ■

Example 4.5

Three students, whom we will call A, B, and C, each write the word "test" on a piece of paper. The papers are given to a psychic who claims he can match the handwriting to the student by his appearance. Suppose the psychic is guessing, so that the students names are randomly assigned to the three papers. What is the probability that there are two or more incorrect assignments? The possible assignments of names to the papers are listed in Table 4.2; and each has probability $1/6$, since they are presumably equally likely. Five of these outcomes have two or more incorrect assignments. Thus

$$P(2 \text{ or more incorrect}) = \frac{5}{6}$$

If the psychic plans to use this experiment to impress audiences, he will be sadly disappointed. About five-sixths of the time, he will make either two or three mistakes. ■

TABLE 4.2   ASSIGNMENTS OF THREE NAMES TO THREE PAPERS

Owner of the Paper		
A	B	C
A	B	C
A	C	B
B	A	C
B	C	A
C	A	B
C	B	A

Note that in using ordered arrangements, their number can be found by multiplying the number of choices for the first entry by the number for the second entry and so forth. For the case of two coins, we could write H or T in the first position; thus there were two choices for it. Once this was done, we again had two choices for the second position. Thus there would be $2 \times 2 = 4$ outcomes. In the case of the assignment of names by the psychic, we could write down any one of the three names in the first position, then any one of the two remaining in the second, and finally the one remaining in the third. Thus the sample space contained $3 \times 2 \times 1 = 6$ outcomes.

The outcomes of a sample space are not always equally likely. When this is the case, it may still be possible to construct an elementary sample space with equally likely outcomes.

Example 4.6   A sample space of sums for the throw of a pair of dice is $S = \{2, 3, 4, 5, 6, 7,$ $8, 9, 10, 11, 12\}$. A sample space of equally likely outcomes is the set of $6 \times 6 = 36$ pairs of the form:

(Number of first die, number on second die)

Each has probability 1/36. The event "sum is 3" is $E = \{(1, 2), (2, 1)\}$ and has probability 2/36. The sums, the pairs that produce these sums, and their probabilities are summarized in Table 4.3. To better illustrate the pattern for forming the probabilities, the fractions have not been reduced. ∎

TABLE 4.3   OUTCOMES IN DICE

					(6, 1)					
				(5, 1)	(5, 2)	(6, 2)				
			(4, 1)	(4, 2)	(4, 3)	(5, 3)	(6, 3)			
		(3, 1)	(3, 2)	(3, 3)	(3, 4)	(4, 4)	(5, 4)	(6, 4)		
	(2, 1)	(2, 2)	(2, 3)	(2, 4)	(2, 5)	(3, 5)	(4, 5)	(5, 5)	(6, 5)	
(1, 1)	(1, 2)	(1, 3)	(1, 4)	(1, 5)	(1, 6)	(2, 6)	(3, 6)	(4, 6)	(5, 6)	(6, 6)

Sum	2	3	4	5	6	7	8	9	10	11	12
Probability	1/36	2/36	3/36	4/36	5/36	6/36	5/36	4/36	3/36	2/36	1/36

A comment regarding notation. Properly speaking, probabilities are assigned to sets of outcomes that we call events. If $e_1, e_2, \ldots$ are outcomes, their probabilities should be denoted by $P(\{e_1\})$, $P(\{e_2\})$, ... We avoid this cumbersome notation by simply writing $P(e_1)$, $P(e_2)$, ... Repeated parentheses are also suppressed. Thus, for the case of two coins, we write $P(H, T)$ for either $P((H, T))$ or $P(\{(H, T)\})$.

As we have seen, with each probability experiment there is a function P that assigns probabilities to the possible events. The foregoing examples and discussion should have made it clear that this function has certain special properties.

> **DEFINITION 4.3**
>
> The set function P is said to be a *probability function* for the discrete sample space $S = \{e_1, e_2, \ldots\}$ if:
> 1. For any event E, $0 \le P(E) \le 1$, that is, the probability of an event is nonnegative and never exceeds 1.
> 2. The probability of an event E is the sum of the probabilities of the outcomes that E comprises.
> 3. $P(S) = 1$, that is, the sum of the probabilities of all the outcomes is 1.

This is, in fact, the definition of a *discrete* probability function. Apart from conditions (1) and (3), there are no restrictions on how the probabilities of the outcomes should be assigned. To be useful, however, they must reflect the realities of the physical situation under investigation.

Example 4.7   What is the probability of exactly two boys in a family with three children? As outcomes we use ordered triples indicating the sex of the first, second, and third child. Thus

$$S = \{(M, M, M), (M, M, F), (M, F, M), (F, M, M),$$
$$(M, F, F), (F, M, F), (F, F, M), (F, F, F)\}$$

The event "exactly two boys" is $E = \{(M, M, F), (M, F, M), (F, M, M)\}$ The probability of E, $P(E)$, depends on how probabilities are assigned to the outcomes. If they are equally likely, each has probability $\frac{1}{8}$ and $P(E) = \frac{3}{8} = .375$. Examination of birth records indicate that $P(E)$ is more nearly .385. Thus the outcomes are not equally likely and their probabilities must be determined empirically. ■

The following example further illustrates the use of the addition property (2) of the above definition.

Example 4.8   Consider the game of dice with the sample space of sums $S = \{2, 3, 4, 5, 6, 7, 8, 9, 10, 11, 12\}$. The associated probabilities are given in Table 4.4. In the game of dice, *craps* is said to occur if a 2, 3, or 12 is thrown. The probability of this event $E = \{2, 3, 12\}$ is $P(E) = P(2) + P(3) + P(12) = \frac{1}{36} + \frac{2}{36} + \frac{1}{36} = \frac{4}{36} = \frac{1}{9}$. The reader should return to Example 4.4 and verify that in fact there are four pairs that can produce a 2, 3, or 12. ■

TABLE 4.4 PROBABILITIES
OF POSSIBLE SUMS IN DICE

Outcome	Probability
2	1/36
3	2/36
4	3/36
5	4/36
6	5/36
7	6/36
8	5/36
9	4/36
10	3/36
11	2/36
12	1/36

In many problems the outcomes of the sample space are too numerous for their listing to be practical. At this point, we could proceed to develop sophisticated counting techniques to deal with such problems. In the interest of conserving time for later statistical development, we choose however, not to pursue this matter.

Part A

1. Two people are to be interviewed as to their political affiliation, with the responses being recorded as Republican, Democrat, or Independent. Using ordered pairs, give a sample space for the experiment.

2. Three draws with replacement between draws are made from a bowl containing one blue, four red, and five white poker chips. Give the sample space if
   a. An outcome is the number of white chips drawn.
   b. An outcome describes both the number of red and the number of white chips drawn.

3. Express the following statements using probabilities:
   a. 12% of the tires produced by this method fail to pass inspection.
   b. 58% of the people believe this new product is superior to the present one.
   c. 30% of all first-year students are on probation after one semester.
   d. 5% of the sample means are more than two standard deviations from the population mean.

4. Express the following in terms of percents or relative frequencies.
   a. The probability that a randomly selected package of bacon will weigh less than 16 ounces is 0.02.
   b. The probability that a randomly selected voter favors the issue is 0.54.
   c. The probability that the sample means exceeds 150 is 0.07.
   d. The probability that the sample mean is within three units of 72 is 0.997.

5. A coin with its faces numbered 1 and 2 and a die are tossed.
   a. Give a sample space of equally likely outcomes.
   b. Find the probabilities of the events representing the possible sums.
   c. If the experiment of tossing the coin and the die is performed 2400 times, about how many times should the sum 4 occur?

6. A pair of dice are tossed. Using Table 4.3 find:
   a. The probability that both dice show the same.
   b. The probability of a sum less than 7.
   c. The probability of an 8 by some other way than (4, 4).
   d. The probability of an even sum.

7. Two stories can appear on any of the four pages of the student paper.
   a. Produce a sample space where the outcomes describe the pages on which the two stories can appear.
   b. What is the probability that the stories appear on different pages?

8. Three people, each with a different occupation, are observed by a person who claims to have psychic powers. The psychic then predicts each person's occupation. Assuming that the psychic knows the three occupations but is simply guessing at which belongs to whom, what is the probability that she guesses:
   a. All three correctly?
   b. At least two correctly?

9. Let $S = \{e_1, e_2, e_3, e_4\}$ with $P(e_1) = \frac{1}{6}$, $P(e_2) = \frac{1}{3}$, $P(e_3) = \frac{1}{3}$, and $P(e_4) = \frac{1}{6}$.
   a. Find $P(E)$ where $E = \{e_1, e_2\}$.
   b. Find $P(S)$.
   c. Devise an experiment of drawing a ball from an urn where there would be four outcomes with the above probabilities.
   d. Devise an experiment that involves sampling a hypothetical population of college students where there would be four outcomes with the above probabilities.

10. Three coins are tossed. A sample space of equally likely outcomes of the form (first coin, second coin, third coin) is to be used.

    a. How many outcomes are in this sample space?

    b. List the sample space.

    c. Find the probability of each of the following events.

        i. No heads.

        ii. Exactly one head.

        iii. Two or more heads.

11. Three dice are tossed. Assuming a sample space of equally likely triples find:

    a. The number of outcomes in the sample space.

    b. The number of outcomes with an even number in each entry. Do not try to list them, please!

    c. The probability that the three dice each show an even number.

12. An urn contains two white and two red balls, designated r_1, r_2, w_1, and w_2. Two balls are to be drawn from the urn without replacement between drawings.

    a. Give a sample space of equally likely pairs.

    b. What is the probability that the first ball drawn is red and the second is white?

    c. What is the probability that in the two draws, one white and one red ball results (any order)?

13. A firm has two positions to be filled and four equally qualified applicants, two men and two women. If two of these applicants are selected by random drawing, what is the probability that both of the women are selected? Hint: Label the men M_1 and M_2, and the women W_1 and W_2, then write out the sample space.

14. The results of classifying 400 students by sex and by major is given below (a two-way classification). If a student is selected at random from these 400, what is the probability that

    a. The student is a psychology major?

    b. The student is a business major?

    c. The student is female?

    d. The student is a female health science major?

    e. The student is a male psychology major?

		Major			
Sex	Psychology	Sociology	Health science	Business	Total
Male	40	30	20	60	150
Female	75	50	35	90	250
Total	115	80	55	150	400

Part B

15. In a marketing test of package design, customers are shown four boxes that will hold different amounts of a laundry detergent and asked to rank the boxes according to the amount they hold. If it is impossible for the customer to discern any difference in the sizes, what is the probability of a customer ranking the boxes correctly?

16. A firm classifies its sales by size (small, medium, or large) and by type (mail order or over-the-counter). Using the classification of 250 sales below, find the (empirical) probability of

    a. A small sale.

    b. A mail order sale.

    c. A large over-the-counter sale.

    d. A large mail order sale.

Sale type	Sale size			Total
	Small	Medium	Large	
Counter	50	80	70	200
Mail	20	20	10	50
Total	70	100	80	250

17. In a study of taste discriminization, a subject's eyes are blindfolded and his nasal passages blocked. He is then presented with four cups of juice: one onion, one tomato, one orange, and one apple. He tastes each and is asked to identify it from the list of four possibilities. Assuming he must guess at each identification and does not name any juice twice, what is the probability of two or more correct identificatons?

18. Suggest a way by which one might estimate:
    a. The probability that a thumbtack will land point up when dropped on a floor.
    b. The probability that an activated burglar alarm will be a false alarm.
    c. The probability that a randomly selected family will have an income exceeding $30,000.

19. $S = \{e_1, e_2, e_3\}$ is the sample space for an experiment. $P(e_1) = \frac{1}{4}$ and $P(e_3) = \frac{3}{8}$. Find $P(e_2)$.

20. A population consists of the four observations 2, 6, 4, and 2. Samples of size 2 are drawn without replacement from this population and their means determined. What is the probability of obtaining a sample with a mean of 3? A mean of 2?

21. A coin is tossed until a head appears. Describe a sample space for this experiment.

22. How is drawing red and white balls from an urn similar to selecting a random sample of women and men from a classroom?

23. A family of four, consisting of parents and two children, is asked whether they like (L) or dislike (D) a certain television show. Let the responses of the father, the mother, the first child, and the second child form an ordered quadruple—thus (L,L,D,L) is one possible outcome.
    a. How many possible outcomes are there if each individual's response is independent of those of the other family members?
    b. How many possible outcomes are there if the husband and wife always give the same response?

4.2   RULES FOR COMPUTING THE PROBABILITIES OF COMPOUND EVENTS

If E and F are events associated with a probability experiment, we often have occasion to form other events from these by modifying and combining their descriptions. These new events are called *compound events*. We conserve space and combine the definitions of three such events into a single statement.

DEFINITION 4.4

Let E and F be events.
1. The event (E or F) is the event consisting of those outcomes that are in E or F or both.

2. The event (E and F) is the event consisting of those outcomes that are in both E and F.
3. The event (not E) or the complement of E is the event consisting of those outcomes of the sample space that are not in E.

Thus the event (E or F) occurs if either E or F or both occur. The event (E and F) occurs only if both E and F occur. The event (not E) occurs if E does not occur. For the reader familiar with set notation, (E or F) is the set union $E \cup F$, (E and F) is the set intersection $E \cap F$, and (not E) is the complement $\bar{E}$. A useful way of representing these events is illustrated in Figure 4.1. The points inside the region E represents the outcomes that are in E, those in the intersection of the regions (E and F), and so forth.

FIGURE 4.1   DIAGRAMS REPRESENTING EVENTS

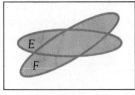

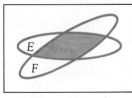
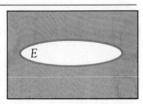

(E or F) shaded   (E and F) shaded   (not E) shaded

Example 4.9

Consider the experiment of selecting a student at random from a high school student body. Let E be the event "student is a senior." Let F be the event "student is a girl." Then (E or F) occurs if the student selected is either a senior, a girl, or both. The event (E and F) occurs if the student selected is both a senior and a girl. And (not E) occurs if the student selected is not a senior. ■

Example 4.10

Consider the experiment of throwing a pair of dice. Let E be the event "sum is less than 9." Thus $E = \{2, 3, 4, 5, 6, 7, 8\}$. Similarly, let $F = \{2, 4, 6, 8, 10, 12\}$ be the event "sum is even." Then

(not E) = {9, 10, 11, 12} (those sums not in E)
(E and F) = {2, 4, 6, 8} (those sums in both E and F)
(E or F) = {2, 3, 4, 5, 6, 7, 8, 10, 12} (those sums in either E or F)

Note that those outcomes which are in common, that is, in both E and F, appear only once in (E or F), even though they appear in both the lists of E and F. ■

A word of reassurance. In most instances we will use expressions such as (E and F) only in formulas and general discussions. In practice, these will be replaced by some specific description, as in the following example.

Example 4.11

P(white rat and red eyes) would be an instance of *P(E and F)*
P(sophomore or psychology major) would be an instance of *P(E or F)*
P(not passing) would be an instance of *P(not E)* ■

Now consider *P*(not *E*). How is this related to *P(E)*? If we know that 60% of the seeds in a culture germinated, then it must be that 100% − 60% = 40% failed to germinate. If ⅖ of the class smokes, then 1 − ⅖ = ⅗ do not smoke. Theorem 4.1 translates this information into a probability statement.

THEOREM 4.1
For an event *E*

$$P(\text{not } E) = 1 - P(E).$$

Example 4.12

If the probability of a customer requiring two or fewer minutes at a bank teller window is .7, then the probability that he will require more than two minutes is immediately recognized as 1 − .7 = .3. Or, if 70% take two or fewer minutes, then 30% take more than two minutes. ■

Intuitively, the probability of the event (*E* or *F*) should be related to the sum of the individual probabilities of *E* and *F*. For example, if 12% of the class received A's and 18% received B's, then clearly 12% + 18% = 30% received an A or B. Of course this works out nicely, since no student can receive both an A and a B as a grade. In this case the two events are pictured as in Figure 4.2.

FIGURE 4.2 *E* AND *F* HAVE NO OUTCOMES IN COMMON

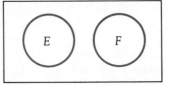

FIGURE 4.3 *E* AND *F* HAVE OUTCOMES IN COMMON

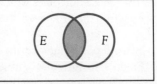

Now suppose we have a class where 40% are women and 12% are psychology majors. Can we now add to find the percent that are women or psychology majors? Not if some of the women are also psychology majors, for the sum 40% + 12% = 52% would count those in both categories. The two events are now as pictured in Figure 4.3. If, for example, 4% of the class were females and psychology majors, then there would in fact be 40% + 12% − 4% = 52% − 4% = 48% who are either females or psychology majors.

In general, when dealing with (E or F), we must be on guard as to whether E and F have any outcomes in common. If so, the sum $P(E) + P(F)$ must be adjusted by $P(E$ and $F)$.

THEOREM 4.2

Let E and F be events. Then

$$P(E \text{ or } F) = P(E) + P(F) - P(E \text{ and } F).$$

This theorem does not often assist in finding $P(E$ or $F)$, because the term $P(E$ and $F)$ may be equally difficult to compute. As we have seen, one exception occurs when the events have no outcomes in common.

DEFINITION 4.5

The events E and F are said to be *mutually exclusive* if they have no outcomes in common.

Note that if two events are mutually exclusive, then the occurrence of one excludes the occurrence of the other. In set terminology, such events are said to be *disjoint*. For mutually exclusive events, we obviously have $P(E$ and $F) = 0$ and Theorem 4.2 simplifies to the following:

THEOREM 4.3

Let E and F be mutually exclusive events. Then

$$P(E \text{ or } F) = P(E) + P(F)$$

Example 4.13

Consider the experiment of selecting a random sample of 50 male college students and determining their mean weight. Suppose the event E, "mean weight less than 138 pounds," has a probability of .05, while the event F, "mean weight greater than 165 pounds," has a probability of .02. Then

P(sample mean is less than 138 pounds or greater than 165 pounds)
$$= P(E \text{ or } F) = P(E) + P(F) = .05 + .02 = .07.$$

In other words, if 5% of all the sample means are below 138 and 2% are above 165, then 7% will be one or the other. ■

The reader should be alert for the use of the connective "or" in the description of an event. It generally means that simpler events are involved; this will be particularly useful if they are mutually exclusive.

Example 4.14

1. In drawing a card from a bridge deck,

$$P(\text{six or ace}) = P(\text{six}) + P(\text{ace}) = \frac{4}{52} + \frac{4}{52} = \frac{8}{52} = \frac{2}{13}$$

2. In selecting a voter from a list in which 35% are Republicans, 12% are Independents, and 53% are Democrats,

$$P(\text{Democrat or Independent}) = P(\text{Democrat}) + P(\text{Independent})$$
$$= .53 + .12 = .65.$$

3. In throwing a pair of dice, $P(\text{sum is less than 9 or even})$ does not involve mutually exclusive events.

This event occurs if the sum is any one of 2, 3, 4, 5, 6, 7, 8, 10, or 12. Summing their probabilities, we obtain $\frac{30}{36} = \frac{5}{6}$. If we wished to use Theorem 4.2, we would need

$P(\text{sum less than 9})$

$$= P(2) + P(3) + P(4) + P(5) + P(6) + P(7) + P(8) = \frac{26}{36}$$

$$P(\text{sum is even}) = P(2) + P(4) + P(6) + P(8) + P(10) + P(12) = \frac{18}{36}$$

$$P(\text{sum is less than 9 and even}) = P(2) + P(4) + P(6) + P(8) = \frac{14}{36}$$

Then

$$P(\text{sum less than 9 or even}) = \frac{26}{36} + \frac{18}{36} - \frac{14}{36} = \frac{30}{36} = \frac{5}{6} \quad ■$$

Our next task is to develop a formula for $P(E \text{ and } F)$. For this, we need the idea of *conditional probability*. As a way of introduction, imagine yourself playing cards. You have a five-card hand with three aces. You plan to throw away the two nondescript cards and draw two additional ones from the deck. With luck, you will receive the fourth ace. Suppose, however, that you are able to peek at an opponent's hand and it contains the remaining ace. Now what? Certainly there is no chance of getting the fourth ace. The point is that you have gained additional information about the state of the deck, and you have thus been able to change the probabilities that you assign to the outcomes.

Generally speaking, the probability of an event F depends on the outcomes which comprise both it and the sample space. If it should happen that we know some related event E has occurred, then this information may be a basis for altering the sample space and thus the probability of the occurrence of F. We then speak of the conditional probability of F subject to the condition that E has occurred. This we denote by $P(F|E)$, which is read "the conditional probability of F given E." Before attempting a formal definition, let us consider two examples. The first illustrates the notation.

Example 4.15

1. Two machines, I and II, produce bolts. Five percent of those from I and 10% of those from II are defective. This we could indicate by:

$$P(\text{defective bolt}|\text{machine I}) = .05$$
$$P(\text{defective bolt}|\text{machine II}) = .10.$$

2. Suppose a mortality table shows that the probability of dying within the year for a 25-year-old male is .00193. This can be stated as:

$$P(\text{male dying within the year}|\text{age} = 25) = .00193. \quad \blacksquare$$

In the next example we try to motivate the formula that will be used in defining $P(F|E)$.

Example 4.16

Suppose 100 written examinations for driver's licenses are tabulated. Of these, 72 are for individuals over 40 years old. Additionally, there are 30 perfect scores, with 18 coming from the over-40 age group. Suppose, now, a paper is selected at random. The probability of obtaining a perfect paper is clearly $30/100 = .3$. Suppose, however, it is known to be from the over-40 age group. Obviously the probability is then $18/72 = .25$. Thus

$$P(\text{perfect paper}|\text{paper is from the over-40 age group}) = \frac{18}{72}$$

Now suppose we identify F with the event "perfect paper" and E with "paper is from the over-40 age group." Thus $P(F|E) = {}^{18}\!/_{72}$. In this fraction, we recognize the denominator as being the count of E (the over-40 age group). What of the numerator? It is the count of those papers that are jointly from the over-40 age group *and* perfect, that is, the count of $(E \text{ and } F)$. Dividing both numerator and denominator of this fraction by 100, the count of the sample space, we obtain

$$P(F|E) = \frac{18}{72} = \frac{18/100}{72/100} = \frac{P(E \text{ and } F)}{P(E)} \quad \blacksquare$$

It should be clear that when we are considering $P(F|E)$, the event E becomes, in essence, a new sample space. This is sometimes called the *reduced sample space*. Thus $P(F|E)$ is the probability of F relative to this reduced space. Our definition is based on this observation and the results of the last example.

DEFINITION 4.6

Let E and F be events with $P(E) \neq 0$. The *conditional probability* of F given E is denoted $P(F|E)$ and is defined as

$$P(F|E) = \frac{P(E \text{ and } F)}{P(E)}$$

For our purposes, the conditional probability formula in the above definition is less interesting than the rearrangement which produces a formula for the probability of $(E \text{ and } F)$.

THEOREM 4.4

Let E and F be events. Then

$$P(E \text{ and } F) = P(E)P(F|E)$$

This formula, sometimes called the *multiplication rule,* states that to find the probability of $(E \text{ and } F)$, we multiply the probability of E by the probability of F given E. Thus, just as "or" often suggests addition, "and" suggests multiplication.

Example 4.17

Two cards are drawn in succession from a bridge deck without replacement between drawings. The event "two aces" can occur only as (ace on the first draw and ace on the second draw). By the multiplication rule,

P(ace on the first draw and ace on the second draw)
 $= P$(ace on the first draw) $\times$ P(ace on the second draw|ace on the first draw)
$= \left(\dfrac{4}{52}\right)\left(\dfrac{3}{51}\right) = \dfrac{1}{221}$

Note that in computing the conditional probability of the ace on the second draw, we are simply taking into account the composition of the deck at this state of the experiment. ∎

The multiplication rule is a very powerful tool, particularly when used to find the probability of events that can be visualized as a sequence or succession of simpler events. One simply multiplies together the probability of each, taking into account the occurrence of the preceding events of the sequence.

Example 4.18

Two birds are selected at random from a cage containing five male (M) and two female (F) finches. What is the probability that both are males? Visualize the experiment as consisting of two stages, at each of which a bird is selected. Then, using M for male,

$P(M$ at first stage and M at second stage)

$$= P(M \text{ at first stage}) \cdot P(M \text{ at second stage}|M \text{ at first stage})$$

$$= \left(\frac{5}{7}\right) \cdot \left(\frac{4}{6}\right) = \frac{10}{21} \quad \blacksquare$$

Example 4.19

A shipment contains 10 items, of which four are defective. Three items are se-lected at random. What is the probability that exactly one of the three is defec-tive? Denoting defective and nondefective by D and N, and visualizing three stages, we see that exactly one defective item occurs by (D, N, N) or (N, D, N) or (N, N, D). Thus we need to add the probability of each of these. Now by the multiplication rule,

$$P(D, N, N) = \left(\frac{4}{10}\right)\left(\frac{6}{9}\right)\left(\frac{5}{8}\right) = \frac{120}{720} = \frac{1}{6}$$

$$P(N, D, N) = \left(\frac{6}{10}\right)\left(\frac{4}{9}\right)\left(\frac{5}{8}\right) = \frac{120}{720} = \frac{1}{6}$$

$$P(N, N, D) = \left(\frac{6}{10}\right)\left(\frac{5}{9}\right)\left(\frac{4}{8}\right) = \frac{120}{720} = \frac{1}{6}$$

Thus

$$P(\text{one defective}) = P(D, N, N) + P(N, D, N) + P(N, N, D)$$

$$= \frac{1}{6} + \frac{1}{6} + \frac{1}{6} = \frac{3}{6} = \frac{1}{2} \quad \blacksquare$$

A very important relation exists between E and F when the occurrence of E does not affect the probability of F and vice versa.

DEFINITION 4.7

The events E and F are said to be *independent* if

$$P(F|E) = P(F) \text{ and } P(E|F) = P(E)$$

For independent events, the multiplication rule becomes $P(E$ and $F) = P(E)P(F|E) = P(E)P(F)$.

THEOREM 4.5

Let E and F be independent events. Then

$$P(E \text{ and } F) = P(E)P(F)$$

Example 4.20

1. Two coins are flipped. The outcome of one in no way affects that of the second. Thus

$$P(2 \text{ heads}) = P(\text{head on first and head on second})$$
$$= P(\text{head on first})P(\text{head on second}) = \frac{1}{2} \cdot \frac{1}{2} = \frac{1}{4}$$

2. Thirty-two percent of all women and 20% of all men in a community are reportedly in favor of a certain bond issue. A reporter interviews a woman and a man (randomly selected) for a newscast.

P(both favor the issue)
= P(the man favors the issue and the woman favors the issue)
= P(the man favors the issue) · P(the woman favors the issue)
= (.20) · (.32) = .064

3. A coin and a pair of dice are tossed. Obviously what happens with the dice is independent of the coin. Thus

$$P(\text{head and a sum of } 7) = \frac{1}{2} \cdot \frac{1}{6} = \frac{1}{12} \quad \blacksquare$$

The concept of independence as well as the multiplication rule can be readily extended to the case of more than two events.

Example 4.21

At a factory producing generators, the three principal components—the armature, the field, and the frame—are independently produced and then assembled. Tests show that 6% of the armatures, 5% of the fields, and 2% of the frames are defective. What is the probability that a randomly selected motor has no defective components? Each component must be nondefective; these events have probabilities of .94, .95, and .98 respectively. The probability of the joint occurrence of these three independent events is

$$P(\text{no defective components}) = (.94) \cdot (.95) \cdot (.98) = .875 \text{ (rounded)}$$

Thus about 87.5% of the production should have no defects.   ■

Example 4.22

A rat runs the maze in Figure 4.4, beginning at A. Assuming it makes a random selection of the exit at each barricade, the probability that the rat will follow a particular path, say the one indicated, is

$$P = \left(\frac{1}{2}\right) \cdot \left(\frac{1}{3}\right) \cdot \left(\frac{1}{4}\right) = \frac{1}{24}$$

If, on the other hand, the rat tends to select an exit reasonably close to its entry point, then this result would be incorrect.

FIGURE 4.4   TEST MAZE

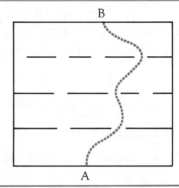

We note in passing that the statement $P(E \text{ and } F) = P(E)P(F)$ could be used to define the notion of independence. As a consequence, this equation finds a great deal of use in statistics when we need to test whether or not two population characteristics are independent. If they are, it must be that the proportion of the population having both characteristics is the product of the respective proportions.

Example 4.23

Of 800 applicants for a law school, 45% are accepted. It is reported that of two acceptable preparatory programs, say I and II, 60% of the applicants used program I. Now if the choice of a preparatory program is independent of the chance of being accepted, then for a randomly selected applicant,

$$P(\text{accepted and used program I}) = P(\text{accepted}) \times P(\text{used program I})$$
$$= (.45) \cdot (.60) = .270 = 27\%.$$

Thus about 27% or 216 of the 800 applicants should have been accepted and used program I. Naturally, if we were performing a statistical test to determine whether the school was biased toward either of the programs, we would not expect to observe exactly 216 students from program I. How to decide when an observed value differs significantly from an expected value (216 in this case) is covered in the chapters on hypothesis testing. ■

Finally, we should mention that Theorem 4.5 can be applied when one is sampling from a large population, where the removal of a few members does not appreciably alter the probabilities of the events involved. When this is the case, the product rule is applied as if the events were independent.

Example 4.24

In a large community, 42% of the adults are over 45 years of age. Two people are selected at random. The probability that both are over 45 years of age is:

P(first is over 45 and second is over 45)
= P(first is over 45) · P(second is over 45|first is over 45)
≈ P(first is over 45) · P(second is over 45) = .42 × .42 = .1764.

The removal of one adult from the population does not appreciably change the proportion of adults who are over 45 years of age. ■

4.3   TREE DIAGRAMS

As we have seen, the multiplication rule is often used to find the probability of a sequence of outcomes which can be regarded as having occurred at different stages of an experiment. A useful illustrative device is the tree diagram. In such a diagram the stages are represented by nodes (circles) and the outcomes by branches (line segments) extending from the nodes at which they occur. On each branch is marked the conditional probability of the corresponding outcome.

The tree diagram for the age problem (Example 4.24) is given in Figure 4.5. The stages correspond to the two selections, the outcomes to the possible age categories. The desired probability is found by simply multiplying the probabilities of the outcomes along the branches of the path representing the event "both over 45 years of age."

FIGURE 4.5   TREE DIAGRAM FOR SELECTING TWO ADULTS FROM A COMMUNITY IN WHICH 42% OF THE ADULTS ARE OVER 45 YEARS OLD

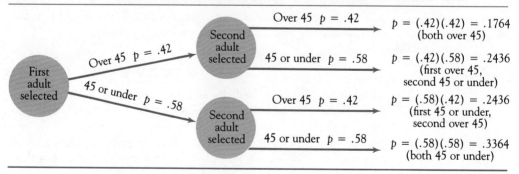

Example 4.25

The tree diagram of Figure 4.6 represents the bird selection experiment of Example 4.18. Use this diagram to find the probability of one male and female finch being selected.

The event "one male and one female" is represented by

$$E = \{(M, F), (F, M)\}$$

Thus

$$P(E) = p_2 + p_3 \text{ (see diagram)}$$
$$= \frac{5}{21} + \frac{5}{21} = \frac{10}{21}$$

**FIGURE 4.6   TREE DIAGRAM FOR SELECTING TWO BIRDS FROM A
CAGE CONTAINING FIVE MALES *(M)* AND TWO FEMALES *(F)***

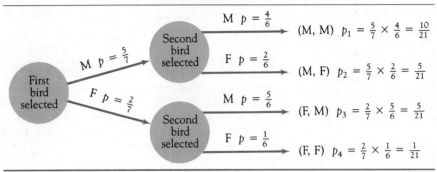

In complex problems there may be more than two branches (outcomes) leading from a particular node. And the nodes to which they lead need not be the
same form.

EXERCISE SET 4.2

Part A

1. Determine in each case whether or not the events E and F are mutually exclusive.
   a. A student is selected at random. E: The student is a history major. F: The
      student is a senior.
   b. A textbook is selected at random at the student bookstore. E: The price ex-
      ceeds \$20. F: The textbook is used.
   c. Two cards are drawn at random from a standard bridge deck. E: Both cards
      are black. F: The second card is the king of hearts.
   d. A pair of dice are tossed. E: The sum is seven. F: Both dice show an even
      number.
   e. Five electrical components are tested. E: At least one component failed. F:
      No component failed.
2. Refer to problem 1. Under what conditions does the event $(E$ or $F)$ occur?
3. Determine in each case whether the events E and F are independent.
   a. A new auto is selected from some particular manufacturer. E: The price of the
      auto exceeds \$15,000. F: The auto is a convertible.
   b. A pair of dice are tossed twice. E: The sum on the first roll is a 7. F: The
      sum on both rolls is a 7.

    c. A die is tossed and a coin is selected from a bowl. *E*: The coin is a quarter. *F*: The die shows a 3.

    d. An individual is selected at random. *E*: The individual's systolic blood pressure exceeds 140. *F*: The individual's hair color is not black.

    e. A student is selected at random. *E*: The student is a male physical education major. *F*: The student's weight exceeds 150 pounds.

4. Refer to problem 3. Under what conditions does the event (*E* and *F*) occur?

5. Consider the experiment of selecting a sample of 25 adults and determining the number who smoke. Give the complement (not *E*) of each of the following events:

    a. *E*: More than 20 smoke.

    b. *E*: At most two smoke.

    c. *E*: Five or more smoke.

    d. *E*: None smoke.

    e. *E*: At least one smokes.

6. If *E* and *F* are mutually exclusive, what is $P(E \text{ and } F)$? What does this say about any possible relationshlip between the concepts of mutually exclusive and independent events?

7. In a study of learning and recognition, let *E* be the event "phrase previously read," *F* "phrase previously heard," and *G* "phrase recognized when encountered a second time." What is meant by:

    a. $P(G|E)$?

    b. $P(G|F)$?

    c. $P(\text{not } G|E)$?

    d. $P(G|\text{not } E)$?

    e. Would you expect $P(G|E) = P(G|F)$?

8. What is the value of $P(F|E) + P(\text{not } F|E)$?

9. The probabilities of the possible outcomes of an experiment are given in the accompanying table. Let $E = \{e_1, e_2, e_3\}$ and $F = \{e_2, e_3, e_5\}$. Find:

    a. $P(E \text{ and } F)$.

    b. $P(E \text{ or } F)$.

    c. $P(\text{not } E)$.

    d. $P(E|F)$.

    e. $P(F|E)$.

Outcome	Probablity
e_1	.05
e_2	.20
e_3	.50
e_4	.10
e_5	.15

10. The betting layout for the game of roulette is pictured below with the red numbers in boldface. The remaining are black, except for 0, which is green. Find:

    a. $P(\text{first or third column number})$.

    b. $P(\text{red or first column number})$.

    c. $P(\text{red and first column number})$.

    d. $P(\text{red number}|\text{third column number})$.

	0	
1	2	**3**
4	**5**	6
7	8	**9**
10	11	**12**
13	**14**	15
16	17	**18**
19	20	**21**
22	**23**	24
25	26	**27**
28	29	**30**
31	**32**	33
34	35	**36**

11. In three tosses of a coin, exactly one head results. What is the probability that the head occurred on the first toss? Solve by using a reduced sample space.

12. A pair of dice are tossed. If both show the same, what is the probability of a sum less than 5?

13. Suppose that 50.3% of the population is female and 49.7% is male. Further, suppose that 6% of the males and 5% of the females are colorblind. Find the probability that a randomly selected individual is:
    a. A colorblind male.
    b. A male with normal vision.
    c. Colorblind.
    Make use of a tree diagram.

14. A newspaper vendor randomly mixes 50 late editions in with 30 early editions of the morning paper. What is the probability that the paper you purchase is a late edition if one paper has already been sold?

15. A business firm estimates that on a scarce item, the probability that supplier A will fill an order is .8, while with company B it is .6. An order is placed with both. What is the probability that the item will be obtained?

16. A test has three multiple-choice questions. The first has four choices, the second two, and the third three. A student guesses at all three answers. Find the probability that:
    a. All answers are correct.
    b. All answers are wrong.
    c. The first two are correct, the last is wrong.
    d. At most two answers are correct.

Part B

17. Of the students taking Statistics I, 70% pass and 30% fail. To enter the course, the students must first pass a screening examination. Sixty percent of the students pass this and go on to Statistics I. The 40% who fail this examination must take a remedial course, A. The 50% who pass this course are permitted to enter Statistics I, while those who fail are not permitted to continue. What percentage of all students who attempt this bureaucratic (albeit academic) jungle succeed in successfully completing Statistics I?

18. A backup system involving a critical component is often formed by placing two of the components in parallel. The accompanying figure shows two valves mounted in parallel serving an oxygen supply system. If the probability of such a valve functioning properly is .90, what is the reliability of this system—that is, what is the probability that at least one of the valves functions properly?

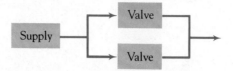

19. Suppose a fast-food chain reports that the probability of a successful first year for a new franchise is .85 and that all franchises with a successful first year remain successful. Further, in the event the first year is not successful, there is a probability of .08 that it will be disastrous and the business will fail before the year is out. Finally, if the first year is not successful but the franchise does not fail during this time, the probability is of .95 that it will ultimately become successful and remain so.
    a. Find the probability of a nonsuccessful first year.
    b. Find the probability of a disastrous first year.
    c. What percentage of all new franchises become successful businesses?

20. A temperature monitoring system in the pasteurization of a dairy product uses three temperature gauges. If each gauge has a probability of .90 of reporting the correct temperature, what is the probability that two or more of the gauges will report the correct temperature?

21. In a study of loneliness at Stanford University, 70 individuals (25 lonely and 45 nonlonely) rated or categorized 100 problems from 1 (least familiar as a problem) to 9 (most familiar as a problem). The most common type of problem for the lonely subjects centered around inhibited sociability. Selected problems and the probability that they would appear in the top five problems of the subjects are given in the table.

Problem: I find it hard to . . .	Probability of being among top five problems		Mean category placement	
	Lonely	Not lonely	Lonely	Not lonely
Make friends in a simple, natural way	.28	.00	6.16	2.84
Introduce myself to O(s) at parties	.24	.18	6.36	5.84
Make phone calls to O to initiate social activity	.20	.02	6.12	4.76
Participate in groups	.16	.04	5.12	4.60
Get pleasure out of a party	.16	.02	5.64	4.64
Get into the swing of a party	.12	.07	6.00	4.89
Relax on a date and enjoy myself	.12	.00	5.84	3.56
Be friendly and sociable with O	.08	.00	5.36	3.78
Participate in playing games with O	.04	.02	5.04	4.49
Get buddy-buddy with O	.04	.00	5.64	4.93
Entertain O at my home	.00	.04	4.84	4.56
Get along with O	.00	.02	6.08	5.04
Extend myself to accept O's friendship	.00	.00	4.48	3.33

Note. O = the other person.
Source: "Interpersonal Problems of People Who Describe Themselves as Lonely," *J. Consul. and Clinical Psychology*, 1979, vol. 47, no. 4, pp. 762–764.

    a. What does a high probability with respect to a particular problem denote? Is a high probability desirable?
    b. Consider the problem: "I find it hard to get into the swing of the party." How

many of the lonely subjects rated this in the top five problems? How many of the nonlonely?

c. The mean of the categories into which these problems were placed is also given. Notice that every one of the problems of inhibited sociability was placed in a higher category (on the average) by the lonely subjects than by the nonlonely. The author states that "the chance probability of such consistency is $(\frac{1}{2})^{13}$." How is this figure arrived at?

22. Let $S = \{e_1, e_2, e_3, e_4\}$ with $P(e_1) = \frac{5}{16}$, $P(e_2) = \frac{3}{16}$, $P(e_3) = \frac{3}{16}$, and $P(e_4) = \frac{5}{16}$. Also let $E = \{e_1, e_2\}$ and $F = \{e_2, e_3\}$.
   a. Are E and F independent?
   b. Are E and F mutually exclusive?

23. A consumer group reports that 25% of all 1-pound packages of a certain product are underweight. Three packages of this product are purchased at random.
   a. What is the probability that all three are underweight?
   b. What is the probability that none are underweight?
   c. What is the probability that exactly one is underweight?

24. A computer manufacturer offers a microcomputer in either a unitized or modular configuration and with a choice of 16K, 64K, or 256K of random access memory. The following table is a classification of sales by both configuration and size of memory.

Configuration	Memory			Total
	16K	64K	256K	
Unitized	280	510	1150	1940
Modular	1460	1590	410	3460
Total	1740	2100	1560	5400

If a sales record is chosen at random, what is the probability that it is for a computer with
   a. A modular configuration?
   b. A 64K memory?
   c. A unitized configuration and a 256K memory?
   d. A 16K or a 256K memory?
   e. A unitized configuration if it has a 16K memory?
   f. A 64K memory given that it has a modular configuration?

25. Sixty percent of the employees of a company live in the town in which the company is located. Suppose that 70% of the employees are women and that 40% are both women and residents of the town. An employee is selected at random. Let R indicate the event "resident" and F the event "female." Find
   a. $P(\text{not } F)$.
   b. $P(\text{not } R)$.
   c. $P(R \text{ or } F)$.
   d. $P(R|F)$.
   e. $P(F|R)$.
   f. $P(\text{not } R|F)$.

26. If the economy should improve, there is a 60% chance that a company's business will improve. If the economy does not improve the company's business will definitely suffer. Suppose a reliable business forecast predicts a 70% chance for an improved economy. What is the probability that the company's business will improve?

27. If E and F are independent, why should you expect (not E) and F to also be independent?

28. Show that the conditional probability formula can be written as:

$$P(F|E) = \frac{P(F) \cdot P(E|F)}{P(E)}$$

4.4   A RESEARCH STUDY

Adapted with permission from *Psychology and Statistics in Psychological Research and Theory*, Donald W. Stilson, Holden Day, 1966.

Example 4.26   The statistician should be a skeptic when interpreting published reports. As an example we consider the abundance of articles concerning the Archimedes Spiral Aftereffect Test. First proposed by E. Freeman and W. Joesy in 1949 ("Quantitative Visual Index to Memory Impairment," *Arch. Neurol. Psychiat.*, 1949, vol. 62, pp. 794–796) this test involves an Archimedes spiral as in Figure 4.7. The spiral is set in motion and observed by the patient. When stopped, the normal response is a continued aftersense of motion. Patients with organic brain damage did not report this aftereffect. Here, then, was a good test for the clinical psychologist to differentiate between patients with functional disorders and those with brain pathology.

FIGURE 4.7   ARCHIMEDEAN SPIRAL

Or so it seemed. Following this publication there were many reports relating to applications and validation of the test. H. A. Page et al. ("Another Application of the Spiral Aftereffect in the Determination of Brain Damage," *J. Consul. Psychology*, 1957, vol. 21, pp. 89–91) supplied data suggesting:

$$P(\text{aftereffect}|\text{functionally ill}) = .85$$
$$P(\text{no aftereffect}|\text{brain damage}) = .60$$

Thus 85% of the functionally ill have the aftereffect, while 60% of those with brain damage do not experience the effect. Two years previously, A. C. Price and H. L. Deabler ("Diagnosis of Organicity by Means of the Spiral Aftereffect," *J. Consul. Psychology*, 1955, vol. 19, pp. 299–302) reported that 96% to 98% of the cases with organic brain syndrome could be detected with this test. This was followed by a similar report in 1957. Also in 1956, A. F. Gallese ("Spiral Aftereffect as a Test of Organic Brain Damage," *J.Clin. Psychology,* 1956, vol. 12, pp. 254–258) reported a somewhat more modest 66% success with the test.

But there were those who were having trouble validating the test. In 1959, Emily Philbrick of the University of California at Berkeley published the results of a study of the use of the spiral aftereffect test to diagnose 81 newly admitted patients at a veteran's center hospital. This report ("The Validity of the Spiral Aftereffect as a Clinical Tool for Diagnosis of Organic Brain Pathology," *J. Consul. Psychology,* vol. 23, no. 1, 1959) differed greatly from previous ones. Table 4.5 shows the results obtained when Dr. Philbrick followed the procedures of Price and Deabler. Similar results were obtained with the other procedures as well.

TABLE 4.5   DISCRIMINATION VALUE OF SPIRAL AFTEREFFECT TESTS USING PRICE AND DEABLER'S SCORING SYSTEM (BEFORE AMYTAL)

Spiral criterion: Price and Deabler (½ scores)	Neurological classification	
	Organic	Nonorganic
Organic scores 0–2		
Number of patients	24	21
Percent of patients	53.3%	46.7%
Nonorganic scores 3–1		
Number of patients	13	14
Percent of patients	48.2%	51.8%

Her conclusions: "Under none of these conditions do the results in any way duplicate the work previously done with the spiral aftereffect test." And for all procedures considered, "in its present form the spiral aftereffect test is not useful for diagnosing organic brain pathology in a general hospital setting. . . . Under the conditions of this study there is no differentiation of organic from nonorganic by the spiral aftereffect test."

Why such contrasting results? First, there is the matter of blinds. In most of the previous investigations, the subjects had been drawn from previously diagnosed hospital patients. Thus according to Dr. Philbrick, "the investigators were working with known groups." In her study, the diagnostic categories were unknown at the time of the tests. These were determined later by two neurologists not involved with the investigation.

Another matter cited by Dr. Philbrick is that "subjects who fall at the extreme ends of the distribution may be correctly diagnosed, while the same test may fail for those closer to the center of the distribution." In her study, the test was applied to patients having irregularities representative of the entire range of the distribution.

Finally, there is the matter of probabilities. Suppose we assume, as is often the case, that roughly 80% of the subjects are functionally ill and the remaining

20% actually have some brain damage. Then using the results of Page and the conditional probability formula, we find

$$P(\text{functionally ill}|\text{no aftereffect}) = .50$$
$$P(\text{brain damage}|\text{no aftereffect}) = .50$$

In short, when there is no aftereffect, then about 50% of the time the patient will be functionally ill and 50% of the time he or she will have brain damage. One might as well flip a coin, except that we know that 80% of the patient population is functionally ill. We can thus get much better results statistically by diagnosing all patients as being functionally ill!

The spiral aftereffect has been discussed in several textbooks. A source from which we have adapted much of the material here is *Probability and Statistics in Psychological Research and Theory* (Donald W. Stilson, Holden Day, 1966). The interested reader will also find (in Chapter 5 of this text) some interesting comments on the subject of detection rates in diagnostic testing. ■

EXERCISE SET 4.3

The following problems all pertain to the research study on the spiral aftereffect. Let E denote the presence of the aftereffect, F the presence of functional illness, and D the presence of brain damage. Further assume that F and D are mutually exclusive.

1. Referring to the study of Page et al.:
   a. Find $P(\text{not } E|F)$.
   b. Find $P(E|D)$.
2. Assuming a population in which 80% of the patients are functionally ill and 20% have brain damage, find $P(E)$. Hint: $P(E) = P(F \text{ and } E) + P(D \text{ and } E)$. Now use the product rule to find these two probabilities. Again use the results of page et al.
3. Continuing with the assumptions of problem (2), find $P(\text{not } E)$.
4. Use the conditional probability formula to compute $P(F|\text{not } E)$. Proceed as follows:

$$P(F|\text{not } E) = \frac{P(F \text{ and not } E)}{P(\text{not } E)} \quad \text{(Why?)}$$
$$= \frac{P(F)P(\text{not } E|F)}{P(\text{not } E)} \quad \text{(Why?)}$$

   Now use the assumptions of problem (2) and the results of problems (1a) and (3) to complete the computation.
5. Find $P(D|\text{ not } E)$ using complementary events and the results of problem (4).

Library

Project

The following problems deal with an optional library project. Most statistical studies utimately end up as a report in some periodical. One locates those within a certain discipline or area through the use of abstracts. These are housed in the library adjacent to the periodicals which they reference and give a summary of each report. They also give the author, title, publication, volume, and year. The abstracts are indexed by subject and author and are collected into yearly volumes. The following problems require the use of *Abstracts in Psychology*.

1. Using the *Abstracts* for 1954, find all topics subsumed under "aftereffects."
2. Using the 1955 *Abstracts*, find three reports dealing with aftereffects. Include titles, author, publication, volume, and year.

3. Refer to problem (2). Look up the call numbers for two of the journals and publications and read the reports. Give three references from one of these reports.
4. Using the 1954 *Abstracts* and the author index, find the title of an article by E. A. Weinstein and S. Malitz.
5. In 1979, Phillip C. Kendall and Julian D. Ford published an article dealing with the reasons for clinical research. Read the article and give the two most frequent replies.

COMPUTER PRINTOUT

Often we simulate a probability experiment by having the computer "perform" an experiment over and over again, using a random number generator. In some cases simulation is used to determine probabilities empirically. In others it is used to test how well the proposed probabilities agree with reality.

Two Minitab simulations are shown. The first is for 90 throws of a single die. The possible outcomes are placed in column C1, their probabilities in C2, and the 90 outcomes generated by the program are placed in C3. The histogram shows reasonably good agreement with the 15 expected occurrences of each of the six faces.

Next, 720 simulated throws of a pair of dice are obtained by generating 720 outcomes for a single die into C3 and then repeating (for a second die) with C4. The corresponding sums are placed in C5 and represent the sums from a pair of dice that have been thrown 720 times. Compare the histogram of the sums in C5 with the results in Table 4.3.

FIGURE 4.8

```
MTB >SET THE FOLLOWING OUTCOMES INTO C1
DATA>1,2,3,4,5,6
DATA>END
MTB >SET THE FOLLOWING PROBABILITIES INTO C2
DATA>.166667,.166667,.166667,.166667,.166667,.166667
DATA>END
MTB >DRANDOM 90 OUTCOMES FROM C1, USE THE
PROBABILITIES OF C2, PUT IN C3
     90 DISCRETE RANDOM OBSERVATIONS

 6.0000  6.0000  1.0000  3.0000  1.0000  6.0000  1.0000
 6.0000  3.0000  3.0000  6.0000  2.0000  3.0000  4.0000
 3.0000  3.0000  1.0000  1.0000  5.0000  6.0000  1.0000
 2.0000  5.0000  6.0000  4.0000  4.0000  5.0000  2.0000
 3.0000  3.0000  3.0000  4.0000  2.0000  6.0000  4.0000
 4.0000  2.0000  1.0000  2.0000  5.0000  5.0000  5.0000
 1.0000  6.0000  6.0000  2.0000  3.0000  1.0000  3.0000
 4.0000  3.0000  4.0000  2.0000  5.0000  5.0000  5.0000
 3.0000  6.0000  3.0000  4.0000  1.0000  6.0000  4.0000
 4.0000  6.0000  4.0000  3.0000  3.0000  5.0000  2.0000
 3.0000  1.0000  5.0000  4.0000  2.0000  6.0000  3.0000
 2.0000  5.0000  2.0000  3.0000  6.0000  1.0000  6.0000
 4.0000  6.0000  6.0000  5.0000  6.0000  1.0000
```

```
          MTB >HISTOGRAM C3

             MIDDLE OF    NUMBER OF
             INTERVAL     OBSERVATIONS
                 1.         13      *************
                 2.         12      ************
                 3.         19      *******************
                 4.         14      **************
                 5.         13      *************
                 6.         19      ******************

          MTB >ERASE C3
          MTB >DIMENSION 720
          MTB >NO PRINT

          MTB >DRANDOM 720 OUTCOMES FROM C1, USE PROBABILITIES
          OF C2, PUT IN C3
          MTB >DRANDOM 720 OUTCOMES FROM C1, USE PROBABILITIES
          OF C2, PUT IN C4
          MTB >LET C5 = C3 + C4
          MTB >HISTOGRAM OF C5
             EACH * REPRESENTS 5 OBSERVATIONS

             MIDDLE OF    NUMBER OF
             INTERVAL     OBSERVATIONS
                 2.         13      ****
                 3.         28      ******
                 4.         63      *************
                 5.         89      *****************
                 6.         98      *******************
                 7.        123      *************************
                 8.        106      *********************
                 9.         66      *************
                10.         63      *************
                11.         40      ********
                12.         26      ******

          MTB >STOP
```

CHAPTER CHECKLIST

Vocabulary

Probability experiment

Outcome

Sample space

Event

Probability

Equally likely outcomes

Probability function

Mutually exclusive events

The event (not *E*)

The event (*E* or *F*)

The event (*E* and *F*)

Conditional probability

Addition rule

Multiplication rule

Independent events

Tree diagram

Techniques

Finding a sample space for a given experiment

Finding a sample space of equally likely outcomes for an experiment

Finding the probabilities of outcomes and events when a probability function is given

Finding the probability distribution for an experiment

Finding the probability $P(\text{not } E)$ from that of E

Finding the probability $P(E \text{ or } F)$ from that of $P(E)$, $P(F)$, and $P(E \text{ and } F)$

Finding the probability $P(E \text{ or } F)$ from $P(E)$ and $P(F)$ when E and F are mutually exclusive events (addition rule)

Finding the probability $P(F|E)$, using a reduced sample space

Finding the probability $P(F|E)$ from that of $P(E \text{ and } F)$ and $P(E)$

Finding $P(E \text{ and } F)$ from $P(E)$ and $P(F|E)$ (multiplication rule)

Find $P(E \text{ and } F)$ from $P(E)$ and $P(F)$ when E and F are independent

Determining if two events are independent

Finding the probabilities of the outcomes of a two- or three-stage experiment using a tree diagram and the multiplication rule

Notation

e_i	The ith outcome of an experiment
S	The sample space of an experiment
E	An event
$P(E)$	The probability of the event E
$(\text{not } E)$	The complement of the event E (set notation $\bar{E}$)
$(E \text{ or } F)$	The union of the event E and F (set notation $E \cup F$)
$(E \text{ and } F)$	The intersection of the events E and F (set notation $E \cap F$)
$P(F|E)$	The conditional probability of F subject to the condition that E has occurred

CHAPTER TEST

1. **a.** Give two examples of mutually exclusive events and state the addition rule for two such events.

   b. Give two examples of independent events and state the multiplication rule for two such events.

2. The effectiveness of an influenza vaccine is to be tested by administering it to randomly selected subjects and then exposing the subjects to the live virus.

   a. What is the probability experiment?

   b. Give a sample space.

   c. When the experiment was tried on 1440 test subjects, 1050 of them did not contract the influenza. Estimate the probability that the vaccine is effective (assuming exposure ordinarily produces influenza).

3. An experiment has the outcomes e_1, e_2, e_3, e_4, and e_5 with probabilities as given in the table below. Let $E = \{e_1, e_2\}$ and $F = \{e_2, e_3, e_5\}$. Find:

   a. $P(E)$.

   b. $P(\text{not } E)$.

   c. $P(E \text{ or } F)$.

   d. $P(E \text{ and } F)$.

Outcome	Probability
e_1	.40
e_2	.25
e_3	.20
e_4	.10
e_5	.05

4. Two students enroll in a statistics course with five sections. Assuming that students are assigned randomly to the different sections, produce a sample space of equally likely outcomes where the outcomes describe the sections in which the two students are enrolled.

5. Let E and F be independent events with $P(E) = .2$ and $P(F) = .5$. Find:
   a. $P(E \text{ and } F)$.
   b. $P(F|E)$.
   c. $P(E \text{ or } F)$.

6. A pair of dice are thrown. Find
   a. $P(\text{sum} < 6)$.
   b. $P(\text{sum is even or } 6)$.
   c. $P(\text{sum} < 5 | \text{sum is even})$.

7. Four cards are drawn in succession from a standard bridge deck with no replacement between draws. What is the probability that all four aces are drawn?

8. A room contains five men and four women. Two people are selected at random with no replacement between selections.
   a. Find the probability that the first person selected is a man and the second is a woman.
   b. Find the probability that all the people selected are men.
   c. Find the probability that the second person selected is a woman, given that the first to be selected was a woman.
   d. Find the probability that one man and one woman are selected.

9. One thousand people were surveyed as to whether or not they used professional help in preparing their income tax return. They were also classified by sex and by age. The results are summarized below:

Category	Men Young	Old	Women Young	Old	Total
Used assistance	40	120	60	180	400
Did not use assistance	180	140	200	80	600
Total	220	260	260	260	1000

If one of these people is selected at random, what is the probability that the person selected:
a. Made use of professional assistance?
b. Is young and made use of professional assistance?

    c. Is old and made use of professional assistance?

    d. Did not use professional assistance, given that the individual is a young female?

10. Five equally qualified firms are competing for two government contracts. Two of the firms are rated as small, the other three as large. Assuming that at most one contract will be awarded to a company, make use of a probability tree to find the probability that:

    a. Both contracts go to small companies.

    b. Both contracts go to large companies.

    c. One contract is to a large company and the other to a small company.

CHAPTER FIVE
Random Variables and Probability Distributions

This chapter develops two important concepts: *random variables* and *probability distributions*. These form the bridge between probability and statistics.

A random variable is a variable whose values depend on outcomes of a probability experiment. Apart from the designation "random," these variables are very much like the population variables we have been using all along. Random variables often originate naturally when we focus on some interesting feature of the outcomes. For example, in the game of dice, it is usually the sums 2, 3, 4, . . ., 12 and not the pairs (1, 1), (2, 1), . . ., (6, 6) that are of interest. These sums are the values of a random variable X.

From a mathematical point of view, random variables are necessary when we attempt to develop a theory covering some broad class of probability experiments. Complex verbal descriptions are too varied to work with in a general way. Numerically valued variables provide a standard format for representing the results of any probability experiment.

Probability distributions are much like the relative frequency distributions that have been used to describe the data of a sample. Only now they predict the distribution of the outcomes when the experiment is performed many times. Depending on the form of the variable, a probability distribution will be either discrete or continuous. For example, the sum X in the roll of a pair of dice is discrete and has a discrete probability distribution. The diameter X of a randomly selected ball bearing from a production run, however, is continuous and has a continuous distribution. Continuous and discrete probability distributions are of distinctly different forms.

This chapter is concerned with the concepts and nomenclature used with probability distributions and not with particular distributions per se. We start with distribution tables, progress to the notions of the mean and standard deviation of a distribution, and finally examine joint distributions of two variables and the notion of independence for two variables.

5.1   RANDOM VARIABLES

As we have seen, the sample space for a probability experiment contains all the possible outcomes of the experiment. A moment's reflection will show that the sample space plays a role like that of the population in a statistical investigation. In Chapter 1, it will be recalled, we found it useful to treat a population as the measurements or observations made on its members and not the members per se. Similarly, mathematical treatments of probability experiments are much simpler if we deal with the outcomes as numbers, not as verbal statements. These numbers constitute the values of a random variable.

DEFINITION 5.1

Let S be the sample space of a probability experiment on which there is defined a probability function. A *random variable* on S is a function (rule) that associates a number with each member of S.

As with population variables, we use a capital letter such as X to denote a random variable and its lower-case counterpart x for a particular value. The set of all possible values assigned by X is called the range of X and is denoted S_X.

Example 5.1

A coin is flipped twice. The outcomes of a sample space and the values of X, where X assigns the number of heads is given in Table 5.1. Thus $S_X = \{0, 1, 2\}$ ■

TABLE 5.1 ASSIGNMENT OF VALUES BY A RANDOM VARIABLE

Outcome	x
(H, H)	2
(H, T)	1
(T, H)	1
(T, T)	0

While it is possible for many random variables to be defined on a single sample space, it is generally best to choose one variable whose values are most useful for the problem being considered. Table 5.2 illustrates some possibilities.

TABLE 5.2

Type of experiment	Description of X
Throwing a pair of dice	Possible sums; $S_X = \{2, 3, 4, \ldots, 12\}$
Throwing a coin until a head occurs	The throw on which the first head occurs: $S_X = \{1, 2, 3, 4, \ldots\}$
Questioning a sample of 100 students about an issue	The number in the sample taking a certain position; $S_X = \{0, 1, 2, \ldots, 100\}$
Measuring the temperature of a beaker of tap water	The temperature observed; if any temperature between, say 0 and 100°C is possible, then S_X is the interval $0 \leq x \leq 100$.
Measuring the weight of a randomly selected student	The weight observed; S_X is some interval, say, $60 \leq x \leq 300$
Finding the average miles per gallon of a sample of 18 autos	The mean of the sample; S_X is some interval, say $5 \leq x \leq 60$

The first three random variables in Table 5.2 are examples of what we have called *discrete variables:* the range S_X is finite or can be enumerated. The remaining three are *continuous:* their ranges when viewed as point sets of the number line are continuous intervals. We first discuss discrete variables.

5.2 DISCRETE PROBABILITY DISTRIBUTIONS

In Chapter 4 we stated the conditions that a probability function must meet in its assignment of probabilities to outcomes and events. Now, if a random variable X is defined on a sample space, then each of its values represents an outcome or an event and we can thereby determine a probability $P(x)$ for it. We thus obtain a probability function P whose range S_X is viewed as another sample space for the experiment.

> **DEFINITION 5.2**
>
> Let X be a discrete random variable with range S_X. A *probability distribution* for the variable X is a probability function P defined on S_X.

In some instances the probability distribution for a random variable X is simply a table that lists the values of the variable and their corresponding probabilities. In others, it is a specific formula from which the probabilities can be calculated.

Example 5.2

Find the distribution of the number of heads X when a coin is tossed twice. As a sample space, we use $S = \{(H, H), (H, T), (T, H), (T, T)\}$, each outcome of which has probability 1/4. The value $x = 1$ corresponds to the event $\{(H, T), (T, H)\}$ and has probability $P(1) = 1/4 + 1/4 = 1/2$. In a similar manner, we find $P(0) = 1/4$ and $P(2) = 1/4$. The distribution is given in Table 5.3.

TABLE 5.3 PROBABILITY DISTRIBUTION FOR THE NUMBER OF HEADS IN TWO TOSSES OF A COIN

x	$P(x)$
0	1/4
1	1/2
2	1/4

The probability histogram of the distribution of the last example is given in Figure 5.1. This is of exactly the same form as the relative frequency histograms used for discrete population variables, which were discussed in Chapter 2. Note that the values of X are the class marks and that the area bounded by the histogram is 1. Further, the area of the rectangle erected over any x is just the probability of x, namely $P(x)$. In fact, if the variable has only integral values, then the height is just the probability $P(x)$.

FIGURE 5.1 PROBABILITY HISTOGRAM

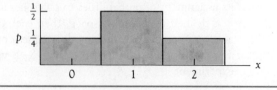

Example 5.3

Each day, a company receives six voltage control units, which are ultimately installed in stationary power plants. As a monitor of quality, two of these units are chosen at random and tested before installation. Find the distribution of the number X of defective units in a sample of two if there are two defective units in the shipment.

In the sample there could be 0, 1, or 2 defective units. Thus $S_X = \{0, 1, 2\}$. Let the units of the shipment be denoted by G_1, G_2, G_3, G_4, D_1, D_2 where G and D denote "good" and "defective" respectively. There are 15 possible samples of size 2. These samples and the corresponding values of X are shown in Table 5.4. If samples are selected at random, they are equally likely to be chosen. Thus since there are eight samples having one defective unit, $P(1) = 8/15$. The complete distribution is given in Table 5.5. It is interesting to note that if the company adopts the policy of refusing a shipment if it detects one or more defective units, then about $6/15 = 40\%$ of the time this sampling plan will lead to the acceptance of shipments containing two defective units. ■

TABLE 5.4 THE NUMBER X OF DEFECTIVE UNITS IN SAMPLES OF SIZE 2

Sample	x
G_1, G_2	0
G_1, G_3	0
G_1, G_4	0
G_1, D_1	1
G_1, D_2	1
G_2, G_3	0
G_2, G_4	0
G_2, D_1	1
G_2, D_2	1
G_3, G_4	0
G_3, D_1	1
G_3, D_2	1
G_4, D_1	1
G_4, D_2	1
D_1, D_2	2

TABLE 5.5 THE DISTRIBUTION OF THE NUMBER X OF DEFECTIVE CONTROL UNITS IN SAMPLES OF SIZE 2

x	$P(x)$
0	6/15
1	8/15
2	1/15

5.3   NOTATION

So far, nothing substantially new has been introduced. We are still concerned with the probabilities of outcomes and events. However, we now have some additional notation to aid us in expressing these ideas.

Consider, for example, that we are usually not as interested in the probability of a particular value x as we are in the fact that the value lies in some set. We use the notation $P(X < a)$ to indicate the probability of an observed value less than a. Similarly $P(a < X < b)$ denotes the probability that the value is between a and b. Although it is somewhat redundant, we can write $P(X = a)$ in place of $P(a)$ if we wish to emphasize the role of the variable.

Example 5.4

Twenty people are randomly selected and asked whether they are registered voters. Let X describe the number of positive responses. Then $S_X = \{0, 1, 2, 3, . . ., 20\}$. $P(X < 8)$ denotes the probability that fewer than eight are registered. $P(6 < X < 14)$ denotes the probability that the sample will show between six and 14 registered voters. Similarly, the probability of finding 15 or more registered voters may be indicated by either $P(X \geq 15)$ or $P(X > 14)$. ■

Example 5.5

A researcher reports that when a certain coal tar is applied to a rabbit's skin, the probability of a reaction is $p = .2$. Find the distribution of the number X of rabbits that need to be tested before the reaction is observed. Letting R and N denote a reaction and no reaction, respectively, the first few values of X and the outcomes to which they are assigned are given in Table 5.6. The probabilities are calculated by using the product rule and assuming each test is independent of the others.

TABLE 5.6 POSSIBLE OUTCOMES OF COAL TAR EXPERIMENT

Outcome	Number of rabbits tested until a reaction (X)	Probability
(R)	1	$P(X = 1) = .2$
(N, R)	2	$P(X = 2) = (.8)(.2) = .16$
(N, N, R)	3	$P(X = 3) = (.8)(.8)(.2)$
		$= (.8)^2(.2) = .128$
(N, N, N, R)	4	$P(X = 4) = (.8)(.8)(.8)(.2)$
		$= (.8)^3(.2) = .1024$

This suggests that the probability of needing to test x rabbits is

$$P(X = x) = (.8)^{x-1}(.2)$$

Thus, for $x = 5$,

$$P(X = 5) = (.8)^{5-1}(.2) = (.8)^4(.2) = .08192$$

The probability histogram for this distribution is given in Figure 5.2. Will we need to test many rabbits to obtain a reaction? Do not let the small probabilities associated with large values of X deceive you. The chance of three or fewer rabbits being needed is

$$P(X \le 3) = P(X = 1) + P(X = 2) + P(X = 3) = .2 + .16 + .128 = .488$$

Thus $P(X \ge 4) = 1 - P(X < 4) = 1 - P(X \le 3) = 1 - .488 = .512 = 51.2\%$. About 51% of the time, four or more rabbits will be needed.

FIGURE 5.2 PROBABILITY HISTOGRAM

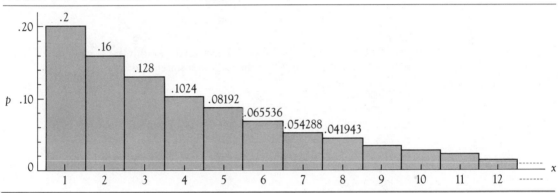

EXERCISE SET 5.1

Part A

1. A coin is tossed three times. Find the distribution (as a table) of the number of heads X.
2. Which of the following variables are discrete and which are continuous?
   a. The number X of incorrect answers on a 10-question true-false test.
   b. The number of sevens when a pair of dice is tossed 15 times.
   c. The mileage X attained by a randomly selected auto of a certain make.
   d. The number of registered voters in a random sample of 60 adults.
   e. The mean weight of a random sample of 25 packages of bacon labeled 1 pound.
3. A wheel of fortune is divided into eight equal sectors, two of which are white, one red, three blue, and two orange. We are interested in the color on which the pointer stops. Define an experiment and an associated random variable X, give its range and distribution, and sketch the histogram.
4. An alternate way to determine the distribution of X in Example 5.3 is to consider two successive draws without replacement from a box containing four units marked G and two marked D. Then

$$P(X = 0) = P(G, G) = \left(\frac{4}{6}\right)\left(\frac{3}{5}\right) = \frac{12}{30} = \frac{6}{15} = .4$$

   In this way find $P(X = 1)$ and $P(X = 2)$.
5. Is the table below a probability distribution?

x	P(x)
0	1/4
1	1/3
2	1/2
3	1/4

6. Does the formula $P(X = x) = x/4$ for $x = 0, 1, 2, 3$, define a probability distribution?

7. Let X be the number of people in a random sample of 40 who favor an increase in the highway speed limit. Describe in words each of the following (as in Example 5.4):
   a. $P(X = 0)$.
   b. $P(X \leq 8)$.
   c. $P(20 \leq X \leq 30)$.
   d. $P(X > 20)$.

8. Let X be the number of kilowatt-hours used monthly by a randomly selected suburban household. Interpret the following statements:
   a. $P(X \leq 1200) = .5$.
   b. $P(800 < X < 1600) = .85$.
   c. $P(X < 400) = .05$.

9. Let X describe the possible sums when a pair of dice are thrown. Find:
   a. $P(X < 4)$.
   b. $P(5 \leq X < 8)$.
   c. $P(X = 10)$.
   d. $P(X > 12)$.

10. A shipment of five calculators contains three defective units. A random sample of size 2 is selected. Give the distribution of the number X of defective units:
   a. By listing the possible samples.
   b. Computing the probabilities directly, as in problem 4.

11. The distribution of family sizes X in a small community is given in the table. Find:
   a. The probability histogram.
   b. $P(X < 5)$.
   c. $P(X \geq 7)$.

x	P(X = x)
2	2/93
3	4/93
4	5/93
5	22/93
6	38/93
7	13/93
8	6/93
9	3/93

12. A proposed probability distribution for the number X of days of sick leave used weekly by each of 100 employees of a large company is given below. Find:
   a. $P(X = 0)$.
   b. $P(X \geq 5)$.
   c. $P(2 < X < 6)$.
   d. c if $P(X \leq c) = .30$.
   e. c if $P(X > c) = .25$.

x	P(x)
0	.01
1	.02
2	.10
3	.17
4	.20
5	.25
6	.17
7	.05
8	.03

Part B

13. A box contains both good and defective bulbs. Three are drawn at random and tested. Let X denote the number of defective bulbs. Find the distribution of X (approximately) if 2% of the bulbs are defective and there are many bulbs in the box, so that we may ignore any change resulting from removal of a bulb. Sketch the histogram.

14. A die is repeatedly thrown. Let X indicate the number of throws until a 4 occurs. Find the distribution of X. (Hint: For, say, $x = 3$, consider non-4 on the first throw followed by non-4 on the second throw and a 4 on the third.

15. A coin is tossed until a head appears. Find the distribution of the number of tosses X that are needed. From this compute the probability that three or more tosses will be needed.

16. If a coin is tossed five times, it can be shown (see the next chapter) that the probability distribution for the number of heads X is given by

$$P(X = x) = \frac{5!}{x!(5 - x)!}\left(\frac{1}{2}\right)^5$$

where $x! = 1 \cdot 2 \cdot 3 \cdot \ldots \cdot x$. For example

$$P(X = 3) = \frac{5!}{3!(5 - 3)!}\left(\frac{1}{2}\right)^5 = \frac{5!}{3!2!}\left(\frac{1}{2}\right)^5 = 10\left(\frac{1}{2}\right)^5 = \frac{10}{32} = \frac{5}{16}$$

   a. Give the distribution of X as a table.
   b. Use this table to compute the probability of three or more heads in five tosses of a coin.

17. A random number between 1 and 10 (inclusively) is selected. Find the probability distribution of the number of divisors X of the number. Note: A number such as 6 has 4 divisors: 1, 2, 3, and 6.

18. A whitewater fishing resort claims that the distribution of the number of fish X in the daily catch of experienced fishermen is given below. What is the average catch if this distribution is followed exactly? (Hint: consider, say, 100 fishermen).

x	P(x)
0	.02
1	.08
2	.10
3	.18
4	.25
5	.20
6	.15
7	.02

5.4   THE MEAN AND STANDARD DEVIATION

In Chapter 3, we saw that the mean and standard deviation play an important role in describing the distribution of a population variable. We should expect there to be similar quantities of equal importance for random variables.

Estimates of a population mean are obtained by sampling the population. If the sample mirrors the population exactly, then its mean is equal to the mean of the population. Similarly, to obtain the mean of a random variable we need a sample of X in which the theoretical frequencies are followed exactly.

Consider the modified wheel of fortune in Figure 5.3. Clearly $P(1) = \frac{1}{6}$, $P(2) = \frac{1}{3}$, and $P(3) = \frac{1}{2}$. If we are interested in an average, we spin the wheel many times and record the 1s, 2s, and 3s as they occur. Suppose there are 600 spins and that the theoretical frequencies are followed exactly. There will be $\frac{1}{6} \times 600 = 100$ ones, $\frac{1}{3} \times 600 = 200$ twos, and 300 threes. Their average is thus:

$$\frac{100 \cdot 1 + 200 \cdot 2 + 300 \cdot 3}{600}$$

or equivalently

$$\frac{100}{600} \cdot 1 + \frac{200}{600} \cdot 2 + \frac{300}{600} \cdot 3 = \left(\frac{1}{6}\right) \cdot 1 + \left(\frac{1}{3}\right) \cdot 2 + \left(\frac{1}{2}\right) \cdot 3 = \frac{7}{3}$$

The last sum we recognize as $1 \cdot P(1) + 2 \cdot P(2) + 3 \cdot P(3)$. The same result is obtained if we try 900 spins, 1500 spins, or whatever. It is natural to refer to this theoretical average, so to speak, as the mean of X and, as with population variables, to denote it by μ or μ_X.

FIGURE 5.3   MODIFIED WHEEL OF FORTUNE

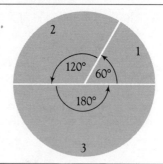

DEFINITION 5.3

Let x_1, x_2, x_3, . . . be the values of a discrete random variable x. The *mean of* X, denoted μ_X, is defined as

$$\mu_X = x_1P(x_1) + x_2P(x_2) + \ldots + = \Sigma xP(x)$$

provided the sum exists.

Example 5.6   Let X indicate the sum when a pair of dice are thrown. If theoretical frequencies are followed, then as many 2s as 12s will occur, as many 3s as 11s, and so forth. The symmetry of the distribution about $x = 7$ suggests that the mean should be 7. By direct computation, we find:

$$\mu_X = 2 \cdot P(2) + 3 \cdot P(3) + 4 \cdot P(4) + 5 \cdot P(5) + 6 \cdot P(6)$$
$$+ 7 \cdot P(7) + 8 \cdot P(8)$$
$$+ 9 \cdot P(9) + 10 \cdot P(10) + 11 \cdot P(11) + 12 \cdot P(12)$$
$$= 2\left(\frac{1}{36}\right) + 3\left(\frac{2}{36}\right) + 4\left(\frac{3}{36}\right) + 5\left(\frac{4}{36}\right) + 6\left(\frac{5}{36}\right) + 7\left(\frac{6}{36}\right) + 8\left(\frac{5}{36}\right)$$
$$+ 9\left(\frac{4}{36}\right) + 10\left(\frac{3}{36}\right) + 11\left(\frac{2}{36}\right) + 12\left(\frac{1}{36}\right) = \frac{252}{36} = 7 \quad \blacksquare$$

Note that the mean of a random variable X is not necessarily the value of X that occurs most often. In fact, as we saw in the wheel of fortune example, it need not even be an outcome. The mean is the value we expect if the outcomes of many runs of the experiment are averaged. It locates the center of the distribution of X in the same manner as the mean of a population variable locates the center of the population.

Example 5.7   The distribution of the number of items purchased by customers in the express line at a supermarket is given in Table 5.7. To find the mean number of items purchased, we have computed the products $xP(x)$ in the third column. This column total is the mean. Thus $\mu_X = 3.55$. See Figure 5.4. $\blacksquare$

TABLE 5.7 COMPUTING THE MEAN NUMBER OF PURCHASES

x	$P(x)$	$xP(x)$
1	.10	.10
2	.20	.40
3	.25	.75
4	.20	.80
5	.10	.50
6	.05	.30
7	.10	.70
	$\Sigma xP(x) =$	3.55

Recall now that the variance of a set of observations is an average of the squares of the deviations of the observations from their mean. To find the average or mean of x_1, x_2, . . . where there is a probability distribution P, we have computed

$$\Sigma xP(x)$$

FIGURE 5.4   PROBABILITY HISTOGRAM FOR NUMBER OF PURCHASES X

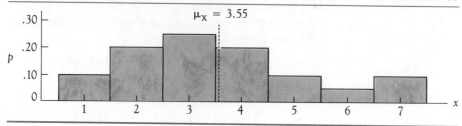

Similarly, for the squares of the deviations $(x_1 - \mu_X)^2$, $(x_2 - \mu_X)^2$, . . ., which also have the respective probabilities $P(x_1)$, $P(x_2)$, . . ., we should compute

$$(x_1 - \mu_X)^2 P(x_1) + (x_2 - \mu_X)^2 P(x_2) + \ldots + = \Sigma(x - \mu_X)^2 P(x)$$

DEFINITION 5.4

Let x_1, x_2, . . . be the values of a discrete random variable with mean μ_X: The *variance* of X, denoted σ_X^2, is defined to be $\sigma_X^2 = \Sigma(x - \mu_X)^2 P(x)$, with the summation extending over all x in S_X.

As would be expected, the square root of the variance is the standard deviation.

DEFINITION 5.5

The standard deviation of a discrete random variable X is defined by

$$\sigma_X = \sqrt{\sigma_X^2} = \sqrt{\Sigma(x - \mu_X)^2 P(x)}$$

Example 5.8

The distribution of the variable X in Table 5.8 is from the wheel of fortune discussion at the beginning of this section. The mean is easily seen to be $\mu_X = 7/3$. The variance is found by:

$$\sigma_X^2 = \left(1 - \frac{7}{3}\right)^2 \frac{1}{6} + \left(2 - \frac{7}{3}\right)^2 \frac{1}{3} + \left(3 - \frac{7}{3}\right)^2 \frac{1}{2}$$

$$= \frac{16}{9} \cdot \frac{1}{6} + \frac{1}{9} \cdot \frac{1}{3} + \frac{4}{9} \cdot \frac{1}{2}$$

$$= \frac{30}{54} = \frac{5}{9} = \cdot 5556 \quad \text{(rounded)}$$

The standard deviation is $\sigma_X = \sqrt{.5556} = .75$ (rounded). ■

TABLE 5.8 COMPUTING THE VARIANCE

x	$P(x)$	$(x - \mu)^2 P(x)$
1	1/6	8/27
2	1/3	1/27
3	1/2	2/9
	$\Sigma(x - \mu)^2 P(x) =$	5/9

In Chapter 3, we gave an alternate formula for computing the variance of a statistical distribution. By rearranging the above formula, we obtain an analogous result.

THEOREM 5.1

Let $x_1, x_2, \ldots x_n$ be the values of a discrete random variable X with mean μ_X and probability distribution P. Then the variance of X is given by:

$$\sigma_X^2 = \sum_{i=1}^{n} x_i^2 P(x_i) - \left[\sum_{i=1}^{n} x_i P(x_i) \right]^2$$

Applying this formula to the distribution of the last example, we obtain

$$\sigma_X^2 = 1^2 \cdot \frac{1}{6} + 2^2 \cdot \frac{1}{3} + 3^2 \cdot \frac{1}{2} - \left[1 \cdot \frac{1}{6} + 2 \cdot \frac{1}{3} + 3 \cdot \frac{1}{2} \right]^2$$

$$= \left(\frac{1}{6} + \frac{4}{3} + \frac{9}{2} \right) - \left(\frac{1}{6} + \frac{2}{3} + \frac{3}{2} \right)^2$$

$$= \frac{5}{9}$$

This is exactly the result previously computed.

Often it is useful to arrange all the computations for the mean and the variance in a single table.

Example 5.9

The distribution of diameters of doll heads (to the nearest inch) produced by a 19th century doll maker is given in Table 5.9. Find the mean and variance. The arrangement of the computations should be self-explanatory. Then $\mu_X = 4.7$ and $\sigma_X^2 = 22.7 - (4.7)^2 = 0.61$. ∎

TABLE 5.9 COMPUTING THE MEAN AND THE VARIANCE OF X

x	$P(x)$	$xP(x)$	$x^2 P(x)$
4	.5	2	8
5	.3	1.5	7.5
6	.2	1.2	7.2
		$\Sigma xP(x) = 4.7$	$\Sigma x^2 P(x) = 22.7$

In Chapter 7, we will investigate how well the mean of a sample can be expected to approximate that of the population from which it was taken. Important to this is the *distribution of the means of samples* of the size being used. The following example anticipates this development.

Example 5.10

A street has five houses H_1, H_2, H_3, H_4, and H_5, with 0, 1, 1, 3, and 2 children respectively. Let X denote the number of children in a randomly selected house and let $\overline{X}$ denote the mean number of children in a random sample of three of these houses. Find the distribution of both X and $\overline{X}$ and compare their means. Since X has the values 0, 1, 2, and 3 with the distribution of Table 5.10, the mean is

$$\mu_X = 0\left(\frac{1}{5}\right) + 1\left(\frac{2}{5}\right) + 2\left(\frac{1}{5}\right) + 3\left(\frac{1}{5}\right) = \left(\frac{7}{5}\right)$$

The 10 possible samples of size 3 are given in Table 5.11.

TABLE 5.10   DISTRIBUTION OF THE NUMBER X OF CHILDREN

x	$P(x)$
0	1/5
1	2/5
2	1/5
3	1/5

TABLE 5.11   SAMPLES OF SIZE $n = 3$ AND THEIR MEANS

Houses	Sample (number of children)	$\overline{x}$
H_1, H_2, H_3	0, 1, 1	2/3
H_1, H_2, H_4	0, 1, 3	4/3
H_1, H_2, H_5	0, 1, 2	3/3 = 1
H_1, H_3, H_4	0, 1, 3	4/3
H_1, H_3, H_5	0, 1, 2	3/3 = 1
H_1, H_4, H_5	0, 3, 2	5/3
H_2, H_3, H_4	1, 1, 3	5/3
H_2, H_3, H_5	1, 1, 2	4/3
H_2, H_4, H_5	1, 3, 2	6/3 = 2
H_3, H_4, H_5	1, 3, 2	6/3 = 2

If the sample is selected randomly, then each of the above samples has probability 1/10 of being selected. Since $\overline{x} = 2$ is produced by two of these samples, it has probability 2/10 = 1/5 of occurring. The distribution of $\overline{X}$ is summarized in Table 5.12. Thus

$$\mu_{\overline{X}} = \left(\frac{2}{3}\right)\left(\frac{1}{10}\right) + 1\left(\frac{1}{5}\right) + \left(\frac{4}{3}\right)\left(\frac{3}{10}\right) + \left(\frac{5}{3}\right)\left(\frac{1}{5}\right) + 2\left(\frac{1}{5}\right)$$

$$= \frac{7}{5} = \mu_X \quad \blacksquare$$

TABLE 5.12 DISTRIBUTION OF SAMPLE MEANS

$\overline{x}$	$P(\overline{x})$
2/3	1/10
1	1/5
4/3	3/10
5/3	1/5
2	1/5

The fact that the mean of the sample variable $\overline{X}$ of the last example turned out to be equal to the mean of the variable X is no accident. In Chapter 7 we shall see that this is always the case. That is, *the mean of the sample means is the population mean*: $\mu_{\overline{X}} = \mu_X$.

5.5 THE EXPECTED VALUE OF FUNCTIONS OF RANDOM VARIABLES

The mean and variance of the random variable X have been computed by formulas that are special cases of the more general expression:

$$\Sigma g(x_i)P(x_i)$$

If $g(x) = x$, then $g(x_1) = x_1$, $g(x_2) = x_2$, and so on, and $\Sigma g(x_i)P(x_i) = \Sigma x_i P(x_i)$, which is the mean of X. If we let $g(x) = (x - \mu)^2$, then $g(x_1) = (x_1 - \mu)^2$, $g(x_2) = (x_2 - \mu)^2$, and so on, and $\Sigma g(x_i)P(x_i) = \Sigma(x_i - \mu)^2 P(x_i)$, which is the variance. In each instance we can regard $g(x_1)$, $g(x_2)$, . . . as values of a *new random variable* $g(X)$. By varying g many other useful quantities can be derived.

DEFINITION 5.

Let X be a discrete random variable with probability distribution P and let g be a function defined on S_X. Then the *expected value* of the random variable $g(X)$ is denoted by $E[g(X)]$ and is defined to be

$$E[g(X)] = \Sigma g(x_i)P(x_i)$$

Often we simply write E rather than $E[g(X)]$.

A useful and interesting application occurs when the experiment is a game of chance with a payoff $g(x)$ for each outcome x. Positive values of $g(x)$ are associated with payoffs by the "house" to the player, negative values with those by the player to the house. Then $E[g(X)]$ is the mean value of the variable $g(X)$ and is the average payoff that will result if the theoretical distribution of X is followed exactly. From a practical point of view, $E[g(X)]$ is what the player expects to win or lose per game if the game is played many, many times.

Example 5.11

Ten thousand raffle tickets at $3.00 each are sold on a $20,000 automobile. An individual purchases five tickets. If he wins, his net gain is $19,985. A loss costs him $15.00. Let X have the values 0 and 1, representing a loss and a win respectively. Then $g(0) = -15$, $g(1) = 19,985$, $P(0) = 9995/10,000$, and $P(1) = 5/10,000$. These values are summarized in Table 5.13. Then

$$\begin{aligned}
E = E[g(X)] &= g(0)P(0) + g(1)P(1) \\
&= -15\left(\frac{9995}{10,000}\right) + 19,985\left(\frac{5}{10,000}\right) \\
&= -5.00
\end{aligned}$$

This expected loss of $5.00 agrees with the observation that if the $10,000 profit is spread over 10,000 tickets, each ticket contributes $1.00 to it. Thus five tickets contribute $5.00. ■

TABLE 5.13   THE DISTRIBUTIONS OF X AND $g(X)$

x	$g(x)$	$P(x)$
0	-15	9995/10,000
1	19,985	5/10,000

There are many applications of the expected value to problems in economics and decision theory. One must simply use some imagination in describing different phenomena as games of chance.

Example 5.12

Each good bolt produced by a machine shop yields a profit of 5¢, each defective one a loss of 10¢. Approximately 2% of the production is defective. What is the average profit per bolt? We can interpret this as a game the machine shop plays with the consumer: The company wins 5¢ each time it produces a good bolt and loses 10¢ for each defective one. Letting $x = 0$ and 1 denote "good" and "defective" respectively, we summarize this information in Table 5.14. Then

$$\begin{aligned}
E &= g(0)P(0) + g(1)P(1) \\
&= 5(.98) + (-10)(.02) \\
&= 4.70 \text{ (cents)}
\end{aligned}$$

Thus the average profit per bolt produced is 4.7¢. ■

TABLE 5.14   THE DISTRIBUTIONS OF X AND $g(X)$

x	$g(x)$	$P(x)$
0	5 (cents)	.98
1	-10 (cents)	.02

Example 5.13

Suppose the IRS has discovered that 67.5% of all income tax returns have no errors, that 24% have a minor error averaging $28, that 8% have an error or fraud averaging $250, and that the remaining 0.5% have an error or gross fraud averaging $6000. If the amount to be spent on processing returns is not to exceed the error or fraud uncovered, what should be the budget for the agency if it expects to process 20 million returns?

Let c be the cost of processing a return and let X have the values 0, 1, 2, and 3, identifying the four categories of returns described above. We can think of the agency playing a game where it loses c dollars for each return it processes with no error, $(28 - c)$ dollars for each with a minor error, and so forth. The necessary facts are presented in Table 5.15. We would like to find the value of c for which the expected value is 0. This will tell us the amount to spend for processing a return.

$$E = -c(.675) + (28 - c)(.240) + (250 - c)(.080) + (6000 - c)(.005)$$
$$= -c + 56.72$$

Thus $E = 0$ if $c = 56.72$. Now, if 20 million returns are to be processed, the budget should not exceed

$$(2 \times 10^7)(56.72) = \$1,134,400,000$$

This should be regarded as a ballpark figure since many simplifying assumptions have obviously been made. ■

TABLE 5.15   THE DISTRIBUTIONS OF X AND $g(X)$

x	$g(x)$	$P(x)$
0	$-c$	.675
1	$28 - c$	.240
2	$250 - c$	.080
3	$6000 - c$	.005

5.6   PROPERTIES OF THE EXPECTED VALUE FUNCTION

When working with expected values, there are several properties that prove useful.

THEOREM 5.2.

1. If C is a constant, then $E[C] = C$.
2. If C is a constant, then $E[Cg(X)] = C \cdot E[g(X)]$.
3. $E[g(X) + h(X)] = E[g(X)] + E[h(X)]$.

In terms of payoffs in a game of chance, these can be interpreted as follows:

1'. If each payoff is the same value C, then C is the average payoff.
2'. If each payoff is multiplied by the same constant C, then the mean payoff is multiplied by C.
3'. If there are two payoff functions operating independently, then the mean payoff of their sum is the sum of the respective payoffs.

To illustrate the usefulness of this theorem, suppose we produce a new random variable Y by subtracting μ_X from each value of X. Thus $Y = X - \mu_X$. Now, since μ_X is a constant, $E(\mu_X) = \mu_X$. Also, by definition, $\mu_X = E(X)$. The mean of Y is then found to be

$$
\begin{aligned}
\mu_Y &= E(X - \mu_X) \\
&= E(X) - E(\mu_X) \qquad \text{by property (3)} \\
&= \mu_X - \mu_X = 0.
\end{aligned}
$$

Thus the mean of Y is 0.

In the next chapter, we shall have need to introduce a new variable Z by setting

$$
Z = \frac{X - \mu_X}{\sigma_X}
$$

Using Theorem 5.2 and proceeding as above, we can show that this new variable has a mean of 0 and a standard deviation of 1.

THEOREM 5.3

Let X have a distribution with mean μ_X and standard deviation σ_X. Then the variable

$$
Z = \frac{X - \mu_X}{\sigma_X}
$$

has mean $\mu_Z = 0$ and standard deviation $\sigma_Z = 1$.

Example 5.14

A Research Study

Does Jet Noise Kill?

Studies that have been widely publicized by the media have linked high noise levels around airports to increased hypertension, increased admissions to mental hospitals, hearing impairment in children, and increased mortality rates. Rarely, however, do the follow-up studies receive the same attention. This is unfortunate, since in many instances the original analysis has been challenged as being faulty. We discuss here a report on one such follow-up: "Los Angeles Airport Noise and Mortality-Faulty Analysis and Public Policy," by Ralph R. Frerichs, DVM, et al. in the April 1980 issue of the *American Journal of Public Health*.

The original report by W. C. Meecham and N. Shaw (*British Journal of Audiology*, vol. 13, 1979, pp. 77–80) compared the mortality rates of an area adjoining the Los Angeles airport with a background noise of 90 adjusted decibels (dba) or higher to a control area three miles south of the airport having a level of some 45 to 50 dba. During 1970, 887 people were reported to have died in the test area, as compared with 672 in the control area. Of course, the populations in the two areas were not the same size, but after adjusting for this and some other differences, Meecham and Shaw reported the death rate to be 19% higher in the airport test area than in the control area.

Frerichs and his colleagues became suspicious when they compared the two mortality rates in the two areas with that of the county as a whole. At that time, the rate for the county was $p = .0088$ or 8.8 per one thousand. Yet the data by Shaw and Meecham showed their test and control areas to have rates of $p = .005$ (five per thousand) and $p = .0043$ (4.3 per thousand). In short, these two areas appeared to have much lower mortality rates than the county as a whole.

Looking further, they also found that the racial compositions of the test and control areas were quite different. Proportionately more black people lived in the test areas and more Asians in the control areas. The significance of this is readily apparent, since the death rate among black people is higher than among Asians. Thus in comparing the two areas, a higher rate in the airport area can be explained by other factors.

But the matter does not end here. Based on these considerations, Frerichs and his colleagues undertook a complete reexamination of the data as outlined by Meechan and Shaw. This required reconstruction of the data using census information. Fortunately, deaths are coded by the census tract of the residence of the deceased and are recorded with information on age, race, and sex and a disease code. With this information, the death rates could be adjusted for these confounding variables. Standard mortality ratios (SMR) were used to derive these rates. An SMR is simply the expected number of deaths for the area divided by the number actually observed and is often used to avoid the insensitivity and instability of mortality rates in small populations.

Table 5.16 summarizes the conclusions of Frerichs et al. Note that when

---

TABLE 5.16 TWO-YEAR MORTALITY, ADJUSTED FOR DIFFERENCES IN AGE, RACE, AND SEX, IN LOS ANGELES AIRPORT AND CONTROL AREAS, 1970–1971

Area	Deaths All causes, 1970–1971	Population, 1970	Two-year mortality rate per 1000 population Crude	Age-, race- sex-adjusted[a]
Airport core	893	64,801	13.78	18.20
Control core	732	58,240	12.57	18.15
Percent difference[b]	22.0%	11.3%	9.6%	0.3%
Airport extended	2,005	124,820	16.06	17.06
Control extended	1,178	92,101	12.79	18.27
Percent difference[b]	70.2%	35.5%	25.6%	−6.8%

[a]Age-, race-, sex-adjusted by direct method to the 1970 Los Angeles County population.
[b][(Airport minus Control)/Control] × 100.

adjusted for age, race, and sex, the two-year mortality rates are practically identical. Moreover, when the test and control areas are extended slightly to compensate for changes in census recording techniques, one finds the test area actually has a lower rate than that of the control.

The authors of this rebuttal make a strong point in their summary: "the findings from population-based observational studies of this nature should be made with great caution. The public believes (in such reports) that an objective experiment was performed in which death rates in noise exposed versus unexposed populations were compared. . . . It is evident that this is not the case. . . . In fact, people who live near an airport may differ from people who live elsewhere. These differences may account for the observed differences in the morbidity and mortality rates."

As part of their report, the authors also summarize other studies of this nature and cite flaws in the design of the investigations. Newspaper reports and public opinion (which is shaped by these reports) notwithstanding, they conclude that "any link between airport noise and mortality must be based on sounder evidence than has been presented to date if causality is to be inferred." ■

EXERCISE SET 5.2

Part A

1. Let X be a random variable with the probability distribution given in the table below.

x	$P(x)$
0	.2
1	.2
2	.4
3	.2

a. Find the mean, variance, and standard deviation of X.
b. Find the probability of X being within one standard deviation of the mean.
c. Find $P(\mu - 2\sigma < X < \mu + 2\sigma)$.

2. A random variable X has the distribution given in the table below.

x	$P(x)$
10	1/12
11	1/6
12	1/2
13	1/6
14	1/12

a. What interval contains all outcomes that are at or within one standard deviation of the mean? At or within two standard deviations of the mean?
b. Find the probability of X being on each of the intervals of part a.

3. A box contains four slips of paper numbered 0 through 3. A slip is drawn and the number x is recorded.
a. Find the distribution of X.
b. Find the mean, variance, and standard deviation of X.

4. A box contains eight balls, numbered 1, 2, 2, 2, 3, 3, 4, and 5. A ball is drawn at random and the number observed, X, is recorded. Suppose the ball is replaced, the balls thoroughly mixed, and the drawing repeated. If this is done many, many times, what is a good prediction of the average of the results recorded?

5. A store has determined that, rounded to the nearest minute, 15% of its customers take 1 minute to check out, 20% take 2 minutes, 30% take 3, 25% take 4, and 10% take 5 minutes.
   a. What is the average time a customer spends at the check-out counter?
   b. If the store has 900 customers per 8-hour day, about how many clerks will be needed if the customers' arrivals at the store are spread out uniformly over the 8-hour period?

6. A die is thrown. Suppose you receive in dollars the number showing on the die. How much would you expect to receive on the average in many plays of this game? What is a fair price to pay for playing the game?

7. One hundred tickets for a lottery are sold by a sorority. The stubs are marked with an identifying number and the amount the holder of the corresponding ticket will win if that stub is drawn. Suppose 50 tickets are marked 0¢, 20 are marked 10¢, 15 are marked 25¢, 10 are marked 50¢, two are marked $1.00, two are marked $5.00, and one is marked $25.00. What is the expected value of a single-ticket purchase?

8. A fire casualty insurance company has determined the distribution below for the loss X on each $1000 of insurance it issues on structures of a certain class.

x	$P(x)$
$1000	.005
600	.008
300	.012
100	.050
0	.925

   a. What is the expected loss for each $1000 of insurance issued on this type of structure?
   b. If the carrier must also assess an additional $3.00 per thousand loading fee to cover expenses, profits, and the like, what should the premium be for a $300,000 policy?

9. In a learning experiment, a monkey is to reach through an opening in a cage, select from one of four objects, and try to bring it into the cage. Suppose that only one of the objects will pass through the opening and that the monkey does not try any object twice. Let X denote the number of objects it tries before succeeding.
   a. Assuming the monkey selects at random, find the probability distribution of X.
   b. Find the expected value of X using the distribution of part a.
   c. What does the expected value of X represent in terms of this experiment?

10. Two investment plans, A and B, each require a funding of $10,000. For plan A the possible returns are $12,000, $14,000, or $18,000 with respective probabilities of .40, .35, and .25. (Note: As used here, a return of $14,000 is a profit of $4000.) For plan B there is a 50-50 chance to break even or double your money. Which is the better plan?

11. A friend offers the following game. A coin is flipped three times. If zero or three heads occur, he pays you $5.00. If one or two heads appear, you pay him $4.00. What is the expected value of the game? What would be your financial position after 100 games if the probability distribution of heads is followed exactly?

12. Let X be a random variable with $\mu_X = 150$ and $\sigma_X = 10$. Find μ_Y for each of the following:
   a. $Y = 3X$.
   b. $Y = \frac{1}{2}X + 20$.
   c. $Y = X - 150$.
   d. $Y = (X - 150)/10$. Also find σ_Y.
13. Let X have the probability distribution below.

x	$P(x)$
0	1/6
2	1/2
3	1/3

   a. Give the probability distribution (as a table) of the variable $Y = 3X$.
   b. Compute $E[3X]$ directly from the table of part a.
   c. Compute $E[X]$ and use this with Theorem 5.2 to find $E[3X]$.
14. In a Bernoulli experiment, there are only two outcomes, failure or success. Let X assign the numbers zero and one respectively to these outcomes. Thus there is either zero or one success. Suppose the probability of success is $\frac{1}{4}$ and that of failure is $\frac{3}{4}$. What is the expected value of X? What is the variance of X?

Part B

15. In the game of roulette, a $1.00 bet on the color black will return $2.00 (the dollar bet and one more) if a black number occurs; otherwise the dollar is lost. Assuming there are 18 red and 18 black numbers and one green one on the board, what is the expected value of a bet on a black number?
16. In the game of roulette, a $1.00 bet on some particular number returns $36.00 (the dollar bet and 35 more) if that number occurs. Assuming the board has 37 numbers, what is the expected value of a bet on a single number?
17. A woman plans to invest in one of two equally priced stocks, A and B, which have respective probabilities of .5 and .2 of a price increase. The increase of A is estimated at $20.00, that of B at $60.00. Both will decline $5.00 if there is a price decrease. Which is the best buy?
18. The probability distribution for X is given by $P(X = x) = x/10$ for $x = 1, 2, 3, 4$.
   a. Construct the probability histogram and estimate the mean.
   b. Compute μ_X and compare it with your estimate in part a.
19. The grades of five students on an examination were A, B, C, B, and A. Let X assign 4 to an A, 3 to a B, and 2 to a C.
   a. Give the probability distribution of X and find its mean.
   b. Determine the distribution of the variable $\overline{X}$ whose values are the means of samples of size 2 selected from these five students.
   c. Compute the mean of $\overline{X}$ and show that it is equal to μ_X.
   d. Give the probability histograms of both distributions.
20. Suppose five individuals, I_1, I_2, I_3, I_4, and I_5, constitute the entire population and that they have the respective positions of for, for, against, for, and against on the issue of lowering the drinking age. Let X have the values 1 (for) and 0 (against).
   a. What proportion of the population favors the issue?

    b. Consider samples of size 4 and let $\hat{P}$ be the proportion of the sample that favors the issue. Find the distribution of $\hat{P}$.

    c. Find $\mu_{\hat{p}}$ and compare with the result of part *a*.

21. A promoter is planning a sports show for a weekend in February. If it does not rain, he figures he will net about $40,000; he will lose about $12,000 if it does. Prior records indicate that the probability of rain on a weekend in February is approximately .3. An insurance company will sell him rain insurance at $400 per thousand.

    a. What is his expected value if he has no insurance?

    b. What is his expected value if he takes out $12,000 worth of rain insurance?

    c. Show that if the hapless promoter purchases *c* thousands of dollars worth of rain insurance that $E \leq 24,400$.

    d. What factors other than the expected value might play a factor in his decision on whether to buy insurance?

22. A neighborhood rent-all store is considering the addition of video recorders to its list of available items. A survey of similar stores has produced the probability distribution shown below for the number of requests on any given day. What is the expected number of requests for video recorder rentals on any given day?

x	P(x)
0	.05
1	.15
2	.30
3	.20
4	.20
5	.10

23. The bakery section of a supermarket always carries large sheet cakes for its customers. The cakes are restocked every two days and any old unsold cakes are thrown away. Each cake costs $4.00 and sells for $9.00. The demand for these cakes for a two-day period has the distribution shown below.

x	P(x)
0	0
1	.3
2	.4
3	.3

    a. What is the expected demand?

    b. What is the expected profit if the store stocks only one cake?

    c. What is the expected profit if the store stocks only two cakes?

    d. What is the expected profit if the store stocks exactly three cakes?

5.7   CONTINUOUS PROBABILITY DISTRIBUTIONS

We have on several occasions considered continuous variables such as weights of males, diameters of bolts, and arrival times of airline flights. It will be recalled that these are characterized by having values that fill out continuous intervals on the number line. We now wish to discuss probability distributions for such random variables.

What are the properties of a continuous probability distribution? To find out, let us begin by considering a particular example, the increase X in the weight of a bacterial culture maintained under controlled conditions for 72 hours. First, there is the matter of the probability of some specific increase, say $x = 23 = 23.0000 . . .$ milligrams exactly. Obviously we cannot weigh to see if this occurs, but we can tell if the increase is somewhere around 23. When one considers the infinitely many other possible increases that are near $x = 23$, we see that the change of an exact increase of 23 milligrams must be quite small, in fact 0.

The probability of any particular value x *of a continuous random variable X is 0.*

It is clear, however, that an increase x must occur and that there are some intervals on which this is more likely to occur than others. Thus:

A probability distribution for a continuous variable X assigns probabilities to intervals of values of X.

The actual assignment of probabilities depends on the concept of a *probability density curve*, which in turn is simply an abstraction of a density histogram. To illustrate, suppose we sample the distribution of the weight increase X for the cultures described above by repeating the growth experiment 40 times under the same conditions, obtaining the density histogram in Figure 5.5.

FIGURE 5.5
HISTOGRAM OF THE DISTRIBUTION OF 40 WEIGHT INCREASES

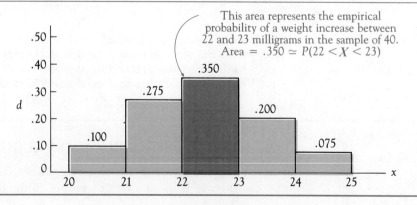

This area represents the empirical probability of a weight increase between 22 and 23 milligrams in the sample of 40.
Area $= .350 \simeq P(22 < X < 23)$

As a brief review, it will be recalled that the density of an interval is found by:

$$d = \frac{\text{relative frequency of the interval}}{\text{length of the interval}}$$

This density is then the height of the rectangle to be erected over the interval. When constructed in this way, the total area of the histogram is 1 and the rectangle over any interval has an area equal to the relative frequency and thus the probability that is to be assigned to this interval. Thus for the interval $22 < X < 23$

$$\text{area} = (\text{base}) \cdot (\text{height}) = (\text{base}) \cdot (\text{density}) = 1 \cdot (.350) = .350$$
$$= P(22 < X < 23).$$

Of course this probability is only an estimate of the actual probability, since it is obtained from a sample.

In an effort to obtain better estimates of the probabilities, suppose we next consider a sample of 100 values of X. Having more observations, we group the data by using smaller intervals and obtain the distribution whose probability histogram is given in Figure 5.6. The probability of X being between 22 and 23 is the area under this histogram and above this interval, and it is found by adding the areas of the four rectangles. Thus

$$P(22 < X < 23) = (.25)(.32) + (.25)(.36) + (.25)(.44) + (.25)(.44)$$
$$= .08 + .09 + .11 + .11 = .390.$$

This should be a better approximation than that obtained with the smaller sample.

FIGURE 5.6   HISTOGRAM FOR 100 WEIGHT INCREASES

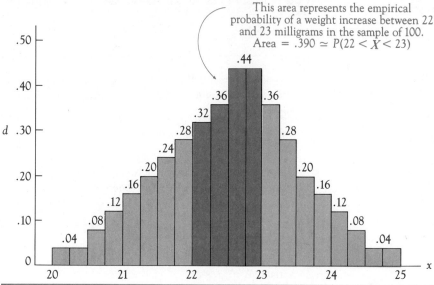

Suppose we now pass a smooth curve through the midpoints of the tops of the rectangles of the last histogram. The areas under this curve are very close to those of the histogram. As we repeat this whole procedure for larger and larger samples of X with smaller and smaller intervals, we begin to sense that these curves might become more and more alike. At the limit we would obtain one smooth curve, which would have the property that the areas beneath it are the basis for assigning theoretical probabilities to intervals. In Figure 5.7, we have indicated $P(22 < X < 23)$.

FIGURE 5.7
DENSITY CURVE FOR THE DISTRIBUTION OF WEIGHT INCREASES

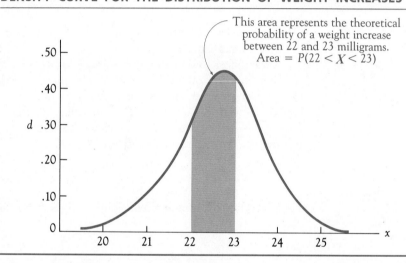

This limiting process, which we have just described, does happen in those cases of most interest to the statistician. The resulting curve is called a probability density curve.

> *A continuous probability distribution is determined by a density curve. The probability assigned to an interval is the area under this curve lying above this interval.*

Any function f whose graph is a density curve is called a probability density function. For the sake of completeness:

DEFINITION 5.7

A nonnegative function f is called a *probability density function* for the random variable X if:

1. The total area bounded by the graph of the function and the *x*-axis is 1.
2. The area under the graph of *f* and above the interval $a < X < b$ is the probability $P(a < X < b)$.

Many important continuous distributions are presented throughout this text as different applications are discussed. Our approach in each case is to give the shape of the density curve and, where needed, the mean and standard deviation of the resulting distribution. The distribution will be given as a table so that probabilities may be easily read. Thus we shall not actually have to find the areas that produce the probabilities. Still, we shall find that it is very useful to think of probabilities as areas under a density curve.

Before leaving the subject, let us consider three simple examples.

Example 5.15

The density (curve) for the distribution of X is given in Figure 5.8. Shade the area representing:
1. $P(2 < X < 3)$
2. $P(X > 5)$.

First, locate the intervals on the number line. We have shaded these boldly. Draw two vertical lines through the endpoints of each interval. The enclosed areas in Figure 5.9 are the desired probabilities. ■

FIGURE 5.8   A DENSITY CURVE

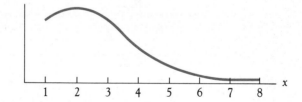

FIGURE 5.9

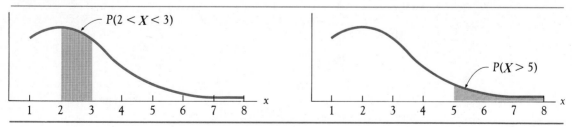

Example 5.16   The distribution of X is given by the density curve of Figure 5.10. Which is most likely to happen, a value of X between 25 and 35 or one between 40 and 45? Comparing the two areas, we see $P(25 < X < 35) < P(40 < X < 45)$, and thus a value of X between 40 and 45 is more likely. ■

FIGURE 5.10 $P(25 < X < 35) < P(40 < X < 45)$

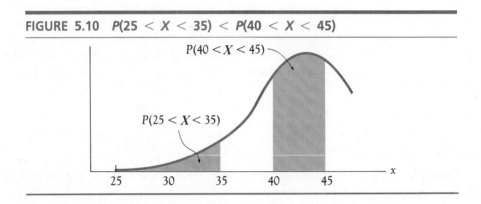

Example 5.17   The probability density curve for the variable X is given in Figure 5.11. All intervals of any particular size have the same probability. A probability distribution of this form is called a *uniform distribution*. Note that the total area under this curve and above the x-axis is 1.

FIGURE 5.11   A UNIFORM DISTRIBUTION

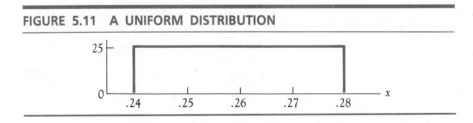

1. Find $P(.25 < X < .27)$. The area above the interval from .25 to .27 is $.02 \times 25 = .50$, which is the desired probability. Thus $P(.25 < X < .27) = .50$. See Figure 5.12.
2. Find $P(X > .265)$. Here the area $= .015 \times 25 = .375$. Thus $P(X > .265) = .375$. See Figure 5.13.
3. Find $P(X \geq .265)$. This is the same as $P(X > .265) = .375$, since $P(X = .265) = 0$; that is, the probability of any particular value of X is 0 for a continuous distribution. See Figure 5.14. ■

FIGURE 5.12   THE SHADED AREA REPRESENTS P (.25 < X < .27)

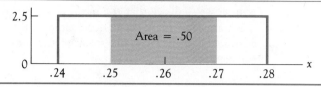

FIGURE 5.13   THE SHADED AREA REPRESENTS P (X > .265)

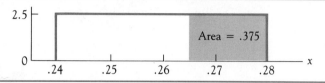

FIGURE 5.14   P (X ≥ .265) = P (X > .265) = .375

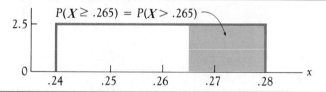

5.8   PARAMETERS OF DISTRIBUTIONS

Many distributions are members of a family of similar distributions which differ only with respect to certain key quantities, called *parameters*. For example, all uniform distributions are of the form shown in Figure 5.15. The interval endpoints a and b are parameters of this distribution, since once they are specified, the distribution is determined completely.

DEFINITION 5.8

A number is called a *parameter* of a distribution if it is one of (possibly) several numbers that determine the distribution.

FIGURE 5.15   THE UNIFORM DISTRIBUTION

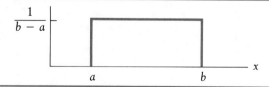

When a distribution is used with a population, we freely interchange the terms *distribution parameter* and *population parameter*. For example, in the next chapter we shall see that the mean μ and standard deviation σ are parameters of the normal distribution. If, for example, IQ scores are approximately normally distributed, then we would speak of the mean and standard deviation as being parameters of the IQ scores.

EXERCISE SET 5.3

Part A

1. The density of X is given.

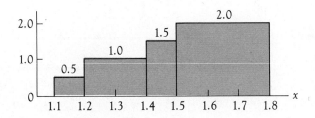

   a. Find $P(1.5 \le X \le 1.8)$.
   b. Find $P(X < 1.4)$.
   c. Find $P(X \le 1.8)$.

2. The distribution for a sample of X is given.

Interval	P
0.0–0.5	.1
0.5–1.0	.1
1.0–1.5	.2
1.5–2.0	.3
2.0–2.5	.2
2.5–3.0	.1

   a. Construct the histogram using areas.
   b. Estimate $P(.5 < X < 2.0)$.
   c. Estimate $P(X \le 1.5)$.

3. The random variable X has values between 2 and 6 and has the probability density given below.

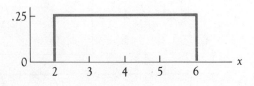

     a. Find $P(X = 3)$.
     b. Find $P(3 < X < 5)$.
     c. Find $P(3 \leq X \leq 5)$.
     d. Find $P(X \leq 6)$.
     e. Find $P(X > 5.3)$.

4. Let X be the weight of a randomly selected person.
     a. Give a plausible explanation of why $P(X = 150) = 0$.
     b. If $P(X < 150) = .90$, find $P(X > 150)$.

5. Let X be a continuous random variable with a probability distribution P. Suppose $P(X \geq 0) = .5$, $P(0 \leq X \leq 3) = .4$, and $P(3 \leq X \leq 5) = .085$.
     a. Find $P(X > 3)$.
     b. Find $P(X = 3)$.
     c. Find $P(0 \leq X \leq 5)$.
     d. Find $P(X > 5)$.
     e. Find c if $P(X < c) = .9$.

6. For each of the following continuous variables, sketch what you believe should be the shape of the density curve.
     a. The distance X from an individual's home to place of employment.
     b. The time lapse between the purchase of a new auto and the discovery of the first flaw or the first failure.
     c. The intensity of the light emitted by a 60-watt bulb.
     d. The volume of wine X in a randomly selected 1-liter bottle from a certain vintner.

Part B

7. The continuous variable X has the distribution below. Uisng the given areas find:
     a. $P(2 \leq X \leq 3)$.
     b. $P(X \leq 2)$.
     c. $P(X \geq 1)$.

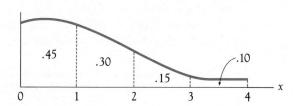

8. The continuous random variable X has values between 0 and 3 and its probability density function is given by $f(x) = (\tfrac{2}{9})x$.
     a. Find $P(0 \leq X \leq 3)$.
     b. Find $P(1 < X < 2)$.
     c. Find $P(X \geq 1.5)$.
     Hint: Sketch the graph of f. Use the formulas for the area of a triangle and the area of a trapezoid to find the needed areas. Note the area of a trapezoid is $A = \tfrac{1}{2}b(h_1 + h_2)$, where b is the length and h_1 and h_2 the heights.

9. Estimate the mean and standard deviation of a probability distribution if the density is as shown.

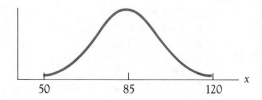

10. Estimate the mean and standard deviation of a probability distribution if the density is as shown.

11. In Example 5.5 $P(X = x) = (.8)^{x-1}(.2)$ was shown to be the distribution of the number X of tests needed for a first reaction by rabbits to a coal tar. This distribution is a member of the family of distributions which are of the form $P(X = x) = p^{x-1}(1 - p)$. What is (are) the parameters of this distribution?

5.9   MULTIVARIATE PROBABILITY DISTRIBUTIONS

It may not always be possible or practical to describe the outcomes of an experiment by a single variable. When several variables are used, a *joint probability distribution* is required.

As an example of a joint probability distribution for two variables, we consider the toss of a coin and a die. Let X be the number showing on the die and let Y assign a value of 0 to a head and 1 to a tail on the coin. The joint probability of (simultaneously) obtaining, say, a 4 on the die and a head on the coin is indicated by $P(X = 4, Y = 0)$ or more simply $P(4, 0)$. The joint distribution of X and Y is found by enumerating the sample space (Table 5.17).

TABLE 5.17   ASSIGNMENT OF VALUES BY TWO RANDOM VARIABLES

Outcome	x	y	P(x, y)
(1, H)	1	0	1/12
(2, H)	2	0	1/12
(3, H)	3	0	1/12
(4, H)	4	0	1/12
(5, H)	5	0	1/12
(6, H)	6	0	1/12
(1, T)	1	1	1/12
(2, T)	2	1	1/12
(3, T)	3	1	1/12
(4, T)	4	1	1/12
(5, T)	5	1	1/12
(6, T)	6	1	1/12

Joint probability distributions of two variables are generally given as a two-way table where the column and row headings are the values of the respective variables. Any table entry shows the joint probability of the pair that head up that column and row which contain the entry. For the die and the coin experiment discussed above, the joint distribution is given in Table 5.18.

TABLE 5.18 A JOINT PROBABILITY DISTRIBUTION

			x				
y	1	2	3	4	5	6	$P(y)$
0	1/12	1/12	1/12	1/12	1/12	1/12	6/12 = 1/2
1	1/12	1/12	1/12	1/12	1/12	1/12	6/12 = 1/2
$P(x)$	$^2\!/_{12}$ = $^1\!/_6$	$^2\!/_{12}$ = $^1\!/_6$	$^2\!/_{12}$ = $^1\!/_6$	$^2\!/_{12}$ = $^1\!/_6$	$^2\!/_{12}$ = $^1\!/_6$	$^2\!/_{12}$ = $^1\!/_6$	

If we sum all the entries in the first row of the above table, we obtain the probability that $Y = 0$ and $X = 1$ or 2 or 3 or 4 or 5 or 6, which is the probability that $Y = 0$ will occur by some means. Thus $P(Y = 0) = \frac{1}{2}$. Similarly $P(Y = 1) = \frac{1}{2}$ is obtained by summing the second row. This probability distribution for Y alone is called the *marginal distribution* of Y.

Similarly, summing the individual columns of the table produces the marginal distribution of X. The statement $P(X = 4) = \frac{1}{6}$ states that the probability of obtaining a 4 on the die by any means when the coin and the die are thrown is $\frac{1}{6}$ which, of course, is as it should be.

DEFINITION 5.9

If $P(X, Y)$ is the joint probability distribution of the discrete variables X and Y, then the *marginal distributions* of X and Y are given by

$$P(X = x) = \sum_i P(X = x, Y = y_i)$$
$$P(Y = y) = \sum_i P(X = x_i, Y = y)$$

In the above experiment, we found the joint probabilities by listing the outcomes of the sample space. Since the coin and the die obviously do not influence one another, we could have used the product rule as developed for independent events to compute these probabilities. Thus, for a head on the coin and a 4 on the die,

$$P(X = 4, Y = 0) = \left(\frac{1}{6}\right) \cdot \left(\frac{1}{2}\right) = \frac{1}{12}$$

Note that the factors here are just the marginal probabilities, that is,

$$P(X = 4, Y = 0) = \left(\frac{1}{6}\right) \cdot \left(\frac{1}{2}\right) = P(X = 4) \cdot P(Y = 0).$$

This relationship suggests the following definition for independence.

> **DEFINITION 5.10**
>
> The variables X and Y are *independent* in the probabilistic sense if and only if their joint distribution is the product of their marginal distributions:
>
> $$P(X, Y) = P(X) \cdot P(Y)$$

This definition extends directly to several variables and applies to continuous as well as discrete variables.

Example 5.18

Instructors are rated 1, 2, or 3 by their personnel committees and 1 or 2 by a composite of their student evaluations. Thus each instructor has a committee ranking X and a student ranking Y. The joint distribution is given in Table 5.19. Verify that X and Y are independent.

TABLE 5.19   THE JOINT DISTRIBUTION OF PERSONNEL COMMITTEE AND STUDENT RATINGS

	x			
y	1	2	3	**P(x)**
1	.10	.10	.20	.40
2	.15	.15	.30	.60
P(y)	.25	.25	.50	1.00

We need to check

$$P(X = x, Y = y) = P(X = x) \cdot P(Y = Y)$$

for the six pairs (1, 1), (1, 2), (2, 1), (2, 2), (3, 1), and (3, 2). For example, $P(X = 3, Y = 1) = .2$ (from the table). The marginal probability $P(X = 3) = .50$ is from the sum of the third column. Similarly from the sum of the first row, we have $P(Y = 1) = .4$. Then $P(X = 3) \cdot P(Y = 1) = (.50) \cdot (.40) = .20$ and this is $P(X = 3, Y = 1)$. All the remaining products may be checked and found to agree. Note that the above checks simply involve comparing the *product* of the row and column sums with the corresponding entries in the table. ■

In actual tests of statistical independence, we do not expect exact agreement between the products and the joint probabilities. We must then cope with the problem of just how much disagreement is caused by the sampling and not the variables themselves.

Finally, we mention the idea of *functional dependence* of two variables. This concept is concerned with a possible relation between the values of the variables rather than with their probabilities. Probabilistic and functional dependence are best thought of as unrelated topics.

DEFINITION 5.11

The variable Y is *functionally dependent* on the variable X if the values of Y are uniquely determined when the values of X are specified.

Example 5.19

Three drawings without replacement are made from an urn containing three white and four black balls. Let X denote the number of white and Y the number of black balls resulting. Then $X + Y = 3$ or equivalently $Y = 3 - X$. If $x = 2$, then $y = 1$. If X is specified, then Y is determined. Thus Y is functionally dependent on X. ■

EXERCISE SET 5.4

Part A

1. A penny and a nickel are tossed. Assigning a 0 to a head and a 1 to a tail, let X and Y represent the outcomes for the penny and nickel, respectively.
   a. Give the joint probability distribution of X and Y.
   b. Determine the marginal probability distributions of X and Y.

2. Consider the joint distribution below. Are X and Y independent (probabilistically)

		x	
y	0	1	2
1	.12	.09	.09
2	.20	.12	.18
3	.08	.09	.03

3. A joint distribution of X and Y is given. Find the marginal distributions of X and Y and use this to check for independence of X and Y (in a probabilistic sense).

		x	
y	1	2	3
0	.10	.10	.05
1	.05	0	.05
2	.15	.05	.05
3	.05	.15	.05
4	.10	0	.05

4. The marginal distributions of two independent variables are given. Find their joint distribution as a table.

x	P(x)	y	P(y)
1	.20	0	.10
2	.40	1	.50
3	.30	2	.40
4	.10		

5. Two coins and a die are tossed. Let X be the number of heads on the two coins, and Y the number showing on the die.
   a. Find the marginal distributions of X and Y. (Remember the marginal distribution of X is just the distribution of X.)
   b. Noting that X and Y are obviously independent, find their joint distribution. (Make use of the marginal distributions in part *a*.)
6. An urn contains four white and five red balls. Four balls are drawn without replacement. Let X be the number of white balls drawn and Y the number of red. Are X and Y functionally dependent?

Part B

7. If X and Y are two random variables, then it can be shown that

$$E[X + Y] = E[X] + E[Y]$$

Thus the expected value of the sum of two random variables is the sum of their respective expected values. Recalling that the expected value may be thought of as an average, interpret the above statement for the following two variables.
   a. X is the number of fish caught by a randomly selected fisher on the opening day of trout season and Y is the number caught on the second day.
   b. X and Y are the amounts won by randomly selected gamblers at Reno and Las Vegas, respectively.
8. A green and a red die are thrown. Let X denote the number showing on the green die and Y that on the red. Then X and Y denotes the sum from the two dice. Find $E[X]$ and $E[Y]$ and then use these to find $E[X + Y]$. (Make use of the result stated in problem 7.)
9. Let $x = 1$ and $x = 0$ correspond respectively to success and failure when a component is tested and suppose $P(X = 1) = \frac{4}{5}$ and $P(X = 0) = \frac{1}{5}$. One easily finds $E[X] = \mu_X = \frac{4}{5}$. Suppose now a second component is tested and we let Y be the number of successes, with values 1 or 0 having the same distribution as X. Then $X + Y$ is the number of successes in the two trials or tests.
   a. Find $E[X + Y]$.
   b. What is the expected number of successes in three trials?
   c. Generalize the above results to n trials.
10. If X and Y are two independent variables, then

$$E[XY] = E[X] \cdot E[Y]$$

Thus the expected value of a product of two independent variables is the product of their independent values. Interpret this result for the product $Z = XY$, where X and Y are the number of heads resulting from throwing two pennies and two nickels, respectively.

COMPUTER PRINTOUT

The printout from a program for computing the mean (K1), the variance (K2), and the standard deviation (K3) of a probability distribution is given in Figure 5.16. Minitab has no direct commands for these. The arrangement of the computations is similar to that used in Example 5.9. Note that labels have been defined and used as the names for the columns C1, C2, C3, and C4. Also note that $^{**}2$ affixed to a variable produces the square of that variable.

FIGURE 5.16 MINITAB COMPUTER PRINTOUT FOR THE MEAN, VARIANCE, AND STANDARD DEVIATION OF A PROBABILITY DISTRIBUTION.

```
MTB >NAME C1 AS 'X', C2 AS 'P(X)', C3 AS 'XP(X)', C4 AS
'XXP(X)'
MTB >SET THE FOLLOWING OUTCOMES INTO 'X'
DATA>2, 3, 4, 5, 6, 7, 8, 9, 10, 11, 12
DATA>END
MTB >SET THE FOLLOWING OUTCOMES INTO 'P(X)'
DATA>.027778, .055556, .083333, .111111, .138889
DATA>.166667
DATA>.138889, .111111, .083333, .055556, .027778
DATA>END
MTB >MULTIPLY 'X' BY 'P(X)', PUT INTO 'XP(X)'
MTB >MULTIPLY 'XP(X)' BY 'X', PUT INTO 'XXP(X)'
MTB >PRINT 'X', 'P(X)', 'XP(X)', 'XXP(X)'
COLUMN        X           P(X)        XP(X)        XXP(X)
COUNT        11             11           11            11
ROW
    1         2.        0.027778     0.05556       0.11111
    2         3.        0.055556     0.16667       0.50000
    3         4.        0.083333     0.33333       1.33333
    4         5.        0.111111     0.55555       2.77778
    5         6.        0.138889     0.83333       5.00000
    6         7.        0.166667     1.16667       8.16668
    7         8.        0.138889     1.11111       8.88890
    8         9.        0.111111     1.00000       8.99999
    9        10.        0.083333     0.83333       8.33330
   10        11.        0.055556     0.61112       6.72228
   11        12.        0.027778     0.33334       4.00003

MTB >LET K1 = SUM('XP(X)')
MTB >LET K2 = SUM('XXP(X)') - SUM('XP(X)')**2
MTB >LET K3 = SQRT(K2)
MTB >PRINT K1,K2,K3
   K1        7.00001
   K2        5.83331
   K3        2.41522
MTB >STOP
```

CHAPTER CHECKLIST

Vocabulary

Random variable
Probability distribution

Density curve
Density function

The mean of a random variable
The variance of a random variable
The standard deviation of a random
   variable
Payoff function
Fair game
Expected value
Continuous random variable

Uniform distribution
Parameter of a distribution
Joint probability distribution
Marginal distribution
Independent random variables
Functionally dependent variables

Techniques

Finding an appropriate random variable for a given probability experiment
Finding the probability distribution of a discrete random variable when the
   experiment is described
Finding the mean, variance, and standard deviation of a discrete random
   variable when the distribution is given; also, interpreting these quantities
Finding the expected value of a game of chance and determining if the
   game is fair
Finding the mean and standard deviation of $Z = (X - \mu)/\sigma$
Finding the probabilities when a simple density curve is given
Finding the joint distribution of X and Y when X and Y are associated with
   an experiment having two components or stages
Finding probabilities of events by using a joint distribution
Finding the marginal distributions of X and Y from their joint distribution
Determining if X and Y are independent when their joint distribution is given

Notation

$P(X)$	The probability function for the random variable X
$P(X = c)$	The probability that X takes on the specific value $x = c$
$P(a < X < b)$	The probability that X. takes on values between a and b
μ_X	The mean of X
σ_X^2	The variance of X
σ_X	The standard deviation of X
$g(X)$	A random variable which is a function of X; also used as the payoff function in games of chance
$E[g(X)]$	The expected value of $g(X)$
E	The expected value (used in place of $E[g(X)]$)
$f(x)$	The density function for the distribution of X
$P(X, Y)$	The probability function of the joint distribution of X and Y
$P(X = x, Y = y)$	The probability that $X = x$ and $Y = y$

**CHAPTER
TEST**

1. The distribution of X is given below. Find:
    a. $P(X = 2)$.
    b. $P(X \geq 3)$.
    c. $P(2 \leq X < 5)$.
    d. c if $P(X \leq c) = .80$.

x	P(x)
0	.10
1	.20
2	.20
3	.30
4	.20

2. A uniform distribution is given below.

   a. Find $P(X \geq 5)$.
   b. Find $P(2 \leq X \leq 3)$.
   c. Find c if $P(X \geq c) = \frac{1}{3}$.

3. Three prospective voters are asked whether they plan to vote for or against a municipal improvement project. Let X be the number who respond that they will vote for the project. Assuming the electorate is equally split on this issue,
   a. Find the distribution of X.
   b. Find the mean and variance of this distribution.
   c. Sketch the histogram and locate the mean.

4. Suppose the probability distribution for the number X of games required in a World Series is as given below. What is the expected number of games for a World Series?

x	P(x)
4	.20
5	.50
6	.20
7	.10

5. A florist believes her daily sales of carnations have the probability distribution below. Find the expected value of the daily carnation sales and interpret the result.

x (Dozen)	P(x)
16	.50
17	.30
18	.20

6. Suppose an insurance company has a rate of $8.00 per thousand for term life insurance for healthy individuals of a certain age group. Suppose that for this age group, the probability of dying within a year is .0018. What can the company expect to make (before expenses) on:
   a. A sale of $1000 worth of this insurance?
   b. Sales of this insurance totaling $100 million?

7. The joint probability distribution of X and Y is given below. Find:
   a. $P(X = 2 \text{ and } Y = 1)$.
   b. The marginal distribution of X.
   c. The marginal distribution of Y.
   d. What is the probability of $X = 2$ by any means?

		y	
x	0	1	2
1	0	1/5	2/15
2	1/6	1/9	1/4
3	1/12	0	1/18

8. The joint distribution of X and Y is given. Are X and Y independent?

		y
x	0	1
0	0	1/4
1	1/3	5/12

9. A man arrives home late at night. In his pocket are three keys, only one of which fits his door. Assume he selects a key at random in the dark (with no replacement).
   a. Find the probability distribution of the number X of keys he must try in order to get the door open.
   b. What is the expected number of keys he will need to try?

10. Suppose X has a distribution whose density curve is given below and let the areas be as indicated.

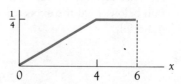

   a. Find $P(X \geq 4)$.
   b. Find $P(X \leq 5)$.
   c. Find $P(2 \leq X \leq 6)$.

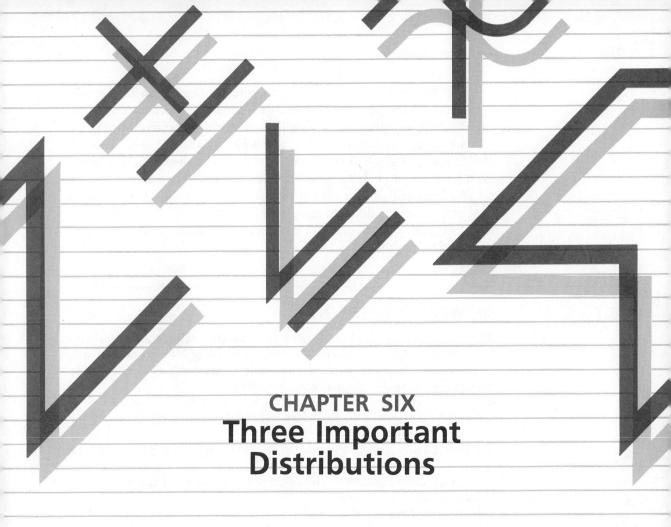

CHAPTER SIX
Three Important Distributions

The concept of a probability distribution was introduced in the preceding chapter. We now turn our attention to three important such distributions, the *binomial,* the *normal,* and the *chi-square.*

The distribution of the number of successes in a fixed number of repeated independent trials is called a binomial distribution. A success may be as simple as a reply of "yes" to a question or as complex as a positive reaction to a medication by a patient. Thus the binomial distribution finds application with opinion sampling and more generally with inferences concerning the proportion of a population exhibiting some characteristic of interest.

The normal distribution is extremely important. There are many naturally occurring variables whose distributions conform more or less to a characteristic bell-shaped form; for these, the normal distribution often proves to be a good approximation. Blood pressures, I.Q. scores, diameters of pine seedlings, and measurement errors are but a few examples of such variables. Additionally, in inferential statistics there are many important sample variables that have approximately normal distributions when the sample size is large.

The chi-square distribution is a continuous distribution that applies to the sum of the squares of several independent, normally distributed variables. We will illustrate one application of this important distribution with the classic "goodness of fit" test for which it was first used. This test offers a way to tell whether a theoretical model or distribution is appropriate for some particular population or experiment. For example, we can test whether a pair of dice are fair or whether some political figure enjoys equal support among different age groups. In Chapter 13 this same "goodness of fit" test is developed into the popular chi-square tests of independence and homogeneity.

6.1   BINOMIAL EXPERIMENTS

The binomial distribution is associated with a *binomial experiment*, which consists of repeated trials with only two possible outcomes—yes or no, right or wrong, for or against, and so forth—which we shall call *success* and *failure*. The specific properties of a binomial experiment are as follows:

1. There is a fixed number n of repeated trials.
2. Each trial is exactly the same and has only two possible outcomes: success and failure.
3. The probability p of success and $q = 1 - p$ of failure are the same at each trial.
4. The trials are independent. The outcome of any particular trial is not influenced by previous outcomes.
5. It is the total number of successes that is of interest, not their order of occurrence.

There may be 0, 1, 2, . . ., n successes in the n trials and these we regard as values of a random variable.

DEFINITION 6.1

The number X of successes in a binomial experiment is called a *binomial random variable*.

The following are examples of binomial experiments.

1. A coin is flipped seven times and the number of heads X recorded. Let success be a head, failure a tail. Each flip represents a trial, and clearly the outcomes of the trials are independent of one another. Thus the number of trials is $n = 7$, the probability of success (a head) on each trial is $p = \frac{1}{2}$, and $S_X = \{0, 1, 2, 3, 4, 5, 6, 7\}$. We are interested in the probability distribution of X, that is, with the probability of obtaining exactly 0 heads, exactly 1 head, and so on up to exactly 7 heads.
2. A pair of dice is thrown 24 times and the number X of sevens recorded. Let success be a seven and any other sum a failure. The successive rolls of the

dice are independent of one another. Thus $n = 24$, $p = \frac{1}{6}$, $q = 1 - \frac{1}{6} = \frac{5}{6}$, and we seek the distribution of X, that is, $P(X = 0)$, $P(X = 1)$, $P(X = 2)$, . . ., $P(X = 24)$.

3. An examination consists of ten multiple choice questions, each with five parts. If a student guesses at each answer, then each question is a trial of a binomial experiment where success is a correct answer. Thus $n = 10$, $p = \frac{1}{5}$, $q = \frac{4}{5}$, and $S_X = \{0, 1, . . ., 10\}$, where X is the total number of correct answers.

As the second example indicates, the term "binomial" is not restricted to those experiments with only two outcomes at each trial. We can always collect the outcome or outcomes of interest into a single event defined as success and define the remaining outcomes as failure.

There are also many experiments with repeated trials that can be approximated as binomial experiments. Typical of these is opinion sampling. Each interview of a subject is a trial. However, since the same subject is not interviewed twice, the probability of success changes from trial to trial. If, however, the population being sampled is large, this change is so slight that it can be ignored and we can treat the sampling as a binomial experiment.

For example, suppose that 70% of all executives have a stock option with their companies and 200 executives are asked whether or not they have this desirable "perk." If we treat the option as a success, this sampling may be viewed as a binomial experiment with the parameters $n = 200$, $p = .7$ and $q = .3$.

6.2   BINOMIAL COEFFICIENTS

In developing the binomial distribution, we need to know the number of ways in which exactly x successes can occur in the n trials of a binomial experiment. Since each trial results in either a success S or failure F, an outcome can be represented as an ordered arrangement of Ss and Fs. In Table 6.1 are listed those outcomes by which exactly $x = 3$ successes can be obtained in $n = 5$ trials. For example, the outcome $(S, S, F, F, S,)$ indicates that the three successes occurred on the first, second, and fifth trials, the failures on the third and fourth.

**TABLE 6.1   OUTCOMES
RESULTING IN THREE
SUCCESSES IN FIVE TRIALS**

(S, S, S, F, F)
(S, S, F, S, F)
(S, S, F, F, S)
(S, F, S, S, F)
(S, F, S, F, S)
(S, F, F, S, S)
(F, S, S, S, F)
(F, S, S, F, S)
(F, S, F, S, S)
(F, F, S, S, S)

It should be clear that to write down an outcome corresponding to x successes in n trials, we need only choose x of the n positions of the outcome in which to place the successes S. The remaining $n - x$ positions must be filled by failures F. The number of ways this can be done is given by the binomial coefficient $\binom{n}{x}$, appropriately read "n choose x." The computation of these coefficients is detailed in Theorem 6.1.

THEOREM 6.1

In an ordered arrangement with n positions, x of the positions may be labeled S and the remaining $n - x$ labeled F in $\binom{n}{x}$ ways where

$$\binom{n}{0} = 1$$

$$\binom{n}{1} = \frac{n}{1}$$

$$\binom{n}{2} = \frac{n(n-1)}{2 \cdot 1}$$

$$\binom{n}{3} = \frac{n(n-1)(n-2)}{3 \cdot 2 \cdot 1}$$

$$\binom{n}{x} = \frac{n(n-1)(n-2) \ldots (n-x+1)}{x \cdot (x-1)(x-2) \cdot \ldots \cdot (1)}$$

A few observations about computing $\binom{n}{x}$. When x is not 0, there are x factors in both the numerator and denominator of the fraction. The first factor of the numerator is n and each succeeding factor is 1 less than the previous until the x factors are obtained. The first factor of the denominator is x and the succeeding factors decrease in the same way (by 1) until the factor 1 is obtained.

Example 6.1

Compute (1) $\binom{9}{4}$; (2) $\binom{5}{3}$; (3) $\binom{8}{0}$.

1. $\binom{9}{4} = \frac{9 \cdot 8 \cdot 7 \cdot 6}{4 \cdot 3 \cdot 2 \cdot 1} = 126$

The coefficient "9 choose 4" has $x = 4$ factors in both its numerator and denominator.

2. $\binom{5}{3} = \frac{5 \cdot 4 \cdot 3}{3 \cdot 2 \cdot 1} = 10$

The coefficient "5 choose 3" has $x = 3$ factors in both its numerator and denominator.

3. $\binom{8}{0} = 1$

Note that "n choose 0" is always 1. ■

As we have stated, there are $\binom{n}{x}$ outcomes corresponding to exactly x successes in a binomial experiment with n trials. We make no attempt to prove this; rather, we shall be content to illustrate it with an example. Note, however, that we already have one confirmation. In the preceding example we found $\binom{5}{3} = 10$, and in Table 6.1 we saw this to be the number of outcomes by which $x = 3$ successes may be obtained in $n = 5$ trials.

Example 6.2

A coin is flipped four times. In how many ways can one obtain two heads? Let success correspond to a head, failure to a tail. The number of outcomes with $x = 2$ heads in $n = 4$ trials is

$$\binom{4}{2} = \frac{4 \cdot 3}{2 \cdot 1} = 6$$

These six outcomes are listed in Table 6.2. ■

TABLE 6.2 OUTCOMES YIELDING TWO HEADS IN FOUR TOSSES OF A COIN

(H, H, T, T)
(H, T, H, T)
(H, T, T, H)
(T, H, H, T)
(T, H, T, H)
(T, T, H, H)

6.3 THE BINOMIAL DISTRIBUTION

We now turn to deriving the binomial distribution. Let us begin by considering a specific problem. From a box containing one white and two red balls, five draws with replacements between draws are made. What is the probability of obtaining exactly three white balls?

One outcome by which three white balls can occur is $(W, W, W, R, R,)$. Since the drawings are independent and the probability of obtaining a white and a red ball at each draw are $\frac{1}{3}$ and $\frac{2}{3}$, respectively, this particular outcome has the probability

$$(\tfrac{1}{3})(\tfrac{1}{3})(\tfrac{1}{3})(\tfrac{2}{3})(\tfrac{2}{3}) = (\tfrac{1}{3})^3 (\tfrac{2}{3})^2$$

Similarly, we find that the outcome (W, R, R, W, W) has the same probability, since

$$(\tfrac{1}{3})(\tfrac{2}{3})(\tfrac{2}{3})(\tfrac{1}{3})(\tfrac{1}{3}) = (\tfrac{1}{3})^3(\tfrac{2}{3})^2$$

In fact, each of the $\binom{5}{3} = 10$ outcomes by which three successes and two failures can occur has this probability $(\tfrac{1}{3})^3(\tfrac{2}{3})^2$. Thus

$$P(X = 3) = 10(\tfrac{1}{3})^3(\tfrac{2}{3})^2 = \binom{5}{3}(\tfrac{1}{3})^3(\tfrac{2}{3})^2$$

If we were to next consider the probability of $X = 4$ successes, we would find

$$P(X = 4) = 5(\tfrac{1}{3})^4(\tfrac{2}{3})^1 = \binom{5}{4}(\tfrac{1}{3})^4(\tfrac{2}{3})^1$$

For x successes we infer

$$P(X = x) = \binom{5}{x}(\tfrac{1}{3})^x(\tfrac{2}{3})^{5-x}$$

In general, for a binomial experiment with n trials, there are $\binom{n}{x}$ outcomes having x successes and $n - x$ failures, and each has probability $p^x q^{n-x}$

THEOREM 6.2

In a binomial experiment with n trials where the probabilities of success and failure on any trial are p and q, respectively, the probability distribution of the number of successes X is given by

$$P(X = x) = \binom{n}{x}p^x q^{n-x}$$

This probability distribution is called the binomial distribution.

Example 6.3

A die is rolled four times. Find the probability of exactly two sixes in the four rolls. This is a binomial experiment with success being a six, failure any other result. Thus $n = 4$, $p = \tfrac{1}{6}$, $q = \tfrac{5}{6}$ and $P(X = x) = \binom{4}{x}(\tfrac{1}{6})^x(\tfrac{5}{6})^{4-x}$. In particular,

$$P(X = 2) = \binom{4}{2}(\tfrac{1}{6})^2(\tfrac{5}{6})^{4-2} = 6(\tfrac{1}{6})^2(\tfrac{5}{6})^2 = \tfrac{25}{216} = .1157 \quad \blacksquare$$

Example 6.4

The probability of recovering from a certain blood disorder is .3. Five patients are selected at random and subjected to a new treatment. If the treatment has no effect, what is the probability that four or more recover?

Four or more occurs as exactly four or exactly five. Thus

$$P(X) \geq 4) = P(X = 4) + P(X = 5)$$

where

$$P(X = x) = \binom{5}{x}(.3)^x(.7)^{5-x}$$

FIGURE 6.1 SELECTED BINOMIAL DISTRIBUTIONS

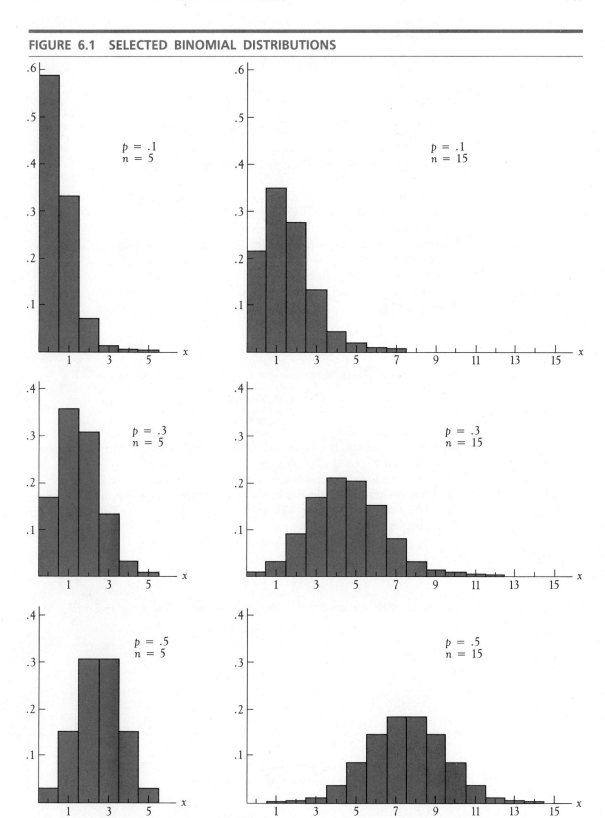

We find

$$P(X = 4) = \tbinom{5}{4}(.3)^4(.7)^1 = .02835$$
$$P(X = 5) = \tbinom{5}{5}(.3)^5(.7)^0 = .00243$$

and thus

$$P(X \geq 4) = .02835 + .00243 = .03078 \ (3.078\%)$$

Thus if the treatment has no effect, then only about 3% of all samples of five patients will have four or more patients who recover. ∎

The histograms of several binomial distributions are given in Figure 6.1. They are almost bell-shaped for values of p near ½, and this appearance becomes more prominent as the number n of trials increases. For large n, the resemblance holds for values of p quite far removed from ½, which suggests that the normal distribution can be used to approximate binomial distributions when n is large. We shall see how to do this shortly.

The mean and variance of a discrete variable X with probability distribution P are given by:

$$\mu_X = \sum_{i=1}^{n} x_i P(x_i) \qquad \sigma_x^2 = \sum_{i=1}^{n} x_i^2 P(x_i) - (\mu_X)^2$$

For the binomally distributed variable X we would substitute

$$P(x) = P(X = x) = \tbinom{n}{x}p^x q^{n-x}$$

Thus for any particular choice of the parameters n, p, and q, these formulas may be evaluated to obtain the mean and variance of the binomial distribution, often, however, with a great deal of effort.

Now the mean is simply the expected number of successes in the n trials. If, say, a die is thrown 300 times and a four is a success, then the number of fours should be about one-sixth of 300, that is, $(\frac{1}{6}) \cdot 300 = 50$. This suggests the mean is $\mu_x = n \cdot p$.

For $n = 1$ trial, the mean is p and the variance is

$$\sigma_X^2 = 0^2 \cdot q + 1^2 \cdot p - p^2$$
$$= p - p^2 = p(1 - p) = pq$$

For several trials it turns out that the variances add. Thus for $n = 2$ trials

$$\sigma_X^2 = pq + pq = 2pq$$

And in general, for n trials

$$\sigma_X^2 = npq$$

THEOREM 6.3

Let X have the binomial distribution $P(X = x) = \tbinom{n}{x}p^x q^{n-x}$. Then the mean, variance, and standard deviation of X are given by:

$$\mu_X = np$$
$$\sigma_X{}^2 = npq$$
$$\sigma_X = \sqrt{npq}$$

Example 6.5

A coin is tossed four times. Let success be a head. Then $n = 4$, $p = \frac{1}{2}$, $q = \frac{1}{2}$, and

$$\mu_X = np = 4(\tfrac{1}{2}) = 2 \qquad \sigma_X{}^2 = npq = 4(\tfrac{1}{2})(\tfrac{1}{2}) = 1$$

These values may also be computed directly from their definitions.

$$\mu_X = 0P(X = 0) + 1P(X = 1) + 2P(X = 2)$$
$$+ 3P(X = 3) + 4P(X = 4)$$
$$= 0\left(\frac{1}{16}\right) + 1\left(\frac{4}{16}\right) + 2\left(\frac{6}{16}\right) + 3\left(\frac{4}{16}\right) + 4\left(\frac{1}{16}\right)$$
$$= \frac{32}{16} = 2$$
$$\sigma_X{}^2 = (0 - 2)^2\left(\frac{1}{16}\right) + (1 - 2)^2\left(\frac{4}{16}\right)$$
$$+ (2 - 2)^2\left(\frac{6}{16}\right) + (3 - 2)^2\left(\frac{4}{16}\right) + (4 - 2)^2\left(\frac{1}{16}\right)$$
$$= \frac{16}{16} = 1 \quad \blacksquare$$

Obviously the formulas of Theorem 6.3 are the simplest way to obtain the mean and variance of the binomial distribution.

6.4 CUMULATIVE BINOMIAL PROBABILITY TABLES

Often we need probabilities such as $P(X \leq k)$ and $P(1 \leq X \leq k)$. To compute the probability of, say, eight or fewer successes in 15 trials would require the evaluation of

$$P(X \leq 8) = P(X = 0) + P(X = 1) + \ldots P(X = 8) = \sum_{x=0}^{8} P(X = x)$$

This is a tedious task at best, and for this reason extensive tables of binomial probabilities are available that allow us to find sums such as these. Tables for $n = 5$, 6, 7, 8, 9, 10, 15, 20, and 25 are provided in the appendix (Table II) and will be used for examples and exercises. Tables for other values of n may be obtained by using Minitab; additionally, we shall soon have the normal distribution available for approximating such probabilities when the number of trials is large. The following examples illustrate the use of Table II.

Example 6.6

It is asserted that 60% of the adults in a certain community are registered voters. If 20 residents are selected at random, what is the probability that eight or fewer will be registered? We regard this as a binomial distribution problem with $n = 20$, $p = .6$ and $q = .4$. Here X is the number of registered voters in a sample of 20. We need

$$P(X \leq 8) = \sum_{x=0}^{8} P(X = x)$$

We enter Table II with $n = 20$ at the row headed $x = 8$ and the column headed $p = .6$. At their intersection we find the entry .0565. Thus

$$P(X \leq 8) = .0565$$

Thus a sample with eight or fewer registered voters will occur in only about 5.65% of all samples of 20 residents. ∎

Example 6.7

The fairness of a coin is to be tested by flipping it 15 times. It should show around seven or eight heads if it is fair. Suppose we decide to reject the coin if it shows fewer than four heads or more than 11. If many *fair* coins are subjected to this test, how often is one rejected? This is a binomial experiment where X is the number of heads in the $n = 15$ trials and $p = .5$ (fair coin). We accept the coin if $4 \leq X \leq 11$. Now

$$P(4 \leq X \leq 11) = P(X \leq 11) - P(X \leq 3)$$
$$= .9824 - .0176 = .9648$$

The probability of accepting a fair coin is .9648, so the probability of rejecting it is $1 - .9648 = .0352$. ∎

Note in this last example that if one is planning to use the coin in some game, it is of no consequence if a good coin is rejected. Also, no matter what type of an acceptance test one may choose (short of accepting the coin regardless of its performance), there will always be a small probability of rejecting a fair coin. It is not inconceivable that a fair coin will show 15 heads or even 0 heads in 15 consecutive throws.

Example 6.8

A test consists of 20 problems, all with five-part multiple choice answers. A student guesses at each answer. What is the probability that she has five or more correct? Here $n = 20$, $p = \frac{1}{5} = .2$, and X is the number of correct answers.

$$P(X \geq 5) = 1 - P(X \leq 4)$$
$$= 1 - .6296$$
$$= .3704 \; \blacksquare$$

It should be noted that the entries in Table II are rounded to four decimal places. Thus, if in the last example we had wanted the probability of 14 or more correct answers, we would have found

$$P(X \geq 14) = 1 - P(X \leq 13) = 1 - 1.0000 = 0.0000$$

This is correct to four decimal places. It is not, however, exact. There is a small probability, less than .00005, that the student will obtain 14 or more correct answers if she guesses.

We conclude this section with an example that shows how we can find the probability of an error when we make a decision based on the results of a sample. These ideas will be explored in much greater detail in Chapter 9, where we examine hypothesis testing.

Example 6.9

Suppose 60% of all automobiles are dark-colored (black, blue, brown, silver-gray, and green) and we suspect that cars with these colors are accident-prone. We plan to examine a sample of 20 automobiles that have been reported in accidents. If these colors are not more accident-prone than others, then about 12 (60% of 20) of the autos should be dark in color. Now to find 13, 14, or even 15 would perhaps not be surprising. Suppose then our decision is to reject the contention that the dark colors are not accident-prone if we find 16 or more dark colors in our sample of 20.

Even if the probability of a dark-colored auto being in an accident is .6 (the same as for the occurrence of this color in the population), it is still possible to find a few samples of 20 with 16 or more dark colors. When this happens we reject the hypothesis and make a mistake. What is the chance of this happening? In probability terms, what is the probability of 16 or more successes in a binomial experiment with $p = .6$ and $n = 20$ trials?

$$P(X \geq 16) = 1 - P(X \leq 15) = 1 - .9490 = .0510 \ (5.1\%)$$

Thus if $p = .6$, then about 5.1% of all samples will lead us to the wrong decision. ■

Note in this last example that if our sample should have only, say, seven or eight (or fewer) dark-colored cars, then we should reexamine our contention that these colors are accident-prone. Statistical studies do indicate, however, that these colors are involved in significantly more accidents than is predicted from their distribution in the population.

EXERCISE SET 6.1

Part A

1. Which of the following are binomial experiments or may be approximated as such?
   a. Four students are selected at random from a class of 20 to form a debate team. Let X be the number of seniors on the team.
   b. An automatic dialing maching places 100 calls to randomly selected numbers in a large metropolitan area. Let X be the number of calls that are answered.
   c. Five draws with replacement are made from a barrel containing 100 ticket stubs, 10 of which bear your name. Let X be the number of times your name is drawn.
   d. Fifty people are selected at random in a large city and asked whether or not they are registered to vote. Let X be the number registered.

2. Devise three binomial experiments where the distribution below would apply.

$$P(X = x) = \binom{5}{x}\left(\frac{1}{3}\right)^x\left(\frac{2}{3}\right)^{5-x}$$

3. Let $P(X = x) = \binom{4}{x}(.4)^x(.6)^{4-x}$. Compute and tabulate this distribution and construct its histogram.

4. Assume that the probabilities of male and female births are the same. Find the probability that in six hospital births:
   a. All will be males.
   b. Exactly one will be a male.

5. Suppose the probability of making eye contact with a stranger you meet on the street is .3. If you walk down a street and meet five different people (all strangers), what is the probability of making eye contact with:
   a. Exactly three of the people?
   b. Two or fewer of the people?
   c. At least one of the people?

6. Let $P(X = x) = \binom{4}{x}(.4)^x(.6)^{4-x}$, as in problem 3. Compute the mean and variance directly and also by using Theorem 6.3. Compare the results.

7. Find μ_X and σ_X for the binomial distribution with the parameters
   a. $n = 100$ and $p = .7$.
   b. $n = 60$ and $p = \frac{1}{2}$.
   c. $n = 400$ and $p = .25$.

8. A lot of 10,000 computer chips contain 500 that are defective. How many defectives should a sample of 1200 contain?

9. A manufacturer claims that a production run of 30,000 stereo needles contains only 900 that are defective. The needles are shipped out in lots of 400. How many defectives should we expect to see in most of these lots? What assumptions about the selection of the lot or the production process are necessary?

10. A box contains 10 balls numbered 0, 1, 2, 3, 4, . . ., 9. Let 10,000 drawings with replacement be made from the box, the numbers on the balls being recorded to form a sequence of digits. If X is the number of zeros in such a sequence, find the mean and standard deviation of X.

11. Let $P(X = x) = \binom{15}{x}(.4)^x(.6)^{15-x}$. Using the cumulative table of binomial probabilities, find:
   a. $P(X \le 6)$.
   b. $P(X \ge 3)$.
   c. $P(4 \le X \le 7)$.

12. Let $P(X = x) = \binom{20}{x}(.3)^x(.7)^{20-x}$. Using the cumulative probability tables:
   a. Find $P(X \le 7)$.
   b. Find $P(X > 8)$.
   c. Find k if $P(X \le k) = .9829$.
   d. Find k if $P(X > k) = .5836$.

13. An office manager has determined that 40% of the calls placed to his office find a busy line. If 15 randomly timed calls are made to the office during business hours, what is the probability that four or more find a busy line?

14. Of the 20,000 students at a certain college, 40% live at home. Ten randomly selected students are to be interviewed. What is the probability that we will find between two and six (inclusively) who live at home?

Part B

15. A grower asserts that the germination level of a certain lot of seed is 80%. Twenty-five seeds are planted and 16 germinate. How often should such poor

performance be observed in samples of 25 seeds from this population, if the germination level is as asserted?

16. A medical researcher knows that the probability of recovering from a certain disease is .3. To test a new drug, she will give it to 20 patients. Unless nine or more recover, the drug will be judged ineffective. If in fact the drug has no effect, what is the probability that this test will lead to a correct conclusion?

17. A company receives from a manufacturer large lots of a certain item and has devised the following lot acceptance sampling plan: Select a sample of 15 items from the lot and test. If three or more are found defective, reject the entire lot. What fractions of the lots are rejected if the entire lot has:
    a. 5% defective?
    b. 10% defective?
    c. 20% defective?
    Does this appear to be an effective acceptance test?

18. Proceed as in problem 17, only now reject the entire lot if a sample of $n = 20$ has two or more defective items. Compare this plan with that in problem 17 for effectiveness.

19. A drunk staggers out of a tavern to the sidewalk. After 25 steps (either forward or backward) he is seven steps ahead of where he started. Is he so drunk that he cannot tell forward from backward?

20. An office has five supervisors and a study has shown that each uses a secretary about 15 minutes of each hour. Think of the day as being broken up into 15-minute time intervals—8:00–8:15, 8:15–8:30, and so on—and let X denote the number of secretaries needed in any one of these time intervals. Find the distribution of X and its expected value. What assumptions are needed?

21. What is the probability of obtaining exactly three red cards in a sequence of five draws from a deck of cards (with replacement) and repeating this result two out of the three times this experiment is performed?

22. In the waiting problem, a coin is tossed repeatedly. What is the probability that the fourth head will occur on the seventh flip? Hint: The fourth head on the seventh flip is the result of three heads in the first six flips and a head on the seventh.

23. Let X be the number of binomial trials needed to produce exactly k successes. The distribution of X is called the *negative binomial* or *Pascal* distribution and is given by

$$P(X = x) = \binom{x-1}{k-1} p^k q^{x-k}$$

Problem 22 above is a specific instance where this distribution applies. There the problem was to determine the probability that $x = 7$ trials would be needed to produce $k = 4$ successes with $p = q = \frac{1}{2}$.
    a. If $k = 8$ and X has a negative binomial distribution, what are the possible values of X?
    b. A pair of dice are thrown repeatedly. What is the probabilty that exactly X throws will be needed to produce two sevens? What is the probability that $X = 3$ throws will be needed?

24. A pollster decides to interview as many people as necessary in order to obtain 10 people who believe that the President is doing a good job. If 40% of the population believe this to be the case, find the distribution of the number X who will need to be interviewed. Use the negative binomial distribution of problem 23.

25. A hypergeometric experiment is much like a binomial one, except that the trials consist of draws without replacement from a population of size N which originally contained k successes. The result is that the probability of success changes from trial (draw) to trial. The distribution of the number of successes in n trials is then

$$P(X = \text{x}) = \frac{\binom{k}{x}\binom{N-k}{n-x}}{\binom{N}{n}}$$

Ten rats, four of which are white and six brown, are available for an experiment. If three are selected at random, what is the probability that all three are brown?

26. Five cards are dealt from a standard bridge deck. What is the probability of obtaining exactly three clubs? Use the hypergeometric distribution of problem 25.

6.5   THE NORMAL DISTRIBUTION

The normal distribution is one of the more important continuous distributions. The graph of its density function, called a *normal curve*, is shown in Figure 6.2. As we have already mentioned, many populations, such as weights of males, reaction times to a medication, and scores on a standardized test, show distributions that are approximately of this form. It also turns out that there are many uses of this distribution in both estimation and hypothesis testing. The normal distribution is often referred to as the *Gaussian distribution* in honor of Karl Friedrich Gauss (1777–1855), who studied it in connection with the analysis of errors that occur in measurements.

FIGURE 6.2. A NORMAL DISTRIBUTION

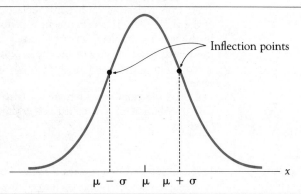

The bell-shaped normal curve is a probability density curve for a continuous variable X, and thus the total area bounded by this curve and the x-axis is 1. The curve is perfectly symmetric about a line perpendicular to the x-axis through the mean μ_X. This line bisects the bell exactly. One-half the area under the curve is to the left of this line, the other half is to its right. Because of the perfect symmetry the mean, median, and mode all coincide. The standard deviation σ_x is the distance from the mean to the point of inflection, that is, to the point where the concavity of the curve changes from down to up.

The normal curve is given in Figure 6.3 for several values of the standard deviation σ_x. Changing the mean μ_X merely translates the curve to the right or the left. Varying the standard deviation changes the shape of the bell. Thus

FIGURE 6.3   THREE NORMAL DISTRIBUTIONS WITH DIFFERENT STANDARD DEVIATIONS

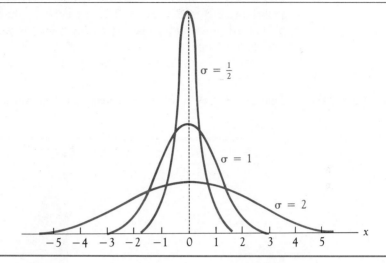

changing the mean and standard deviation may change the location and the shape of the curve, but it still remains a normal curve. It is important to note that if the mean and standard deviation are specified for a normal distribution, then the normal curve is determined completely.

6.6   THE STANDARD NORMAL DISTRIBUTION

The above discussion suggests that we will perhaps need a different normal curve for each choice of the mean and standard deviation. It turns out that we can always standardize the variable of interest so that we work with a variable having a mean of 0 and a standard deviation of 1. This variable, which is commonly denoted by Z, has a distribution called the *standard normal distribution*. For the sake of completeness we shall include its density function in its definition.

DEFINITION 6.2

The variable Z having a mean of 0 and a standard deviation of 1 is said to have the *standard normal distribution* if its probability density function is given by

$$f(z) = \frac{1}{\sqrt{2\pi}}\, e^{-z^2/2}$$

Here e is an irrational number whose approximate value is 2.71828.

We now turn to the probability distribution of the standard normal variable Z, which is given in Table III of the appendix. This table is constructed so that one may readily find the area under the normal curve between 0 and z_1 and thus the probability $P(0 \leq Z \leq z_1)$ of Z falling between these limits. From these entries all needed probabilities may be readily found as we show in the following sequence of examples.

Example 6.10   Find $P(0 \leq Z \leq 1.26)$. The desired area is shaded in Figure 6.4.

FIGURE 6.4 $P(0 \leq Z \leq 1.26) = .3962$

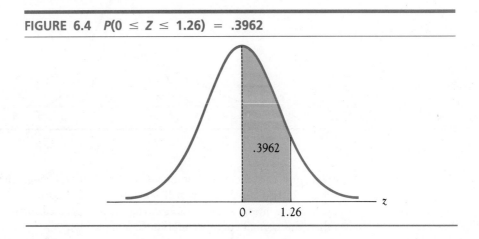

Entering the table in the row headed $z = 1.2$, we find in the column headed .06 the entry .3962. Thus $P(0 \leq Z \leq 1.26) = .3962$. ∎

Example 6.11   Find $P(Z \geq 2.00)$ (see Figure 6.5).

FIGURE 6.5 $P(Z \geq 2.00) = .0228$

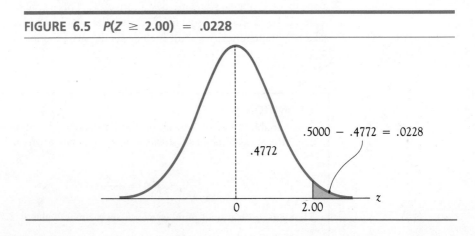

The total area to the right of $z = 0$ is .5. The desired area is thus the difference between .5 and the table entry .4772. Thus

$$P(Z \geq 2.00) = .5 - P(0 \leq Z \leq 2.00)$$
$$= .5 - .4772$$
$$= .0228 \quad \blacksquare$$

Example 6.12   Find $P(.3 \leq Z \leq 1.5)$ (see Figure 6.6).

FIGURE 6.6 $P(.3 \leq Z \leq 1.5) = .3153$

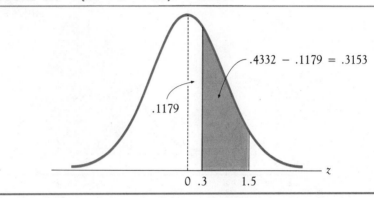

We seek the difference of the areas from 0 to 1.50 and from 0 to .30. Thus

$$P(.3 \leq Z \leq 1.5) = P(0 \leq Z \leq 1.5) - P(0 \leq Z \leq .3)$$
$$= .4332 - .1179$$
$$= .3153 \quad \blacksquare$$

Example 6.13   Find $P(-2.34 \leq Z \leq 0)$ (see Figure 6.7).

FIGURE 6.7 $P(-2.34 \leq Z \leq 0) = .4904$

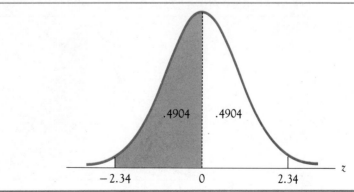

By symmetry the area needed is equal to that of the corresponding region to the right of center. We find

$$P(-2.34 \le Z \le 0) = P(0 \le Z \le 2.34)$$
$$= .4904 \quad \blacksquare$$

Example 6.14   Find $P(-1.3 \le Z \le 1.3)$ (see Figure 6.8.

FIGURE 6.8 $P(-1.3 \le Z \le 1.3) = .8064$

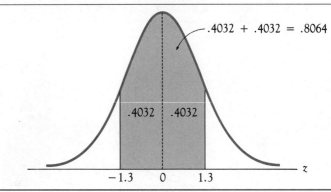

All we need to do is double the area between the limits of 0 and 1.3.

$$P(-1.3 \le Z \le 1.3) = P(-1.3 \le Z \le 0) + P(0 \le Z \le 1.3)$$
$$= .4032 + .4032$$
$$= .8064 \quad \blacksquare$$

Often in applications we know the probability $P(0 \le Z \le z_1)$ and need to find the limit z_1. We then work from the table entry back to the z limit.

Example 6.15   Find z_1 if $P(0 \le Z \le z_1) = .4881$ (see Figure 6.9).

FIGURE 6.9 $P(0 \le Z \le 2.26) = .4881$

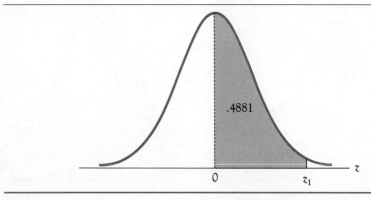

We are now given that the area under the curve is .4881. We find the entry .4881 in the body of the table in the row headed 2.2 and the column headed .06. Thus $z_1 = 2.26$. ■

One rarely needs to interpolate in this table. The nearest entry is sufficient in nearly all cases.

Example 6.16   Find z_1 if $P(-z_1 \leq Z \leq z_1) = .9486$ (see Figure 6.10).

FIGURE 6.10 $P(-z_1 \leq Z \leq z_1) = 2 \cdot P(0 \leq Z \leq z_1)$

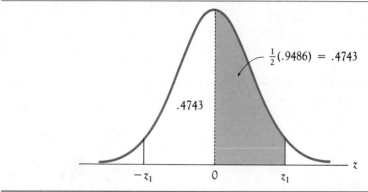

By symmetry the area from 0 to z_1 must be one-half of this, that is, $P.(0 \leq Z \leq z_1) = .4743$. The nearest table entry is .4744 and corresponds to $z = 1.95$. ■

The reader will find that the use of the normal distribution table is greatly simplified if one sketches a graph, shades the desired region, and reasons from it. It should also be noted that the entries in this table are exact only to four decimal

---

FIGURE 6.11   THE EMPIRICAL PREDICTIONS

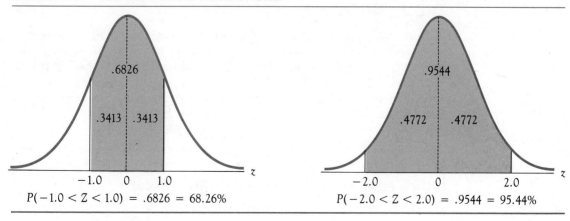

places and that the table has been discontinued at $z = 3.49$, since the area beyond this point is quite small. While rarely needed, more extensive tables are available.

The empirical prediction of Chapter 3 for the distribution of a population about its mean has its origin in the normal distribution. Since the standard deviation is 1, the probability of Z being within one standard deviation of its mean $\mu_Z = 0$ is $P(-1 \leq Z \leq 1) = .6826$ or about 68%. Similarly, for two standard deviations, we find the familiar prediction $P(-2 \leq Z \leq 2) = .9544$ or about 95%. We illustrate these results graphically in Figure 6.11.

Example 6.17   Assuming a population with an approximately normal distribution, within how many standard deviations of the mean are found 50% of its members? We seek z such that $P(-z \leq Z \leq z) = .5$ or equivalently $P(0 \leq Z \leq z) = .2500$. The nearest table entry is .2486, corresponding to $z = .67$. Thus approximately 50% of the population is within .67 standard deviations of the mean (see Figure 6.12).

FIGURE 6.12 $P(-.67 < Z < .67) = .5000 = 50.00\%$

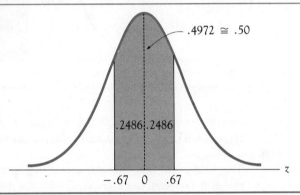

6.7   THE z_α NOTATION

In future applications it proves useful to have a notation for the z value that corresponds to an area α under the extreme right tail of the standard normal distribution curve. We refer to this value by writing the area as a subscript, z_α. Thus since $P(Z > 1.28) = .10$, we can write $z_{.10} = 1.28$. Such numbers are referred to as *percentage points* of the distribution or, in hypothesis testing, as *critical values*. The relationship of z_α to α is shown in Figure 6.13.

DEFINITION 6.3

If α is a given probability, then z_α is the number such that

$$P(Z > z_\alpha) = \alpha$$

FIGURE 6.13   THE AREA α ASSOCIATED WITH z_α

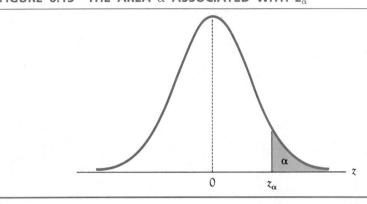

Example 6.18

Find $z_{.05}$. In Figure 6.14 we seen that the area to the right of the desired value of z is .05. Thus the entry in Table III is $.5000 - .05 = .4500$; for this we find $z = 1.64$ to be the nearest entry. Thus

$$z_{.05} = 1.64 \quad \blacksquare$$

FIGURE 6.14   THE AREA ASSOCIATED WITH $z_{.05}$

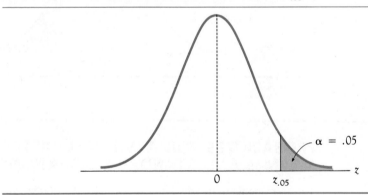

Example 6.19

Find $z_{.20}$ (see Figure 6.15). The area to the right of $z_{.20}$ is .20 and the table entry is $.5000 - .2000 = .3000$. For this $z = .84$ (nearest entry). Thus

$$z_{.20} = .84 \quad \blacksquare$$

Example 6.20

Find the 90th percentile of the variable Z having the standard normal distribution. From Figure 6.16 we see that the 90th percentile is $z_{.10}$. The table entry is the area $.5000 - .1000 = .4000$. for this we find $z_{.10} = 1.28$ (nearest entry). The 90th percentile is thus located 1.28 standard deviations above the mean.

FIGURE 6.15   THE AREA TO THE RIGHT OF $z_{.20} = .84$ IS $\alpha = .20$

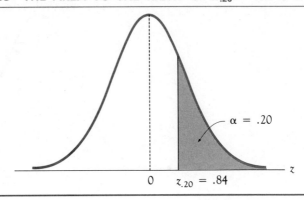

FIGURE 6.16   90% OF THE AREA IS BELOW THE 90TH PERCENTILE AND 10% IS ABOVE

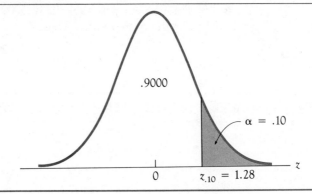

6.8   PROBABILITES FOR A NORMAL DISTRIBUTION WITH MEAN μ AND STANDARD DEVIATION σ

In most applications involving a normally distributed variable X, the mean μ_X and standard deviation σ_X are different from 0 and 1. To find probabilites associated with this distribution, we transform the variable X to the variable Z with which we have been working.

THEOREM 6.4

Let X have a normal distribution with mean μ_X and standard deviation σ_X. Then the variable

$$Z = \frac{X - \mu_X}{\sigma_X}$$

has the standard normal distribution.

The number z produced from X by this transformation is often called a z *score*. A z score, first introduced in Chapter 3, is simply the distance of the corresponding x value from the mean μ_X measured in standard deviation units.

The probability of the normally distributed variable X being between the limits x_1 and x_2 is, of course, given by the area under the normal density curve for X between these limits. But more important, it is also given by the area under the standard normal curve between the corresponding z scores z_1 and z_2. To be more specific:

$$P(x_1 \leq X \leq x_2) = P(z_1 \leq Z \leq z_2)$$

where z_1 and z_2 are computed by

$$z_1 = \frac{x_1 - \mu_X}{\sigma_X} \quad \text{and} \quad z_2 = \frac{x_2 - \mu_X}{\sigma_X}$$

The relationship between the areas involved is shown in Figure 6.17. Thus we need only the standard normal table to deal with all normal distributions.

FIGURE 6.17   AREAS UNDER A NORMAL DENSITY CURVE OBTAINED FROM THOSE UNDER THE STANDARD NORMAL DENSITY

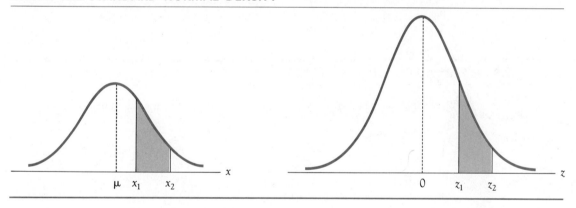

Example 6.21

The effective time X of a certain medication is normally distributed with mean $\mu_X = 144$ minutes and a standard deviation $\sigma_X = 30$ minutes. Approximately what percent of patients will experience an effective time of 2 hours (120 minutes) or less? We seek $P(X \leq 120)$. Using

$$z = \frac{x - 144}{30}$$

the z score corresponding to $x = 120$ is $z = (120 - 144)/30 = -.80$. We readily find $P(Z \leq -.80) = .5 - .2881 = .2119$. Thus $P(X \leq 120) = .2119$. Using the probability as a predictor of the relative frequency, we see that about 21% of the patients will experience an effective time of 2 hours or less. The corresponding areas are shown in Figure 6.18.

FIGURE 6.18 $P(X \leq 120) = P(Z \leq -.80)$

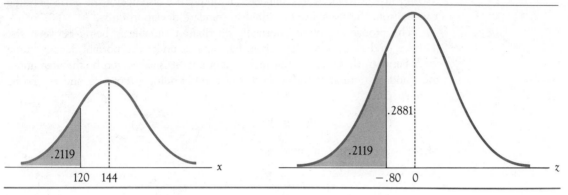

Example 6.22

The diameter X of a steel bearing produced by a company has a normal distribution with a mean of 3.76 centimeters and a standard deviation of .03 centimeters. Acceptable tolerances for the bearings are 3.75 ± .03 centimeters. What is the probability that a randomly selected bearing will be acceptable? We need $P(3.72 \leq X \leq 3.78)$. The z scores are found by using $z = (x - 3.76)/.03$ (see Figure 6.19).

FIGURE 6.19 $P(3.72 \leq X \leq 3.78) = P(-1.33 \leq Z \leq .67)$

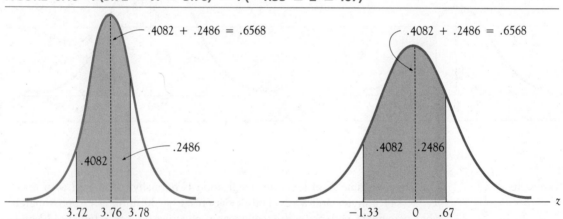

$$z_1 = \frac{3.72 - 3.76}{.03} = -1.33 \text{ (rounded)}$$

and

$$z_2 = \frac{3.78 - 3.76}{.03} = 0.67 \text{ (rounded)}$$

Thus

$$P(3.72 \le X \le 3.78) = P(-1.33 \le Z \le 0.67)$$
$$= .4082 + .2486$$
$$= .6568 \ (65.68\%) \ \blacksquare$$

Example 6.23

The lifetime X of a battery pack designed for use in hand-held calculators has a distribution that is approximately normal with a mean of 640 minutes and a standard deviation of 46 minutes. The manufacturer wishes to guarantee a certain life for the battery packs and wishes no more than 10% to fail to meet the guarantee. For what length of time should the battery packs be guaranteed? Interpreting 10% as the probability .10, we are to find the time x_1 such that $P(X \le x_1) = .10$ (see Figure 6.20). If

$$z_1 = \frac{x_1 - 640}{46}$$

is the corresponding z score, we have $P(Z \le z_1) = .10$. The nearest entry to $.4000 = .5000 - .1000$ in Table III is .3997, from which we obtain $z_1 = -1.28$. We obtain x_1 by solving

$$-1.28 = \frac{x_1 - 640}{46}$$

and find $x_1 = 581.1$. Thus a guarantee of, say, 9 hours (540 minutes) should be more than adequate. $\blacksquare$

FIGURE 6.20 AREAS UNDER NORMAL CURVES FOR EXAMPLE 6.23

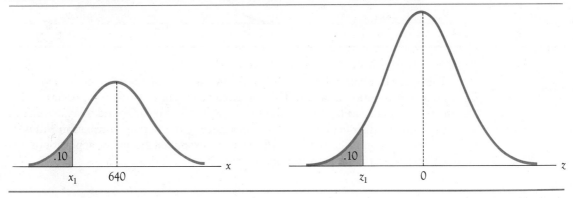

To reiterate a previous statement, a z score is simply the number of standard deviations the corresponding x value lies from its mean. Thus in the last example, once we find $z = -1.28$ we can immediately state that x is 1.28 standard deviations below its mean (of 640) and write

$$x = 640 - 1.28(46) = 581.1.$$

Example 6.24

Suppose the salaries of part-time workers in a metropolitan area have a distribution that is approximately normal with a mean of $4.32 and a standard deviation of $0.50 per hour. Find the 70th percentile (see Figure 6.21). We need the salary x such that

$$P(X > x) = .30 \ (30\%)$$

To this corresponds $z_{.30} = .52$ (nearest entry to $.5000 - .3000 = .2000$). Thus the 70th percentile is located $.52$ standard deviations above the mean at

$$x = 4.32 + (.52)(.50) = 4.58$$

This result can also have been obtained by substituting $z_{.30} = .52$ for z in

$$z = \frac{x - 4.32}{.50}$$

and solving for x. Again we find

$$x = 4.32 + (.52)(.50) = 4.58 \quad \blacksquare$$

FIGURE 6.21   TO THE 70TH PERCENTILE CORRESPONDS $z_{.30}$

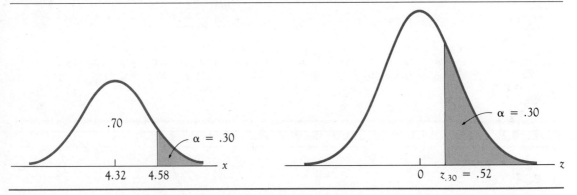

The preceding examples should give some insight into the importance of the normal distribution as a model that approximates many distributions occurring in nature, science, and business. Shortly, we will show how it is used to approximate the binomial distribution. The full importance of the normal distribution will not become apparent, however, until we have examined its many applications to estimation and hypothesis testing.

Example 6.25

A Research Study

The MCAT

It is almost an axiom in testing circles that the distribution of scores on an examination will be approximately normal when there are a large number of more or less uniformly prepared participants. As a specific example of how good the normal approximation actually is, we consider the Medical College Aptitude Test (MCAT).

TABLE 6.3 PERCENTAGES OF NEW MCAT EXAMINEES ACHIEVING SCALED SCORE LEVELS AND ASSOCIATED PERCENTILE RANK RANGES BY AREA OF ASSESSMENT

Scaled score	Biology Percent achieving score	Biology Percentile rank range	Chemistry Percent achieving score	Chemistry Percentile rank range	Physics Percent Achieving Score	Physics Percentile rank range	Science problems Percent achieving score	Science problems Percentile rank range	Skills analysis: reading Percent achieving score	Skills analysis: reading Percentile rank range	Skills analysis: quantitative Percent achieving score	Skills analysis: quantitative Percentile rank range	Scaled score
15	0.0	99.9	0.1	99.9	0.4	99.7–99.9	0.3	99.8–99.9	0.0	99.9	0.0	99.9	15
14	0.2	99.9	0.8	99.2–99.9	1.3	98.4–99.6	1.1	98.7–99.7	0.1	99.9	0.4	99.7–99.9	14
13	2.1	97.9–99.8	3.0	97–99.1	2.6	97–98.3	2.6	97–98.6	2.8	98–99.9	1.3	98.4–99.6	13
12	5.3	94–97.8	4.9	92–96	5.3	91–96	3.6	93–96	6.4	92–97	4.2	95–98.3	12
11	10.6	83–93	7.4	85–91	8.8	83–90	7.2	86–92	15.9	76–91	7.6	87–94	11
10	13.8	69–82	11.8	73–84	10.0	73–82	11.7	74–85	20.7	55–75	10.6	77–86	10
9	12.7	56–68	12.8	60–72	11.1	62–72	12.9	62–73	15.9	39–54	10.9	66–76	9
8	15.8	41–55	12.8	47–59	16.0	46–61	13.5	48–61	10.5	29–38	15.3	51–65	8
7	11.7	29–40	14.9	33–46	12.7	33–45	14.9	33–47	9.3	19–28	14.5	36–50	7
6	10.5	18–28	13.4	19–32	12.0	21–32	13.3	20–32	5.9	13–18	13.5	23–35	6
5	6.8	12–17	9.9	09–18	11.6	09–20	9.8	10–19	4.7	09–12	9.8	13–22	5
4	5.4	06–11	6.0	03–08	6.4	03–08	6.2	04–09	2.8	05.0–08	6.6	06–12	4
3	3.1	02.2–05	1.8	00.5–02	1.6	00.3–02	2.5	00.6–03	2.0	03.0–04.9	3.2	02.2–05	3
2	1.7	00.5–02.1	0.4	00.1–00.4	0.2	00.1–00.2	0.5	00.1–00.5	2.9	00.0–02.9	1.7	00.5–02.1	2
1	0.4	00.0–00.4	0.0	00.0	0.0	00.0	0.0	00.0			0.4	00.0–00.4	1
Scaled score Mean	8.0		7.9		7.9		7.8		7.7		7.5		Scaled score
Std. Deviation	2.58		2.50		2.57		2.51		2.54		2.53		

a Combined April and October 1980 administrations; N = 49,646.

Source: Association of American Medical Colleges.

The MCAT, which is given twice yearly, covers topics from biology, chemistry, and physics and the ability to apply such information to problems in medical science. The test takes 6½ hours and contains 327 questions, producing scores in six areas. These raw scores are then converted to a scale from 1 to 15 and reported. Thus if there were 45 biology questions, a 43, 44, or 45 would be reported as a 15, a 40, 41, or 42 as a 14, and so forth.

Table 6.3 contains the distributions of scaled scores for the 49,646 participants in 1980. For the moment, we concentrate primarily on the chemistry scores. In Figure 6.22, both the (density) histogram for their distribution and the normal density curve for a variable with a mean of 7.9 and a standard deviation of 2.5 are drawn. The results are typical of what is regarded as a good approximation by the normal distribution.

FIGURE 6.22   HISTOGRAM OF CHEMISTRY SCORES FITTED WITH A NORMAL CURVE HAVING $\mu = 7.9$ and $\sigma = 2.5$

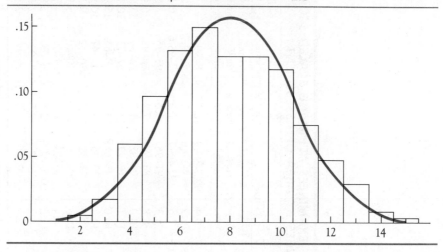

To further investigate the quality of the approximation, let us compare a particular prediction by the normal distribution with what was actually observed. For example, we see that 8.8% of the scores were a 12 or better. Using

$$z = \frac{x - 7.9}{2.5}$$

with $x = 12$ we find $z = 1.64$ and $P(X \geq 12) = P(Z \geq 1.64) = .5000 - .4495 = 5.05\%$. While in the right ballpark, this does not agree well with the observed 8.8%, but this is typical.

Actually, in this case we can obtain an improved approximation. The scaled scores are discrete and it is actually the area under the histogram between 11.5 and 12.5 that corresponds to a score of 12. Thus a better approximation should be furnished by the area to the right of $x = 11.5$ (see Figure 6.23). We then find

$$z = \frac{11.5 - 7.9}{2.5} = 1.44$$

and $P(X \geq 11.5) = P(Z \geq 1.44) = .5000 - .4251 = .0749 = 7.49\%$. Certainly this is an improvement.

FIGURE 6.23   APPROXIMATING $P(X \geq 12)$ BY THE AREA UNDER THE NORMAL CURVE

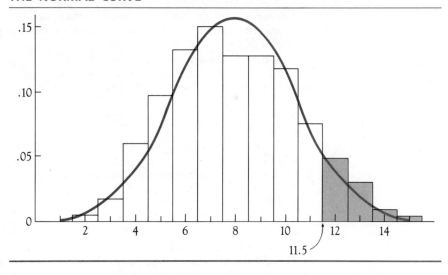

By design, the components of the MCAT should all have a mean of around 8 and a standard deviation of 2.5. Knowing this, schools and individuals are able to make predictions about admission cutoff scores before the test results are available. In a sense, one has a theoretical distribution, of which each year's results are a sample.

For example, suppose a school plans to consider only those applicants who score in the top 15%. Since $P(Z \geq z) = .1500$ yields $z = 1.04$, the cutoff is around 1.04 standard deviations above the mean and this is

$$x = 8 + (1.04)(2.5) = 10.6$$

A scaled score of $x = 11$ would probably be used. We see from Table 6.3 that 18.2% of the biology, 16.2% of the chemistry, and 18.4% of the physics scores were actually at this level.

The results illustrated in this study are typical, if not better, than most that are obtained when the normal approximation is used. Certainly when it is used carefully, the normal approximation is a valuable tool. When used to approximate naturally occurring distributions, one must recognize that 2 to 5% errors in predictions are not uncommon. ■

EXERCISE SET 6.2

Part A

In each of the following problems you should make a rough sketch or sketches of the normal curves that are used and mark off the boundaries of the intervals involved.

1. Find the area under the standard normal density curve.
   a. Between $z = 0$ and $z = 1.24$.
   b. Between $z = 1.00$ and $z = 2.00$.
   c. Within $z = \pm 1.40$.
   d. To the right of $z = 2.34$.
   e. To the left of $z = 1.50$.
   f. To the left of $z = -1.68$.
2. Suppose the distribution of X may be approximated by the normal distribution. Interpreting probabilities as relative frequencies, find the percentage of the population that is.
   a. Within ½ standard deviation of the mean.
   b. More than 2.5 standard deviations above the mean.
3. Let Z have the standard normal distribution. Find:
   a. $P(0 \leq Z \leq 1.43)$.
   b. $P(Z \geq 2.2)$.
   c. $P(.7 \leq Z \leq 1.4)$.
   d. $P(-1.6 < Z < 1.6)$.
   e. $P(Z \leq -2.4)$.
4. Let Z have the standard normal distribution. Find:
   a. $P(1.5 \leq Z \leq 2.5)$.
   b. $P(Z = 1.6)$.
   c. z_1 if $P(0 < Z < z_1) = .3508$.
   d. z_1 if $P(-z_1 \leq Z \leq z_1) = .9500$.
   e. z_1 if $P(Z \geq z_1) = .05$.
5. Find the following percentage points. In each case, sketch the normal distribution and shade the area α corresponding to z_α.
   a. $z_{.01}$.
   b. $z_{.34}$.
   c. $z_{.05}$.
   d. $z_{.10}$.
   e. $z_{.50}$.
6. Suppose the grades on an examination are distributed approximately normally with a mean of 800 and a standard deviation of 200. Using percentage points of Z, find:
   a. The 90th percentile.
   b. The 99th percentile.
   c. The upper quartile.
7. Let X have a normal distribution with a mean of 123 and a standard deviation of 12. Find:
   a. $P(105 \leq X \leq 141)$.
   b. $P(X > 136)$.
   c. $P(X \leq 148)$.
   d. k if $P(123 - k \leq X \leq 123 + k) = .8230$.
   e. x_1 if $P(X > x_1) = .0250$.
8. At a reforestation project, the trunk diameters of a certain species of tree at five years of age are normally distributed with a mean of 6 inches and a standard deviation of ½ inch when measured at a point 18 inches above ground level. What percent of such trees should have trunk diameters between 5¼ and 6¾ inches? Assume a normal distribution.
9. The daily production of a cracking station at a certain refinery is approximately normally distributed, with a mean of 10,000 gallons and a standard deviation of 1200 gallons. What is the probability that on a randomly selected day the production will be below 8000 gallons?

10. Gasoline mileages for a fleet of similarly equipped rental autos are said to be normally distributed, with a mean of 17.2 and a standard deviation of 2.1 miles per gallon. You randomly select one of these autos, fill up its 20-gallon gas tank, and start off of a 300-mile trip. What is the probability that you will run out out of gas before reaching your destination?

11. The lives of a certain type of dry cell battery used in solid-state portable radios are normally distributed with a mean of 110 hours and a standard deviation of 14 hours. If the batteries are guaranteed for 80 hours, what fraction of the production will have to be replaced under the guarantee?

12. A student stated that he scored at the 95th percentile on a test where the mean was 70 and the standard deviation was 10. If the scores are normally distributed, what is his score?

13. The length of time X needed to complete a state teachers examination is normally distributed with $\mu_X = 86$ and $\sigma_x = 14$ minutes. How much time should be allotted to be certain that 90% of the examinees complete the test?

14. In a study of the dimensions of respirable fiber glass particles as found in certain manufacturing processes, the fiber diameters were established as having a mean of 1.2 and a standard deviation of 0.6 micrometers (μm). What percent of all such fibers would be filtered out by a filter having a porosity of 0.4 μm? Assume a normal distribution.

Part B

15. Two machines produce threaded rods of the same length. The first has a mean diameter of 0.25 inches, the second a mean of 0.28 inches. Both have a standard deviation of 0.008 inches. Rods having diameters between 0.235 and 0.265 and between 0.274 and 0.286 inches are acceptable. Assuming that the number of rods produced by both machines is the same and assuming their diameters are both approximately normal, estimate the percentage of the production that is not accepted.

16. An allergy test for a certain pollen requires that two drops of a concentrated solution be placed under the patient's tongue. Individuals sensitive to the pollen have a reaction time that is approximately normally distributed, with a mean of 14 seconds and a standard deviation of 4 seconds. How often should one observe a reaction time of less than 5 seconds with such individuals?

17. A magazine reported that 62% of its subscribers are of age 35 or greater. Assuming that the standard deviation is about 10 years, what is the mean age of the subscribers? Assume the ages have a distribution that is approximately normal.

18. The scores X on a military three-month physical fitness review are normally distributed, with a mean of 71 and a standard deviation of 8. Any individual whose score exceeds the 80th percentile does not have to undergo the next review. To what score does this correspond?

19. Thermocouples used in certain gas furnaces have service lives that are approximately normal with a mean of 35 and a standard deviation of four months. How often should these be replaced if not more than 3% should fail while in service?

20. Family incomes in a certain region are normally distributed, with a mean and standard deviation of $20,000 and $2200, respectively. If two families are selected at random, what is the probability that both will have incomes below $18,000?

21. The gestation period X in humans has a distribution that is approximately normal, with a mean of 266 days and a standard deviation of 16 days. What percentage of all births are at least 300 days after conception?

22. If X is normally distributed, then for any two constants a and b ($a \neq 0$), the

variable $Y = aX + b$ is also normally distributed. Suppose X is normally distributed with $\mu_X = 70$ and $\sigma_X = 20$ and let $Y = 5X - 50$. Find $P(300 < Y < 470)$.

23. Suppose that raw scores on a certain national test have a normal distribution, with a mean of 125 and a standard deviation of 20. These are to be transformed to a new Y scale using a formula of the form $Y = aX + b$. Find a and b if the variable Y is to have a mean of 500 and a standard deviation of 100. Hint: If the variables were both standardized, then corresponding values of x and y would produce the same z.

24. A teacher announces that she gives A's to those whose class totals are more than 1.6 standard deviations above the mean, B's to those between .6 and 1.6 standard deviations (inclusively) above the mean, and C's to those within .6 standard deviations of the mean, while failing the rest. If the class totals can be approximated by a normal distribution, what percentage of A's, B's, and C's does she give?

25. The specification limits for a fine wire heating element are 2200 ± 150 ohms. If the resistances are normally distributed with a mean and standard deviation of 2200 and 80 ohms, approximately what percent of the heating elements do not meet specifications?

26. Research evidence supports the hypothesis that bees fly minimum distances between successively visited flowers (i.e., they do not tend to fly to the nearest flower). In one study the minimum distances were found to have a mean of 1.32 and a standard deviation of 0.72 meters. Assuming a normal distribution for such distances, what percent of all flights between successively visited flowers are at least ⅓ meters? (Source: "Comparison of Distances Flown by Different Visitors to Flowers of the Same Species," N. M. Waser, *Oecologia*, vol. 55, 1982).

27. The application rate for a medfly control pesticide is 15.4 pounds per acre and the safe reentry level for workers into a sprayed field is 6.0 pounds per acre. Ten days after application, the residual concentration of the pesticide is approximately normal, with a mean of 3.2 and a standard deviation of 1.4 pounds per acre. What is the probability that it is safe to reenter a field at this time?

6.9   APPROXIMATING THE BINOMIAL DISTRIBUTION

As previously mentioned, if the parameters n, p, and q are such that the binomial distribution is not overly skewed, then it may be approximated by the normal distribution.

To see how this approximation is made, consider the problem of finding $P(X = 14)$ when $n = 20$, $p = .6$, and $q = .4$. The mean and standard deviation of the binomal distribution are $\mu_X = 20(.6) = 12$ and $\sigma_X = \sqrt{20(.6)(.4)} = 2.19$ (rounded). We start with a normal distribution having this same mean and standard deviation (see Figure 6.24).

Now imagine the histogram for the binomial distribution constructed on this same set of axes. The probability of $X = 14$ is represented by the area of that rectangle having boundaries of $x_1 = 13.5$ and $x_2 = 14.5$. Thus we want the area under the normal curve between these limits.

Next we transform $x_1 = 13.5$ and $x_2 = 14.5$ to z scores using

$$z = \frac{x - 12}{2.19}$$

FIGURE 6.24 APPROXIMATING $P(X = 14)$ USING THE NORMAL DISTRIBUTION

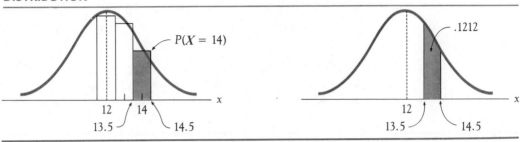

and obtain $z_1 = .68$ and $z = 1.14$. From Table III in the Appendix we then find

$$P(.68 \leq Z \leq 1.14) = .3729 - .2517 = .1212$$

Thus for the binomial distribution $P(X = 14) = .1212$. This compares very favorably with the table value of

$$P(X = 14) = P(X \leq 14) - P(X \leq 13) = .8744 - .7500 = .1244.$$

THEOREM 6.5

For large values of both np and nq, the binomial distribution is approximately normal.

A good rule of thumb is that both np and nq should exceed 5. Additionally, one should be very cautious when p is near 0 or 1, as there will then be a high degree of skewness to contend with.

To summarize the process, the first step is to adjust the values of X by adding or subtracting $\frac{1}{2}$ to obtain the adjusted limits. Next we transform these limits to z scores, using

$$z = \frac{x - \mu_X}{\sigma_X} = \frac{x - np}{\sqrt{npq}}$$

Finally, the desired probability is found using the standard normal distribution.

Example 6.26

A coin is tossed 300 times. What is the probability that the number of heads is between 140 and 160 (inclusively)? A direct computation would require evaluating:

$$P(140 \leq X \leq 160) = P(X = 140) + P(X = 141) + \ldots + P(X = 160)$$

with $P(X = x) = \binom{300}{x}(\frac{1}{2})^{300}$ (since $p = q = \frac{1}{2}$ and $n = 300$). We use instead the area of the histogram bounded by $x_1 = 139.5$ and $x_2 = 160.5$. The z scores are computed from these using

$$z = \frac{x - np}{\sqrt{npq}} = \frac{x - 300\left(\frac{1}{2}\right)}{\sqrt{300\left(\frac{1}{2}\right)\left(\frac{1}{2}\right)}} = \frac{x - 150}{\sqrt{75}} = \frac{x - 150}{8.660}$$

Thus $z_1 = (139.5 - 150)/8.660 = -1.21$ and $z_2 = (160.5 - 150)/8.660 = 1.21$. Then $P(-1.21 \le Z \le 1.21) = .3869 + .3869 = .7738$. The approximation is $P(140 \le X \le 160) = .7738$. This area is shown in Figure 6.25.

FIGURE 6.25 $P(140 \le X \le 160|X \text{ BINOMIAL}) \approx$
$P(139.5 \le X \le 160.5|X \text{ NORMAL}) = P(-1.21 \le Z \le 1.21) = .7738$

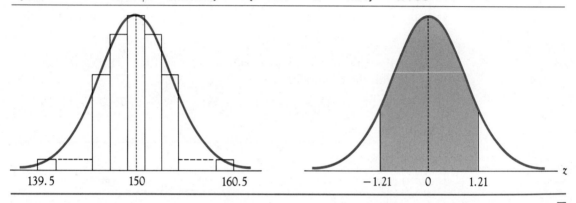

In this last example, both np and nq are quite large (150) and the approximation should be quite good.

Example 6.27

Ten thousand sterile fruit flies are released in an agricultural area to combat propagation of the species through interruption of the breeding cycle (the flies mate only once). It is estimated that the area initially contained 90,000 fruit flies. To ascertain the uniformity of the mixing of the flies, a sample of 500 is collected a short time later and examined. Under ideal conditions, $\frac{1}{10}$ of this sample would be sterile flies. What is the probability that the sample will have between 40 and 60 (inclusively) sterile flies? A trial is the random selection of a fly. There are $n = 500$ trials and the probability of success on each is approximately $\frac{1}{10}$ if there is a uniform mixing. (We say approximate, since the probabilities change very slightly at each trial since there would be no replacement). We seek $P(40 \le X \le 60)$. If $x_1 = 39.5$ and $x_2 = 60.5$,

$$z_1 = \frac{39.5 - 500\left(\frac{1}{10}\right)}{\sqrt{500\left(\frac{1}{10}\right)\left(\frac{9}{10}\right)}} = \frac{39.5 - 50}{\sqrt{45}} = -1.57$$

$$z_2 = \frac{60.5 - 50}{\sqrt{45}} = 1.57$$

Thus $P(40 \leq X \leq 60) = P(-1.57 \leq Z \leq 1.57) = .8836$. Approximately 88% of all samples of 500 should show between 40 and 60 of the sterile flies if a homogeneous mixing has occurred. ∎

Often we wish to work with the fraction or proportion X/n of successes in an experiment. For example, if 160 business firms are surveyed as to their projection of the business climate for the coming year and 100 respond positively, then we would ordinarily report the proportion of positive responses to be $100/160 = .625$ or the equivalent 62.5%. For many people, proportions and percents are easier to interpret than the raw numbers from which they are derived.

If X represents the number of successes in a binomial experiment, then we may regard the proportion X/n as a new variable which we denote by $\hat{P}$. Since

$$Z = \frac{X - np}{\sqrt{npq}} = \frac{X/n - p}{\sqrt{npq/n}} = \frac{\hat{P} - p}{\sqrt{\dfrac{pq}{n}}}$$

it follows that the distribution of $\hat{P}$ may be approximated by the normal distribution as long as np and nq are sufficiently large.

Example 6.28

Suppose that alcohol is a contributing factor in 65% of all auto fatalities. If 500 accidents involving a fatality are investigated, what is the probability that between 60% and 70% of the sample will show alcohol as a factor? We seek $P(.6 \leq \hat{P} \leq .7)$ given that $n = 500$ and $p = .65$. If $\hat{p}_1 = .6$ and $\hat{p}_2 = .7$, the corresponding Z scores are found by

$$z_1 = \frac{\hat{p}_1 - .65}{\sqrt{\dfrac{(.65)(.35)}{500}}} = \frac{.60 - .65}{\sqrt{\dfrac{(.65)(.35)}{500}}} = -2.34$$

$$z_2 = \frac{\hat{p}_2 - .65}{\sqrt{\dfrac{(.65)(.35)}{500}}} = \frac{.70 - .65}{\sqrt{\dfrac{(.65)(.35)}{500}}} = 2.34$$

Thus $P(.60 \leq \hat{P} \leq .70) = P(-2.34 \leq Z \leq 2.34) = .9808$. About 98% of all random samples of 500 should show this behavior. ∎

Example 6.29

It is reported that the probability of a parolee from a state prison being returned to prison within two years is $\frac{2}{3}$. A random sample of 1600 parolee records are examined. What is the probability that 70% or more of the parolees will be found to have returned to prison within two years after their release? Here $n = 1600$ and $p = \frac{2}{3}$, and we seek $P(\hat{P} \geq .7)$ (see Figure 6.26). We find

$$z_2 = \frac{.7 - \dfrac{2}{3}}{\sqrt{\dfrac{\left(\dfrac{2}{3}\right)\left(\dfrac{1}{3}\right)}{1600}}} = 2.83 \text{ (rounded)}$$

and thus

$$P(\hat{P} \geq .7) \approx P(Z \geq 2.83) = .5000 - .4977 = .0023$$

This probability is quite small and simply states that we should expect to see better agreement between the proportion in the sample (.7) and in the population (.667) when working with such a *large* sample.

FIGURE 6.26 $P(\hat{P} \geq .70) = P(Z \geq 2.83)$

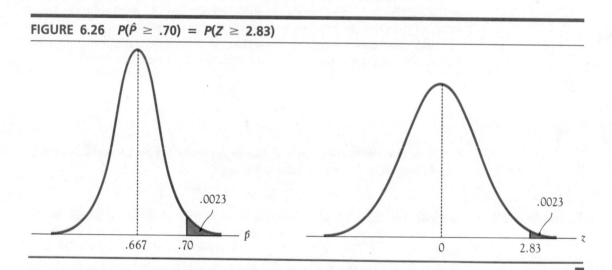

The variable $\hat{P}$ will play an important role in testing hypotheses about population proportions. It will be reintroduced in the next chapter as a proportion of a sample of size n possessing a given property and will be called the *sample proportion variable*. At that time we will find the following result useful.

THEOREM 6.6

Let X have the binomial distribution $P(X = x) = \binom{n}{x}p^x q^{n-x}$ and let the variable $\hat{P}$ be defined by $\hat{P} = X/n$. Then the mean and standard deviation of $\hat{P}$ are given by

$$\mu_{\hat{P}} = p \qquad \text{and} \qquad \sigma_{\hat{P}} = \sqrt{\frac{pq}{n}}$$

The first of these results follows readily from Theorem 5.1 by observing

$$\mu_P = E[\hat{P}] = E[X/n] = (1/n)E[X] = (1/n)np = p$$

The result for the standard deviation can also be obtained from Theorem 5.1, but not so simply. We omit the details.

EXERCISE SET 6.3

Part A

1. Find the mean and standard deviation and sketch the graph of the normal distribution used in approximating the binomial distribution for parameters:
   a. $n = 800$ and $p = .65$.
   b. $n = 200$ and $p = \frac{1}{2}$.
   c. $n = 400$ and $p = \frac{1}{4}$.

2. Find the normal distribution (mean, standard deviation, and graph) used to approximate the distribution of $\hat{P}$ when the binomial distribution has parameters:
   a. $n = 225$ and $p = .4$.
   b. $n = 150$ and $p = .7$.

3. Let X be binomially distributed with $n = 500$ and $p. = .4$. Find:
   a. $P(190 \leq X \leq 210)$.
   b. $P(X \leq 195)$.
   c. $P(X > 215)$.

4. Let X be binomially distributed with $n = 20$ and $p = .3$. Find $P(X \geq 8)$ using
   a. The cumulative binomial tables.
   b. The normal approximation with the end point correction for continuity.
   c. The normal approximation without the end point correction.

5. Let X be binomially distributed with $n = 200$ and $p = .35$.
   a. What is meant by the statement $\hat{P} \geq .38$?
   b. Find $P(\hat{P} \geq .38)$.

6. A recent survey on a college campus revealed that 40% of the students live at home and commute to the campus. If a random sample of 320 students is questioned, what is the probability of finding at least 130 students who live at home?

7. An insurance company estimates that 70% of the apartment dwellers in a certain area have no insurance on their personal possessions. A random sample of 400 apartments reveals 260 with no insurance. What is the probability of such poor agreement? (Here poor agreement represents 260 or less.) Comment on the insurance company's estimate.

8. Suppose that 65% of the undocumented workers in this country receive a wage below the minimum scale. What is the probability that 70% or more of a random sample of 120 such workers do not receive the minimum wage?

9. A political candidate states that 60% of the voters support her. If this is the case and 500 prospective voters are questioned, what is the probability that at least 58% of the sample will support her?

10. Assuming that male and female births are equally likely, what is the probability that 50.4% of a sample of 20,000 birth records will show male births?

Part B

11. An airline has determined that 12% of its reservations are no-shows. On a flight where the plane has 270 seats, 300 reservations are sold. What is the probability that all passengers who show will have a seat?

12. A drug company is considering the marketing of a new medication that its research division believes to be 88% effective. Before its introduction, a large-scale test is to be made of this claim, using a sample of 400 people. If the medication is effective on 82% of the sample, the claim will be considered substantiated and the product marketed. If in fact the medication is 88% effective, what is the probability that the medication will fail this test?

13. A consumers group claims that 20% of a company's soft drink bottles are not filled to the advertised 2 liters. If this is the case, what is the probability that a random sample of 100 bottles will show less than 15% to be underfilled?

14. Consider a large population where the proportion of "successes" is .5. How large a sample is needed in order to be 90% certain that $\hat{P}$ is within 2% of $p = .5$ (50%), that is between $p = .5 \pm .02$?

15. A test contains 120 true-false questions. For each correct answer you receive three points; for each incorrect or omitted answer a point is deducted. A passing score is 160. What is the probability of passing this test if you guess at every answer?

16. Assuming a binomial distribution with parameters n and p, justify each step of the following derivation. What is being established?

a. $\quad \sigma_{\hat{P}}^2 = E[(\hat{P} - \mu_{\hat{P}})^2]$

b. $\quad = E\left[\left(\dfrac{X}{n} - p\right)^2\right]$

c. $\quad = E\left[\left(\dfrac{X - np}{n}\right)^2\right]$

d. $\quad = E\left[\dfrac{1}{n^2}(X - np)^2\right]$

e. $\quad = \dfrac{1}{n^2} E[(X - np)^2]$

f. $\quad = \dfrac{1}{n^2} E[(X - \mu_{\hat{X}})^2]$

g. $\quad = \dfrac{1}{n^2} \sigma_X^2$

l. $\quad = \dfrac{1}{n^2} npq$

i. $\quad = \dfrac{pq}{n}$

j. $\quad \therefore \sigma_{\hat{P}} = \sqrt{\dfrac{pq}{n}}$

6.10   THE CHI-SQUARE DISTRIBUTION

Sums of squares have many applications in statistics and consequently their distributions are of interest. One of the most important of these sums occurs when the variables included are independent and have the standard normal distribution. This sum is a variable which is commonly denoted by the square of the Greek letter chi, χ^2, read "chi-square." The distribution of this variable is accordingly called the chi-square distribution.

The distribution of $\chi^2 = Z_1^2 + Z_2^2 + \ldots + Z_n^2$ depends on the number of independent variables n in the sum. For each n there is in fact a different distribution. To identify the distribution of interest, we use a parameter, ν (the Greek letter nu), whose value is n. For historical reasons ν is called the number of *degrees of freedom*. Thus $\nu = 2$ for $\chi^2 = Z_1^2 + Z_2^2$ while $\chi^2 = Z_1^2 + Z_2^2 + Z_3^2$ has $\nu = 3$ degrees of freedom.

> **DEFINITION 6.4**
>
> Let $Z_1, Z_2, \ldots, z_n$ be independent random variables, each normally distributed with a mean of 0 and a standard deviation of 1. Then the sum of their squares has the chi-square distribution with $\nu = n$ *degrees of freedom*. Accordingly we set
>
> $$\chi^2 = Z_1^2 + Z_2^2 + \ldots + Z_n^2$$

The density curves for the chi-square distribution is sketched in Figure 6.27 for several degrees of freedom. Observe that χ^2 is nonnegative and that the mean and standard deviation depend on the number of degrees of freedom. While we shall not need to know them, it can be shown that the mean is in fact the number of degrees of freedom, that is $\mu = \nu$, and that the standard deviation is $\sigma = \sqrt{2\nu}$. When the number of degrees of freedom is small, there is consequently a definite skewness to the right. This, however, diminishes rapidly with increasing ν; for $\nu > 30$, the distribution is approximately normal.

FIGURE 6.27   CHI-SQUARE DENSITY CURVES

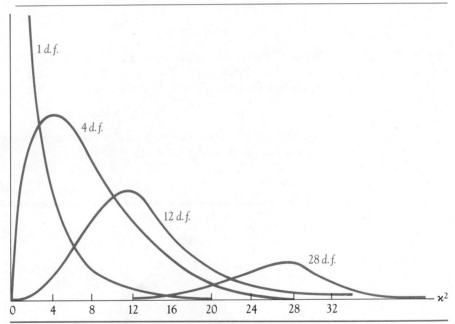

Since the chi-square distribution is continuous, probabilites are found as areas under its density curve. In Figure 6.28, the shaded area represents the probability $P(0 \le \chi^2 < 16.9) = .95$ for the case $\nu = 9$. In this respect, a distribution table could be presented, as with the normal distribution. Each value of ν would,

FIGURE 6.28 $P(0 < \chi^2 < 16.9) = .95$

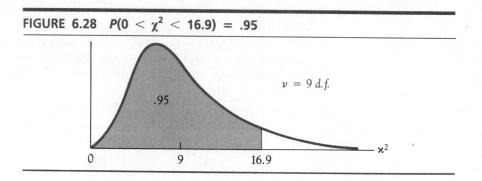

however, require a different table. To avoid this, some condensing and rearranging is necessary.

The chi-square distribution is given in Table IV of the Appendix. The entries of the first row (the column headings) are the degrees of freedom. The row headings are selected probabilities and the entries in the body of the table are the corresponding values if χ^2. Notice how this differs from the distribution table for Z. There the tabular entries were the probabilities. Also there is no attempt to give the probabilities for the full range of values of χ^2.

Example 6.30

For $\nu = 3$, find the probability of a chi-square value less than 4.64. The desired area is shaded in Figure 6.29. Entering table IV along the column headed $\nu = 3$, we find the entry $\chi^2 = 4.64$ in the row headed .80. Thus

$$P(0 \le \chi^2 \le 4.64) = .80$$

See Figure 6.29. Since the values of χ^2 are nonnegative there is no ambiguity in writing this simply as $P(\chi^2 \le 4.64) = .80$ ■

FIGURE 6.29 $P(\chi^2 \le 4.64) = .80$

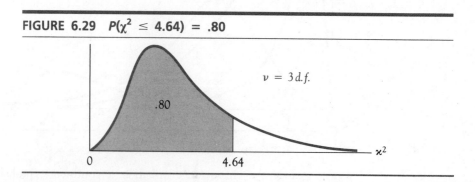

Rarely will one be able to find the chi-square value as an exact tabular entry. For our purposes it will be adequate to round to the nearest entry.

Example 6.31

For $v = 15$, find $P(\chi^2 \geq 22)$. The nearest table entry in the column headed $v = 15$ is $\chi^2 = 22.3$, which is in the row headed .90. Thus $P(0 \leq \chi^2 \leq 22) = .90$, approximately. Now refer to Figure 6.30. Since the total area under the curve is 1,

$$P(\chi^2 \geq 22) = 1 - .90 = .10 \text{ (approximately)} \quad \blacksquare$$

FIGURE 6.30 $P(\chi^2 \geq 22.3) = .10$

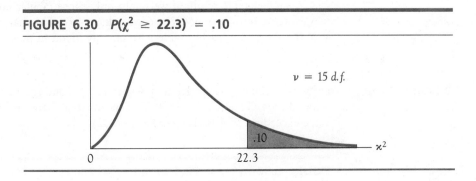

Example 6.32

For $v = 8$, find the chi-square value c such that

$$P(\chi^2 \geq c) = .05$$

In this case, the probability is given. The corresponding area is shaded in Figure 6.31. Obviously the area to the left of $\chi^2 = c$ is .95. Entering Table IV in the row headed .95, we find in the column $v = 8$ the entry 15.5. Thus $c = 15.5$. This value is so large that there is only a 5% chance that it will be exceeded. $\quad \blacksquare$

FIGURE 6.31 $P(\chi^2 \geq 15.5) = .05$

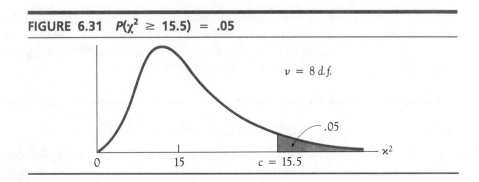

In most applications of the chi-square distribution, we are interested in what constitutes *large* values of the variable. These are understood to be those that lie under the extreme right tail of the density curve and collectively have a small probability of occurrence, say .1, .05, .01, and so forth. As with the standard normal distribution, the lower boundaries of these intervals are denoted by $\chi^2_{.1}$, $\chi^2_{.05}$, $\chi^2_{.01}$. The general case is given in Definition 6.5.

> **DEFINITION 6.5**
>
> χ_α^2 is that number such that $P(\chi^2 \geq \chi_\alpha^2) = \alpha$.

Thus the total area under the density curve to the right of χ_α^2 is α. Returning to the last example, we see that $\chi_{.05}^2 = 15.5$, since $P(\chi^2 \geq 15.5) = .05$.

A simpler table results if we retain only those tabular entries corresponding to χ_α^2 for the most commonly needed values of α. This has been done in Table V of the Appendix; such a table is commonly found in most introductory texts.

Example 6.33

For $\nu = 20$, find $\chi_{.10}^2$. In the column headed $\chi_{.10}^2$ and the row $\nu = 20$ of table V is found 28.4. Thus $\chi_{.10}^2 = 28.4$. This means that the probability of a chi-square value exceeding 28.4 is .10 (see Figure 6.32). ■

FIGURE 6.32 $\chi_{.10}^2 = 28.4$ when $\nu = 20$

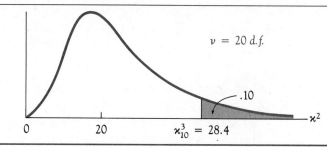

Table V also contains extreme values under the left tail of the chi-square density curve. These serve as boundaries for "small" values of the variable. Loosely speaking, only 5% of the values of chi-square distribution are less than $\chi_{.95}^2$. These are so "small" as to occur only rarely.

6.11   THE GOODNESS-OF-FIT TEST

Long before hypothesis testing was perfected to its present state, the brilliant statistician Karl Pearson (1857–1936) developed a test for comparing a theoretical distribution with that of a population. This test and its many variations employ the chi-square distribution and are referred to as both *goodness-of-fit* and *chi-square* tests.

The problem is to test whether a variable X has some particular distribution in a population. This is done by grouping its values into a certain number of classes, say m, selecting a sample from the population, and comparing the observed frequencies $o_1, o_2, \ldots, o_m$ for these classes with the expected frequencies $e_1, e_2, \ldots, e_m$ predicted by the distribution.

The quantities $(o_1 - e_1)$, $(o_2 - e_2)$, . . ., $(o_m - e_m)$ represent the differences between the observed and predicted frequencies. Some will be positive, others negative; their sum will be 0. Thus they are not independent. As soon as $m - 1$ of them are known, the remaining one is determined. Pearson avoided this cancelling of positive and negative differences by using their squares and succeeded in keeping their size in check by dividing each by the corresponding expected frequency. Thus he considered

$$\frac{(o_1 - e_1)^2}{e_1} + \frac{(o_2 - e_2)^2}{e_2} + \ldots + \frac{(o_m - e_m)^2}{e_m} = \Sigma \frac{(o - e)^2}{e}$$

Each sample yields a new value for the above sum. Thus the sum is a random variable; a little algebraic manipulation shows it to be the sum of the squares of $m - 1$ independent variables. If these variables were each normally distributed, then the chi-square distribution with $v = m - 1$ degrees of freedom would apply to the sum. The interesting point is that when the sample sizes are such that all the expected frequencies are at least 5, this is indeed the case. Accordingly, although it is somewhat inaccurate, it is common to denote this sum by χ^2. To summarize:

THEOREM 6.7

If the expected frequencies e_1, e_2, . . ., e_m are at least of size 5, then the sum

$$\chi^2 = \sum_{i=1}^{m} \frac{(o_i - e_i)^2}{e_i}$$

has approximately the chi-square distribution with $v = m - 1$ degrees of freedom.

Example 6.34

It is asserted that politically, 18% of college students are liberals, 60% are in the middle, and 22% are conservatives. This is a theoretical distribution for a variable X with three values, say, 1, 2, and 3, each of which determines a class. Now suppose we want to test whether this distribution describes or *fits* the students at some large university. A sample of, say, 200 students is selected. The expected frequencies are 18%, 60%, and 22% of 200. Thus

$$e_1 = .18 \times 200 = 36 \qquad e_2 = .60 \times 200 = 120 \qquad e_3 = .22 \times 200 = 44$$

Suppose the observed frequencies are $o_1 = 44$, $o_2 = 128$, and $o_3 = 28$. These frequencies are summarized in Table 6.4.

TABLE 6.4   OBSERVED AND EXPECTED FREQUENCIES FOR THREE POLITICAL AFFILIATIONS

Frequency	1(Liberal)	2(Middle)	3(Conservative)
o_i	44	128	28
e_i	36	120	44

We then compute

$$\chi^2 = \frac{(o_1 - e_1)^2}{e_1} + \frac{(o_2 - e_2)^2}{e_2} + \frac{(o_3 - e_3)^2}{e_3}$$
$$= \frac{(44 - 36)^2}{36} + \frac{(128 - 120)^2}{120} + \frac{(28 - 44)^2}{44}$$
$$= 1.78 + .53 + 5.82 = 8.13$$

For $v = 2$, the nearest entry less than this in Table V is 7.38; this corresponds to a probability of .975. The probability of a chi-square value exceeding $\chi^2 = 7.38$ is .025. Thus $\chi^2 = 8.13$ is indeed large, indicating a poor agreement between the proposed distribution and that of the student population at this school (see Figure 6.33).

FIGURE 6.33   THE χ^2 DISTRIBUTION FOR $v = 2$ DEGREES OF FREEDOM

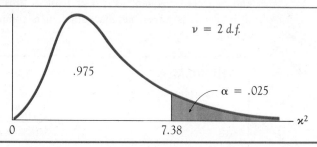

In practice the chi-square goodness of fit test is carried out by comparing the computed χ^2 value with χ^2_α for some small α, say .05 or .01. The number α is called the *level of significance* of the test and χ^2_α is called its *critical value*. The critical value lies under the extreme right tail of the chi-square density curve and is a lower limit for what is considered large values of the variable. Thus if the critical value is exceeded by the computed χ^2 value, a large disagreement is indicated, which is grounds for rejecting the proposed distribution for the population in question.

Returning to the last example, suppose we had selected a significance level of 5%. The critical value would then have been $\chi^2_{.05} = 5.99$. Since $\chi^2 = 8.13 > 5.99$, the results are significant at a level of $\alpha = .05$ and we reject the distribution as being suitable for the university from which the sample was taken.

The steps in performing a goodness-of-fit test are summarized in the following procedure.

PROCEDURE FOR GOODNESS-OF-FIT TEST

Given: A proposed distribution for a variable X in a population and a sample from the population.

1. Partition the range of X into m disjoint class intervals in such a way that their expected frequencies $e_1, e_2, . . ., e_m$ are at least 5.

2. Tally the observations of the sample to obtain the observed class frequencies $o_1, o_2, \ldots, o_m$.
3. Select a significance level α such as $\alpha = .05$ or $\alpha = .01$ and find the critical value χ_α^2, using $\nu = m - 1$ degrees of freedom.
4. Compute

$$\chi^2 = \sum_{i=1}^{m} \frac{(o_i - e_i)^2}{e_i}$$

5. Compare χ^2 and χ_α^2. If $\chi^2 > \chi_\alpha^2$, then reject the proposed distribution.

Example 6.35

We wish to test whether a certain population is normally distributed, with a mean of 160 and a standard deviation of 10. As classes we use intervals with end points of 1 and 2 standard deviations on both sides of the mean, as shown in Figure 6.34.

FIGURE 6.34   A PARTITION OF THE RANGE OF X

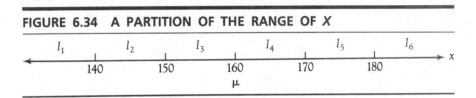

Next we obtain a sample of 300 observations of X, classify them according to these intervals and obtain their frequencies $o_1, o_2, \ldots, o_6$ (see Table 6.5). Now, using the normal distribution, we compute the probabilities and expected frequencies for these intervals. For the first two intervals I_1 and I_2,

$$P(X \leq 140) = P(Z \leq -2) = 0.0228 \qquad e_1 = (0.0228)(300) = 6.84$$
$$P(140 < X \leq 150) = P(-2 < Z \leq 1) = 0.1359 \qquad e_2 = (0.1359)(300) = 40.77$$

The expected frequencies for the remaining intervals are computed similarly and are given in the second row of Table 6.5.

TABLE 6.5   OBSERVED AND EXPECTED FREQUENCIES

Frequency	I_1 $X \leq 140$	I_2 $140 < X \leq 150$	I_3 $150 < X \leq 160$	I_4 $160 < X \leq 170$	I_5 $170 < X \leq 180$	I_6 $X > 180$
o_i	11	50	91	98	38	12
e_i	6.84	40.77	102.39	102.39	40.77	6.84

Suppose we decide on a significance level $\alpha = .05$. Using $\nu = 6 - 1 = 5$ degrees of freedom, we find the critical value $\chi_{.05}^2 = 11.1$. Next we compute

$$\chi^2 = \frac{11 - 6.84)^2}{6.84} + \frac{50 - 40.77)^2}{40.77} + \frac{(91 - 102.39)^2}{102.39}$$
$$+ \frac{(98 - 102.39)^2}{102.39} + \frac{(38 - 40.77)^2}{40.77} + \frac{(12 - 6.84)^2}{6.84} = 10.16$$

Since $\chi^2 = 10.16$ does not exceed $\chi^2_{.05} = 11.1$, there is not sufficient evidence to reject the proposed normal distribution as being suitable for this population. Using Table IV, we note that $\chi^2 = 10.16$ is somewhere around the 90th percentile. This means that a population with this normal distribution will have this large a chi-square value for about 10% of its samples. ∎

The goodness-of-fit test is not restricted to numerically valued variables. The important consideration is to obtain classes or categories for which observed and expected frequencies can be obtained.

Example 6.36

A certain genetic theory predicts that the relative proportion of members of a population in three classes A, B, and C is 3:5:4. A sample of 180 members from this population produced 54 in class A, 57 in B, and 69 in C. Is the sample consistent with the theory? The ratio 3:5:4 means that of each 12 population members we expect to see three in class A, five in class B, and four in C. More precisely:

$$P(\text{Class A}) = \frac{3}{12} = \frac{1}{4}$$
$$P(\text{Class B}) = \frac{5}{12}$$
$$P(\text{Class C}) = \frac{4}{12} = \frac{1}{3}$$

Then $e_1 = (\frac{1}{4}) \cdot 180 = 45$, $e_2 = (\frac{5}{12}) \cdot 180 = 75$, and $e_3 = (\frac{1}{3}) \cdot 180 = 60$. The observed and expected frequencies are summarized in Table 6.6.

TABLE 6.6 OBSERVED AND PREDICTED FREQUENCIES FOR THREE GENETIC CATEGORIES

Frequency	A	B	C
o_i	54	57	69
e_i	45	75	60

Using $\alpha = .05$ and $\nu = 3 - 1 = 2$, we find the critical value $\chi^2_{.05} = 5.99$. Next we compute

$$\chi^2 = \frac{(54 - 45)^2}{45} + \frac{(57 - 75)^2}{75} + \frac{(69 - 60)^2}{60} = 7.47 \text{ (rounded)}$$

This far exceeds the critical value $\chi^2_{.05} = 5.99$, indicating that this genetic model should be rejected for this population (see Figure 6.35).

FIGURE 6.35 THE CHI-SQUARE DISTRIBUTION FOR $v = 2$ DEGREES OF FREEDOM

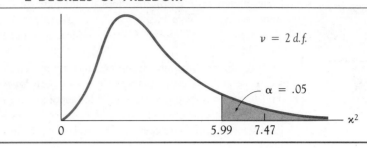

As the next example shows, if the sample size is not large, it may be necessary to condense the classes in order to keep the expected frequencies at five or larger.

Example 6.37

We want to test a pair of dice for fairness and plan to use $\chi^2_{.05}$ as the critical value. Since the probability of a 2 and also a 12 from an honest pair of dice are each $\frac{1}{36}$, we would need at least 180 throws to have all expected frequencies be 5 or larger. An alternate approach is to consider only two outcomes: "sum is 7" and "sum is not 7" (see Table 6.7). These have probabilities of $\frac{1}{6}$ and $\frac{5}{6}$ for honest dice. Now we need only 30 (or more) throws. Suppose we actually use 36 throws and obtain 8 sevens. We then find

$$e_1 = (\frac{1}{6}) \cdot 36 = 6 \qquad e_2 = (\frac{5}{6}) \cdot 36 = 30$$

Thus $\chi^2 = \dfrac{(8 - 6)^2}{6} + \dfrac{(28 - 30)^2}{30} = .8$ (rounded). With $v = 2 - 1 = 1$, we find $\chi^2_{.05} = 3.84$, so there are no grounds for rejecting the fairness of the dice. ∎

TABLE 6.7 EXPECTED AND OBSERVED FREQUENCIES IN 36 THROWS OF A PAIR OF DICE

Frequency	7	Not 7
o_i	8	28
e_i	6	30

We have not tried to be too precise with concepts and terms in the last two sections. Our main goal was to introduce the chi-square distribution and the notion of degrees of freedom. A great deal of what we have done with the goodness-of-fit test will be dealt with more precisely when the various aspects of hypothesis testing are taken up in Chapters 9, 10, and 13.

Example 6.38

A Research
Study

When all else fails, is there a better strategy than guessing at the answers on a multiple choice test? If so, the clues would have to come from the form of the possible responses. This is the subject of a study; "Response Biases in Multiple Choice Item Files," Thomas Mentzer, *Educational and Psychological Measurement*, vol. 2, 1982.

Using 35 files of multiple choice test items supplied with psychology textbooks, three possible types of response biases were investigated:

1. The letter bias. Suppose the possible responses are denoted by the letters A, B, C, and D. Is it the case that one of these letters is overutilized or underutilized for the correct response? The author hypothesized that "A" is underutilized.

2. The format bias. Some test items have three explicit answers A, B, and C, and for D use either "all of the above" or "none of the above." Is it the case that D is the correct answer more often than one might expect? If so, is either of the forms "all" or "none" preferred? A similar question was posed for items where only A and B are explicit answers and C and D provide a choice between "both A and B" and "neither A nor B."

3. The length-of-answer bias. If the possible responses are of varying lengths, is the longest one correct more often than 25% of the time? (Obviously questions involving the format bias would not be considered here because of the length of the C and D responses.)

The analysis was carried out by using the chi-square goodness of fit test on each of the 35 files, the categories being the four possible responses A, B, C, and D. Thus $\nu = 4 - 1 = 3$; when there is no bias, each letter should be the correct answer about 25% of the time.

When all questions of each file were considered, a letter preference bias was found in only seven of the 35 files; the letter "A" was underutilized in only three of these. When, however, those questions that contained a format type of clue were dropped from consideration, a significant letter bias was found in 17 of the files. In nearly all of these, the bias resulted from underutilizing "A" as the correct answer.

The format analysis was more interesting. Twenty-four of the 35 files showed a format bias, due almost exclusively to the overutilization of "all of the above" as the correct response. No bias was found in the use of "none of the above." In 10 of the files, a significant bias toward "both A and B" was found. Finally, a length-of-answer bias was found in 17 of the 35 files, due to overusing the longest answer as the correct one.

Biases such as those uncovered here could possibly be of use in taking a test. Unfortunately, they do not appear to be so great as to permit a strategy that will appreciably improve an otherwise poor performance. ∎

**EXERCISE SET
6.4**

Part A

1. Using Table IV, find:
   a. $P(0 \le \chi^2 \le 10.6)$ when $\nu = 6$ degrees of freedom (df).
   b. $P(\chi^2 > 25.0)$ when $\nu = 15$ df.
2. Using Table V, find:
   a. $\chi^2_{.05}$ when $\nu = 6$ df.
   b. $\chi^2_{.01}$ when $\nu = 10$ *df*.

3. Using Table IV, find:
   a. $P(\chi^2 > 14.9)$ when $v = 4$ df. Compare with $\chi^2_{.005}$ from Table V.
   b. $P(\chi^2 > 18.3)$ when $v = 10$ df. Compare with $\chi^2_{.05}$ from Table V.
4. Compute the χ^2 value for the goodness-of-fit test associated with the following observed and predicted frequencies. Compare the result with $\chi^2_{.05}$. Does there appear to be significant disagreement?

i	1	2	3
o_i	6	16	28
e_i	10	15	25

5. A die is tested for fairness by tossing it 120 times. If it is fair, each face should have a relative frequency close to ⅙. The actual observations were 20 ones, 25 twos, 18 threes, 19 fours, 23 fives, and 15 sixes. Using $\alpha = .05$, test for fairness of the die.
6. In a goodness-of-fit test using $n = 15$ classes, what is the probability of a computed chi-square value exceeding 31.3 if in fact the theoretical distribution is the distribution of the population?
7. In one of his classical studies, Mendel observed and recorded the shape and size of peas as follows:

Round and yellow:	315
Round and green:	108
Oblong and yellow:	101
Oblong and green:	32

Mendel had predicted that these variables would occur in the ratio $6:2:2:1$. Test his prediction using $\chi^2_{.10}$ as the critical value.
8. We wish to test whether shortsightedness is independent of the sex of the individual. If so, the proportion should be the same in both sexes. A sample of 479 adults with 233 males and 246 females showed 7 males and 14 females with this sight defect. Using $(7 + 14)/479 = .044$ as the proportion that should be in each class, test the assertion of independence.
9. We have been told that the bells on the respective dials of a slot machine appear with the same frequency. Test that this is the case if in 300 plays, 39 bells appeared on dial 1, 28 on dial 2, and 30 on dial 3. (Hint: First decide on what is the best estimate for the probability of a bell on any dial if it is the same for all three dials.)

Part B

10. The I.Q. scores of 79 third grade students were reported as follows:

Class	Frequency
80–90	8
90–100	15
100–110	39
110–120	12
120–130	5

Test that these scores came from a normal distribution. Suggestion: Use the *sample* mean and standard deviation as that of the population.

11. Extremely small values of χ^2 indicate a strong agreement, perhaps even a rigged sample. It is reported that of the registered voters in a district, 10% are independents, 50% are Democrats, and 40% are Republicans. A pollster reported that a sample of 300 voters showed 33 independents, 151 Democrats, and 116 Republicans. If the reported distribution is correct, what is the probability (approximately) of obtaining a sample which shows such good agreement?

12. A company has 500 daily employees. The number absent during one week showed the following distribution by days. Does absenteeism appear to be equally distributed over the days of the week? Use $\alpha = .01$.

Day	Monday	Tuesday	Wednesday	Thursday	Friday
Number absent	40	25	20	30	45

13. A study of 400 three-children families produced the distribution of boys shown below. Test the hypothesis that boys and girls are equally likely in three-children families. Use $\alpha = .05$.

Number of boys	0	1	2	3
Number of families	40	180	120	60

14. When the number v of degrees of freedom is large, the normal distribution can be used to find approximate values of χ^2 by using the following formula:

$$\chi_\alpha^2 = \frac{1}{2}[z_\alpha + \sqrt{2v - 2}]^2$$

  a. Compute $\chi_{.05}^2$ when $v = 25$, using this formula, compare the result with the entry in Table V.

  b. Compute $\chi_{.95}^2$ when $v = 400$.

15. The chi-square distribution looks more and more like a normal distribution as the number of degrees of freedom v increases. One would expect that the chi-square distribution could be approximated by the normal distribution having the same mean and standard deviation. This is in fact the case. Using

$$z = \frac{\chi^2 - \mu_{\chi^2}}{\sigma_{\chi^2}}$$

with $v = 50$ degrees of freedom, compute $\chi_{.05}^2$.

16. Using the formula in problem 15, show the formula relating χ^2 to z is

$$\chi^2 = (\sqrt{2v})z + v$$

COMPUTER
PRINTOUT

**FIGURE 6.34 MINITAB PRINTOUT OF BINOMIAL DISTRIBUTION;
PARAMETERS:** $n = 20$, $p = .5$

```
MTB > GENERATE INTEGERS 0--20 IN C1
MTB > BINOMIAL PROBABILITIES N = 20, P = .5, PUT
INTO C5

   BINOMIAL PROBABILITIES FOR N =  20  AND P =
0.500000

        K            P(X = K)           P(X LESS or = K)
        0             0.0000                  0.0000
        1             0.0000                  0.0000
        2             0.0002                  0.0002
        3             0.0011                  0.0013
        4             0.0046                  0.0059
        5             0.0148                  0.0207
        6             0.0370                  0.0577
        7             0.0739                  0.1316
        8             0.1201                  0.2517
        9             0.1602                  0.1119
       10             0.1762                  0.5881
       11             0.1602                  0.7483
       12             0.1201                  0.8684
       13             0.0739                  0.9423
       14             0.0370                  0.9793
       15             0.0148                  0.9941
       16             0.0046                  0.9987
       17             0.0011                  0.9998
       18             0.0002                  1.0000
MTB > PLOT C5 VS. C1
      C5
 0.200-
      -
      -
      -                              *
      -
      -                       *      *
 0.150-
      -
      -
      -                    *            *
 0.100-
      -
      -
      -              *               *
      -
 0.050-
      -
      -           *               *
      -
      -
 0.000-  * * * * *                        * * * * *
      -           *                     *
        +---------+---------+---------+---------+---------+C1
        0.0       5.0      10.0      15.0      20.0      25.0
MTB >STOP
```

Minitab contains provisions for accessing most of the important probability
distributions. In Figure 6.34 are given the program and printout for
generating the binomial distribution for the case $n = 20$ and $p = .5$. A plot
of this distribution is also included. Following this (Figure 6.35 and 6.36) are
the plots of the binomial distributions corresponding to $n = 20$ and $p = .3$

and $p = .2$. Obviously one should expect some difficulty in approximating either of these last two distributions with the normal distribution.

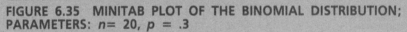

FIGURE 6.35   MINITAB PLOT OF THE BINOMIAL DISTRIBUTION; PARAMETERS: $n = 20$, $p = .3$

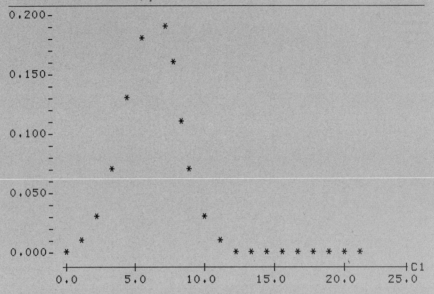

FIGURE 6.36   MINITAB PLOT OF THE BINOMIAL DISTRIBUTION; PARAMETERS: $n = 20$, $p = .2$

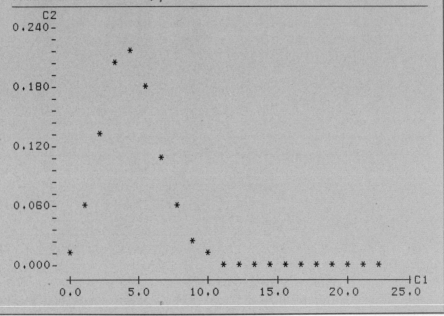

CHAPTER CHECKLIST

Vocabulary

Binomial experiment
Trial
Binomial coefficient
Binomial distribution
Normal distribution

Standard normal distribution
Percentage points or critical values
Chi-square distribution
Degrees of freedom
Goodness-of-fit test

Techniques

Determining if an experiment is a binomial experiment
Using binomial coefficients to find the number of ways to choose x objects from among n
Finding binomial probabilities, using both the binomial formula and the cumulative table of binomial probabilities
Computing the mean, variance, and standard deviation of the binomial distribution
Finding probabilities associated with the normal distribution
Finding critical values z_α of the normal distribution
Approximating the binomial distribution with the normal one when n is large
Approximating the distribution of the proportion $\hat{P}$ with the normal distribution
Finding probabilities associated with the chi-square distribution
Finding critical values χ_α^2 of the chi-square distribution
Performing goodness-of-fit tests

Notation

$\binom{n}{k}$ The binomial coefficient "n choose k"
z_α Critical value or percentage point of the normal distribution
χ^2 The chi-square variable
ν The number of degrees of freedom ("df" is also used)
χ_α^2 A critical value of the chi-square distribution
$\hat{P}$ The sample proportion variable

CHAPTER TEST

1. Which of the following are binomial experiments or can be approximated as such?
   a. Surveying 300 corporate executives as to whether they believe a Democrat will be the next president to be elected.
   b. Giving a written drivers' examination test to 40 randomly selected applicants where the passing rate is 80%.
   c. Testing a sample of four quartz watches from a shipment of 25 watches of which 10 are defective.
   d. Surveying 1000 families as to whether they belong to a dental plan, when it is known that 35% of all families belong to such a plan.
2. Let $P(X = x) = \binom{3}{x}(.4)^x(.6)^{3-x}$.
   a. Find the mean and standard deviation of this distribution.

b. Compute the probabilities $P(X = x)$ for $x = 0, 1, 2,$ and 3 and graph the distribution.

3. Let X be binomially distributed with $n = 20$ and $p = .4$.
   a. Using the cumulative binomial probabilities table, find $P(X \geq 10)$.
   b. Using the normal approximation to the binomial distribution, compute $P(X \geq 10)$ and compare with the results in part a.

4. A television production company knows from past experience that the probability of a new pilot being accepted by a network is $p = .3$. If 10 pilots are produced, what is the probability that four or more will be accepted?

5. A company in bankruptcy estimates that of its many outstanding accounts receivable, only 75% are collectible. What is the probability that in a random sample of 50 of these accounts, less than 70% will be collectible?

6. Find (a) $z_{.05}$, (b) $z_{.01}$, (c) $z_{.75}$.

7. The charge card accounts of a bank have balances with a mean of $340 and a standard deviation of $80. Assuming these balances are normally distributed,
   a. What is the probability that a randomly selected account will have a balance exceeding $500?
   b. Find the 90th percentile for these balances.

8. There are three popular surgical procedures, A, B, and C, for dealing with an occlusion of the carotid arteries. When 90 surgeons were surveyed as to which procedure they preferred, the results were as follows:

Technique	A	B	C
Preferred	24	34	32

Does there appear to be any difference in the popularity of these three procedures? (Are they equally popular?) Use $\alpha = .05$.

9. Four hundred workers were surveyed as to whether they were satisfied with their occupations. The workers were then classified into three categories: white collar, blue collar, and professional. The results were as follows:

Satisfaction	Professional	White collar	Blue collar
Satisfied	96	80	92
Unsatisfied	44	64	24

Using $\alpha = .05$, test whether there is the same degree of job satisfaction by workers in these three categories. Hint: first find an estimate for the proportion of satisfied workers when the categories are combined.

10. A standard I.Q. test produces scores that are normally distributed with a mean of 100 and a standard deviation of 15. Suppose that the top 10% of the scores are used to identify individuals with superior intelligence. What is the lower cutoff score for the superior classification?

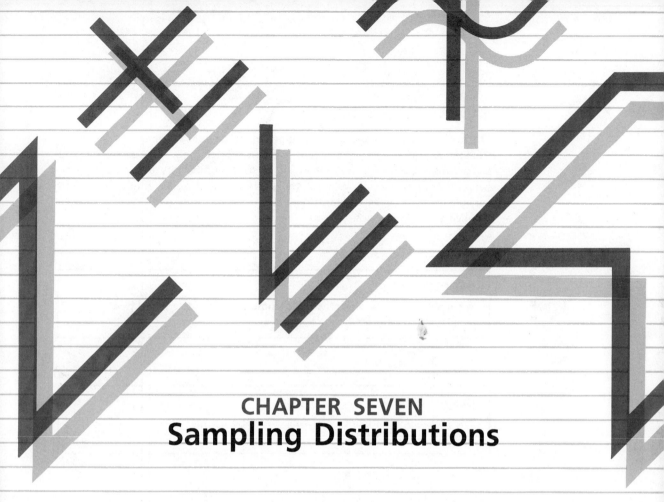

CHAPTER SEVEN
Sampling Distributions

Inferential statistics is concerned with making decisions about populations based on information obtained from random samples. In most problems the decision ultimately becomes a choice among one of several alternatives. Whether we are selecting from several competing hypotheses or choosing the form of an estimate, the challenge is to make a decision that has only a small probability of being in error, without being so conservative as to be useless. To do this we need to understand how such quantities as sample means, variances, and proportions are distributed relative to their counterparts in the population. These sampling distributions are the subject of this chapter.

7.1   SAMPLE STATISTICS

Our first goal is to relate the sampling process to random variables. As we saw in Chapter 5, this is easily done. The process of selecting a random sample and computing some descriptive statistic is in fact a probability experiment. Each sample furnishes another value of the statistic and these determine a random variable having a probability distribution. In sampling theory this type of variable is called a *sample statistic*.

DEFINITION 7.1

A *sample statistic* is a random variable whose values are determined by all samples of a certain size from a population.

DEFINITION 7.2

The probability distribution of a sample statistic is called a sampling distribution.

The computation of three such statistics, the mean $\overline{X}$, the variance S^2, and the standard deviation S, was studied in Chapter 3. Additionally, the sample proportion $\hat{P}$ was introduced with the binomial distribution in Chapter 6. Note that we have followed the usual practice of using capital letters for variable names.

We shall soon see that the distribution of $\overline{X}$ is centered at the population mean μ_X. This proves to be an extremely useful relationship, since otherwise the sample means would have a built-in bias with which we would have to constantly contend. Similar relationships apply to other statistics as well.

DEFINITION 7.3

A sample statistic is said to be an *unbiased estimator* of a population parameter if its expected value is this parameter.

Once we show that $\mu_{\overline{X}} = \mu_X$ and $\mu_{\hat{P}} = p$ we can conclude that the mean $\overline{X}$ and proportion $\hat{P}$ are unbiased estimators respectively of the population mean μ_X and proportion p. And while we shall have no occasion to do so, it can be shown that the use of $n - 1$ rather than n in the definition of the sample variance makes S^2 an unbiased estimator of the population variance σ_X^2.

7.2   THE DISTRIBUTION OF $\overline{X}$

In this section we consider the more important aspects of the distribution of the mean $\overline{X}$. The major result, the *central limit theorem*, states roughly that this distribution is approximately normal when the sample size is large. Before examining this relationship in depth, let us consider a special case that will shed some light on some details of this remarkable theorem.

We begin by considering a population consisting of the equally likely outcomes

$$x_1 = 1, \; x_2 = 2, \; x_3 = 3, \; x_4 = 4$$

The distribution and probability histogram are given in Table 7.1 and Figure 7.1.

TABLE 7.1
DISTRIBUTION OF X

X	f(x)
1	¼
2	¼
3	¼
4	¼

FIGURE 7.1   PROBABILITY HISTOGRAM FOR X

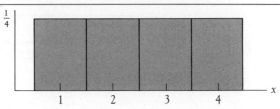

The symmetry of the distribution shows the mean to be $\mu_X = 5/2 = 2.5$, which is easily verified by direct computation. For the variance

$$\sigma_X^2 = \sum_{i=1}^{4} (x_i - \mu_X)^2 P(x_i)$$

$$= \left(1 - \frac{5}{2}\right)^2\left(\frac{1}{4}\right) + \left(2 - \frac{5}{2}\right)^2\left(\frac{1}{4}\right) + \left(3 - \frac{5}{2}\right)^2\left(\frac{1}{4}\right) +$$

$$\left(4 - \frac{5}{2}\right)^2\left(\frac{1}{4}\right)$$

$$= \frac{5}{4} = 1.25$$

We now want to examine the distribution of $\overline{X}$ for samples of size $n = 2$ from an infinite population having the above distribution. To avoid having to deal with an infinite number of each of the values 1, 2, 3, and 4 and the resulting samples we use the device of sampling with replacement. Note that this permits samples such as (1, 1) as well as both (2, 3) and (3, 2), which would occur with an infinite population.

The $4 \times 4 = 16$ distinct samples that can arise are detailed in Table 7.2. Each of these occurs with the same frequency and has probability $\frac{1}{16}$. Since the sample mean $\overline{x} = \frac{3}{2}$ occurs twice, it has probability $\frac{2}{16} = \frac{1}{8}$. Continuing in this manner we obtain the distribution of $\overline{X}$ in Table 7.3.

TABLE 7.2   MEANS OF SAMPLES OF SIZE 2

Observations	Sample	$\overline{x}$
x_1, x_1	1, 1	1
x_1, x_2	1, 2	$\frac{3}{2}$
x_1, x_3	1, 3	2
x_1, x_4	1, 4	$\frac{5}{2}$
x_2, x_1	2, 1	$\frac{3}{2}$
x_2, x_2	2, 2	2
x_2, x_3	2, 3	$\frac{5}{2}$
x_2, x_4	2, 4	3
x_3, x_1	3, 1	2
x_3, x_2	3, 2	$\frac{5}{2}$
x_3, x_3	3, 3	3
x_3, x_4	3, 4	$\frac{7}{2}$
x_4, x_1	4, 1	$\frac{5}{2}$
x_4, x_2	4, 2	3
x_4, x_3	4, 3	$\frac{7}{2}$
x_4, x_4	4, 4	4

TABLE 7.3   PROBABILITY DISTRIBUTION OF $\overline{X}$

$\overline{x}$	$f(\overline{x})$
1	1/16
3/2	2/16 = 1/8
2	3/16
5/2	4/16 = 1/4
3	3/16
7/2	2/16 = 1/8
4	1/16

The mean and variance of $\overline{X}$ are easily found:

$$\mu_{\overline{X}} = 1\left(\frac{1}{16}\right) + \frac{3}{2}\left(\frac{1}{8}\right) + 2\left(\frac{3}{16}\right) +$$
$$\frac{5}{2}\left(\frac{1}{4}\right) + 3\left(\frac{3}{16}\right) + \frac{7}{2}\left(\frac{1}{8}\right)$$
$$+ 4\left(\frac{1}{16}\right) = \frac{5}{2} = 2.5$$

$$\sigma_{\overline{X}}^2 = \left(1 - \frac{5}{2}\right)^2 \frac{1}{16} + \left(\frac{3}{2} - \frac{5}{2}\right)^2 \frac{1}{8} + \left(2 - \frac{5}{2}\right)^2 \frac{3}{16} +$$
$$\left(\frac{5}{2} - \frac{5}{2}\right)^2 \frac{1}{4} + \left(3 - \frac{5}{2}\right)^2 \frac{3}{16} + \left(\frac{7}{2} - \frac{5}{2}\right)^2 \frac{1}{8} + \left(4 - \frac{5}{2}\right)^2 \frac{1}{16}$$
$$= \frac{5}{8}$$

How are these related to the population mean $\mu_X = 5/2$ and variance $\sigma_X^2 = 5/4$? Just as in previous examples, the mean of the sample means is the population mean, that is:

$$\mu_{\overline{X}} = \mu_X$$

The variance of the mean is seen to be the population variance divided by 2. It is not just an accident that 2 is also the sample size. In fact, when the sample size is n,

$$\sigma_{\overline{X}}^2 = \frac{\sigma_X^2}{n}$$

For the standard deviation this becomes

$$\sigma_{\overline{X}} = \frac{\sigma_X}{\sqrt{n}}$$

In words, the standard deviation of the mean is the standard deviation of the population divided by the square root of the sample size.

Now compare the histograms of X and $\overline{X}$ in Figure 7.2. They have very little in common apart from the symmetry. The symmetry of X has been passed along to the mean $\overline{X}$. It turns out that had the distribution of X been normal, then this would also have been passed along to the mean. Thus we have the Theorem 7.1.

FIGURE 7.2 THE DISTRIBUTIONS OF X AND $\overline{X}$

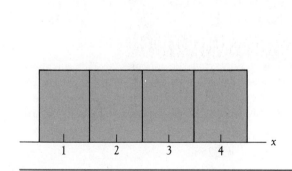 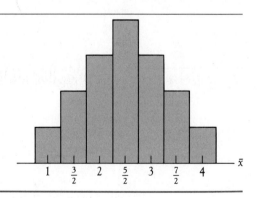

THEOREM 7.1

Let X be normally distributed with mean μ_X and standard deviation σ_X. Then the mean $\overline{X}$ associated with samples of size n is normally distributed, with mean and standard deviation

$$\mu_{\overline{X}} = \mu_X$$

$$\sigma_{\overline{X}} = \frac{\sigma_X}{\sqrt{n}}$$

But there is still more to the story. Even if we start with a population that is not normal, we find that for moderately large samples, the distribution of $\overline{X}$ resembles the normal distribution and becomes increasingly more so as the sample size increases. This is the Central Limit Theorem.

THEOREM 7.2   (CENTRAL LIMIT THEOREM)

Let X have a distribution with mean μ_X and standard deviation σ_X. As the sample size n becomes large, the distribution of the mean $\overline{X}$ approaches the normal distribution, with mean and standard deviation

$$\mu_{\overline{X}} = \mu_X$$

$$\sigma_{\overline{X}} = \frac{\sigma_X}{\sqrt{n}}$$

Experience indicates that the sample size n should be at least 30 before this approximation can be used.

The Central Limit Theorem assumes that the population is infinite. One may well ask what happens when this is not the case. If we make the additional assumption that the population is large in comparison to the large samples that are needed, then the theorem may still be used. In many applications this additional assumption is no restriction, since the populations are large relative to the samples.

Example 7.1

Statistical data for a large labor union show that the incomes X of its members have a mean of $22,580 and a standard deviation of $4200. What is the distribution of sample means for samples of size $n = 100$? Since $n = 100$ is considerably larger than the minimal size of $n = 30$, the distribution of $\overline{X}$ will be approximately normal, with mean

$$\mu_{\overline{X}} = \mu_X = 22580$$

and standard deviation

$$\sigma_{\overline{X}} = \frac{\sigma_X}{\sqrt{n}} = \frac{4200}{\sqrt{100}} = 420$$

The distributions of both X and $\overline{X}$ are shown in Figure 7.3.

FIGURE 7.3   THE DISTRIBUTIONS OF X AND $\overline{X}$

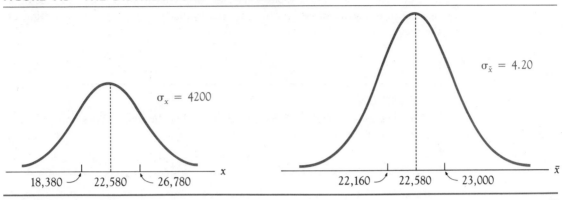

Example 7.2   Suppose that aptitude test scores for entering college students have a mean of 250 and a standard deviation of 75 and are approximately normally distributed. Discuss the distribution of samples of size $n = 9$. Since the aptitude test scores X are approximately normal, the same will be true of the sample means. The mean and standard deviation of $\overline{X}$ are

$$\mu_{\overline{X}} = \mu_X = 250$$

$$\sigma_{\overline{X}} = \frac{\sigma_X}{\sqrt{9}} = \frac{75}{3} = 25$$

The distributions of both X and $\overline{X}$ are given below (Figure 7.4).

FIGURE 7.4   THE DISTRIBUTION OF X   THE DISTRIBUTION OF $\overline{X}$

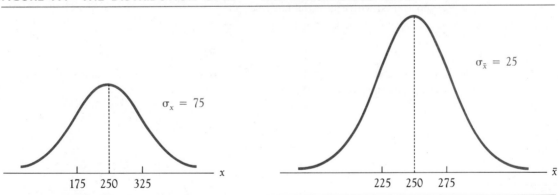

In any discussion involving samples, it soon becomes obvious that there is a problem of terminology. For any particular sample there is the standard deviation s. This sample is from a population having a standard deviation σ_X. Moreover, the mean of this sample belongs to a population of sample means having a standard deviation $\sigma_{\overline{X}}$. We extricate ourselves somewhat from this by introducing a new term, the *standard error*, which is used in place of the standard deviation of the mean.

DEFINITION 7.4

The *standard error* (*S.E.*) of the mean is defined as the standard deviation of the distribution of $\overline{X}$. Thus

$$S.E. = \sigma_{\overline{X}}$$

When we are dealing with large or infinite populations, this standard error will be computed by

$$S.E. = \frac{\sigma_X}{\sqrt{n}}$$

Example 7.3

Sacks of sugar filled by an automatic loading machine have a distribution of weights that is approximately normal with a mean of 5.35 and a standard deviation of .16 pounds. For samples of size $n = 64$, the standard error of the mean is

$$S.E. = \frac{\sigma_X}{\sqrt{n}} = \frac{.16}{\sqrt{64}} = .02 \text{ (pounds)} \quad \blacksquare$$

When the mean $\overline{X}$ is normal or approximately so, the standard normal distribution is used with the variable Z:

$$Z = \frac{\overline{X} - \mu_{\overline{X}}}{\sigma_{\overline{X}}}$$

Of course $\mu_{\overline{X}} = \mu_X$ and $\sigma_{\overline{X}} = S.E. = \sigma_X/\sqrt{n}$. Thus in computing the z score corresponding to a value x we use

$$z = \frac{\overline{x} - \mu_X}{S.E.} = \frac{\overline{x} - \mu_X}{\sigma_X/\sqrt{n}}$$

Example 7.4

A federal highway inspector is told that the speeds of autos on a certain stretch of highway have a mean of 56 and a standard deviation of 6 miles per hour (mph). A sample of $n = 36$ autos are observed. What is the probability that their mean speed exceeds 58.5 mph? The distribution of $\overline{X}$ is approximately the normal distribution shown in Figure 7.5. The desired area is to the right of $\overline{x} = 58.5$. Using $\mu_X = 56$ and $S.E. = 6/\sqrt{36} = 6/6 = 1$, the z score corresponding to $\overline{x} = 58.5$ is found to be

$$z = \frac{\overline{x} - \mu_X}{S.E.} = \frac{58.5 - 56}{1} = \frac{2.5}{1} = 2.5$$

FIGURE 7.5 THE DISTRIBUTION OF $\overline{X}$

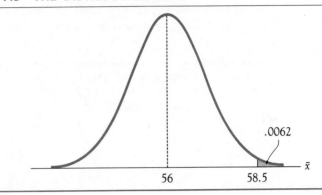

Thus $P(\overline{X} > 58.5) = P(Z > 2.5) = .5000 - .4938 = .0062$. What would you think if you were the inspector and drew such a sample? Either you were extremely unlucky, receiving a sample that occurs less than 1% percent of the time, or there is something wrong with the asserted distribution. ■

The next example will help clear up any uncertainty regarding the use of the distribution of X as opposed to that of $\overline{X}$.

Example 7.5

Suppose the concentration of glucose in the blood of patients with a certain condition is approximately normal, with a mean of 100 milligrams per deciliter (mg/dl) and a standard deviation of 6 mg/dl.
1. What is the probability that a randomly selected patient with this condition will produce a glucose concentration less than 98 mg/dl?
2. What is the probability that a random sample of 16 of these patients will produce a mean concentration less than 98 mg/dl?

The first question deals with the occurrence of certain values of X, not of $\overline{X}$. Thus the distribution of X is used. The desired area is shaded under the normal density curve in Figure 7.6. Using $z = (x - \mu_X)/\sigma_X = (x - 100)/6$ we find $z = -.33$ for $x = 98$. Thus $P(X < 98) = P(Z < -.33) = .3707$.

FIGURE 7.6 THE DISTRIBUTION OF X

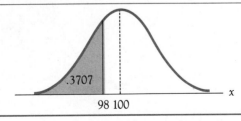

The second question involves the probability of observing certain values of the mean $\overline{X}$. Thus the distribution of $\overline{X}$ is to be used. Since $z = \dfrac{\overline{x} - \mu_X}{S.E.} = \dfrac{\overline{x} - 100}{6/\sqrt{16}} = \dfrac{\overline{x} - 100}{1.5}$ we find that for $\overline{x} = 98$, $z = (98 - 100)/1.5 = -1.33$. Thus $P(\overline{X} < 98) = P(Z < -1.33) = .0918$. The corresponding area is shaded under the density curve for $\overline{X}$ in Figure 7.7. ■

FIGURE 7.7   THE DISTRIBUTION OF $\overline{X}$

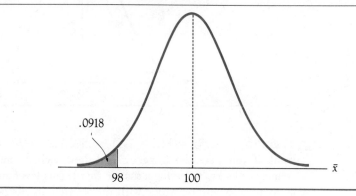

In using the formula

$$z = \frac{\overline{x} - \mu_X}{S.E.}$$

it should be noted that the resulting z value is the number of standard errors at which $\overline{x}$ is located from the population mean. Thus the standard error is the statistician's unit of measure for locating the sample mean relative to the population mean. Roughly speaking, the central limit theorem states that most large samples have means that are within two or three standard errors of the population mean.

FIGURE 7.8   DISTRIBUTION OF $\overline{X}$

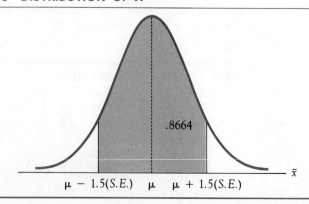

Example 7.6

What is the probability that a sample of size $n = 200$ will have a mean that is within 1.5 standard errors of the population mean? The mean $\overline{X}$ will have the distribution shown in Figure 7.8. The desired area is shaded and corresponds to $P(-1.5 < Z < 1.5) = .8664$. Thus about 86.6% of all samples will have means within 1.5 standard error of the population mean. ■

EXERCISE SET 7.1

Part A

1. Consider the experiment of selecting a random sample of $n = 64$ apples from an orchard where the individual weights X are normally distributed with a mean of 8.4 and a standard deviation of 2.4 ounces. Each sample has a mean and these are the values of the random variable $\overline{X}$.
   a. Find the mean of $\overline{X}$.
   b. Find the standard deviation of $\overline{X}$.
   c. What is another name for the quantity found in part b?
   d. Sketch the distribution of $\overline{X}$.
2. Random samples of size n are being investigated from a population with $\mu = 80$ and $\sigma = 30$.
   a. Sketch the sampling distribution of the mean $\overline{X}$ if $n = 36$.
   b. Sketch the sampling distribution of the mean $\overline{X}$ if $n = 100$.
   c. Draw several sketches showing the shape of the distributions as the sample size becomes progressively larger and larger.
   In each case mark off the mean and the points that are 1 S.E. from the mean.
3. A normal population has a mean of 84 and a standard deviation of 28. For each case describe the distribution of $\overline{X}$ and then find the number of standard errors the given sample mean $\overline{x}$ is located above or below $\mu_{\overline{x}}$.
   a. $n = 16, \overline{x} = 91.0$.
   b. $n = 196, \overline{x} = 79.5$.
   c. $n = 1600, \overline{x} = 86.5$.
   d. $n = 6, \overline{x} = 100$.
4. A population has a mean of $\mu = 150$ and a standard deviation $\sigma = 50$. What percent of all samples of size 100 have means that:
   a. Are between 140 and 160?
   b. Exceed 165?
5. Let $\overline{X}$ be the sample mean associated with samples of size 64 from a population with $\mu = 128$ and $\sigma = 12$. Find
   a. $P(\overline{X} > 132)$.
   b. $P(126 < \overline{X} < 130)$.
6. What is the probability that the mean of a sample of size n will be within:
   a. 2 standard errors of the population mean? Assume $n \geq 30$.
   b. 1.5 standard errors of the population mean?
7. A large engineering school claims that its graduating electrical engineers had starting salaries which are normally distributed with a mean of $25,600 and a standard deviation of $1840.
   a. What is the probability that a randomly selected graduate had a starting salary below $25,000?
   b. What is the probability that a sample of 16 graduates has a mean starting salary below $25,000?
   c. Interpret the answers to parts a and b as percentages.
8. A manufacturer claims that a certain candy bar is packaged to have a mean weight of .8 ounces and that the standard deviation of these packaged weights is

.12 ounces. What is the probability that a sample of 36 of these bars will have a mean below .75 ounces?

9. In a study of tumor growth rates, the volume doubling time for a certain type of tumor was found to have a mean $\mu = 202$ days and a standard deviation $\sigma = 12.5$ days. If these values are correct, what is the probability that a random sample of 36 of such tumors will have a volume doubling time exceeding 205 days?

10. Midlevel management personnel with major corporations are reported to have stock portfolios whose values have a mean of $13,400 and a standard deviation of $4480. What is the probability that the mean portfolio value of a sample of 300 such managers will be within $1000 of the stated mean?

Part B

11. David Templer's Death Anxiety Scale is a popular instrument in psychological studies. In one study of adults between 18 and 83 years of age by researchers at the University of Connecticut ("Age Norms for Templer's Death Anxiety Scale," S. Stevens et al., *Psychology Report*, 46, 1980, 205–206), $\mu = 6.89$ and $\sigma = 3.20$ were established for scores on this test. In a related study of 68 elderly adults, a mean of 5.74 was found. What is the probability of obtaining such a low mean if in fact there is no difference in anxiety between the elderly and adults in general?

12. In a sample of 1760 hospitals across the nation it was found that the average length of stay for patients under 65 years of age was approximately 6 days. (For those 65 and over, the average was 10.6 days.) Assuming a normal distribution with a standard deviation of 2.8 days for those under 65,
    a. What is the probability that an entering patient (randomly selected of age 65 or less) will be hospitalized less than 8 days?
    b. What is the probability that a random sample of 60 patients (again 65 or under) will have a mean stay of between 5 and 7 days?
    (Source: *Hospitals*, June 1981, "Selected Hospital Statistics for February 1981".)

13. Why would one expect the standard error of the mean to become small as the sample size becomes large? How is this confirmed mathematically?

14. A population consists of 0, 2, 4, 6, and 8. Assuming samples of size $n = 2$ are drawn with replacement between each selection,
    a. Find the sampling distribution of the mean $\overline{X}$ for such samples by examining all possible samples.
    b. Verify that $\mu_{\overline{X}} = \mu_X$.
    c. Verify that $\sigma_{\overline{X}} = \sigma_X/\sqrt{n}$.

15. If X is a continuous variable, is $\overline{X}$ continuous or discrete?

16. What is meant by the statement that S^2 is an unbiased estimator of σ^2?

17. Interpret the following statement: For a symmetric distribution, the median is an unbiased estimator of the mean.

18. Suppose a population has a mean of 500 and a standard deviation of 150. Many samples of size $n = 100$ are selected and the mean of each is found.
    a. What would you predict for the mean of all these means?
    b. What would you predict for the standard deviation of this distribution of sample means?
    c. Sketch what the histogram of the distribution of these means should look like.
    d. What would you predict for the mean of all the sample variances?

19. A population has a mean of 100 and a standard deviation of 30. The distribution

of the means of samples of size $n = 100$ is being discussed. Give each of the following (a problem of notation).

a. μ_X.
b. $\mu_{\bar{X}}$.
c. σ_X.
d. $\sigma_{\bar{X}}$.
e. $S.E.$

7.3   THE *t* DISTRIBUTION

In most applications involving the mean $\bar{X}$, the standard deviation σ_X of the population is not known. When the sample size n is 30 or larger, good results are obtained by estimating σ_X by the standard deviation s of the sample. The standard error is then approximated by

$$S.E. = \frac{s}{\sqrt{n}}$$

The normal distribution is still used with the variable

$$Z = \frac{\bar{X} - \mu_X}{S/\sqrt{n}}$$

For samples of size less than 30, the sample standard deviations have so much variability that they are no longer reliable estimates of σ_X. More to the point, the variable

$$T = \frac{\bar{X} - \mu_X}{S/\sqrt{n}}$$

can no longer be approximated by the normal distribution.

The distribution of this new variable T when the underlying population is normal was determined in 1908 by W. S. Gossett, a chemist at Guinness' Brewery in Dublin. Apparently company policy prohibited publication of research, so he used the pseudonym "Student;" as a result the distribution is often referred to as the *Student t distribution*.

The t distribution is not just a single distribution. For each sample size there is a different distribution. In their development, Gossett used the chi-square distribution; consequently each has an associated degree of freedom ν. This parameter, it turns out, is one less than the sample size.

THEOREM 7.3

Let X have a normal distribution with mean μ_X and standard deviation σ_X. Then if $\bar{x}$ and s are the mean and standard deviation of a sample of size n from this distribution,

$$t = \frac{\bar{x} - \mu_X}{s/\sqrt{n}}$$

is the value of a variable T having the t distribution with $\nu = n - 1$ degrees of freedom.

The t distributions for several degrees of freedom are shown in Figure 7.9. Each is symmetric about $t = 0$. They resemble the standard normal distribution but become increasingly more spread out with decreasing n, because of the greater uncertainty induced by the variability of the sample standard deviation. For the t distribution, the standard deviation σ_t is greater than 1 and increases in size as the sample size decreases.

FIGURE 7.9
t DISTRIBUTIONS FOR SELECTED DEGREES OF FREEDOM

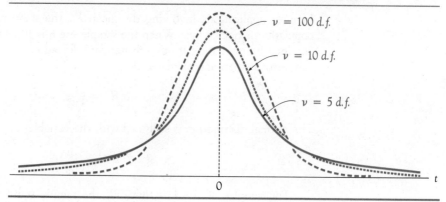

As with the chi-square distribution, it is not practical to tabulate the t distribution for each degree of freedom ν. Table VI of the appendix contains the percentage points or critical values t_α for the most commonly used probabilities α. The critical value notation is exactly the same as with the standard normal and chi-square distributions (see Figure 7.10). For completeness we include Definition 7.5.

DEFINITION 7.5

If α is a given probability, then t_α is that number such that $P(T > t_\alpha) = \alpha$.

FIGURE 7.10   THE CRITICAL VALUE t_α

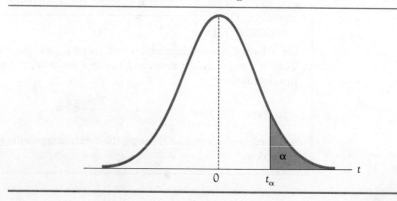

Example 7.7   In Table VI, using $\nu = 8$ and $\alpha = .05$, we find $t_{.05} = 1.860$. Thus $P(T > 1.860) = .05$ (Figure 7.11).

FIGURE 7.11 $P(T > 1.860) = .05$ **WHEN** $\nu = 8$

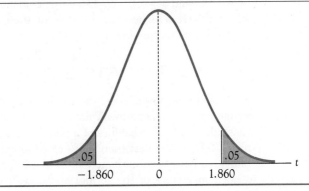

Because of the symmetry, we see also that $P(T < -1.860) = .05$ or, equivalently, $P(T \geq -1.860) = .95$. Thus $t_{.95} = -1.860$. ■

Example 7.8   For $\nu = 20$, find t_1 if $P(-t_1 < T < t_1) = .99$. The area to the right of $t = t_1$ is obviously half the remaining area, that is, $\alpha = (1 - .99)/2 = .005$ (see Figure 7.12).

FIGURE 7.12 $P(-2.845 < T < 2.845) = .99$ **WHEN** $\nu = 20$

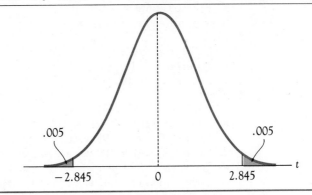

Entering Table VI in the column headed $\alpha = .005$ and row $\nu = 20$, we find $t_{.005} = 2.845$. Thus $P(-2.845 < T < 2.845) = .99$. ■

The t distribution proves to have many applications in estimation and hypothesis testing.

Example 7.9

The temperature X of warm water springs in a basin is reported to have a mean of 39°C. A sample of $n = 16$ springs from the west end of the basin produced $\bar{x} = 41.5$°C and $s = 4.5$°C. Does this evidence indicate that the springs at the west end have a higher mean? Assuming their mean is $\mu = 39$, we first compute the t value corresponding to $\bar{x} = 41.5$:

$$t = \frac{41.5 - 39}{4.5/\sqrt{16}} = 2.222$$

For $v = 15$ degrees of freedom, we find $t_{.025} = 2.131$. Thus $t = 2.222 > t_{.025}$. Assuming the temperatures X are approximately normal with a mean of 39°C, we conclude that the probability of a sample having a mean exceeding 41.5°C is less than .025; that is, less than 2.5% of all samples should show a mean this high. Either we have been extremely unlucky in obtaining one of these samples or the mean temperature of the springs at the west end exceeds 39°C. We adopt the latter position. ■

The reader will find that the t distribution is often used even when the sample size far exceeds 30. This seems to be particularly true in psychology and sociology. In fact it sometimes seems that the t distribution is used almost exclusively in place of the normal one. There is no criticism in this observation. Whenever σ_X is replaced by s, the t distribution will always furnish better results. For large samples, however, this may hardly be noticeable. As an example, for $v = 60$, $t_{.025} = 2.00$ while $z_{.025} = 1.96$.

EXERCISE SET 7.2

Part A

1. Under what conditions may the t distribution be used with $\overline{X}$?
2. Sketch the t distribution, find t_α, and shade the corresponding area when:
   a. $\alpha = .05$ and $v = 20$.
   b. $\alpha = .01$ and $v = 6$.
3. Find:
   a. $t_{.05}$ when $v = 21$.
   b. $t_{.10}$ when $v = 16$.
   c. $t_{.005}$ when $v = 10$.
4. Find t:
   a. If for $v = 8$, $P(T > t) = .01$
   b. If for $v = 12$, $P(T > t) = .005$
   c. $P(-t < T < t) = .95$ when $v = 4$.
   d. $P(-t < T < t) = .90$ when $v = 20$.
   e. $P(T < t) = .99$ when $v = 9$.
5. A genetic strain of dwarf apple trees is reported to have mature heights whose distribution is approximately normal with a mean of 13 feet. A sample of 16 of these trees produced heights with a mean of 15.2 feet and a standard deviation of 3.4 feet. How likely is a mean this large if the population mean is in fact 13 feet?

6. The stopping distance of a test auto equipped with low pressure radial tires on a specially prepared test surface at 50 miles per hour is reported as being approximately normal with a mean of 262 feet. A new high-pressure version of this same tire (same compound and same tread pattern) in 16 tests had stopping distances with a mean of 240 feet and a standard deviation of 44 feet. If there is no difference in the distribution of stopping distances for these two types of tires, what is the probability of obtaining such a small sample mean (that is, 240 or below)? Simply give some estimate of how small the probability is.

Part B

7. Why is it natural to expect $t_{.05} > z_{.05}$, $t_{.01} > z_{.01}$, and so on?
8. The variance of the t distribution is given by

$$\sigma^2 = \frac{v}{v - 2} \qquad (v > 2)$$

Investigate the behavior of the variance for large values of v, say $v = 25$, $v = 50$, $v = 100$, and $v = 1000$. What does the variance approach as the number of degrees of freedom becomes large? Why should this be expected?

9. If a random sample of size $n = 16$ is selected from a normal population with mean μ, what is the probability that the sample mean will fall on the interval $\mu \pm 1.75$ S.E., where $S.E. = s/\sqrt{n}$?

10. In some texts a double subscript notation is used with percentage points to indicate the corresponding area α and the number of degrees of freedom v. Thus for $\alpha = .05$ and $v = 8$ we see $t_{(.05,8)} = 1.860$.
   a. Find $t_{(.01,15)}$.
   b. Find $t_{(.005,7)}$.

7.4

THE DISTRIBUTION OF THE SAMPLE PROPORTION $\hat{P}$

We begin by examining two occurrences of the sample proportion $\hat{P}$ and then review how it is related to the binomial and the normal. As will soon become apparent, the details of the distribution of $\hat{P}$ have already been established.

Consider first the simple binomial experiment of tossing a coin, say, 40 times and observing the number of heads X. If $x = 18$, then the proportion or fraction of heads in the $n = 40$ trials is

$$\hat{p} = \frac{x}{n} = \frac{18}{40} = .45 = 45\%$$

Each time this experiment is repeated another value $\hat{p}$ of $\hat{P}$ results. If we think of the 40 trials as a sample from a binomial population with $p = \frac{1}{2}$, then $\hat{p}$ is a sample proportion. And it is not unreasonable to call $p = \frac{1}{2}$ the population proportion.

Next consider a sample of $n = 400$ that asks residents of a large city whether they are home owners. Suppose $x = 280$ owners are found. The proportion or fraction of home owners in the sample is

$$\hat{p} = \frac{x}{n} = \frac{280}{400} = .70 = 70\%$$

Each time this sampling experiment is performed, values of both the number X and proportion $\hat{P}$ of home owners in the sample are obtained. And as was pointed out in the last chapter, this variable X has essentially the binomial distribution. Thus we can think of $\hat{P}$ as being associated with samples from a binomial population.

Now recall that we approximated the binomial distribution with the normal one when the number of trials was large. More precisely, we approximated the distribution of

$$Z = \frac{X - np}{\sqrt{npq}} \qquad \text{(where } q = 1 - p\text{)}$$

with the standard normal distribution. Further, we saw that when both the numerator and denominator of this fraction were divided by n and X/n was replaced with $\hat{P}$, the same could be said of

$$Z = \frac{\hat{P} - p}{\sqrt{\dfrac{pq}{n}}}$$

Since Z is approximately normal, the same must be true of $\hat{P}$. Finally, comparing this fraction with the standard formula

$$Z = \frac{\hat{P} - \mu_{\hat{P}}}{\sigma_{\hat{P}}}$$

we inferred that the mean and standard deviation of $\hat{P}$ are

$$\mu_{\hat{P}} = p \qquad \sigma_{\hat{P}} = \sqrt{\frac{pq}{n}}$$

THEOREM 7.4

Let $\hat{P}$ represent the proportion of a sample of size n that possesses a certain characteristic and let p be the corresponding proportion in the population. When the sample size is large, the distribution of $\hat{P}$ is approximately normal, with mean

$$\mu_{\hat{P}} = p$$

and standard deviation

$$\sigma_{\hat{P}} = \sqrt{\frac{pq}{n}}$$

where $q = 1 - p$.

In most applications a large sample size may be translated to be one for which both np and nq exceed 5. The one exception is when either p or q is extremely close to 0.

Example 7.10 Discuss the distribution of the proportion $\hat{P}$ associated with samples of size $n = 1600$ from a population with $p = .4$. Since $np = 1600(.4) = 640$ and $nq = 1600(.6) = 960$ far exceed the minimum of 5, the distribution of $\hat{P}$ is approximately normal, with mean and standard deviation

$$\mu_{\hat{P}} = p = .4$$

$$\sigma_{\hat{P}} = \sqrt{\frac{pq}{n}} = \sqrt{\frac{(.4)(.6)}{1600}} = .012$$

The graph of this distribution is given in Figure 7.13.

FIGURE 7.13 DISTRIBUTION OF $\hat{P}$

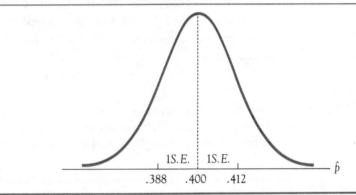

As with the mean, we will now introduce the standard error.

DEFINITION 7.6

The *standard error* S.E. of the proportion $\hat{P}$ is defined as the standard deviation of the distribution of $\hat{P}$.

Thus for large samples, $S.E. = \sqrt{(pq)/n}$ and we may write $Z = (\hat{P} - p)/S.E.$. It may seem confusing that the standard error of both the mean $\overline{X}$ and the proportion $\hat{P}$ are denoted by S.E. No problems arise from this, however since it will always be clear from the discussion which sampling distribution is involved.

Example 7.11 In support of a bid, a large aerospace company stated that 67% of its engineering staff had a bachelor's degree in engineering. Suppose a sample of 300 personnel files showed that 183 had engineering degrees. Thus $\hat{p} = 183/300 = .61 = 61\%$.

Is a random sample with a proportion this small (or smaller) very likely? The required probability is $P(\hat{P} \leq .61)$. Since

$$z = \frac{\hat{p} - p}{\sqrt{\dfrac{pq}{n}}} = \frac{.61 - .67}{\sqrt{\dfrac{(.67)(.33)}{300}}} = -2.21$$

we see that these values of $\hat{P}$ are more than 2.2 standard errors below the proportion $p = .67$. The probability will be quite small. In fact, $P(Z \leq -2.21) = .0136$ and thus only slightly more than 1% of all samples should have a proportion this small. ∎

Inferences based on sample proportions generally require the use of much larger samples than would be needed with means. The reason for this is that small differences translate into large percentages. More specifically, each unit in the second decimal place of the difference $\hat{p} - p$ represents a 1% difference when $\hat{p}$ and p are expressed as percentages. Thus if $\hat{p} - p = .08$ there is an 8% difference between $\hat{p}$ and p.

Example 7.12

Suppose the population proportion is estimated to be around $p = .5$. What sample size is needed to be 95% certain that the sample proportion differs from that of the population by at most $.04 = 4\%$? Refer to Figure 7.14.

FIGURE 7.14 $P(p - .04 < \hat{P} < p + .04) = .95$

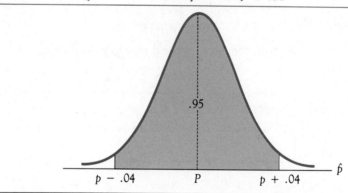

Since $P(-1.96 < Z < 1.96) = .95$ and $\hat{p} - p = \pm.04$, we have

$$1.96 = z = \frac{\hat{p} - p}{S.E.} = \frac{.04}{\sqrt{\dfrac{(.5)(.5)}{n}}}$$

On solving for n, we obtain $n = 600.25$. Thus the sample size should be at least 601. ∎

In this last example, if we reduce the difference to $.02 = 2\%$, the required sample size becomes 2401. In opinion polling, for example, sampling sizes in the range of 1200 to 2500 are commonly used in order to keep the error in the estimate or prediction in the 2 to 3% range.

The sampling distribution of $\hat{P}$ that we have been discussing applies to large samples and to populations where p is not close to 0 or 1. When small samples are used, the binomial distribution is an alternative regardless of the size of p. If, however, p is quite small, say .05, .01, .001, and so on, as in communicable disease studies, it may be necessary to survey large segments of the population, involving literally tens of thousands of subjects. In these cases another distribution, the Poisson distribution, is frequently used. Information on this distribution can be found in more advanced texts.

EXERCISE SET 7.3

Part A

1. Describe the distribution of $\hat{P}$ for samples of size $n = 400$ from a population with $p = .6$. Give the mean and standard error and sketch the distribution.

2. Assuming a population with $p = \frac{1}{2}$, what is the difference between the standard errors of the proportions associated with samples of size $n = 400$ and $n = 1600$?

3. Consider the distribution of $\hat{P}$ for samples of size 625 from a population with $p = .4$.
   a. Find $P(.38 < \hat{P} < .42)$.
   b. Find $P(\hat{P} > .44)$.

4. Suppose large samples are drawn from a population. Approximately what percentage of all samples of size n will have proportions that:
   a. Are within one standard error of p?
   b. Are within two standard errors of p?
   c. Are within 1.64 S.E. of p?
   d. Exceed p by two standard errors?

5. Suppose public opinion is split 60–40 in favor of increasing taxes to balance the budget. If a random sample of 300 people are interviewed, what is the probability that the proportion favoring a tax reduction for this purpose will be less than 54%?

6. A director of nursing personnel of a large metropolitan hospital estimates that three-eighths of the applicants for nursing-related jobs fail to attend the scheduled interview. Assuming this also holds for other hospitals in the area, what is the probability that in a random sampling of 800 scheduled interviews, fewer than 35% of the applicants will fail to attend?

7. In a 1970 New York study on the propagation of the subculture surrounding drug users, it was found that of marijuana users who had occasionally tried heroin, 45% had a heroin-using friend. Assume $p = .45$ holds for the population of such users. If a random sample of 500 occasional marijuana users are interviewed, what is the probability that the proportion who have a heroin-using friend will exceed 50%?

8. Suppose a nationwide survey reveals that 35% of all men smoke. You wonder if this applies to your area and interview a sample of 625 men. What is the probability that the proportion of smokers in the sample exceeds 41% if indeed $p = .35$? If you actually obtained such results, what would you think about the suitability of $p = .35$ for your population?

9. In a study of college students by the American Council for Education and by researcher Prof. Alexander Astin at the University of California at Los Angeles,

more than 67% of the students interviewed stated that a very important reason in their decision to attend college is "to be able to make more money." Assuming this percentage to be true in your area, what is the probability that a random sample of 300 students would have between 64% and 70% agreeing with this reason (assuming a similar questionnaire, etc.)?

10. The proportion of registered voters who favor an incumbent governor is believed to be around 50%. To estimate the true proportion, a random sample will be selected and interviewed. How large should the sample be if its proportion is to be within 2% of that of the population with a probability of .90?

Part B

11. In a study of immunocompetence and survival rates for women with operable breast cancer, it was found that the probability of disease recurrence within 48 months following surgery was $p = .26$ for those with optimal immune response capability and $p = .61$ for those who were judged suboptimal. Let us accept $p = .61$ for the suboptimal population and suppose that a treatment to increase the immune response is tried on 124 such patients. If the treatment is ineffective, what is the probability that the proportion having a recurrence in the next 48 months is less than 55%?

    Source: Immunocompetence, Immuno Suppression, and Human Breast Cancer, A. Adler, *Cancer*, Vol. 45, April 15, 1980.

12. In a study of anticoagulant therapy following heart attacks (myocardial infarctions) by researchers at the Johns Hopkins Medical School, it was found that the fatality rate for those not using the follow-up anticoagulant therapy was somewhere around 31%. They also found a fatality rate of only 18% for a sample of 597 patients who were treated with anticoagulants (heparin and coumadin derivatives). What is the probability of a sample this size producing such a small proportion if the anticoagulants had been ineffective? (Assume a population with $p = .31$.)

    Source: Additional Data Favoring Use of Anticoagulant Therapy in Myocardial Infraction, M. Szklo, M.D., et al.; *Journal of the American Medical Association*, Vol. 242, No. 12, Sept. 21, 1979.

13. In a federally funded study conducted between 1978 and 1981 by two University of Massachusetts researchers (J. D. Wright and P. Rossi of the Social and Demographic Research Institute), it was found that the proportion of American families acknowledging gun ownership remained constant at about 50%. Suppose this proportion holds true for your community. What is the probability that a random sample of 900 families would have 54% or more who acknowledge owning guns?

14. In a study of cigarette smoking and peptic ulcers at the Cracow Medical School in Poland, 26.7% was established as the norm rate for peptic ulcer pain syndrome in men. If this syndrome is not related to smoking, what is the probability that in a random sample of 600 smokers more than 28% will have this syndrome?

15. According to the Mendelian law of selection or segregation in genetics, a flower such as a snapdragon carries two types of genes (alleles): R (red) and w (white). The possible matings of these genes can be described by the following table:

Gene from ovule	Gene from pollen	Offspring
R	R	RR (red)
R	w	Rw (pink)
w	R	wR (pink)
w	w	ww (white)

These four cases are equally likely: each has probability ¼. Thus if we ignore variations caused by such factors as mutations, 25% of all the flowers should be white. What is the probability that of 400 of these flowers, at least 26.5% are white?

16. How large a sample is needed from a population with $p = .3$ in order to have

$$P(.26 < \hat{P} < .34) = .90$$

17. It is reported that about 40% of taxpayers believe it would be easy to cheat on their income tax. If this is true, how large a sample of taxpayers is needed to produce a proportion with this belief that falls on the interval $.40 \pm .03$ with a probability of $.99$?

7.5   THE DISTRIBUTION OF THE SAMPLE VARIANCE S^2

The formula for the variance of a sample with n observations is

$$s^2 = \frac{\sum_{i=1}^{n} (x_i - \bar{x})^2}{n - 1}$$

Multiplying this equation by $(n - 1)$ and dividing by the population variance σ^2 yields

$$\frac{(n - 1)s^2}{\sigma^2} = \sum_{i=1}^{n} \frac{(x_i - \bar{x})^2}{\sigma^2}$$

The quantity of the right looks a great deal like the value of a chi-square variable. This, in fact, is the case. The variable, however, has $n - 1$ rather than the n degrees of freedom one might expect from such sums.

THEOREM 7.5

Let S^2 be the variance associated with samples of size n from a normal distribution with variance σ^2. Then the variable

$$\chi^2 = \frac{(n - 1)S^2}{\sigma^2}$$

has the chi-square distribution with $\nu = n - 1$ degrees of freedom.

Example 7.13

A sample of size $n = 10$ is selected from a normal population with variance $\sigma^2 = 12$. What is the probability that the sample variance exceeds 18? The χ^2 distribution with $\nu = 10 - 1 = 9$ degrees of freedom is given in Figure 7.15.

FIGURE 7.15   THE χ^2 DISTRIBUTION WITH $\nu = 9$ DEGREES OF FREEDOM

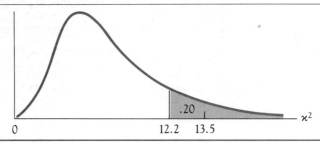

For $s^2 = 18$ and $\sigma^2 = 12$, the computed chi-square value is

$$\chi^2 = \frac{(n-1)s^2}{\sigma^2} = \frac{9(18)}{12} = 13.5$$

Thus $P(S^2 \geq 18) = P(\chi^2 > 13.5)$. Because of the limited nature of the chi-square tables this cannot be determined exactly. We do see from Table IV, however, that $P(\chi^2 \geq 12.2) = 1 - .80 = .20$ and $P(\chi^2 > 14.7) = 1 - .90 = .10$. Thus the probability is somewhere between .2 and .1. ■

7.6

THE DISTRIBUTION OF THE RATIO OF TWO SAMPLE VARIANCES: THE F DISTRIBUTION

The ratio of the variances of two independent samples from normal populations is of extreme importance in tests involving the means of several populations. The distribution of this ratio was first investigated by Sir Ronald Fisher (1890–1962) and it is called the F distribution in his honor. The derivation of this distribution depends on the fact that if two independent chi-square variables are divided by their respective degrees of freedom, then their ratio has the F distribution with these degrees of freedom.

DEFINITION 7.7

Let χ_1^2 and χ_2^2 be two independent chi-square variables with ν_1 and ν_2 degrees of freedom, respectively. Then the variable

$$F = \frac{\chi_1^2/\nu_1}{\chi_2^2/\nu_2}$$

has the F *distribution* with ν_1 and ν_2 degrees of freedom.

The F distribution has two parameters, ν_1 and ν_2; ν_1 is associated with the numerator and ν_2 with the denominator. To identify a particular distribution we shall write $F(\nu_1, \nu_2)$. It is not important which of the two chi-square variables one chooses for the numerator, since the ratio of the two will have the F distribution. It is important, however, that the degrees of freedom should match those of the chosen numerator and denominator.

The form of the distribution $F(\nu_1, \nu_2)$ is shown in Figure 7.16. As is customary, the lowercase f is used for values of F. These are nonnegative and, depending on the number of degrees of freedom, can be quite large. The distribution is continuous and has a mean or expected value $\mu_F = \nu_2/(\nu_2 - 2)$. For large values of ν_2, this ratio is approximately 1. And as with any continuous distribution, probabilities are represented by areas under this density curve.

FIGURE 7.16 THE *F* DISTRIBUTION

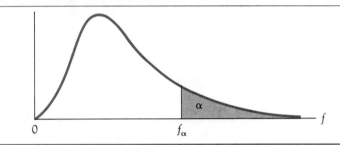

The tabulation of the distribution of $F(\nu_1, \nu_2)$ must be abbreviated considerably because of the two separate sets of degrees of freedom. Table VII of the Appendix contains the percentage points corresponding to $\alpha = .05, .025,$ and $.01$. These we denote by f_α, or by $f_\alpha(\nu_1, \nu_2)$ if we wish to stress the number of degrees of freedom.

Example 7.14

Find $f_{.05}(6, 8)$. In Table VII first locate the subtable headed $\alpha = .05$. Then find the number of degrees of freedom of the numerator, $\nu_1 = 6$, as a column heading and that of the denominator, $\nu_2 = 8$, as a row heading. The entry at the intersection of this row and column is $f_{.05}(6, 8)$. Thus $f_{.05}(6, 8) = 3.58$ and $P(F > 3.58) = .05$ when $\nu_1 = 6$ and $\nu_2 = 8$. This area is shown in Figure 7.17. ■

FIGURE 7.17 $f_{.05}(6, 8) = 3.58$

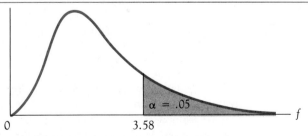

More extensive tabulations for the F distribution are available. Most do not include percentage points for large α, since it can be shown that

$$f_{1-\alpha}(\nu_1, \nu_2) = \frac{1}{f_\alpha(\nu_2, \nu_1)}$$

Example 7.15

Find $f_{.99}(5, 4)$. We use

$$f_{.99}(5, 4) = \frac{1}{f_{.01}(4, 5)}$$

Refer to Table VII with $\alpha = .01$, noting that the degrees of freedom are interchanged. At the intersection of the column headed $\nu_1 = 4$ and the row headed $\nu_2 = 5$ we find $f_{.01}(4, 5) = 11.4$. Thus

$$f_{.99}(5, 4) = \frac{1}{11.4} = .0877$$

The two areas are shown in Figure 7.18 ∎

FIGURE 7.18   F DISTRIBUTIONS FOR EXAMPLE 7.15

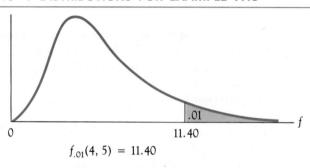

$f_{.01}(4, 5) = 11.40$

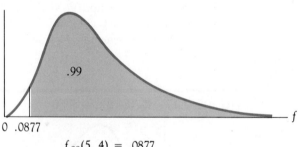

$f_{.99}(5, 4) = .0877$

In most applications we avoid having to use this formula by working with f values that are greater than 1.

To see how the F distribution applies to the ratio of two variances, consider two populations having variances σ_1^2 and σ_2^2. Let S_1^2 be associated with samples of size n_1 from the first population and S_2^2 with samples of size n_2 from the second. In the previous section we saw that the variables

$$\chi_1^2 = \frac{(n_1 - 1)S_1^2}{\sigma_1^2} \qquad \chi_2^2 = \frac{(n_2 - 1)S_2^2}{\sigma_2^2}$$

have chi-square distributions. Dividing these equations by their respective degrees of freedom and forming their ratio yields

$$F = \frac{\chi_1^2/(n_1 - 1)}{\chi_2^2/(n_2 - 1)} = \frac{S_1^2/\sigma_1^2}{S_2^2/\sigma_2^2}$$

THEOREM 7.6

Let S_1^2 and S_2^2 be the variances associated with independent samples of size n_1 and n_2 from two normal populations with variances σ_1^2 and σ_2^2. Then

$$F = \frac{S_1^2/\sigma_1^2}{S_2^2/\sigma_2^2}$$

has the F distribution with $\nu_1 = n_1 - 1$ and $\nu_2 = n_2 - 1$ degrees of freedom.

Theorem 7.6 is a remarkable result. It applies to large and small samples alike. In the case of large samples it will be recalled that the expected value of F is near 1. For such samples there is a high probability that the ratios S_1^2/σ_1^2 and S_2^2/σ_2^2 do not differ too greatly, regardless of the normal populations involved.

In most applications we work under the assumption that both samples are from the same normal population and attempt to show the contrary. When $\sigma_1^2 = \sigma_2^2$ we see that Theorem 7.6 also applies to the ratio

$$F = \frac{S_1^2}{S_2^2}$$

Again for large samples, there is a high probability that this ratio should be close to 1. Thus just as we expect the means of large samples to be approximately the same, so it is for their variances.

Example 7.16

Two samples of sizes $n_1 = 10$ and $n_2 = 30$ from the same normal distribution have variances $s_1^2 = 54$ and $s_2^2 = 38$. Are such differences in variances unusual? For these variances the value of F is

$$f = \frac{s_1^2}{s_2^2} = \frac{54}{38} = 1.42$$

We note that for $\nu_2 = 30 - 1 = 29$, the expected value of F is

$$\mu_F = \frac{\nu_2}{\nu_2 - 2} = \frac{29}{29 - 2} = \frac{29}{27} = 1.07$$

FIGURE 7.19   THE F(9, 29) DISTRIBUTION

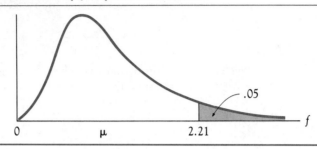

(see Figure 7.19). The computed f value is quite close to the expected value of the F(9, 29) distribution. Thus this result is not unusual. Additionally we note that f values beyond $f_{.05}(9, 29) = 2.21$ are ordinarily considered large and are the basis for questioning whether the two samples had come from the same population. These results are more or less typical of what one might expect when using two samples from the same population. ■

In this last example, the variances $s_1^2 = 54$ and $s_2^2 = 38$ might appear to be in strong disagreement. The F distribution leads us to conclude otherwise. The reader should become accustomed to having this happen in working with variances. Since variances are squares of standard deviations, large differences may be deceiving. The standard deviations for these two samples are $s_1 = \sqrt{54} = 7.35$ and $s_2 = \sqrt{38} = 6.16$, which seem to be in reasonably good agreement. If you agree with this, then you must also agree that the variances of 38 and 54 compare favorably.

Example 7.17

Scores on two forms, of an achievement test, 1 and 2, are supposed to have the same distribution, which is approximately normal. However, some evidence suggests that the variance of the form 1 scores may exceed that of the form 2 scores. Suppose samples of $n_1 = 25$ form 1 scores and $n_2 = 40$ form 2 scores are selected and found to have respective variances of $s_1^2 = 68.2$ and $s_2^2 = 31.6$. Discuss the assertion that test scores on these two test forms have the same variance.

   If the distributions for both forms are identical, then the two samples are from the same population. Then

$$f = \frac{s_1^2}{s_2^2} = \frac{68.2}{31.6} = 2.16$$

is the value of a variable having the F distribution. For $v_1 = 25 - 1 = 24$ and $v_2 = 40 - 1 = 39$, we find $f_{.05}(24, 39) = 1.79$ (using the nearest table entry). Thus $P(F > 1.79) \le .05$. That is, the probability that the ratio of the sample 1 to sample 2 variance exceeds 1.79 is at most .05. Certainly f values as large as $f = 2.16$ rarely occur (see Figure 7.20). These samples strongly suggest that the form 1 test scores have a variance exceeding the variance of the form 2 scores.

FIGURE 7.20 $f_{.05}(24, 39) = 1.79$

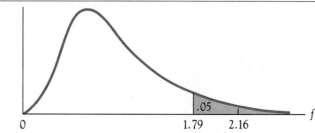

EXERCISE SET 7.4

Part A

1. Samples of size $n = 16$ are being selected from a normal population with variance $\sigma^2 = 23$. Estimate the probability that the sample variance will exceed 30.

2. The scores X on an assertiveness evaluation test are reported to be approximately normal with a variance of 26. If S^2 is the variance of samples of size 9, find the number c such that $P(S^2 < c) = .95$. In other words, what is the 95th percentile for the sample variances?

3. An I.Q. test is scaled or normalized to have scores with a mean of 100 and a standard deviation of 15. Estimate the probability that the standard deviation of a sample of 15 scores will exceed 20.

4. Find c so that:
   a. $P(F < c) = .95$ when $v_1 = 6$ and $v_2 = 12$.
   b. $P(F < c) = .99$ when $v_1 = 10$ and $v_2 = 10$.

5. Find:
   a. $f_{.05}(7, 10)$.
   b. $f_{.01}(20, 17)$.

6. Two samples of size $n_1 = 16$ and $n_2 = 25$ from the same normal population had variances of $s_1^2 = 29.6$ and $s_2^2 = 9.8$. Are such great differences in sample variances unusual?

7. A hydrocarbon emission control system is installed on 21 steam-powered electric generating plants and tested. The emission levels had a mean of 40.6 and a standard deviation of 11.2 parts per million (ppm). Following the six-month service of the emission control equipment, emission tests were again conducted on 16 of the plants. This time the mean and standard deviation were 39.1 and 18.9 ppm. Is such an increase in variation in sample data (as reflected in the standard deviations) unusual, if in fact the distribution of the emission levels is the same from the serviced equipment as from the original?

8. At the beginning of a three-hour laboratory class the students were given a test of basic arithmetic skills. Two weeks later the same test was given again. This time however, the first two hours of the laboratory involved distractions that repeatedly put the students in stressful situations (overheated classroom, power mower operating outside the classroom creating a loud noise, etc.). The question, of course, is whether this stress changed the distribution of the grades. The results are given below. Is such an increase in variances unusual if the stress produced no change in the distribution of grades?

First test	Second test
$\bar{x} = 152$	$\bar{x} = 153$
$s^2 = 171$	$s^2 = 900$
$n = 26$	$n = 26$

Part B

9. Find μ_F and indicate what might be considered "large" values of F.
   a. $F(5, 3)$
   b. $F(5, 6)$
   c. $F(5, 40)$
10. The distribution $F(4, 20)$ is given at the right. What values of F would be considered large for this distribution?

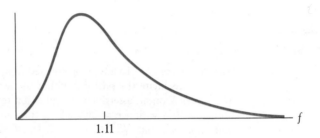

1.11

11. Does $f_{.05}(10, \nu_2)$ increase or decrease as ν_2 becomes large? Does $f_{.05}(\nu_1, 10)$ increase or decrease as ν_1 gets large? What happens to $f_{.05}(\nu_1, \nu_2)$ as both ν_1 and ν_2 become large? What does this imply about the ratio of two sample variances as the sample sizes become larger and larger?
12. Let S_1^2 and S_2^2 be the variances of samples of 25 ACT scores. Assuming the samples are independent and the ACT scores are normal, find $P(S_1^2/S_2^2 > 2)$ approximately.
13. Find:
   a. $f_{.95}(4, 6)$.
   b. $f_{.99}(10, 12)$.
   Sketch both distributions and show the areas involved.
14. It can be shown that t^2 with ν degrees of freedom has the $F(1, \nu)$ distribution. Check this for selected values of α and ν by comparing the square of $t_{\alpha/2}(\nu \ d.f.)$ with $f_\alpha(1, \nu)$.

Example 7.17

A Research Study

Data: The Weak Link

In the course of designing a study, concern with such factors as defining the objectives, selecting the samples, and choosing the proper tests often overshadow the gathering of data. We should not forget that unreliable data invariably lead to worthless results, regardless of the amount of analysis to which they are subjected. And we should not assume that accepted techniques by trained technicians will produce good data.

In cancer studies the worth of a therapy for treating tumors is often determined by whether it produces a reduction in the tumor size or a slowing of its growth rate. In either case the tumor must be measured beneath the skin and body tissue of the patient. To simulate what might be encountered under ideal conditions of measurement, two Mayo Clinic researchers, Dr. C. G. Moertel and Dr. J. A. Hanley, placed spheres of varying size under layers of foam and asked 16 oncologists (cancer specialists) to measure their diameters. The results are presented in "The Effect of Measuring Error on the Results of Therapeutic Trials in Advanced Cancer," *Cancer*, vol. 38, July 1976, pp. 388–394.

The diameter of the 12 spheres are the column headings of Table 7.4. The tabular entries are the measurements by the 16 oncologists. Unknown to these specialists, spheres 5 and 6 were a matched pair, as were 7 and 8. (The slight differences in their diameters were due to the manufacturing process.) These pairs were used to see how well a specialist could reproduce an observation.

The variation in the 16 entries in any column results from errors caused by the bias of the specialists and random errors that arise when any individual repeats an observation of some specific quantity.

The bottom rows of this table show that the group tended to underestimate tumor sizes. For example, the first column (sphere 1) has a mean of 1.49 centimeters, which is 0.31 centimeters below the actual diameter of the sphere. Thus for this sphere there is a group bias of $-.31$ centimeters, which is approximately 18% of the true diameter. For all 12 spheres, these errors varied from 1% to 18%, with a mean of about 12%.

As we remarked above, spheres 5 and 6 are essentially the same size. Thus the entries of any specific row in column 5 and 6 would be the same if that specialist could reproduce his observations. Obviously this is not the case. Based on the differences in these observations, Moertel and Hanley were able to predict a 9% error in an estimated diameter resulting from an individual's inability to reproduce an observation.

Combining these two sources of variation, the researchers were then able to predict an average error of 17% in the measured diameter of a tumor. Contrasting the ideal conditions of this experiment with those encountered in practice, one must conclude that this 17% figure is an extremely optimistic error estimate. In reality, it is more like a lower bound for the error.

The next step was to examine the effects of such errors on the diagnosis of a growth or reduction in tumor size. Specifically, the researchers asked how often an erroneous diagnosis would result if either a 25% or a 50% change in the crosssectional area of the tumor was taken as a basis for diagnosing growth or shrinkage of the tumor.

Consider the matched pair of spheres 5 and 6. The measurement of 5 may be regarded as that of the initial or baseline tumor, while that of 6 is for the unchanged tumor at some later time. Or their roles could be interchanged, wtih sphere 6 serving as the baseline tumor. Similar considerations apply to spheres 7 and 8. Thus there are 64 pairs of observations of tumors which have not changed size.

Now in how many of these pairs of observations would the measurements of these specialists lead to the wrong diagnosis? These are indicated in Table 7.5 on pages 270 and 271 by a dagger †. We shall concentrate on the false positive or erroneous shrinkage diagnosis. Using the 25% area change criterion, we see that 12 (18.75%) of the pairs would result in a false positive diagnosis of a shrinkage in the tumor size. And remember, this is when both observations are made by the same specialist. When different individuals were involved in the "before" and "after" measurements, the error rate for the false positive diagnosis jumped to 24.8%.

TABLE 7.4 TUMOR DIAMETERS, ACTUAL AND AS MEASURED (IN CENTIMETERS) BY 16 INVESTIGATORS

	Tumor number and diameter (cm)											
Investigator	1 1.8	2 2.3	3 2.7	4 3.5	5 4.3	6 4.4	7 5.3	8 5.5	9 6.4	10 7.5	11 8.6	12 14.5
1	1.3	1.8	2.5	3.0	4.0	4.0	5.0	6.5	5.0	7.0	7.5	15.0
2	1.5	2.0	3.0	4.0	4.5	3.0	5.0	4.5	5.0	6.5	7.5	13.0
3	1.5	2.0	2.5	3.5	4.0	4.0	5.5	5.0	6.0	7.0	8.0	15.0
4	1.5	1.8	2.5	3.3	4.0	3.8	4.7	4.5	5.0	6.4	7.0	13.0
5	1.8	2.0	3.0	4.0	4.5	5.0	5.5	6.0	6.0	6.5	9.0	15.0
6	1.8	2.3	3.0	4.0	4.5	5.0	5.5	5.5	6.0	7.5	7.5	15.0
7	1.0	2.0	2.5	3.0	4.0	3.5	4.0	5.0	4.0	6.0	6.0	15.0
8	1.5	2.0	3.0	4.0	5.0	3.5	7.0	5.5	6.5	6.5	4.0	19.0
9	1.6	2.0	2.5	3.5	4.0	4.5	5.5	5.5	5.5	7.5	9.0	14.0
10	1.2	1.5	2.0	3.0	4.0	5.4	4.5	6.0	6.0	7.0	7.5	15.0
11	1.5	2.2	3.0	4.3	4.2	5.2	5.5	5.4	6.0	7.5	10.5	18.0
12	1.5	1.5	2.0	3.0	4.5	3.0	6.0	4.0	4.5	4.5	8.5	14.0
13	2.0	2.0	3.0	3.5	4.0	4.5	5.5	5.5	6.0	7.0	8.0	15.0
14	1.6	2.2	2.8	3.8	4.5	5.6	5.3	5.4	6.2	6.8	8.1	14.0
15	0.8	1.0	1.5	2.0	3.0	2.0	4.0	3.0	4.0	5.0	6.0	12.0
16	1.7	1.8	2.2	3.5	3.8	3.8	4.8	4.6	5.0	6.0	6.0	13.0
Group mean	1.49	1.88	2.56	3.46	4.16	4.06	5.21	5.12	5.42	6.61	7.5	14.69
Group bias	−.31	−.42	−.14	−.04	−.14	−.34	−.09	−.38	−.98	−.89	−1.09	.19

On the basis of these findings the researchers concluded that "a 25% regression in tumor area is inadequate to assure a reasonable degree of accuracy or specificity." Even the 50% criterion is questionable.

However, when these results were coupled with known volume doubling times for tumors, Moertel and Hanley were able to conclude that a positive diagnosis for a particular therapy would be statistically sound if there was a 50% reduction in the largest diameter of the tumor after at least 60 days following the onset of the therapy.

As a result of this study, cancer researchers now require substantially more evidence of tumor shrinkage before they are willing to accept a therapy. ■

EXERCISE SET 7.5

1. Refer to Table 7.4. Compare the number of specialists who underestimated the diameters of the smallest and largest spheres.
2. The measurements of the spheres were made under what have been described as ideal conditions. What are some problems that might be encountered in measuring the diameters of a tumor? In particular, what would be the diameter of a tumor if it was not spherical?
3. Refer to Table 7.4. Compute the average bias using all 12 spheres. What percent is this of the average diameter of all the spheres?
4. Refer to Table 7.4. Which specialists appear to round to the nearest half centimeter? Which to the nearest tenth?
5. Refer to Table 7.5 (page 270). What percent of the 64 pairs produced an erroneous false negative diagnosis of tumor growth when the 25% criteria was applied? What percent produced a false negative when the 50% criteria was used?
6. Find the article in *Cancer* and read it. Answer the following:
   a. What references are cited?
   b. The graph of the standard deviation of the measurements of the individual spheres is given in the article. Copy it and describe what appears to be the trend in the change of the standard deviations as the diameters increase in size.

Library

Project

The following problems refer to a Library Project involving *Index Medicus*. This index is the most complete source of published research in the medical field. Currently, a year's index comprises some 12 to 14 volumes. The first six or so of these are indexed by the authors' names, the remainder by subject and title. The subject classification is somewhat difficult to use because of the many specialized subcategories. Also, some of the journal titles are abbreviated, so that a reference of titles (separate volume) must be occasionally used. In spite of these minor annoyances *Index Medicus* is the standard guide to publications in the medical and health science fields.

1. In 1976, C. G. Moertel, W. F. Taylor, and A. Roth published "Who Responds to Sugar Pills?" Find the source of this publication using both the author and subject index of *Index medicus*. Where was the article published?
2. C. G. Moertel and other researchers at the Mayo Cancer Clinic investigated Linus Pauling's vitamin C therapy for cancer. Locate the article. (This is not easy. First, one must decide at roughly what year to start searching.
3. Check the 1980, 1981, and 1982 editions of *Index Medicus* for articles on laetrile therapy for cancer. Locate one of the articles and summarize its contents.

TABLE 7.5 ERRONEOUSLY DECLARED OBJECTIVE RESPONSE AND OBJECTIVE PROGRESSIONS WHEN THE SAME INVESTIGATOR REPEATS A MEASUREMENT ON A TUMOR WHICH HAS REMAINED THE SAME SIZE

	Erroneous Classifications if B is Baseline Tumor				Tumor "area" Measurements (cm²)		Erroneously Classified Responses if A is Baseline Tumor			
Investigator	≥50% Growth	≥25% Growth	≥25% Shrinkage	≥50% Shrinkage	Tumor A	Tumor B	≥50% Shrinkage	≥25% Shrinkage	≥25% Growth	≥50% Growth
Tumors 5 and 6										
1					16.0	16.0				
2	+	+			20.2	9.0	+	+		
3					16.0	16.0				
4					16.0	14.4				
5					20.2	25.0				
6					20.2	25.0				
7		+			16.0	12.2				
8	+	+			25.0	12.2	+	+		
9					16.0	20.2			+	
10					16.0	20.2			+	
11			+		17.6	27.0			+	+
12	+	+			20.2	9.0	+	+		
13					16.0	20.2			+	
14			+		20.2	31.4			+	+
15	+	+			9.0	4.0	+	+		
16					14.4	14.4				
Tumors 7 and 8										
1			+		25.0	42.2			+	+
2					25.0	20.2				
3					30.2	25.0				
4					22.1	20.2				
5					30.2	36.0				
6					30.2	30.2				
7			+		16.0	25.0			+	+
8	+	+			49.0	30.2		+		
9					30.2	30.2				
10			+		20.2	36.0			+	+
11					30.2	29.2				
12	+	+			36.0	16.0	+	+		
13					30.2	30.2				
14					28.1	29.2				
15	+	+			16.0	9.0		+		
16					23.0	21.2				

Baseline Tumor	Tumor Measured Subsequently	Number of Same-Investigator Evaluations	Number of Investigators Who Reported Objective Responses		Number of Pairings Who Report Objective Progression	
			≥25% Shrinkage	≥50% Shrinkage	≥25% Growth	≥50% Growth
5	6	16	4	4	2	2
6	5	16	2	0	4	4
7	8	16	3	1	3	3
8	7	16	3	0	3	3
TOTAL		64	12 (18.8%)	5 (7.8%)	12 (18.8%)	12 (18.8%)

COMPUTER PRINTOUT

The process of sampling a population with a known distribution may be readily simulated with the aid of Minitab. Since sample means can also be easily computed, we thus have a way to illustrate the distribution of sample means. An example of this is given in Figure 7.21. Here 125 samples of size $n = 4$ are simulated from a normal population with mean 500 and standard deviation 100. Note the technique. First, four samples of size 125 are generated using the NRANDOM command. The first members of each of these four samples then yield a sample of size $n = 4$. Similarly, the second members furnish a second sample, and so forth. The distribution of all samples of size $n = 4$ from this normal population is, of course, normal with mean $\mu = 500$ and standard deviation $\sigma = 100/\sqrt{4} = 50$. Already we begin to see evidence of this result in the distribution of these 125 sample means.

Corresponding results for 600 samples of size $n = 4$ are shown in Figure 7.22. The printing of the means has been omitted. The resulting distribution of these 600 means is quite close to the theoretical distribution of $\overline{X}$.

FIGURE 7.21   MINITAB PRINTOUT: SIMULATION OF 125 SAMPLE MEANS FROM A NORMAL POPULATION

```
MTB >NRANDOM 125, MU = 500, SIGMA = 100, PUT INTO C1
MTB >NRANDOM 125, MU = 500, SIGMA = 100, PUT INTO C2
MTB >NRANDOM 125, MU = 500, SIGMA = 100, PUT INTO C3
MTB >NRANDOM 125, MU = 500, SIGMA = 100, PUT INTO C4
MTB >ADD C1-C4, PUT SUMS INTO C5
MTB >DIVIDE C5 BY 4, PUT RESULTS INTO C6
MTB >PRINT C6
COLUMN          C6
COUNT          125
      517.881    473.790    516.165    501.302    538.591
      476.616    483.562    466.721    506.631    579.350
      545.112    434.594    517.572    416.067    458.516
      457.309    567.999    497.665    469.267    531.657
      602.668    541.109    507.912    508.569    537.807
      487.538    520.088    494.344    451.797    497.764
      568.186    494.411    464.356    517.232    416.084
      467.527    442.982    501.147    501.269    508.623
      562.895    555.030    459.891    460.773    563.647
      552.750    548.284    473.248    496.521    530.634
      448.126    434.371    532.984    478.108    525.938
      552.034    570.485    500.033    550.557    450.381
      564.016    442.541    543.369    545.705    511.780
      487.119    452.710    445.931    481.267    420.626
      502.226    566.815    489.395    529.359    547.807
      438.538    509.480    562.934    458.392    527.800
      535.453    555.847    529.175    534.846    576.700
      609.621    511.221    520.997    509.966    445.509
      547.076    483.964    522.205    471.516    434.033
      470.428    501.011    486.434    567.906    485.598
```

FIGURE 7.21   MINITAB PRINTOUT: SIMULATION OF 125 SAMPLE MEANS FROM A NORMAL POPULATION—cont.

```
     577.655    361.155    488.294    475.877    516.602
     484.309    491.261    488.126    478.993    480.707
     637.474    491.048    545.102    565.126    488.570
     480.834    537.988    448.087    546.036    486.049
     455.027    621.459    452.262    455.058    396.586

MTB >AVERAGE C6

     AVERAGE  =        504.25
MTB > STANDARD DEVIATION C6
     ST.DEV.  =        47.880
MTB >HISTOGRAM C6

     MIDDLE OF      NUMBER OF
     INTERVAL       OBSERVATIONS
        360.            1      *
        380.            0
        400.            1      *
        420.            3      ***
        440.           10      **********
        460.           15      ***************
        480.           22      **********************
        500.           20      ********************
        520.           14      **************
        540.           17      *****************
        560.           14      **************
        580.            4      ****
        600.            2      **
        620.            1      *
        640.            1      *


MTB >STOP
```

FIGURE 7.22   MINITAB PRINTOUT: SIMULATION OF 600 SAMPLE MEANS FROM A NORMAL POPULATION

```
MTB >DIMENSION THE COLUMNS TO 600 ENTRIES
MTB >NOPRINT
MTB >NRANDOM 600 OBSERVATIONS MU = 500, SIGMA = 100,
     PUT INTO C1
MTB >NRANDOM 600 OBSERVATIONS MU = 500, SIGMA = 100,
     PUT INTO C2
MTB >NRANDOM 600 OBSERVATIONS MU = 500, SIGMA = 100,
     PUT INTO C3
MTB >NRANDOM 600 OBSERVATIONS MU = 500, SIGMA = 100,
     PUT INTO C4
MTB >ADD C1-C4, PUT ROW SUMS INTO C5
MTB >DIVIDE C5 BY 4, PUT RESULTS INTO C6
MTB >AVERAGE C6
```

FIGURE 7.22   MINITAB PRINTOUT: SIMULATION OF 600 SAMPLE
MEANS FROM A NORMAL POPULATION—cont.

```
      AVERAGE  =          501.86
MTB >STANDARD DEVIATION C6
      ST.DEV.  =          49.493
MTB >HISTOGRAM C6
      EACH * REPRESENTS    2 OBSERVATIONS


MIDDLE OF    NUMBER OF
INTERVAL    OBSERVATIONS
    360.        3     **
    380.        4     **
    400.       13     *******
    420.       28     **************
    440.       47     ***********************
    460.       57     *****************************
    480.       88     ********************************************
    500.       99     **************************************************
    520.       93     ***********************************************
    540.       63     ********************************
    560.       56     ****************************
    580.       22     ***********
    600.       24     ************
    620.        1     *
    640.        2     *

MTB >STOP
```

CHAPTER CHECKLIST

Vocabulary

Sample statistics

Sampling distributions

Unbiased estimator

Sample mean (variable)

Sample standard deviation (variable)

Sample proportion (variable)

Central limit theorem

Standard error of the mean

t distribution

Critical value

Standard error of the proportion

F distribution

Techniques

Finding the distribution of $\overline{X}$ for a given "small" population by listing all samples

Describing the distribution of $\overline{X}$ when the sample size is large; including the mean and standard error.

Finding probabilities associated with the distribution of $\overline{X}$ when the sample size is large or the sampled population is normal

Describing the distribution of $\overline{X}$ when the population is normal and σ is approximated by s.

Finding critical values t_α of the t distribution.

Describing the distribution of $\hat{P}$ when the sample size or the number of trials n is large; including the mean and standard error

Finding probabilities associated with the distribution of $\hat{P}$ when n is large

Describing the distribution of the sample variance S^2 when the population is normal

Describing the distribution of the ratio of two sample variances from the same normal population

Finding critical values $f_\alpha(\nu_1, \nu_2)$ of the F distribution

Notation

$\overline{X}$	The mean. (A sample statistic)
S^2	The Variance (A sample statistic)
S	The Standard Deviation (A sample statistic)
$\mu_{\overline{X}}$	The mean of $\overline{X}$
$\sigma_{\overline{X}}$	The standard error of $\overline{X}$. Also the standard deviation of $\overline{X}$.
T	The variable of the T distribution
t_α	A critical value of the T distribution
F	The variable of the F distribution
$f_\alpha(\nu_1, \nu_2)$	The critical value of the f distribution which has ν_1 and ν_2 degrees of freedom.

CHAPTER TEST

1. Find (a) $t_{.05}$ when $\nu = 15$; (b) $f_{.05}(8, 10)$; (c) $f_{.95}(6, 4)$.

2. A population has a mean of 78 and a standard deviation of 18. Discuss the distribution of the means of samples of size $n = 36$ from this population. (Give the mean and standard error and make a sketch.)

3. Let $\hat{P}$ be the proportion of students who have a driver's license in a random sample of $n = 200$. Suppose 65% of all students have a driver's license.
   a. Discuss the distribution of $\hat{P}$, giving the mean and standard error and making a sketch.
   b. Find the probability that less than 55% of the students in a randomly selected sample of 200 will have a license.

4. A normal population has a mean of 186 and a standard deviation of 28. What is the probability that the mean of a random sample of size $n = 9$ will be within .5 standard errors of the population mean?

5. A medical study reported that a certain hormone improved the learning ability of 68% of the rats on which it was tested. Granting this to be true in general, suppose an experimenter tries to duplicate the results using 100 rats. What is the probability that less than 50% of the rats will show an improvement in their learning ability?

6. The outstanding account balances of a widely used credit card have a mean of $743 and a standard deviation of $250. In an audit, the 400 accounts of a random sample are examined. What is the probability that the sample mean exceeds $760?

7. Suppose third grade children have I.Q.s which are normally distributed with a mean of 100 and a standard deviation of 10. A sample of $n = 10$ third grade children who have been exposed to stimulating experiences from infancy on (color, sound, music, etc.) are tested and found to have a mean

I.Q. of 108. What is the probability of such a large mean if in fact the "stimulating experiences" have no effect on I.Q. as measured by the test employed?

8. Scores on a standardized mathematics test have a mean of 800 and a standard deviation of 100. Assuming a normal distribution,
    a. Approximately what percentage of all people who take the test score above 840?
    b. Approximately what percentage of all samples of 36 people who take this test will have a mean above 840?

9. A random sample of size $n = 23$ is selected from a normal population having a mean of 50 and a variance of 42. Is a sample variance exceeding 60 very likely?

10. Twenty-five young seedling plants were randomly assigned to two groups, with 12 plants in group A and 13 in B. These groups were then placed in an elementary classroom and the students were instructed to speak kindly and soothingly to those in group A and harshly and loudly to those in B. Otherwise the two groups received the same treatment. The height increases at the end of a four-week period for these two samples are summarized below:

Group A	Group B
$n_1 = 12$	$n_2 = 13$
$\bar{x}_1 = 8.63$ inches	$\bar{x}_2 = 8.00$ inches
$s_1^2 = .49$ (inches)^2	$s_2^2 = 2.63$ (inches)^2

    a. Is a difference in variances of this magnitude unusual if the treatment by the children has no effect?
    b. What assumptions about the distribution of the height increases are necessary?

11. Sketch $F(6, 10)$. (Include the mean.)

12. A population consists of 1, 3, and 4. Find the distribution of the mean $\bar{X}$ associated with samples of size $n = 2$. (Assume the sampling is done with replacement.)

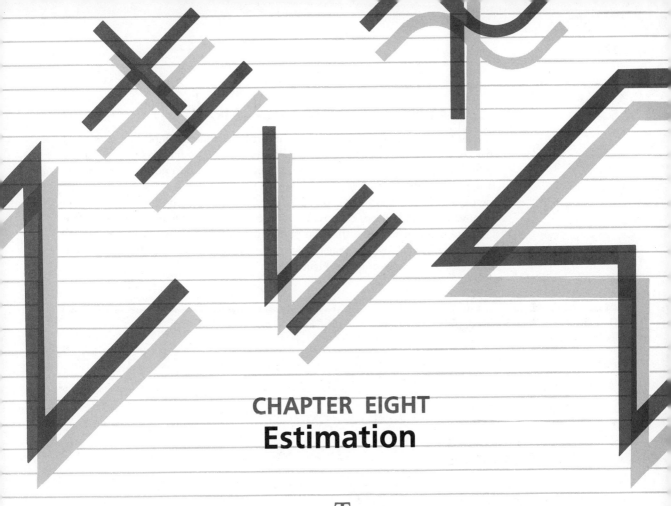

CHAPTER EIGHT
Estimation

The sampling distributions outlined in Chapter 7 are the basis of inferential statistics. We now begin the development of this important subject with the study of estimation.

Problems of estimation seemingly occur everywhere. The health scientist must estimate the pesticide level in a food product. The city controller and the tax assessor must estimate tax rates. The advertising executive must estimate the percentage of the population who will react favorably to a proposed advertising campaign. On a more personal level, there are such routine matters as estimating incidental expenses, travel times, and examination averages. In some of these only crude estimates or guesses are needed. In others, the consequences are of such importance that both good estimates and measures of their reliability are required.

Two types of estimates are considered for both the mean and proportion of a population. The first, called a *point estimate*, is a single number used when a definite value of the parameter is required. The second, a *confidence interval*, is an interval whose end points should be upper and lower bounds for the parameter of interest. The confidence level is a measure of how certain we are that this is in fact the case.

8.1   ESTIMATING THE POPULATION MEAN

Besides being intuitively satisfying, the mean of a random sample is a good estimate of the population mean for several reasons. First, we have seen that $\overline{X}$ is an unbiased estimator of μ: its distribution is centered at μ. Second, this distribution is approximately normal when the sample size is large. As a result nearly all sample means are within two or three standard errors of the population mean. Finally, the standard error decreases as the sample size increases. When these facts are combined we see that the probability of a sample mean being close to the population mean is extremely high, growing even greater as the sample size increases. This is illustrated in Figure 8.1.

FIGURE 8.1   THE STANDARD ERROR OF $\overline{X}$ DECREASES AS THE SAMPLE SIZE INCREASES

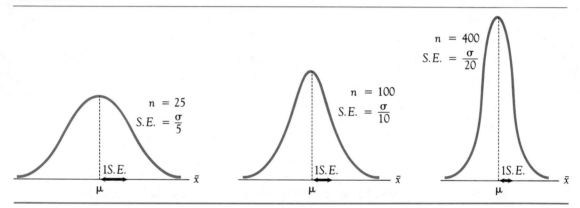

Such considerations suggest that the mean of a randomly selected sample should be a good estimate of the corresponding population mean. Since this mean represents a single point on the number line, hopefully close to μ, it is called a point estimate.

DEFINITION 8.1

The mean $\overline{x}$ of any random sample from a population is called a *point estimate* of the population mean μ.

Unfortunately, the point estimate $\overline{x}$ rarely turns out to be the population mean. In fact, the probability of this happening is zero for any continuous distribution. Still we expect the population mean to be close to the point estimate, in the sense that it lies on some interval centered about $\overline{x}$, extending, say, from $\overline{x} - E$ to $\overline{x} + E$. The longer this interval, the more confident we are that it contains the population mean. This gives rise to the notion of a confidence interval, (Figure 8.2) and the need for a measure of this confidence. This measure, which is called the *level of confidence,* is typically expressed as a percent. Thus when we speak of, say, a 95% confidence interval, the level of confidence is 95%.

FIGURE 8.2   FORM OF A CONFIDENCE INTERVAL FOR μ

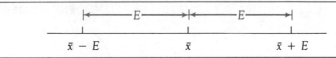

Before attempting to define a confidence interval, we consider the construction of a 95% confidence interval for the mean μ of a population. We shall assume that the sample size n is so large that the sample means may be considered to be normally distributed. We then know the probability is .95 that

$$\mu - 1.96(S.E.) < \overline{X} < \mu + 1.96(S.E.)$$

We interpret this relation as stating that 95% of the sample means lie within 1.96 standard errors of the population mean μ. From this we conclude that μ must be within 1.96 standard errors of 95% of the sample means. Thus, if it were possible to examine all intervals extending 1.96 standard errors on both sides of the mean of each sample of size n, then 95% of these intervals would contain μ and 5% would not. This can be demonstrated more rigorously by showing that the above inequality can be rearranged to yield

$$\overline{X} - 1.96(S.E.) < \mu < \overline{X} + 1.96(S.E.)$$

We omit the details but do note that this inequality is of the form $\overline{X} - E < \mu < \overline{X} + E$ where $E = 1.96(S.E.)$.

Suppose now that we have only a single sample, as is usually the case. Using its mean $\bar{x}$, we form the interval with end points $\bar{x} \pm 1.96(S.E.)$ This interval either contains μ or it does not. Thus the probability that it contains the population mean is either 1 or 0, not .95. On the other hand, 95% of all such intervals do contain μ, so there is a basis for a subjective optimism or *confidence* that this will be one of those that does. For this reason we describe this interval as a 95% confidence interval.

In Figure 8.3, three intervals are shown around the population mean μ. Each extends 1.96(S.E.) on either side of the sample mean. Two of these contain μ and one does not. All are 95% confidence intervals.

The above development began with the problem of constructing a 95% confidence interval with endpoints $\bar{x} \pm E$ and concluded that $E = 1.96(S.E.)$ It should be clear that if an interval with end points $\bar{x} \pm E$ had been given, then the probability $P(\mu - E < \overline{X} < \mu + E)$ and thus the level of confidence could have easily been found, as noted in Definition 8.2.

DEFINITION 8.2

Let $\bar{x}$ be the mean of any random sample from a population and E a positive number. The numbers lying between $\bar{x} - E$ and $\bar{x} + E$ constitute an interval on the number line that is called a *confidence interval* for the population mean μ. The *level of confidence* associated with this interval is the number $P(\mu - E < \overline{X} < \mu + E)$, expressed as a percent.

FIGURE 8.3   THREE 95% CONFIDENCE INTERVALS FOR μ

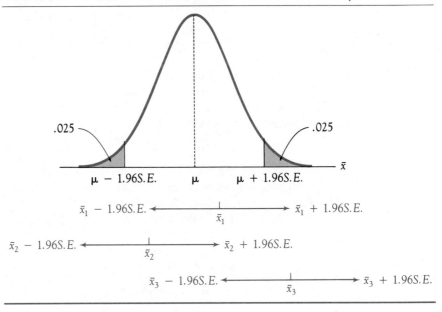

A confidence interval is indicated by writing

$$\overline{x} - E < \mu < \overline{x} + E$$

One must be careful in interpreting such statements, since as we have seen, the interval need not contain μ.

The following terms will prove useful as we proceed.

DEFINITION 8.3
The number E of the confidence interval $\overline{x} - E < \mu < \overline{x} + E$ is called the *maximum error* of the estimate.

DEFINTION 8.4
The numbers $\overline{x} \pm E$ are called the *upper* and *lower limits* of the confidence interval $\overline{x} - E < \mu < \overline{x} + E$.

Example 8.1

Find a 95% confidence interval for the mean of a population with $\sigma = 100$ if a sample of $n = 400$ has a mean of $\bar{x} = 492$. The maximum error of the estimate is

$$E = 1.96(S.E.) = 1.96\,\frac{\sigma}{\sqrt{n}} = 1.96\,\frac{100}{\sqrt{400}} = 9.8$$

The upper and lower limits are

$$\bar{x} \pm E = 492 \pm 9.8 \text{ or } 501.8 \text{ and } 482.2$$

The 95% confidence interval is

$$482.2 < \mu < 501.8 \quad \blacksquare$$

Realistically, we note that if the population mean is not known, the same is probably true of its standard deviation. In most constructions it will be necessary to approximate σ by the sample standard deviation s. As we saw in the last chapter, this introduces no serious error as long as the sample size is large.

Example 8.2

A random sample of semester textbook expenditures by 64 fulltime university students had a mean of \$98.00 and a standard deviation of \$32.00. Find a 95% confidence interval for the mean expenditure for textbooks by students at this university. The maximum error of the estimate is approximated as

$$E = 1.96\,\frac{s}{\sqrt{n}} = 1.96\left(\frac{32.00}{\sqrt{64}}\right) = 7.84$$

$$\bar{x} \pm E = 98.00 \pm 7.84 \text{ or } 105.84 \text{ and } 90.16$$

The confidence interval is

$$90.16 < \mu < 105.84 \quad \blacksquare$$

Now consider how this procedure may be generalized to any confidence level. Note that in terms of the percentage point notation, a 95% confidence interval is of the form

$$\bar{x} - z_{.025}(S.E.) < \mu < \bar{x} + z_{.025}(S.E.)$$

Setting $\alpha = 1 - .95 = .05$ and $\alpha/2 = .025$, we obtain

$$\bar{x} - z_{\alpha/2}(S.E.) < \mu < \bar{x} + z_{\alpha/2}(S.E.)$$

This is in fact the general form. For, if a $(1 - \alpha) \times 100\%$ confidence interval is constructed to the 95% interval, we immediately find that $E = z_{\alpha/2}(S.E.)$ (Figure 8.4).

FIGURE 8.4 **(1 − α) × 100% OF THE SAMPLE MEANS ARE WITHIN** $z_{\alpha/2}(S.E.)$ **of** μ

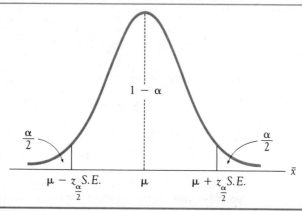

PROCEDURE FOR CONSTRUCTING A CONFIDENCE INTERVAL FOR μ

Given: The size n, the mean $\bar{x}$, and the standard deviation s of a random sample from a population whose mean is μ.
Assumption: The sample size $n \geq 30$.

1. Select a confidence level and express it as a decimal c.
2. Compute $\alpha = 1 - c$ and $\alpha/2$.
3. Find $z_{\alpha/2}$.
4. Compute the maximum error of the estimate $E = z_{\alpha/2}\,(s/\sqrt{n}.)$.
5. Compute the upper and lower limits $\bar{x} \pm E$ and write the confidence interval as

$$\bar{x} - E < \mu < \bar{x} + E$$

Discussions are often simpler if we describe the above interval as having a $(1 - \alpha) \times 100\%$ confidence level.

Example 8.3

Find a 99% confidence interval for the mean textbook expenditure μ, again using the data $n = 64$, $\bar{x} = 98.00$, and $s = 32$. The steps are as follows:
1. $c = .99$
2. $\alpha = 1 - .99 = .01$ and $\alpha/2 = .005$
3. From the standard normal table, $z_{.005} = 2.58$.
4. The maximum error of the estimate is

$$E = z_{.005}\,\frac{s}{\sqrt{n}} = 2.58\,\frac{32}{\sqrt{64}} = 10.32$$

5. The upper and lower limts are $\bar{x} \pm E = 98 \pm 10.32$. Thus the confidence interval is

$$87.68 < \mu < 108.32 \quad \blacksquare$$

Both the 95% and 99% confidence intervals for the mean textbook expenditure from the last two examples are shown in Figure 8.5. Note that the 99% interval is the longer of the two, as would be expected.

FIGURE 8.5   THE 99% AND 95% CONFIDENCE INTERVALS FOR THE MEAN TEXTBOOK EXPENDITURE

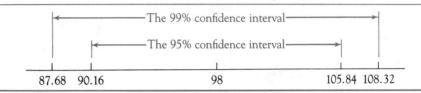

In illustrating the procedure for constructing a confidence interval in the above example, all steps were carefully detailed and enumerated. We shall not be so formal in the future.

Example 8.4

Forty healthy middle-aged women participated in a running program that eventually culminated in their jogging one mile three times a week. At this time their pulse rates at rest had declined on the average by 14 beats per minute. If these decreases had a standard deviation of $s = 4$ beats per minute, find a 90% confidence interval for the mean decrease in pulse rates for similar subjects in this program.

The confidence level as a decimal is $c = .90$. Thus $\alpha = 1 - .90 = .10$ and $\alpha/2 = 10/2 = .05$. From the standard normal table, $z_{.05} = 1.64$. The maximum error of the estiamte is

$$E = 1.64 \frac{s}{\sqrt{n}} = 1.64 \frac{4}{\sqrt{40}} = 1.0 \text{ (rounded)}$$

The upper and lower limits are

$$\bar{x} \pm E = 14 \pm 1 = 15 \text{ or } 13$$

The 90% confidence interval is

$$13 < \mu < 15 \quad \blacksquare$$

How does one select the confidence level? This, it turns out, is a matter of individual choice; nothing indicates that one level should be preferred over another. Nevertheless, a survey of the literature reveals an overwhelming preference for 95%, 99%, and 90% in that order. Unless otherwise indicated, the reader may use 95%.

In planning an experiment it often happens that there is a bound on the error which can be tolerated in the estimate of the mean. If an approximate value of the standard deviation σ is available, then a sample size can be computed that will hold the maximum error within this bound.

Example 8.5

A city wishes to develop a change-out schedule for its street lights based on the expected life of the bulbs. For this an estimate is needed of the mean life of the bulbs that is accurate to within five days at a 95% confidence level. Suppose that in similiar studies the standard deviations of the burning lives have all been around 30 days.

The limits of the 95% confidence interval are $\bar{x} \pm E = \bar{x} \pm 5$. Thus $E = 5$. We know that

$$E = 1.96 \frac{\sigma}{\sqrt{n}} = \frac{1.96(30)}{\sqrt{n}} = \frac{58.8}{\sqrt{n}} \text{ (approximately)}$$

Solving

$$\frac{58.8}{\sqrt{n}} = 5$$

we obtain $n = 138.3$. A sample of at least 139 bulbs should produce a confidence interval where the error of the estimate of the mean μ by the sample mean $\bar{x}$ will be at most 5 days.

Example 8.6
A Case Study

In a report entitled "Offshore Wind Power Potential" (Paul Ossenbrugen et al., Journal of the Energy Division, American Society of Civil Engineers), Vol. 105, No. EY1, January, 1979, wind speeds at Boston's Logan Airport and Boston Lightship, a lighthouse in Massachusetts Bay, are compared for different seasonal groupings (see Table 8.1). Note that a season is determined here by similar wind speeds and not necessarily by months of the year. The report summarizes observations dating to 1945.

TABLE 8.1 SEASONAL DISTRIBUTION OF WIND SPEED, IN MILES PER HOUR (METERS PER SECOND)

| Seasonal group | Logan airport | | Lightship | |
	Mean hourly wind speed	Standard deviation	Mean hourly wind speed	Standard deviation
January, February, March	14.3 (6.4)	6.5 (2.9)	18.1 (8.1)	9.6 (4.3)
April, December	13.9 (6.2)	6.2 (2.8)	16.8 (7.5)	8.6 (3.8)
November	13.2 (5.9)	6.3 (2.8)	16.3 (7.3)	7.7 (3.4)
May, October	12.3 (5.5)	5.5 (2.5)	12.7 (5.7)	7.4 (3.3)
June, September	11.5 (5.1)	4.9 (2.2)	11.8 (5.3)	6.6 (3.0)
July, August	10.9 (4.9)	4.5 (2.0)	10.5 (4.7)	5.5 (2.5)

Assuming the January, February, and March seasonal results are based on 22,600 observations at Logan Airport and 1264 at Lightship, show that the 95% confidence intervals for the mean wind speeds at these two sites (for this season) do not overlap.

The 95% confidence interval for the mean speed μ_1 at Logan Airport is easily found to be

$$14.3 - .1 < \mu_1 < 14.3 + .1 \text{ or } 14.2 < \mu_1 < 14.4$$

Here we have used

$$E = \frac{1.96(6.5)}{\sqrt{22600}} = .1 \text{ (rounded)}$$

Similarly, for the mean μ_2 at Lightship we use

$$E = \frac{1.96(9.6)}{\sqrt{1264}} = .5 \text{ (rounded)}$$

obtaining

$$18.1 - .5 < \mu_2 < 18.1 + .5 \text{ or } 17.6 < \mu_2 < 18.6$$

Clearly, the two intervals do not intersect. While there are better tests for comparing the means of two populations, these results can be accepted as evidence that the two sites, while closely located, have different mean wind speeds. Can such a difference be traced to the less homogeneous topographical conditions on land? And further, is the difference sufficiently large to influence the location of wind-powered generating systems? In this report, the authors do not believe the difference in wind speeds would justify the increased costs of locating power plants offshore. And it is not clear that the land topography is responsible for the differences in wind speeds; there are six months in which the wind speeds are virtually the same at both the airport and the lighthouse. ■

In Figure 8.6 we show how the two confidence intervals for the last example

FIGURE 8.6 95% CONFIDENCE INTERVALS FOR MEAN WIND SPEED AT AIRPORT AND LIGHTSHIP SITES

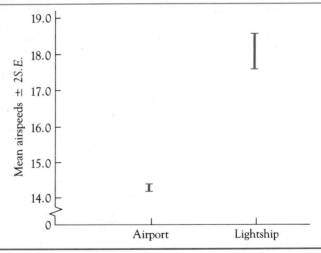

might be graphed. The means are plotted relative to the vertical axis as points, and the confidence intervals are indicated by bracketed vertical segments through these points. This method of displaying confidence intervals is particularly useful when several means are being compared.

Example 8.7
A Case Study

In Figure 8.7 are presented the confidence intervals for mean embryonic weights of winkles (marine snails) for two populations, one living in crevices and the other on boulders along the shores of Holy Island, Anglesey, Wales. It is easy to compare the growth of the embryos at each stage. The confidence intervals do not overlap and there is clearly a definite difference in the mean embryonic weights for the two populations at each stage of the development. It is the authors' hypothesis that there is initially a greater energy expenditure or parental investment in each egg on the part of the crevice females, a difference that may well be due to the different demands of the two habitats. (Source: "The Status of General Reproductive-Strategy Theories, Illustrated in Winkles," Andrew Hart and Michael Begon, *Oecologia*, vol. 52, 1982.) ■

FIGURE 8.7 EMBRYO WEIGHTS BY POPULATION AND BY DEVELOPMENTAL STAGE. (Means are Given with 95% Confidence Limits; Only Batches that Contained More than 50 Embryos are Included.)

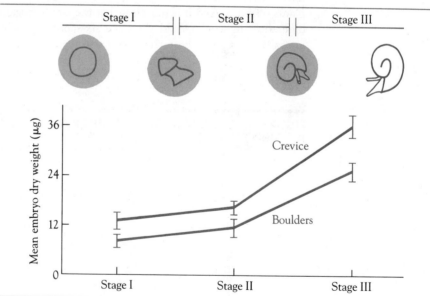

As a variation, many authors use bar graphs, with the heights of the individual bars representing the means of the individual samples. Confidence intervals are still represented by bracketed vertical segments. It is also common to give the mean plus or minus one standard error (a 68% confidence interval). Presumably, from one standard error the knowledgeable reader can construct any confidence interval he or she might need. For example, by doubling the length of the bracketed segment, a 95% confidence interval is produced. Increasing its length by a factor of 1.65 produces a 90% confidence interval, and so forth.

Example 8.8

A Research Study

In a study of learned taste aversion, several litters of rat pups were each divided into two groups, one serving as a control (CS-Sal) and one as a test group (CS-Li). The test group nursed from foster mothers whose milk had a distinctively different flavor from that of their mothers. After nursing they were given a slight dose of lithium chloride by injection to produce nausea. Would they associate the illness with the differently flavored milk? The mean weight gain immediately after nursing would show how much milk they consumed and, when compared with the control group, would show whether they had developed a taste aversion to the milk that "created" their previous illness. The mean weight gains plus or minus one standard error are shown for both control and test samples at four different age levels in Figure 8.8. It is clear that both the experimental and control groups consumed essentially equivalent amounts of milk. In short, the pups failed to develop a taste aversion to the milk of the foster mothers. (Source: "Taste Aversions to Mother's Milk: The Age-Related Role of Nursing in Acquisition and Expression of a Learned Association," Louise Martin and Jeffrey Alberts, *Journal of Comparative Physiological Psychology*, vol. 93, no. 3, 1979.) ∎

FIGURE 8.8   MEAN WEIGHT GAIN (1 *S.E.*) BY PUPS TRAINED AND TESTED IN A NURSING SITUATION

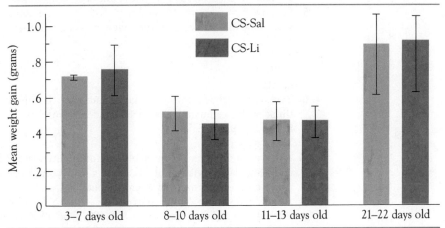

Part A

1. What is the confidence level of each of the following intervals? (Assume a large sample).

   a. $\bar{x} - 1.96 \dfrac{s}{\sqrt{n}} < \mu < \bar{x} + 1.96 \dfrac{s}{\sqrt{n}}$

b. $\bar{x} - 2.33 \dfrac{s}{\sqrt{n}} < \mu < \bar{x} + 2.33 \dfrac{s}{\sqrt{n}}$

c. $\bar{x} - 1.64 \dfrac{s}{\sqrt{n}} < \mu < \bar{x} + 1.64 \dfrac{s}{\sqrt{n}}$

d. $\bar{x} - \dfrac{s}{\sqrt{n}} < \mu < \bar{x} + \dfrac{s}{\sqrt{n}}$

2. Give the formula for computing the maximum error of the estimate E for each of the following confidence intervals. Use the percentage point notation.
   a. A 95% confidence interval.
   b. A 90% confidence interval.
   c. A 92% confidence interval.
   d. A 99.7% confidence interval.

3. Give the confidence level for each of the following intervals:

   a. $\bar{x} - z_{.01} \dfrac{s}{\sqrt{n}} < \mu < \bar{x} + z_{.01} \dfrac{s}{\sqrt{n}}$

   b. $\bar{x} - z_{.05} \dfrac{s}{\sqrt{n}} < \mu < \bar{x} + z_{.05} \dfrac{s}{\sqrt{n}}$

   c. $\bar{x} - z_{.02} \dfrac{s}{\sqrt{n}} < \mu < \bar{x} + z_{.02} \dfrac{s}{\sqrt{n}}$

4. The upper and lower limits of a confidence interval along with the information used in its construction are given. Find the confidence level.
   a. 50 ± 4, $\bar{x} = 50$, $s = 12$, and $n = 36$.
   b. 200 ± 4.96, $\bar{x} = 200$, $s = 24$, and $n = 64$.
   c. 125 ± 1.88, $\bar{x} = 125$, $s = 14$, and $n = 100$.
   d. 1235 ± 3, $\bar{x} = 1235$, $s = 87$, and $n = 128$.

5. Suppose $\bar{x} = 70$, $s = 18$, and $n = 100$. Find the confidence interval associated with these data for each of the following confidence levels:
   a. 90%.
   b. 95%.
   c. 99%.
   d. 80%.

6. You are told that a certain make of auto uses 18 miles per gallon. What do you think of this estimate if the 95% confidence interval is:
   a. $17.5 < \mu < 18.5$?
   b. $12 < \mu < 24$?

7. A survey of the repair costs of 140 microcomputers showed a mean cost of $38.50 and a standard deviation of $12. Find a 90% confidence interval for the average repair costs of this make of microcomputer.

8. A random sample of the start-up costs of 36 franchises of a hardware chain showed $\bar{x} = \$154{,}200$ with $s = \$18{,}400$. Find a 95% confidence interval for the mean start-up cost for these franchises.

9. A sample of routine emergency room treatment costs for 49 patients at a large general hospital had a mean of $72.00 and a standard deviation of $8.40. What confidence level is associated with the interval estimate $69 < \mu < 75$ for the mean cost of a routine emergency room treatment?

10. A business report stated that the inventory value of an auto parts store will be somewhere between $23,000 and $45,000. If this is a confidence interval for the mean inventory value of auto parts stores, and it is derived from a sample of 30 stores with $\bar{x} = \$34{,}000$ and $s = \$25{,}000$, what is the confidence level?

11. In order for an airline to compute the contribution to the total on-board weight from the carry-on luggage of passengers, an estimate of the luggage's mean weight per passenger is needed that will be accurate to within 5 pounds with a 99.7% certainty. How large a sample is needed if it is estimated that the carry-on luggage weights have a standard deviation of around 20 pounds?

12. It is desired to estimate the mean checkout time of customers at a supermarket to within 1 minute with a 95% certainty. How large a sample is needed if it is estimated that the standard deviation of the checkout times is around 5 minutes?

Part B

13. A craft shop randomly surveyed 36 suppliers of a certain size planting pot for the unit cost in lots of 100. The unit costs had a mean of $3.40 and a standard deviation of $.53. Find a 99% confidence interval for the mean unit price of these pots. Also, give upper and lower bounds on the expected cost of a lot of 100 of these pots.

14. A speedy oil and lubrication chain surveyed its stores as to the time needed to process a customer. Six customers from each of its 17 stores were selected at random and monitored. The processing times had a mean of 11 minutes with a standard deviation of 5.8 minutes. Find a 95% confidence interval for the mean time needed to process a customer.

15. Pyrometric cones of different composition are supplied to the ceramic industries to use in measuring temperatures in kilns. Different cones are designed to melt at specific temperatures. Suppose a random sample of 40 No. 17 cones had melting temperatures with a mean of 1845°F and a standard deviation of 54°F. Find a 95% confidence interval for the mean melting temperature of the No. 17 cones.

16. The weights of 30 bronze cast copies of a certain statue had a mean of 17.54 pounds and a standard deviation of 2.46 pounds. Give upper and lower limits on the amount of bronze needed for 100 copies of this statue.

17. The systolic blood pressure of 524 white males and 308 black males between 45 and 74 years of age in the famous Evans county study showed means of 148.9 and 165.0 millimeters of mercury (mm Hg), respectively. Assuming a standard deviation of 18 mm Hg for each group, find 95% confidence intervals for the mean systolic blood pressures of both of these groups and display the results graphically as in Figure 8.6.

18. In a study of immunoresponse to rabies virus, a control group of 30 mice were injected with the virus but not treated. The times to the onset of paralysis for this group had a mean of 10.9 days with a standard deviation of 1.9 days. In one treatment group of 36 mice which received the drug cyclophosphamide the times to the onset of paralysis had a mean of 21.7 days and a standard deviation of 2.8 days.
    a. Find 95% confidence intervals for the mean time to the onset of paralysis for both the treatment and the control groups.
    b. Display the confidence intervals graphically as in Figure 8.6.
    (Adapted from: "Dual Role of the Immune Response in Street Rabies Virus Infection of Mice," Jean S. Smith, Catherine McClelland, et al., *Infection and Immunity*, vol. 35, no. 1, Jan. 1982.)

19. The accompanying table shows the release in milligrams of reducing substances from conditioned leaves 1 to 5 hours after ingestion. Show these confidence intervals graphically as in Figure 8.7. (Source: "The Contribution of Fungal Enzymes to the Digestion of Leaves," Felix Barlocher, *Oecologia*, vol. 52, 1982, pp. 1–4.)

RELEASE OF REDUCING SUBSTANCES*

Time	1	2	3	4	5
Control leaves	4.0 ± 0.3	3.8 ± 0.2	2.5 ± 0.3	1.5 ± 0.1	1.0 ± 0.2
Conditioned leaves	7.3 ± 0.3	6.1 ± 0.2	4.0 ± 0.2	2.1 ± 0.2	1.2 ± 0.2

*In micrograms of glucose equivalents per entire gut, $\pm95\%$ confidence limit from conditioned leaves (4 weeks, *T. splndens*) recovered from G. *fossarum* guts 1 to 5 hours after ingestion. Control: leaves with enzymes inactivated before ingestion.

20. For the rating tag of a gas furnace, the manufacturer needs to know the mean output in British thermal units (Btu) with an error not exceeding 1250 Btu per hour. It is estimated that the standard deviation of the outputs of these furnaces will be around 4000 Btu per hour. How large a sample of these furnaces should be tested?

8.2   SMALL-SAMPLE CONFIDENCE INTERVALS FOR THE MEAN

As we have seen, the approximation of σ by s is often necessary in the construction of confidence intervals for μ. In the last chapter we saw that for a normal population with small samples, such an approximation requires the t distribution rather than the normal distribution for $\overline{X}$. Using the t-distribution, we can now develop a small-sample technique for confidence intervals.

From the similarity of the formulas

$$t = \frac{\overline{x} - \mu}{s/\sqrt{n}}$$

and

$$z = \frac{\overline{x} - \mu}{s/\sqrt{n}}$$

it should be clear that for a $(1 - \alpha) \times 100\%$ confidence interval for μ based on the mean of a small sample, the maximum error is now computed by

$$E = t_{\alpha/2}\frac{s}{\sqrt{n}}$$

rather than

$$E = z_{\alpha/2}\frac{s}{\sqrt{n}}$$

Apart from this the steps are identical to those for large samples.

> ## PROCEDURE FOR CONSTRUCTING A SMALL-SAMPLE CONFIDENCE INTERVAL FOR μ
>
> *Given:* The size n, the mean $\bar{x}$, and the standard deviation s of a sample from the population in question.
> Assumption: The population is normal.
> 1. Select a confidence level and express it as a decimal c.
> 2. Compute $\alpha = 1 - c$ and $\alpha/2$.
> 3. Using $\nu = n - 1$ degrees of freedom, find $t_{\alpha/2}$.
> 4. Compute the maximum error of the estimate $E = t_{\alpha/2}(s/\sqrt{n})$.
> 5. Compute the upper and lower limits $\bar{x} \pm E$ and write the confidence interval as
>
> $$\bar{x} - E < \mu < \bar{x} + E$$

Experience has shown that the normality requirement may be relaxed as the sample size increases and may be dropped completely for large samples ($n \geq 30$). As a consequence one finds that researchers also use the t distribution with large samples whenever σ is approximated by s.

Example 8.9

The sound level (in decibels) on the playground of an elementary school adjacent to a freeway was measured on 15 occasions. These measurements had a mean of 53 and a standard deviation of 12. Find a 99% confidence interval for the mean sound level μ on this playground.

Here $c = .99$, $\alpha = 1 - .99 = .01$ and $\alpha/2 = .005$. Using $\nu = 15 - 1 = 14$ d.f. we find $t_{.005} = 2.977$. The error of the estimate is

$$E = \frac{(2.977)(12)}{\sqrt{15}} = 9.2$$

Finally, the lower and upper limits are

$$\bar{x} \pm E = 53 \pm 9.2 \text{ or } 62.2 \text{ and } 43.8$$

and the confidence interval is $43.8 < \mu < 62.2$. This interval is shown in Figure 8.9; the distribution used in its construction is given in Figure 8.10. ■

FIGURE 8.9   THE CONFIDENCE INTERVAL $43.8 < \mu < 62.1$

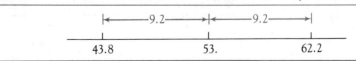

|←——9.2——→|←——9.2——→|

43.8   53.   62.2

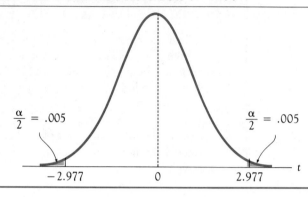

FIGURE 8.10   THE t DISTRIBUTION FOR $\nu = 14$

$\frac{\alpha}{2} = .005$ $\frac{\alpha}{2} = .005$

-2.977 0 2.977 t

Example 8.10
A Case Study

Can You Do Anything About Your Height?

In spite of the tall, statuesque appearance of fashion models and Miss America types, many young girls are distressed at the prospect of being tall. The details of attempts to control the heights of predictably tall girls with stilbestrol, an estrogen hormone, is detailed in the report "Tall Girls: 15 Years of Management and Treatment" by Dr. Norman Wettenhall, et al. in the *Journal of Pediatrics*, vol. 4, 1975.

In this report the treatment and monitoring of 87 girls with an initial mean height of 172.9 centimeters (68.1 inches) and an estimated mature mean height of 180.2 centimeters (70.9 inches) are detailed. The estimates of the final statures were determined by tables that use the individual's present age and a skeletal-specific age based on measurements of the wrists, hands, and knees. This is a well-established procedure producing reliable results.

The experimental subjects were selected from some 450 cases that the doctor treated between 1959 and 1973. To be eligible for treatment a girl had to be seriously concerned about her projected height, which had to be at least 177 cm in nearly all cases.

TABLE 8.2   APPARENT EFFECT OF TREATMENT ON FINAL STATURE IN RELATION TO MENARCHE

Time of commencement	N	Mean (cm)	SD	Range Low	Range High
Before menarche	38	3.8	2.4	−0.5	10.1
After menarche					
0.00–0.50 years	21	3.6	1.9	−1.2	7.8
0.51–1.00 years	12	2.5	1.2	0.8	4.6
1.01–2.00 years	12	2.4	1.6	−0.6	5.6

Table 8.2 summarizes the differences between the estimated and final heights of the subjects. The results are impressive, particularly for the premenarche group.

If we could regard these 38 girls as a random sample, then a 95% confidence interval for the mean reduction in height μ of tall girls who begin treatment before menarche is

$$3.8 - \frac{(2.02)(2.4)}{\sqrt{38}} < \mu < 3.8 + \frac{(2.02)(2.4)}{\sqrt{38}}$$

or

$$3.0 < \mu < 4.6$$

Here we have used $t_{.025} = 2.02$ with $\nu = 37$. Thus the mean reduction in height for the premenarche subjects is somewhere between 3 and 4.6 centimeters. The results become predictably less noteworthy with advancing age. ◼

EXERCISE SET 8.2

Part A

1. A random sample of size $n = 15$ from a normal population has a mean $\bar{x} = 41.2$ and a standard deviation $s = 5.62$. Find a 95% confidence interval for the population mean μ.
2. Suppose $\bar{x} = 23$ and $s = 6$. Assuming a normal distribution for X,
   a. Find a 95% confidence interval for μ if the sample size is $n = 16$.
   b. Find a 99% confidence interval for μ if the sample size is $n = 9$.
3. A sample of nine 1 pound packages of bacon from a certain meat packer had a mean weight of 17.1 ounces and a standard deviation of 2.3 ounces. Find a 95% confidence interval for the mean weight of 1 pound packages of bacon from this packer.
4. Assuming the same confidence level and the same standard deviation s, which would be longer: the 95% confidence interval for μ constructed from a random sample of $n = 16$ observations or from $n = 9$ observations?
5. The pH levels of a stream below a power dam were measured on 10 occasions and found to have a mean pH of 7.9 with a standard deviation of .5. Find a 99% confidence interval for the mean pH level of this stream.
6. An aircraft company needs an estimate of the time for a supplier to fill a reorder on a critical part. In a random sample of nine previous reorders, the resupply times had a mean of 23 days and a standard deviation of seven days. Assuming the resupply times have a distribution that is approximately normal, find a 99% confidence interval for the mean resupply time.
7. A department store wishes to estimate the number of customers who daily pass a particular location in their store. A sampling of the traffic on 18 randomly selected days at this location showed a mean of 1246 customers and a standard deviation of 145. Find a 90% confidence interval for the mean number of customers who daily pass this location. Assume the distribution of the number X that daily pass this location is approximately normal.
8. A random sample of 16 university students are questioned on how much they spend monthly for their personal entertainment. If $\bar{x} = \$56.45$ and $s = \$14.50$, find a 90% confidence interval for the mean expenditure for entertainment using the t distribution. What assumption is necessary?

Part B

9. Assuming a sample of size $n = 30$, which would be larger, the 95% confidence interval constructed using $z_{.025}$ or the one using $t_{.025}$? Generalize.

10. The mean times spent by mother rats in licking their young male and female offspring at various days after birth are shown below.
   a. Estimate the confidence intervals for the mean times for both the male and female pups at 18 days postpartum.
   b. Does the maternal behavior appear to be related to the pups' gender?
   c. Does the licking time appear to vary appreciably with the age of the pups (within each gender)?

   (Source: "Mother Rats Interact Differently with Male and Female Offsprings," Celia L. Moore and Gilda Morelli, *Journal of Comparative and Physiological Psychology*, vol. 93, no. 4, 1979.)

MEAN TIME SPENT BY MOTHER IN ANOGENITAL LICKING OF MALE (SOLID LINE) AND FEMALE (BROKEN LINE) PUPS ON DAYS 2, 3, 5, 7, 10, 14, AND 18 POSTPARTUM. (VERTICAL LINES ARE *S.E.*)

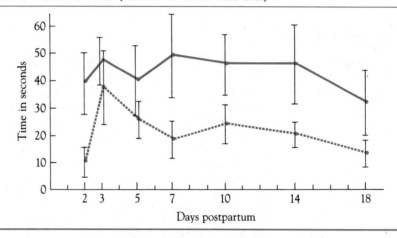

11. Using the data from the case study of girls' heights (Example 8.10), compute 95% confidence intervals for the mean estimated reduction in height for each of the postmenarche groups and display the results graphically along with those of the premenarche group. Use a layout similar to the lower half of the graph of problem 10 above.

12. In a study of absorption of the tranquilizer Colbazam into the blood, 15 subjects were given a 40 milligram tablet. The time to peak concentration and the concentration for each subject are given below. Find a 95% confidence interval for both the mean peak time and mean peak concentration for similar subjects.
   (Source: "Plasma Levels of Clobazam after Three Oral Dosage Forms in Healthy Subjects," J. J. Vallner, T. E. Needham, et al., *The Journal of Clinical Pharmacology*, July 1978.)

SUMMARY OF INDIVIDUAL SUBJECT DATA AFTER ADMINISTRATION OF 40 MG CLOBAZAM AS TABLET DOSAGE FORM

Subject	Peak time (hours)	Peak concentration (nanograms per milliliter)
A	2.00	488
B	1.50	565
C	2.50	741
D	1.50	438
E	2.00	539

Continued

SUMMARY OF INDIVIDUAL SUBJECT DATA AFTER ADMINISTRATION OF 40 MG CLOBAZAM AS TABLET DOSAGE FORM—cont.

Subject	Peak time (hours)	Peak concentration (nanograms per milliliter)
F	1.50	431
G	1.00	630
H	1.50	600
I	0.33	735
J	1.50	683
K	2.00	438
L	2.50	493
M	1.00	609
N	0.67	953
P	0.67	561
Mean	1.48	593
Standard deviation	0.63	137

13. Some phenolic cinnamates in white wines are reponsible for bitterness and browning or presence of color. In a study of their presence and effects, the following measurements were obtained for cinnamates in 10 Chardonay wines. Find 95% confidence intervals for the mean amounts of the two tartaric acids. (Source: "Determination of Phenolic Cinnamates in White Wine and Their Effects on Wine Quality," S. Okamura and M. Watanabe, *Journal of Agricultural and Biological Chemistry*, Vol. 45, no. 4, 1981.)

CONTENTS OF PHENOLIC CINNAMATES IN CHARDONAY WINE, MILLIGRAMS PER LITER.

Wines*	Caffeoyl tartaric acid	p-Coumaroyl tartaric acid	Caffeic acid	p-Coumaric acid
1	26	7	1.0	0.5
2	16	4	1.5	0.8
3	20	7	1.6	0.5
4	26	6	0.7	0.6
5	34	10	1.0	0.6
6	30	8	1.8	1.0
7	19	6	1.1	0.3
8	16	6	0.7	1.2
9	75	21	5.7	2.0
10	24	26	1.9	3.4

*No. 1–No. 8: French Bourgogne wine; No. 9, No. 10; American California wine.

Library

Project

The following research articles either are concerned with finding confidence intervals or employ graphic techniques for reporting such intervals. Select an article and report on it. The report should include information on selection of experimental subjects, gathering of data, and conclusions. Pay particular attention to the topics in this chapter. Note that some of these papers may involve procedures yet to be discussed. If so, limit your discussion to the use of estimation techniques.

a. "A Randomized Study of Sodium Intake and Blood Pressure in Newborn Infants," A Hofman et al., *Journal of the American Medical Association*, vol. 250, no. 3, July 1983.

b. Physical Health Status as a Consequence of Health Practices," W. Reed, *Journal of Community Health*, vol. 8, no. 4, Summer 1983.

c. "Office Laboratory Tests," A. Epstein et al., *Medical Care*, vol. 22, no. 2, February 1984.

d. "Physiologic Effects of a Self-Contained Self Rescuer," E. Petsonk et al., *American Industrial Hygiene Association Journal*, vol. 44, no. 5, May 1983.

e. "Epidermal Growth Factors in Human Milk," J. Moran et al., *Journal of Pediatrics*, vol. 103, no. 3, September 1983.

f. "Comparison of Nutrient Intake in Middle-Aged Men and Women Runners and Controls," S. Blair, *Medicine and Science in Sports and Exercise*, vol. 13, no. 5, 1981.

g. "Quadriceps Function and Training After Knee Ligament Surgery," G. Grimby et al., *Medicine and Science in Sports and Exercise*, vol. 12, no. 1, 1980.

h. "Rewarding Safety Belt Usage at an Industrial Setting," E. Geller, *Journal of Applied Behavorial Analysis*, vol. 16, no. 2, Summer 1983.

i. "HPLC Determination of Caffeine and Theobromine in Coffee, Tea, and Instant Hot Cocoa Mixes," J. Blauch and S. Tarka, *Journal of Food Science*, vol. 48, no. 3, May-June 1983.

j. "Dietary Induced Obesity (in Mice)," F. Bourgeous et al., *British Journal of Nutrition*, vol. 49, no. 1, January 1983.

8.3   ESTIMATING A PROPORTION

The proportion of a population having a certain characteristic is the focus of many statistical studies. The political pollster samples registered voters to find what percentage favor some position or candidate. The consumer advocate is interested in what fraction of the boxes of cereal from a certain manufacturer are underweight. The psychologist tries to estimate the proportion of low-income people who suffer from mental depression. The possibilities abound and each has been or will at some time be the subject of an investigation.

The basis for constructing estimates of a population proportion p is the distribution of the sample proportion $\hat{P}$. We begin by recalling that if we have a sample of size n of which x members possess the characteristic of interest, then the value of the sample proportion is $\hat{p} = x/n$. This we use as an estimate of the proportion p.

DEFINITION 8.5

The proportion $\hat{p}$ of any random sample from a population is a *point estimate* of the corresponding population proportion p.

When we read or hear a statement such as "46% of all teenagers smoke," we are being presented with a point estimate. To judge the quality of a point estimate or use it as the basis for a decision, we need to examine it in conjunction with the size of, say, a 95% confidence interval.

DEFINITION 8.6

A *confidence interval* for the proportion p of a population is an interval of the form

$$\hat{p} - E < p < \hat{p} + E$$

The confidence level for such an interval is the number

$$P(p - E < \hat{P} < p + E)$$

expressed as a percent.

The terminology is exactly the same as with the mean. The number E is called the maximum error of the estimate (of the proportion) and the numbers $(\hat{p} \pm E)$ are again termed the upper and lower limits of the confidence intervals. We omit the formal definitions.

Now suppose we want a 95% confidence interval for p. First recall that the sample proportion $\hat{P}$ has mean $\mu_{\hat{P}} = p$ and standard error $S.E. = \sqrt{(pq)/n}$ and is approximately normal for large sample size n. More precisely,

$$Z = \frac{\hat{P} - p}{\sqrt{\dfrac{pq}{n}}}$$

may be approximated by the standard normal distribution for large n. Thus, paralleling the development for the mean, we should begin the construction of the 95% confidence interval by considering

$$p - z_{.025}\sqrt{\frac{pq}{n}} < \hat{p} < p + z_{.025}\sqrt{\frac{pq}{n}}$$

Exactly as with the mean, this would rearrange to

$$\hat{p} - z_{.025}\sqrt{\frac{pq}{n}} < p < \hat{p} + z_{.025}\sqrt{\frac{pq}{n}}$$

The reader should compare this confidence interval with the corresponding one for the mean (see Figure 8.11).

FIGURE 8.11 95% CONFIDENCE INTERVALS FOR μ AND p

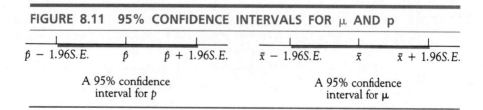

$\hat{p} - 1.96S.E.$	$\hat{p}$	$\hat{p} + 1.96S.E.$	$\bar{x} - 1.96S.E.$	$\bar{x}$	$\bar{x} + 1.96S.E.$

A 95% confidence interval for p A 95% confidence interval for μ

In both cases the maximum error of the estimate is given by

$$E = z_{.025}(S.E.) = z_{.025}\sqrt{\frac{pq}{n}}$$

Of course, the standard errors are computed in different ways.

It should now be clear that for a $(1 - \alpha) \times 100\%$ confidence interval, the maximum error of the estimate should be computed by

$$E = z_{\alpha/2}\sqrt{\frac{pq}{n}}$$

However, since p and q are unknown, the product pq must be approximated by the product $\hat{p}\hat{q}$. We summarize the construction of this confidence interval in the following procedure.

PROCEDURE FOR CONSTRUCTING A LARGE-SAMPLE CONFIDENCE INTERVAL FOR p

Given: The size n and the proportion $\hat{p}$ of a random sample from the population of interest.

Assumptions: Both $np > 5$ and $nq > 5$ where $q = 1 - p$.

1. Select a confidence level and express it as a decimal c.
2. Compute $\alpha = 1 - c$ and $\alpha/2$.
3. Find $z_{\alpha/2}$.
4. Compute the maximum error of the estimate

$$E = z_{\alpha/2}\sqrt{\frac{\hat{p}\hat{q}}{n}}$$

5. Compute the upper and lower limits $\hat{p} \pm E$ and write the confidence interval as

$$\hat{p} - E < p < \hat{p} + E$$

There should be no trouble in remembering these steps. Apart from the computation of the standard error, they are exactly the same as for the mean.

Example 8.12

A random sample of 600 homes produced 318 home owners who admitted to owning at least one gun. Find a 98% confidence interval for the proportion of home owners who admittedly own a gun.

The confidence level as a decimal is $c = .98$. Thus $\alpha = 1 - .98 = .02$ and $\alpha/2 = .01$. From the standard normal tables $z_{.01} = 2.33$. Using $\hat{p} =$

$318/600 = .53$ and $\hat{q} = 1 - .53 = .47$, the maximum error of the estimate is

$$E = 2.33 \sqrt{\frac{(.53)\,(.47)}{600}} = .047$$

The upper and lower limits are

$$\hat{p} \pm E = .53 \pm .047$$

Finally, the confidence interval is

$$.483 < p < .577$$

Thus roughly speaking, somewhere between 48% and 58% of the home owners own a gun to which they will admit. ■

Example 8.12

A random sample of 125 college graduates showed that 62% had changed their major at least once. Find a 95% confidence interval for the proportion of students who change their majors before graduation.

   Here $\alpha = .05$, $\alpha/2 = .025$ and $z_{.025} = 1.96$. The error of the estimate is computed by using $\hat{p} = .62$ and $\hat{q} = .38$. Thus

$$E = 1.96 \sqrt{\frac{(.62)\,(.38)}{125}} = .085$$

The 95% confidence interval is

$$.62 - .085 < p < .62 + .085$$

or

$$.535 < p < .705 \quad ■$$

   Do not let the poor result in this last example go unnoticed. We have learned only that somewhere between 53% and 71% of the students change their major, which does not reveal much. This result highlights the inadequacy of small samples for estimating proportions. For this reason, sample sizes exceeding 1200 are commonplace in opinion sampling, when it is desired to hold the error to around 2 or 3%.

Example 8.13

Consider again the estimation of the proportion of students who change their major. Only now let us assume that $\hat{p} = .62$ was obtained with a sample of $n = 1500$ students. Then

$$E = 1.96 \sqrt{\frac{(.62)\,(.38)}{1500}} = .025 = 2.5\%$$

Notice the difference. The 95% confidence interval is now

$$.595 < p < .645 \quad ■$$

As with the mean, it is possible to determine a sample size that will keep the error in the estimate of p within given limits at any specific confidence level. This requires an estimate for the product pq appearing in E. In Figure 8.12 we have plotted $y = pq = p(1 - p)$ for p between 0 and 1. The maximum occurs at $p = \frac{1}{2}$; here $pq = (\frac{1}{2})(1 - \frac{1}{2}) = \frac{1}{4}$. Thus, lacking any better estimates for pq, we use $pq = \frac{1}{4}$. This is illustrated in the next example.

FIGURE 8.12 $y = p(1 - p)$ **HAS A MAXIMUM AT** $p = \dfrac{1}{2}$

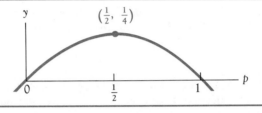

Example 8.14

An investment institution needs to know the percentage of working people who have a tax-sheltered annuity plan. How large a sample will ensure with 95% certainty that the error in the estimate of p by the sample proportion $\hat{p}$ is at most 4%? The upper and lower limits of the confidence interval are

$$\hat{p} \pm E = \hat{p} \pm .04$$

Thus $E = .04$. Also, $E = 1.96 \sqrt{(pq)/n}$ and in this we substitute $pq = \frac{1}{4}$ to obtain

$$E. = 1.96 \sqrt{\frac{1/4}{n}} = \frac{1.96(1/2)}{\sqrt{n}} = \frac{.98}{\sqrt{n}}$$

Equating these two values of E we obtain the equation

$$\frac{.98}{\sqrt{n}} = .04$$

This simplifies to $\sqrt{n} = .98/.04 = 24.5$ and from this we obtain $n = (24.5)^2 = 600.25$. Thus $n = 601$ will suffice. This, of course, is the maximum sample size that would need be used, since we have used a worst-case choice for the value of pq. ■

Example 8.15
A Case Study

In a 20-year study by the National Heart and Lung Institute of coronary heart disease in the town of Framingham, Massachusetts, the following data (Figure 8.13) were amassed in an electrocardiograph study of people who have an unrecognized myocardial infarction (unrecognized possibly because of mild symptoms, misinterpretation of symptoms, or simply disinterest on the part of the patient). Find a 95% confidence interval for the proportion of myocardial infarctions that go unrecognized in both sexes.

FIGURE 8.13 PROPORTION OF ECG-DOCUMENTED MYOCARDIAL INFARCTIONS UNRECOGNIZED (16 YEARS) MEN AND WOMEN 30–62 AT ENTRY; FRAMINGHAM STUDY

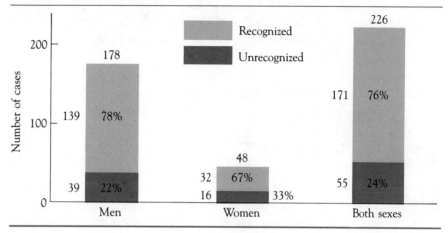

There were in all $n = 226$ documented cases of an infarction. Of these $x = 55$ or $\hat{p} = .243 = 24.3\%$ were previously unrecognized. Using $z_{.025} = 1.96$, $\hat{p} = .243$ and $\hat{q} = .757$, the error of the estimate is

$$E = 1.96 \sqrt{\frac{(.243)\,(.757)}{226}} = .056$$

The upper and lower limits are $.243 \pm .056 = .299$ or $.187$ and the 95% confidence interval is

$$.187 < p < .299.$$

FIGURE 8.14 SUBSEQUENT SURVIVAL IN PERSONS SURVIVING ONE YEAR FOLLOWING INITIAL MYOCARDIAL INFARCTION MEN AND WOMEN 30–79: FRAMINGHAM HEART STUDY

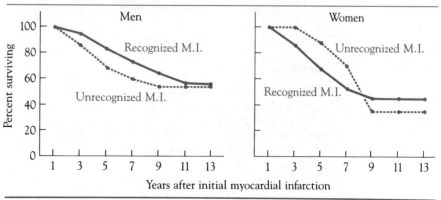

Thus somewhere between 18.7% and 29.9% of all myocardial infarctions are unrecognized. It turns out that these mild attacks are, however, as serious as those that are diagnosed and treated. In Figure 8.14 are plotted the percentage surviving in the years after the initial attack. (Source: "Risk Factors in Young Adults," Thomas Dawber, M.D., Director, Boston University Medical Center-Framington Study, *College Health*, vol. 22, 1973.) ■

Example 8.16

A Research Study

The Nielsen Ratings

As is well known, ratings may make or break a television show, regardless of its quality or character. The ratings are the basis for advertising rates; poor ratings translate to low revenues and a cancelled show. The most influential of the ratings, the Nielsen Ratings, are based on a sample of approximately 1200 homes scattered around the country. Each home has a meter that monitors all television sets in the home; these results are transmitted daily by telephone to computers. On any given day some 950 to 1050 reports are received; vacationing families, broken sets, and broken meters account for the rest.

From these reports Nielsen develops a rating number for each show. This is the proportion p of the reporting sample who had their sets tuned to the show in question for at least 6 minutes. Thus a 15 rating translates to 15% of the sample. Another number reported is the share of the audience, which is the proportion of all sets monitored that are tuned into the show in question.

Nielsen ratings are reported ± 1 S.E., which for most shows is between 1 and 2 points (1% and 2%). An unfortunate aspect of this is that producers and account executives accept a rating such as 15 ± 1.5 as meaning that the true proportion of sets tuned in to the show is somewhere between 13.5% and 16.5%, when in reality this is only a 68% confidence interval. This becomes an important consideration when the difference between, say, the top four shows is less than 3 points. There may then be no significant difference between the sizes of the four audiences. Yet this is not how the results are interpreted.

A criticism that is often made of the Nielsen service is that the sampling process does not include the population for which the ratings are interpreted. In essence there are strata that are never sampled. On the positive side, when the Nielsen ratings are compared with other ratings, they are found to give very good results. In the final analysis, Nielsen is simply marketing a service. The interpretation of the results is up to the executives and producers who subscribe. ■

EXERCISE SET 8.3

Part A

1. Find a 95% confidence interval for the population proportion p if a random sample from this population of size $n = 400$ produced $\hat{p} = .575$.
2. Suppose that in a sample of 1000 students, 374 planned to continue with graduate school immediately after graduation. Find a 99% confidence interval for the proportion of all students who plan on graduate school immediately after graduation.
3. A survey of 435 graduating high school seniors revealed that 31.5% did not know how to write a check. Find a 99% confidence interval for the proportion of all graduating seniors who are similarly ill prepared.

4. Patients fail to take drugs as prescribed for many reasons. Suppose a sample of 500 hypertensive patients had 46% who failed to comply with their physicians' directions on medications. Find a 99% confidence interval for the percentage of noncompliance within the hypertensive population. (The data are contrived, but the percentages are close to actual findings in several studies.)

5. An instructor studied a random sample of 450 enrollments in her course over the preceding three years and found that after four weeks, 22% of these had withdrawn from the course(s). Find a 95% confidence interval for the proportion of students who can be expected to drop her course by the 4th week. Then use this result to compute upper and lower limits on the number of vacant seats she will have after four weeks in a course which presently has 120 students and is fully enrolled.

6. Suppose a Nielsen report showed that of 920 homes reporting, 201 of these tuned in to a certain show. Find a 95% confidence interval for the Nielsen rating number for this show. How would this rating be reported by Nielsen?

7. In a job turnover study, 1400 of 2000 college graduates who left their first employment within two years reported that they did so because the prospects for advancement had been misrepresented. Find a 95% confidence interval for the proportion of graduates who change jobs for the reason stated.

8. A random sample of 500 absences by state workers showed that 73% were due to illness. If the proportion of state workers who are absent because of illness is estimated as .73, what can be claimed with a 90% certainty about the possible size of the error?

9. A random sample of 400 entering kindergarten children showed that 65% could recite their ABCs. Find a 90% confidence interval for the proportion of all entering kindergarten children who are so prepared.

10. A political pollster needs an estimate of the percentage of the electorate who favor a certain candidate with an error of at most 3%. What size sample will ensure this accuracy with a 95% confidence?

11. The Department of Transportation needs an estimate of the percentage of vehicles that exceed 58 miles per hour along a certain stretch of interstate highway. How large a sample is needed if the point estimate is not to be in error by more than 1% with a 95% confidence?

12. How large a sample is needed to estimate the proportion p to within .01 (1%) with a 99% confidence?

Part B

13. It is estimated that somewhere around 60% of the adult population drink coffee on a regular basis. How large a sample is needed if we wish to approximate the proportion of coffee drinkers in the population with an error not exceeding .02 (2%) with a 95% certainty?

14. In a study of the reasons for delay between a woman first suspecting a breast abnormality and consulting her physician, the following reasons were uncovered:

REASONS FOR DELAY (66 WOMEN)

Reasons for Delay	Number of Women (%)	Percentage of Women in Each Category Waiting Longer Than 6 Weeks	Percentage of Women in Each Category Waiting Longer Than 12 Weeks
Assumed symptoms not serious, nothing to worry about, would disappear	29 (43.9)	76	31

REASONS FOR DELAY (66 WOMEN)—cont.

Reasons for Delay	Number of Women (%)	Percentage of Women in Each Category Waiting Longer Than 6 Weeks	Percentage of Women in Each Category Waiting Longer Than 12 Weeks
Developed a rational explanation other than breast cancer	10 (15.2)	30	10
Social or domestic reasons	10 (15.2)	50	30
Initial symptoms confusing	4 (6.1)	75	—
Previous benign breast disease	4 (6.1)	50	25
Too worried to approach GP	2 (3.0)	100	—
Other	5 (7.6)	80	20
No information	2 (3.0)	100	50

Assuming we can treat these subjects as a random sample, find:

a. The 95% confidence interval for the percentage of all women who would wait longer than 6 weeks before consulting their physician because they assume the symptoms are not serious.

b. Proceeding as in (a), find the 95% confidence interval for those who wait longer than 12 weeks.

(Source: "Delay in Treatment for Breast Cancer," S. A. Adams, J. K. Horner, and M. P. Vessey, *Community Medicine,* 1980, no. 2).

15. In a study of methods for lowering appointment failures, the following data were obtained for different types of preappointment reminders.

RATES FOR KEPT, CANCELLED, AND FAILED APPOINTMENTS BY REMINDER GROUP

Group	Number of Appointments	Percent Kept	Percent Cancelled	Percent Failed
No reminder	109	55.0	7.0	38.0
Letter reminder	82	83.7	6.5	9.8
Telephone reminder	80	80.0	11.3	8.8

Find 95% confidence intervals for the percentage of appointments kept by these three populations and graph the results.

(Source: "Lowering Appointment Failures in a Neighborhood Health Center," Susan J. Gates and D. Kathleen Colborn, *Medical Care,* vol. 14, no. 3, 1976).

16. To estimate the proportion of auto owners with no insurance with an error not exceeding .03 (3%), what size sample is needed, assuming a 99% confidence interval?

17. In a sample of 2943 skull films of inpatients at a psychiatric facility over a five-year period, 17.3% ($n = 509$) were found to have some abnormality. Find a 90% confidence interval for the proportion of all psychiatric patients having some skull abnormality.

(Source: "Routine Skull Films in Hospitalized Psychiatric Patients," John Delaney, M.D., *American Journal of Psychiatry,* vol. 133, no. 1, January 1976.)

COMPUTER PRINTOUT

Confidence intervals for the mean are easily produced with the aid of the Minitab computer program. An example is given in Figure 8.15. The task times are the times needed by 15 randomly selected senior management students to solve a hypothetical inventory control problem. The TINTERVAL command uses the t distribution.

FIGURE 8.15 COMPUTER PRINTOUT: CONFIDENCE INTERVAL FOR μ

```
MTB >SET THE FOLLOWING TASK TIMES (MINUTES) INTO C1
DATA> 23,34,29,43,34,32,39,27,45,36,35,38,41,26,29
DATA>END
MTB >TINTERVAL WITH 95 PERCENT CONFIDENCE LEVEL USING
THE DATA OF C1
    C1          N =   15     MEAN =        34.067    ST.
DEV. =      6.46

    A 95.00  PERCENT C.I. FOR MU IS
       (30.4862,     37.6471)
MTB >STOP
```

CHAPTER CHECKLIST

Vocabulary
Point estimate
Interval estimate
Confidence interval
Maximum error of the estimate
Upper and lower limits of a confidence interval
Level of confidence

Techniques
Finding both large and small sample confidence intervals for the mean μ
Finding a large sample confidence interval for the proportion p
Finding the confidence level of an interval of the form $\bar{x} \pm E$ or $\hat{p} \pm E$ when the sample size is large.
Finding the sample size n that will ensure μ does not differ from $\bar{x}$ (or p from $\hat{p}$) by more than E with a given confidence or certainty.

Notation
c	Confidence level
α	α = 1 − confidence level
E	Maximum error of the estimate
$\bar{x} \pm E$	The upper and lower limits of a confidence interval for μ
$\hat{p} \pm E$	The upper and lower limits of a confidence interval for p

1. A sample of size $n = 400$ from a population with mean μ has a mean $\bar{x} = 82.0$ and standard deviation $s = 14.3$. Find a 90% confidence interval for μ.

2. The survival times for 14 laboratory hamsters after exposure to a respiratory virus had a mean of 18.3 days and a standard deviation of 5.7 days. Find a 95% confidence interval for the mean survival time of hamsters exposed to this virus. What assumptions are necessary?

3. A random sample of 250 high school teachers revealed that 34% had a master's degree. Find a 99% confidence interval for the proportion of all high school teachers with this degree.

4. The mean weekly salary of a nurse's aide in a metropolitan area is reported as 346.80 ± 20 (dollars). Suppose this is a confidence interval based on a sample of 144 salaries having a mean and standard deviation of 346.80 and 114.75 respectively. What is the confidence level of the reported interval?

5. When a certain drug was administered, the blood sugar levels of 20 diabetics exhibited decreases with a mean of 138 and a standard deviation of 65. Find a 95% confidence interval for the mean decrease of similar diabetics if these decreases are approximately normal.

6. What sample size is required to estimate the proportion of retired people in California if it is specified that there should be a 95% confidence that the error of the estimate does not exceed .03 (3%)?

7. We wish a 99% confidence interval for the mean number of units completed by graduating seniors in engineering. What size sample will be needed if the maximum error of the estimate is specified as three units and the standard deviation of the units completed is estimated to be 12?

8. Suppose 1144 adults of a certain city are stopped and questioned as to the amount of pocket money they carry. If these amounts have a mean of $38.40 and a standard deviation of $16.20, find a 95% confidence interval for the total amount of pocket money carried by the 1,430,000 adults of the city.

9. Fourteen hundred adults are classified by their age and whether or not they drink coffee on a regular basis (see below). Find a 95% confidence interval for the percentage of coffee drinkers in each age group, show these intervals graphically, and discuss whether it is likely that the percentage of drinkers is the same for each group. Finally, combine all the data into one sample and obtain a 95% confidence interval for the percentage of adult coffee drinkers.

	Ages		
Regularly drink coffee	18–35	35–55	Over 55
Yes	210	289	380
No	190	211	120

10. Six 8 ounce jars of instant coffee from a production lot are randomly selected and their contents weighed. The results (in ounces) are as follows:

$$8.2, \ 8.0, \ 8.4, \ 8.5, \ 8.3, \ 8.2$$

Find a 99% confidence interval for the mean weight of coffee in jars of this lot. Assume such weights are normally distributed.

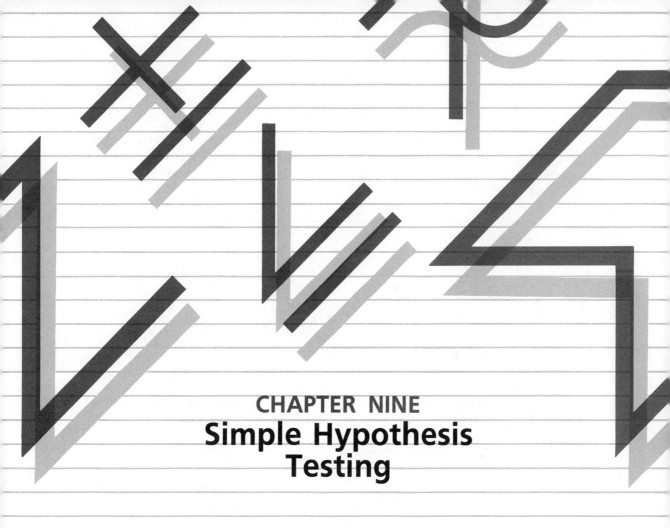

CHAPTER NINE
Simple Hypothesis Testing

Much of inferential statistics is concerned with one form or another of hypothesis testing. In this chapter we take up the subject of simple hypothesis testing, so named because the tests involve some parameter such as the mean or proportion of a single population.

We readily admit that the use of simple hypothesis testing is somewhat limited. Most statistical testing is of a comparative nature, involving several populations represented by treatment and control groups. Nevertheless, the experience one gains in this particularly simple setting proves invaluable when the subject is extended to multiple populations, beginning in the next chapter.

Among the topics to be developed are the nature of the decision rules for choosing between two competing hypotheses, the types of errors that can occur, the significance level of a test, and the use of p values in reporting results.

9.1   TYPES OF HYPOTHESES

The first question is, what is a hypothesis? This is not an easy question to answer. A hypothesis can be a many-faceted idea or a simple statement. It will always contain some uncertainty that is to be resolved. For our purposes Definition 9.1 will suffice.

DEFINITION 9.1

A *hypothesis* is an assertion whose truth is in question.

In this chapter a hypothesis will be a statement about the mean μ of a population or the proportion p that exhibits some property; it will be identified by a letter such as H, H_0, or H_1.

Example 9.1

Suppose the mean weight of male Air Force recruits is thought to be around 154 pounds. The following hypotheses could be derived from this:

Verbal Statement	Mathematical Statement
■ The mean weight is 154 pounds	$H: \mu = 154$
■ The mean weight is less than 154 pounds	$H: \mu < 154$
■ The mean weight is greater than 154 pounds	$H: \mu > 154$
■ The mean weight is not equal to 154 pounds	$H: \mu \neq 154$

Most situations involve two competing hypotheses.

Example 9.2

A political candidate claims that 62% of the registered voters support him. His opposition feels that this percentage is too high but does not know what figure would be correct. In this case there are two obvious hypotheses:

$$H_0: \quad p = .62$$
$$H_1: \quad p < .62$$

Here p is the proportion of registered voters who support the candidate. ■

In Example 9.2, the candidate's statement that 62% of the voters support him may have been interpreted too strictly. He probably meant that at least 62% supported him, which leads to another hypothesis—$H: p \geq .62$. If, however, we are able to test the hypothesis $H_0: p = .62$ against $H_1: p < .62$ and conclude that the former is not a reasonable possibility, then this will also be the conclusion when $H: p \geq .62$ is tested against $H_1: p < .62$.

The tests we will develop require two hypotheses. First, there is the basic or primary one that necessitated the test in the first place. This is called the *null hypothesis* and is denoted H_0. The null hypothesis will embody some precise statement about the mean μ or proportion p, such as H_0: $\mu = 154$ or H_0: $p = .62$. Generally, this hypothesis represents some well-established position that should not be rejected unless there is considerable evidence to the contrary.

We do not test to accept the null hypothesis; it is assumed to be true. Rather, we test to see if it should be rejected. If we should decide to reject it there must be some other hypothesis that we are willing to accept. This is called the *alternate hypothesis* and is denoted by H_1. Note that we do not establish the alternate hypothesis on its own merits, but rather by rejecting the null hypothesis. The alternate hypothesis is often a very nebulous statement such as H_1: $\mu \neq 154$ or H_1: $p < .62$.

In testing hypotheses we speak of testing the null hypothesis against the alternate hypothesis. We either reject or do not reject the null hypothesis. If the null hypothesis is rejected, we may say that the alternate hypothesis is accepted.

Example 9.3

Continuing the discussion of mean weights of Air Force recruits from Example 9.1, we could consider testing any one of the following pairs of hypotheses:

$$\begin{cases} H_0: & \mu = 154 \\ H_1: & \mu \neq 154 \end{cases} \qquad \begin{cases} H_0: & \mu = 154 \\ H_1: & \mu < 154 \end{cases} \qquad \begin{cases} H_0: & \mu = 154 \\ H_1: & \mu > 154 \end{cases} \blacksquare$$

Note that the null hypothesis always involves an equality.

9.2   DECISION RULES

How do we test hypotheses? More precisely, what is the basis of the process by which we decide whether or not to reject the null hypothesis H_0 in favor of the alternate hypothesis H_1? Let us begin by considering a test of

$$\begin{aligned} H_0: & \quad \mu = 154 \\ H_1: & \quad \mu \neq 154 \end{aligned}$$

If $\mu = 154$ is the population mean, then the sampling distribution of the mean $\overline{X}$ for large samples is approximately the normal distribution of Figure 9.1. Thus if we examine a large random sample from the population there is only a very small chance that its mean $\overline{x}$ will be more than, say, two or three standard errors from the assumed $\mu = 154$.

Now, if a sample should produce an $\overline{x}$ so extreme as to be more than two standard errors from $\mu = 154$, there are only two possibilities. Either we have been so unfortunate as to select one of those rarely occurring unrepresentative samples or the population mean does not have the asserted value. We choose the latter alternative and use this result as the basis for rejecting the null hypothesis.

In the above example, the decision on whether to reject the null hypothesis is based on the value of the mean $\overline{x}$ of the sample. In tests concerning the population proportion p, the proportion $\hat{p}$ of the sample will be used. More generally,

FIGURE 9.1 FEWER THAN 5% OF ALL SAMPLE MEANS ARE MORE THAN TWO STANDARD ERRORS FROM THE POPULATION MEAN

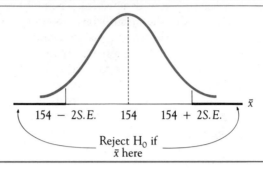

DEFINITION 9.2

A *test statistic* is a sample statistic on which a decision rule is based for a test of hypotheses.

Thus the mean $\overline{X}$ and the proportion $\hat{P}$ are test statistics. In most cases, the decision rule will be to reject the null hypothesis if the sample mean $\overline{x}$ or proportion $\hat{p}$ is more than two or three standard errors from the corresponding μ or p postulated by H_0.

Consider now Z and T scores derived from sample means and proportions using

$$z = \frac{\overline{x} - \mu}{\sigma/\sqrt{n}} \qquad t = \frac{\overline{x} - \mu}{s/\sqrt{n}} \qquad z = \frac{\hat{p} - p}{\sqrt{\dfrac{pq}{n}}}$$

These values are the number of standard errors that $\overline{x}$ is located from μ or $\hat{p}$ from p. Thus if these values are around ± 2 or ± 3 we would in all likelihood reject the null hypothesis.

The point is simply that Z and T are also test statistics when they derive their values from $\overline{X}$ and $\hat{P}$ in the usual way. They are generally easier to use than $\overline{X}$ and $\hat{P}$. Until this fact is fully appreciated we will discuss both Z and $\hat{P}$ or Z and $\overline{X}$ and use samples of such a size that the normal distribution can be assumed.

Now consider again the test of H_0: $\mu = 154$ against H_1: $\mu \neq 154$ and assume the population standard deviation is known to be $\sigma = 15$. Suppose we adopt the decision rule to reject the null hypothesis if the mean $\overline{x}$ of a sample of size $n = 100$ is more than two standard errors from $\mu = 154$. Since one standard error is

$$\frac{\sigma}{\sqrt{n}} = \frac{15}{\sqrt{100}} = 1.5$$

we reject H_0 if

$$\overline{x} < 154 - 2.0(1.5) = 151 \text{ or } \overline{x} > 154 + 2.0(1.5) = 157$$

These two numbers, 151 and 157, are called *critical values* of $\overline{X}$ for this test. If we were to use the test statistic Z, then rejection occurs if $z < -2.0$ or $z > 2.0$ where

$$z = \frac{\overline{x} - 154}{\frac{15}{\sqrt{100}}}$$

Thus $z = \pm 2.0$ are the critical values of Z. Any value of the test statistic more extreme than a critical value forces a rejection of the null hypothesis. These values constitute what is called the *critical region* of the test. In Figure 9.2 the critical regions of our weight test are shown for both $\overline{X}$ and Z.

DEFINITION 9.3

The *critical region* for a test is that set of values of the test statistic for which the null hypothesis will be rejected. The boundary or boundaries of this region are the *critical value(s)* of the test statistic.

FIGURE 9.2   THE CRITICAL REGIONS OF A TWO-TAILED TEST

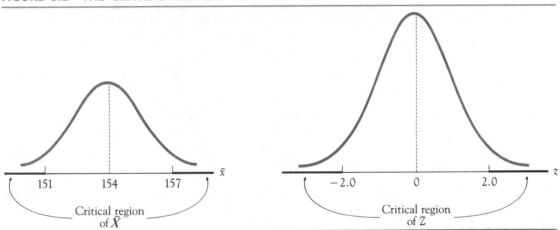

In Figure 9.2 we see that regardless of the statistic used, the critical region consists of two intervals lying under the extreme tails of the distribution of the test statistic. For this reason this test is referred to as a *two-sided* or *two-tailed* test. Note that the two intervals resulted from the lack of specific direction in the alternate hypothesis H_1: $\mu \neq 154$. With this alternate hypothesis, H_0 is rejected if the value of $\overline{X}$ or Z is too extreme in either direction.

Now consider how we would test the two competing hypotheses:

$$H_0: \quad \mu = 154$$
$$H_1: \quad \mu < 154$$

Again we assume $\sigma = 15$. The alternate hypothesis states that the population mean is some number less than 154. A good indicator that this might be the case would be a sample mean considerably less than the asserted $\mu = 154$, say at least two standard errors less. Accepting this as the decision rule, there is then only one critical value of $\overline{X}$ and it is located two standard errors below $\mu = 154$ at

$$\bar{x} = 154 - 2.0 \frac{15}{\sqrt{100}} = 154 - 3 = 151$$

As in the previous discussion, we are assuming a sample size of $n = 100$ and a standard deviation $\sigma = 15$.

Thus when $\overline{X}$ is used as the test statistic, the critical region of this test is made up of all numbers $\bar{x} < 151$. For the Z statistic, the test is even more straightforward. The critical value is $z = -2.0$; the critical region consists of all numbers $z < -2.0$. Both these regions are shown in Figure 9.3 and are seen to lie under the extreme left tails of the corresponding sampling distributions. For this reason the test is called a *one-sided* or *one-tailed test to the left*.

In Figure 9.4 the critical region is shown for a one-tailed test of the hypotheses:

$$
\begin{array}{ll}
H_0: & \mu = 154 \\
H_1: & \mu > 154
\end{array}
\quad (\sigma = 15)
$$

FIGURE 9.3   THE CRITICAL REGION OF A ONE-TAILED TEST TO THE LEFT

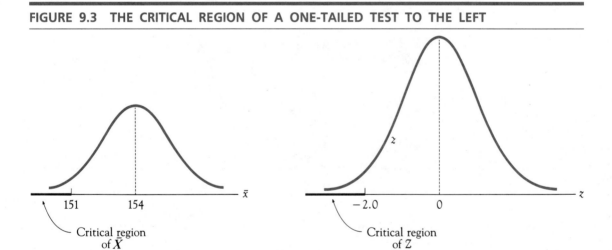

Here the decision rule is to reject H_0 if the mean $\bar{x}$ of a random sample is more than two standard errors greater than $\mu = 154$. This test for obvious reasons would be called a one-sided or one-tailed test to the right.

When we speak of either a two-tailed or a one-tailed test, it should be clear that we are describing the location of the critical region. Moreover, which type is used is determined by the nature of the alternate hypothesis. If there is an absence of direction, as in $H_1: \mu \neq 154$ or $H_1: p \neq .62$, then a two-tailed test is used. On the other hand, when there is a specific direction, as in $H_1: \mu > 154$ or $H_1: p < .62$, then a one-tailed test is called for.

FIGURE 9.4   THE CRITICAL REGION OF A ONE-TAILED TEST TO THE RIGHT

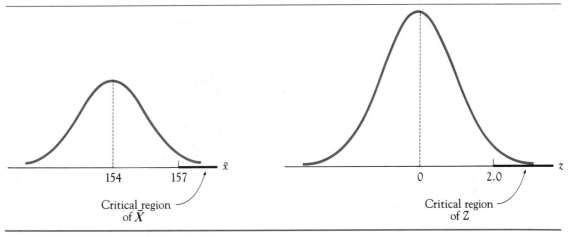

So far, our examples have focused on critical regions of simple hypothesis tests of the population mean μ. The next example shows that these ideas apply equally well to tests of the population proportion p.

Example 9.4

The mathematics competency exam at a university is failed by $p = .24 = 24\%$ of all students who take it. A shorter form of the exam is being considered that some critics claim will be more difficult. As a result, the university plans to test

$$H_0: \quad p = .24$$
$$H_1: \quad p > .24$$

where p is the proportion of all students who fail the new exam. Suppose a random sample of $n = 150$ new students will take the new exam and the decision rule is to reject the null hypothesis if the proportion $\hat{p}$ of the sample failing the exam exceeds $p = .24$ by three standard errors. Find the critical region of the test both in terms of $\hat{p}$ and Z.

The standard error of $\hat{p}$ is

$$S.E. = \sqrt{\frac{pq}{n}} = \sqrt{\frac{(.24)\,(.76)}{150}} = .035$$

The alternate hypothesis $H_1: p > .24$ implies a one-sided test to the right. The critical value of $\hat{p}$ is thus three standard errors to the right of the asserted $p = .24$ at

$$\hat{p} = .24 + (3.0)\,(.035) = .345$$

Thus H_0 is rejected if 34.5% or more of the sample fail the exam. For the test statistic $Z = (\hat{P} - .24)/.035$ the critical value is simply $z = 3.0$. The two critical regions are shown in Figure 9.5. ∎

FIGURE 9.5   CRITICAL REGION OF THE SINGLE-TAILED TEST OF INCREASED EXAM DIFFICULTY

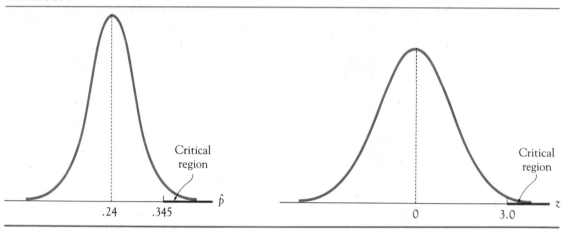

Suppose in Example 9.4 the proportion failing the test turned out to be $\hat{p} = .38$. Since this exceeds the critical value .345, the null hypothesis would be rejected. Had Z been used as the test statistic, we would have computed

$$z = \frac{.38 - .24}{.035} = 4.0$$

Since this value far exceeds the critical value of three standard errors, this result would also produce a rejection of H_0. The Z statistic is obviously easier to use. We do not have to compute a critical region, since it is immediately obvious. And the computed z score (here $z = 4.0$) tells us in terms of the standard error just how extreme a value was produced by the sample. Eventually we will use z scores exclusively.

EXERCISE SET 9.1

Part A

1. In each of the following problems, formulate the null and alternate hypotheses both verbally and in mathematical form. Then decide whether a one- or two-tailed test is appropriate.

   a. A federal highway inspector is told that the average speed of vehicles on a certain federally supported highway is 57.5 miles per hour. The inspector believes this is too low.

   b. 37% of the adult population smoke. It is believed that more teenagers than adults smoke.

   c. Two years ago an extensive study of the activities of students showed that 67% have never skied. It is not known whether or not there has been any change in this figure.

   d. An insurance company reported that 67% of all licensed drivers have never had a ticket nor been in a major accident. The state highway safety commission wants to verify this report but has no evidence to indicate whether the reported figure is too high or too low.

e. A large clothing chain found, in surveying year-end inventories, that 63.5% of all leftover jeans were in small sizes and tried to adjust their ordering patterns accordingly. A test is planned of the effectiveness of the new patterns by examining a random sample of inventories at several stores.

f. The mean grade for all third grade students nationwide on a standardized examination is 125. It is believed that students who are not well nourished will perform more poorly on this test.

2. Past performances of U.S. Marine graduates on an overall proficiency examination have averaged 254 out of a possible score of 300. Give the null and alternate hypotheses for testing:

   a. The present graduates do not differ appreciably from those in the past.
   b. The present graduates are better than previous classes.
   c. The present class does not measure up to previous graduates.

3. The null hypotheses of a test is H_0: $\mu = 560$. Give an alternate hypothesis for:

   a. A two-tailed test.
   b. A one-tailed test to the left
   c. A one-tailed test to the right.

4. In a test, the null hypothesis H_0: $p = .35$ will be rejected if the proportion $\hat{p}$ of a random sample of size $n = 245$ is more than two standard errors from $p = .35$. Sketch the critical regions for both Z and $\hat{P}$ if the test is:

   a. A two-sided test.
   b. A one-sided test to the left.
   c. A one-sided test to the right.

5. A company claims that its accounts receivable have an average age of 45 days. You believe this figure is too low and plan a test. Give suitable null and alternate hypotheses.

6. We plan to test H_0: $\mu = 245$ (with $\sigma = 24$) against H_1: $\mu \neq 245$, using the mean of a sample of size $n = 100$. Find the critical values and regions for both Z and $\overline{X}$ if the decision is to reject H_0 when:

   a. $\bar{x}$ is more than two standard errors from 245.
   b. $\bar{x}$ is more than three standard errors from 245.

7. Individual purchases in a large shopping mall are reported to have a mean of $18.70 and a standard deviation of $4.30. You believe this figure is too high and will reject the claim if the mean of a random sample of 40 purchases is more than 2 standard errors below 18.70. For what sample means will the claim be rejected?

8. In a test of

$$H_0: \quad p = .4$$
$$H_1: \quad p > .4$$

the critical value of the test statistic Z is 1.65. What is the critical value of $\hat{P}$ if the sample size is $n = 800$?

9. A certain medication is known to reduce blood pressure in 60% of all patients with extreme hypertension. An improved version having fewer side effects has been developed. Formulate null and alternate hypotheses and a three-standard-error decision rule based on a random sample of $n = 576$ subjects to test whether:

   a. The new drug is more effective in reducing blood pressure.
   b. The new drug is less effective in reducing blood pressure.
   c. There is no difference in the effectiveness of this drug and its predecessor.

10. An advertising agency claims that only 35% of a newspaper's readers saw a full-page advertisement. The newspaper believes this figure is too low and plans a test using a two-standard-error decision rule. What is the conclusion if 340 of a sample of 900 readers of the paper are found to have seen the ad?

11. The critical region of a test of the mean is the interval $\bar{x} < 125$.
   a. Give a suitable alternative hypotheses if the null hypotheses is H_0: $\mu = 112$.
   b. Assuming $\sigma = 28$ and $n = 30$, find the critical region for Z.
   c. If H_0 is true, what is the probability that the mean of a random sample of size 30 falls in the critical region?
12. The critical region for a test of proportions consists of two intervals $\hat{p} > .34$ and $\hat{p} < .28$.
   a. Give a suitable alternative hypotheses if the null hypotheses is H_0: $p = .31$.
   b. Assuming a sample size of $n = 900$, find the critical region for Z.

Part B

13. Last year the starting salaries for graduates in a certain field had an average of $24,580 and a standard deviation of $3350. A test is planned to determine whether there will be any change this year by examining the starting salaries of $n = 100$ early job offerings. Assuming the null hypothesis will be rejected if the sample mean differs from that of the null hypothesis by at least two standard errors, find the critical region both in terms of $\overline{X}$ and Z if:
   a. It is believed that the salaries have increased.
   b. It is believed that that the salaries have decreased.
14. In a test of

$$
\begin{aligned}
H_0: & \quad \mu = 100 \\
H_1: & \quad \mu \neq 100
\end{aligned}
$$

the critical values of $\overline{X}$ are 100 ± 8.75. What are the critical values of Z if the sample size is $n = 36$ and the population standard deviation is $\sigma = 21$?
15. Suppose we are testing

$$
\begin{aligned}
H_0: & \quad \mu = 180 \\
H_1: & \quad \mu > 180
\end{aligned}
\qquad (\sigma = 21)
$$

and the decision rule is to reject H_0 if the mean of a sample of size $n = 64$ is more than 2.4 standard errors above the mean. What is the decision if $\bar{x} = 188$?
16. A detergent manufacturer has regulated a machine to fill boxes so that the mean amount dispensed is 16.2 ounces. Suppose the standard deviation of the dispensed amounts is around .05 ounces. Does it appear that the machine is properly adjusted if the mean weight of 36 randomly selected packages is 16.28 ounces?
17. Suppose in a one-tailed test with a two-standard-error decision rule, the null hypothesis is rejected. Would this also have been the conclusion for a two-tailed test with the same decision rule?
18. A weight control clinic claims that over a three-month period its patients show weight losses that average 18 pounds and have a standard deviation of 4.3 pounds. State appropriate null and alternate hypotheses for testing this claim if:
   a. It is believed that the claimed weight losses are too high.
   b. It is believed that the stated weight losses may not apply to the category of patients in question.
19. You are testing H_0: $\mu = 120$ with $\sigma = 8$ against H_1: $\mu > 120$. The mean of a sample of size $n = 144$ turns out to be $\bar{x} = 112$. What should you do, in all honesty?

9.3   TYPES OF ERRORS

Any decision based on a sample always has the possibility of being in error. Occasionally the results go against the probabilities. For example, while it is highly unlikely, it is possible to select a random sample of 100 people from the general population, all of whom smoke. Likewise, there are samples of 100 in which none of the members smoke. Fortunately, the chances of such occurrences are small, but when they do happen we are led to an error.

In simple hypothesis testing there are two states of affairs: Either the null hypothesis is true or else the alternate hypothesis is true. Of course, we don't know which prevails; otherwise there would be no need for the test. If it is the case that the null hypothesis is true and our decision is to reject it, then we have made what is called a type I error.

DEFINITION 9.4

A *type I error* occurs if the null hypothesis is rejected when in fact it is true.

Of course, it could be that the alternate hypothesis is true yet our sample does not lead us to reject the null hypothesis. A type II error then results.

DEFINITION 9.5

A *type II error* occurs if the null hypothesis is not rejected when in fact it is false.

A type II error thus occurs when the test does not detect that it is the alternate hypothesis which is true.

The possible results of our testing are summarized in Table 9.1.

TABLE 9.1

	State of affairs	
Decision	H_0 *True*	H_1 *True*
Accept H_0	correct decision	type II error
Reject H_0 (accept H_1)	type I error	correct decision

Example 9.5

Suppose that several years ago an extensive study of the number of hours worked per week by college students had a mean of 18 hours and a standard deviation of 8. This subject is now under study again. Having no indication as to the possible direction of the change, if any, we plan to test

$$H_0: \quad \mu = 18$$
$$H_1: \quad \mu \neq 18$$

A two-tailed test is appropriate for this alternate hypothesis. Suppose the decision rule is to reject the null hypothesis if the mean of a sample of 64 students is more than two standard errors from the asserted $\mu = 18$. Then for $\overline{X}$ the critical values are 16 and 20 ($18 \pm 2.0\ [8/\sqrt{64}]$). See Figure 9.6.

FIGURE 9.6 THE TWO-TAILED CRITICAL REGION FOR TESTING μ = 18 AGAINST μ ≠ 18

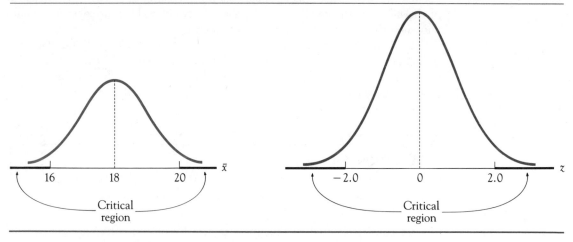

A type I error will be made if the following conditions hold:

1. The mean number of hours worked by the student population is $\mu = 18$.
2. The sample mean falls in the critical region, that is $\bar{x} < 16$ or $\bar{x} > 20$. Or, equivalently, $z < -2.0$ or $z > 2.0$ where

$$z = \frac{\bar{x} - 18}{8/\sqrt{64}}$$

A type II error results if:

1. The mean number of hours worked by the student population is not 18 hours, $\mu \neq 18$.
2. The sample mean $\bar{x}$ is somewhere between 16 and 20, that is $16 \leq \bar{x} \leq 20$ and thus does not fall in the critical region. Or equivalently $-2.0 \leq z \leq 2.0$. ■

The probabilities of the type I and type II errors are denoted by the Greek letters α and β, alpha and beta. Making use of the conditional probability notation we make the following definitions.

DEFINITION 9.6

The probability of a type I error is

$$\alpha = P(H_0 \text{ is rejected}|H_0 \text{ is true})$$

The probability of a type II error is

$$\beta = P(H_0 \text{ is not rejected}|H_1 \text{ is true}).$$

Example 9.6   Consider again the two-tailed test for

$$H_0: \quad \mu = 18$$
$$H_1: \quad \mu \neq 18$$

with the two-standard-error decision rule. Find the probability of a type I error.

The critical regions for both $\overline{X}$ and Z are shown in Figure 9.7. A type I error is made when the distribution is as shown and $\bar{x}$ is in the critical region. Thus

$$\alpha = P(\overline{X} < 16 \text{ or } \overline{X} > 20 | \mu = 18)$$
$$= P(Z < -2.0 \text{ or } Z > 2.0) = .0456$$

Note that this area is under the two tails of the distribution and above the critical region of either $\overline{X}$ or Z. ■

FIGURE 9.7   THE PROBABILITY α OF A TYPE I ERROR IS DETERMINED BY THE CRITICAL REGION

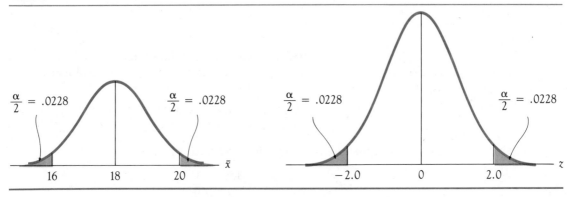

In this last example we interpret the error $\alpha = .0456$ to mean that in repeated applications of this test with a true null hypothesis, about 4.56% of the tests will result in the erroneous decision to reject the null hypothesis.

As a result of the rather vague nature of most alternate hypotheses, the probability β of a type II error can rarely be determined. The following example is contrived to illustrate its computation.

Example 9.7

Suppose we plan to test

$$H_0: \quad \mu = 150$$
$$H_1: \quad \mu = 161$$

with $\sigma = 20$ and the decision rule is to reject H_0 if the mean of a random sample of size $n = 36$ exceeds 156. Find the probabilities α and β.

There are two possible distributions for $\overline{X}$, depending on which of the hypotheses is true (Figure 9.8). The critical region of this test is to the right of $\bar{x} = 156$ under the tail of the first distribution. Thus

$$\alpha = P(X > 156 | \mu = 150)$$
$$= P(Z > 1.8) = .0359$$

Here $z = 1.8$ was obtained from

$$z = \frac{156 - 150}{20/\sqrt{36}}$$

FIGURE 9.8 THE DISTRIBUTION OF $\overline{X}$ UNDER H_0 AND H_1

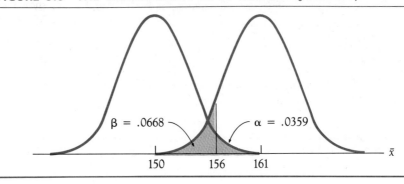

A type II error occurs if the second distribution centered at $\mu = 161$ is the true state of affairs and the sample mean does not fall in the critical region, that is $\bar{x} \leq 156$. Thus

$$\begin{aligned}\beta &= P(\overline{X} \leq 156 | \mu = 161) \\ &= P(Z \leq -1.5) = .0668\end{aligned}$$

Note that in the case $z = -1.5$ was computed by

$$z = \frac{156 - 161}{20/\sqrt{36}}$$

This area, $\beta = .0668$, is located under the extreme left tail of the second distribution. ∎

9.4 THE SIGNIFICANCE LEVEL OF A TEST

It is natural to ask whether it is possible ensure small probabilities α and β of type I and type II errors. Unfortunately the answer is no, if this end is to be accomplished by adjusting the critical region. For, as the critical region is changed to decrease one of these errors, the other increases.

In Figure 9.9 is sketched the two competing distributions of $\overline{X}$ from the last example. If the critical value is moved farther to the right, say, from 156 to 158, then the area α under the right tail of the first distribution decreases from .0359 to .0082. But notice that this has the effect of increasing the area β under the left tail of the second distribution. In fact, β increases from .0668 to .1841. In a similar manner, it can be shown that a decrease in β produces an increase in α.

In most tests of hypotheses, the nature of the alternate hypothesis precludes computation of the probability β of a type II error. Thus it is only natural to use tests where the probability α of a type I error is kept small. This approach has the effect of weighting tests somewhat in favor of the null hypothesis, since considerable evidence will always be required to reject it.

**FIGURE 9.9   THE DISTRIBUTION OF $\overline{X}$ UNDER H_0 AND H_1:
IF α IS DECREASED, β IS INCREASED**

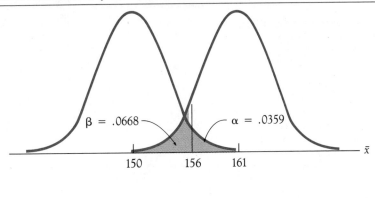

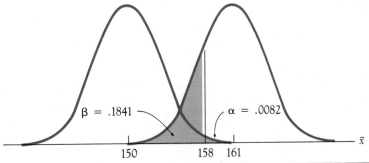

DEFINITON 9.7

The *significance level* of a test is the probability α of a type I error, expressed as a percent.

Thus if $\alpha = .02$ we would speak of a 2% significance level test. As might be suspected, significance levels of 5% and 1% are the most popular.

For a two-tailed test the significance level α is the total area under the extremes of the two tails above the critical region. Thus the area under each tail is $\alpha/2$; for the test statistic Z, the critical values are simply $z_{\alpha/2}$ and $-z_{\alpha/2}$. With a one-tailed test the area α is concentrated solely under one tail and the critical value is either z_α or $-z_\alpha$, depending on whether the direction of the test is to the right or to the left (see Figures 9.10 and 9.11).

Once the significance level of a test is specified, the critical values of Z are easily found. Critical values for the mean $\overline{X}$ requires a computation that involves σ. As with confidence intervals, σ must generally be approximated by the standard deviation of the sample. Thus in most instances the critical value(s) of $\overline{X}$ cannot be found until after the data are gathered. This is simply another argument in favor of using Z as the test statistic.

FIGURE 9.10   CRITICAL VALUES OF Z, $z_{\alpha/2}$ AND $-z_{\alpha/2}$, FOR A TWO-TAILED TEST AT SIGNIFICANCE LEVEL α

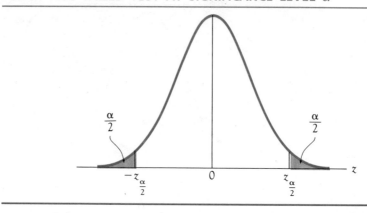

FIGURE 9.11   CRITICAL VALUES OF Z AT SIGNIFICANCE LEVEL α: z_α FOR A ONE-TAILED TEST TO THE RIGHT AND $-z_\alpha$ FOR A ONE-TAILED TEST TO THE LEFT

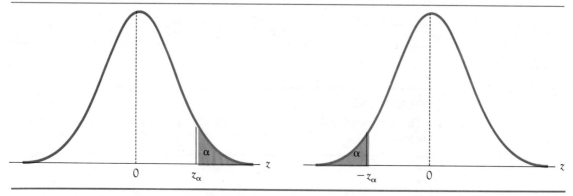

Example 9.8

A 10% significance level will be used to test

$$H_0: \quad p = .25$$
$$H_1: \quad p < .25$$

Assuming a sample large enough to justify using the normal distribution, find the critical value of Z.

The direction $p < .25$ of H_1 indicates a one-tailed test to the left. The area under the extreme left tail of the normal distribution is $\alpha = .10$ (see Figure 9.12). Thus the critical value is $-z_{.10} = -1.28$ and the critical region is $z < -1.28$. ■

FIGURE 9.12 CRITICAL VALUE OF *Z* FOR A ONE-TAILED TEST TO THE LEFT WITH α = .10

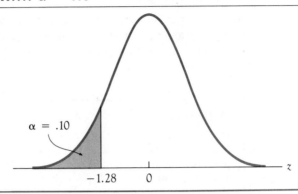

$\alpha = .10$

$-1.28 \qquad 0$

z

Example 9.9

A 5% level of significance is to be used in testing

$$H_0: \quad \mu = 28$$
$$H_1: \quad \mu \neq 28 \qquad \sigma = 6$$

Find the critical values of both $\overline{X}$ and Z, assuming a sample size of $n = 64$.

The lack of a specific direction in the alternate hypothesis dictates a two-tailed test. Thus the area under each tail of standard normal distribution is $\alpha/2 = .05/2 = .025$ and the critical values of Z are $z_{.025} = 1.96$ and $-z_{.025} = -1.96$. The critical values of $\overline{X}$ are now seen to be

$$\overline{x}_1 = 28 - z_{.025} \frac{6}{\sqrt{64}} = 28 - 1.47 = 26.53$$

$$\overline{x}_2 = 28 + z_{.025} \frac{6}{\sqrt{64}} = 28 + 1.47 = 29.47$$

The critical values of both these variables are shown in Figure 9.13. ∎

FIGURE 9.13 THE TWO-TAILED CRITICAL REGION WITH α = .05 FOR TESTING μ = 28 AGAINST μ ≠ 28

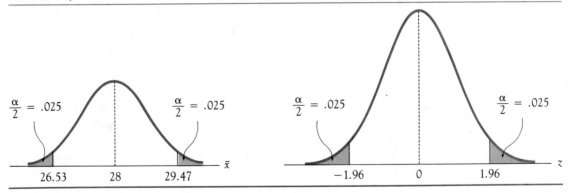

$\frac{\alpha}{2} = .025$ $\frac{\alpha}{2} = .025$

$26.53 \qquad 28 \qquad 29.47$ $\overline{x}$

$\frac{\alpha}{2} = .025$ $\frac{\alpha}{2} = .025$

$-1.96 \qquad 0 \qquad 1.96$ z

9.5   THE *p* VALUE

In the literature one rarely finds a statement that some hypothesis was or was not rejected at a certain level of significance. Rather a summary of the test statistic and a number called the *p* value are given.

> **DEFINITION 9.8**
>
> In a test of hypotheses, the *p value* is the probability that the test statistic being used will have a value at least as extreme (and in the same direction) as that produced by the sample. This probability is computed under the assumption that the null hypothesis is true.

Example 9.10

Consider a test of the two hypotheses:

$$H_0: \quad \mu = 140$$
$$H_1: \quad \mu > 140 \qquad \sigma = 10$$

Suppose a sample of size $n = 64$ produces $\bar{x} = 143.0$. The distribution of $\overline{X}$ is shown in Figure 9.14. Those values that are at least as extreme (as large, in this case) as $\bar{x} = 143$ lie under the right tail of this distribution. Thus the *p* value is

$$p = P(\overline{X}) > 143 | \mu = 140)$$

The corresponding Z score is

$$z = \frac{\bar{x} - 140}{10/\sqrt{64}} = \frac{143 - 140}{1.25} = 2.4$$

Thus the *p* value is $p = P(Z > 2.4) = .0082$ ■

FIGURE 9.14   THE PROBABILITY OF OBTAINING A SAMPLE MEAN MORE EXTREME THAN $\bar{x} = 143$ IS $p = .0082$

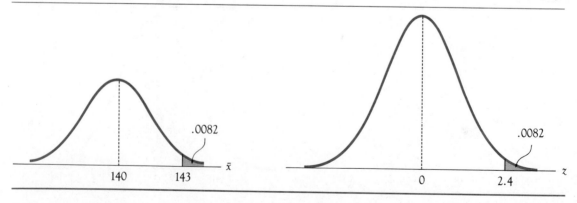

Suppose in the last example we had by chance chosen $\bar{x} = 143.0$ as the critical value for this test. The critical region would then have been exactly the interval of extreme values indicated in Figure 9.14, and the significance level α of the test would have been the corresponding area. That is, we would have $\alpha = .0082$. Further, if the significance level were any smaller than .0082, we would not reject the null hypothesis.

For a single-tailed test we now see that the p value is the smallest significance level at which the data justify rejecting the null hypothesis (assuming that the value $\bar{x}$ or $\hat{p}$ of the test statistics is located on the same side of the mean μ or the proportion p as the critical region). For a two tailed test, this smallest significance level is twice the p value. We illustrate this in the next example.

Example 9.11

In a study of carry-on luggage weights an airline tested:

$$H_0: \quad \mu = 46$$
$$H_1: \quad \mu \neq 46$$

Assuming $\sigma = 9$ find the p value of the test and interpret the result if a sample of $n = 36$ luggage weights had a mean of 50.4 pounds.

The p value of the test is the probability of obtaining a sample mean as large or larger than 50.4 when $\mu = 46$. Thus

$$p = P(\overline{X} > 50.4 | \mu = 46)$$
$$= P(Z > 2.93) = .0017$$

Here $z = 2.93$ was computed by

$$z = \frac{50.4 - 46}{9/\sqrt{36}}$$

Now to the interpretation. Suppose the airline had chosen $\alpha = .0034$ (twice the p value) as the significance level. Then the area under each tail of the distribution of $\overline{X}$ is $\alpha/2 = .0017$, which is precisely the p value (see Figure 9.15).

FIGURE 9.15 FOR A TWO-TAILED TEST THE MINIMAL SIGNIFICANCE LEVEL FOR REJECTING THE NULL HYPOTHESIS IS TWICE THE p VALUE

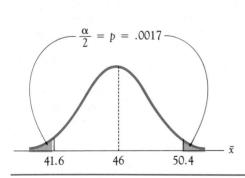

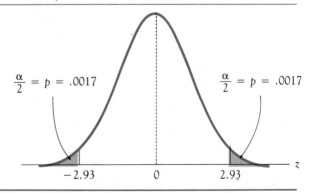

Thus for a two-tailed test, the smallest significance level that would lead to a rejection of H_0 is twice the p value. ∎

The p value is always included along with any published summary of a test, without exception. From it the reader can immediately determine the range of significance values for which the null hypothesis would have been rejected. In those instances where the complete distribution of the test statistic is not available (e.g., T, F, χ^2), only a bound such as $p. < .01$ can be given. For the t distribution this would mean that the p value is between .01 and .005, since $t_{.01}$ and $t_{.005}$ are adjacent entries in the table of this distribution.

Although this convention is by no means universal, the term *significant* is sometimes used to describe the value of the test statistic if the p value is less than .05. *Highly significant* is used if the p value is less than .01.

A word of caution about the definition of p values: Some authors relate the p value to the type of test. For a single-tailed test the p value will be the same by either convention. For a two-tailed test, however, some people report twice the single-tailed value, which we have used. It is easy to argue against this practice. For one thing, this relates the p value to the form of the test, without knowing the form, one cannot determine how extreme a value was observed. Also, p values apply to statistics having both symmetric and asymmetric distributions. With an asymmetric distribution, the doubling of a single-tailed p value could conceivably produce a value greater than 1. There are many other excellent arguments in favor of a one-sided p value; the interested reader many pursue the subject in "p Values: Interpretation and Methodology," Jean Gibbons and John Pratt. *The American Statistician*, vol. 29, 1975.

EXERCISE SET 9.2

Part A

1. Formulate null and alternate hypotheses in non-mathematical terms for each of the following and describe the type I and type II errors. Also, discuss which (if either) of the probabilities α and β should definitely be kept small.
   a. A technical school will retain its certification if its graduates demonstrate a reasonable level of proficiency, say above 65% on an examination.
   b. Driving speeds are being clocked and a driver will be cited if several radar readings of her speed average more than 57.5 miles per hour.
   c. A patient is diagnosed as having hypertension if the average of several systolic blood pressure readings exceeds 147 millimeters of mercury.
   d. An instructor will be deemed ineffective if the mean of a sample of his students' evaluations is appreciably below the mean for the department.
   e. An area is classified as economically depressed and eligible for additional aid if more than 15% of its work force is unemployed.
   f. A child is classified as mentally deficient if its I.Q., as measured by the average of several intelligence tests, is below 80.
2. The null hypothesis H_0: $\mu = 24$ is being tested against H_1: $\mu \neq 24$. Suppose the decision rule is to reject H_0 if $\bar{x}$ is more than two standard errors from $\mu = 24$. Assuming $\sigma = 6$ and $n = 64$, what type of error if any will be made if:
   a. The null hypothesis is true and $\bar{x} = 22.9$?
   b. The null hypothesis is true and $\bar{x} = 26.2$?
   c. H_1 is true and $\bar{x} = 24$.
   d. H_1 is true and $\bar{x} = 25.75$.

3. Consider a test of H_0: $p = .44$ versus H_1: $p > .44$ where the decision rule is to reject H_0 if the proportion $\hat{p}$ of a sample of size $n = 900$ is more than two standard errors greater than $p = .44$. What type of error, if any, is made when:
   a. H_0 is true and $\hat{p} = .48$?
   b. H_0 is true and $\hat{p} = .45$?
   c. H_1 is true and $\hat{p} = .48$?
   d. H_1 is true and $\hat{p} = .45$?

4. A test is being made of H_0: $\mu = 100$ against H_1: $\mu < 100$. Find the probability of a type I error if the decision rule is to reject H_0 when the mean of a sample of size 35 is:
   a. More than 1.64 standard errors below $\mu = 100$.
   b. More than 2.33 standard errors below $\mu = 100$.
   c. More than three standard errors below $\mu = 100$.

5. Find the probability of a type I error in testing H_0: $p = .5$ against H_1: $p \neq .5$ if the decision rule is to reject H_0 if the proportion $\hat{p}$ of a sample of size $n = 625$ is:
   a. More than 1.64 standard errors from $p = .5$.
   b. More than 2.0 standard errors from $p = .5$.
   c. More than 2.33 standard errors from $p = .5$.

6. A metropolitan area's promotional brochure stated that the price of a breakfast in their area averaged $4.05 and had a standard deviation of $.90. A tourism rating service believes that breakfasts are actually much more expensive and plans to make a test of the advertised assertion, using a sample of 36 restaurant prices. For what sample means will the asserted mean be rejected if the significance level of the test is:
   a. 5%?
   b. 1%?
   c. 2%?

7. An import auto distributor stated that the mean sticker price of all cars of a certain model is $9240 and further that these prices have a standard deviation of $820. A domestic producer believes that the distributor is understating the prices and plans a test using a sample of $n = 30$ of the import autos. Find the critical values of both Z and the mean $\overline{X}$, assuming the significance level is:
   a. 5%?
   b. 1%?
   c. 10%?

8. A sample of size $n = 100$ is to be used in testing H_0: $\mu = 38$ against H_1: $\mu < 38$. Assuming $\sigma = 9$, find the critical values of both Z and $\overline{X}$ when the significance level is:
   a. 5%.
   b. 2.5%.
   c. 3%.
   d. 10%.

9. Five years ago a very extensive study showed that 67% of the cars on a freeway at rush hours had only one occupant. A test will investigate whether there has been any change in this figure by observing 1200 randomly selected autos over a five-day period during rush hours.
   a. State the null and alternate hypothesis.
   b. Find the critical regions for both Z and $\hat{P}$ when a 1% significance level test is used.

10. In testing H_0: $p = .5$ against H_1: $p \neq .5$ a sample of $n = 400$ yielded $\hat{p} = .55$. Find the p value of the test.

11. In testing H_0: $\mu = 118$ with $\sigma = 12$ against H_1: $\mu > 118$, a sample of size $n = 36$ yielded $\bar{x} = 123$. Find the p value of the test.

12. In a large-sample hypothesis test of the mean, the p value turned out to be .028. Would the null hypothesis be rejected in:
   a. A single-tailed test with $\alpha = .05$?
   b. A single-tailed test with $\alpha = .01$?
   c. A two-tailed test with $\alpha = .05$?
   d. A two-tailed test with $\alpha = .01$?

Part B

13. A population has a standard deviation of 18 and a mean of either 146 or 156.
   a. Sketch the possible distributions of $\overline{X}$ when the sample size is $n = 36$.
   b. Suppose a test is planned in which we will reject H_0: $\mu = 146$ in favor of H_1: $\mu = 156$ if for a random sample of size 36, $\bar{x} > 153$. What is the probability of a type II error?
   c. Now suppose $\mu = 156$ is taken as the null hypothesis and the decision rule is to reject this in favor of $\mu = 146$ if the sample mean is at or below 153. Find both α and β.

14. What happens to the size of the critical region if the significance level of a test is decreased?

15. Suppose H_0: $\mu = 160$ is rejected in favor of H_1: $\mu \neq 160$ when the significance level is $\alpha = .01$.
   a. Would H_0 have been rejected at a significance level $\alpha = .05$?
   b. Would H_0 have been rejected at a significance level $\alpha = .005$?
   c. Would H_0 have been rejected at $\alpha = .05$ if the alternate hypothesis had been $\mu > 160$? (Assume $\bar{x} > 160$).
   d. Would H_0 have been rejected at $\alpha = .01$ if the alternate hypothesis had been $\mu > 160$? (Assume $\bar{x} > 160$).

16. The p value of a test was reported as .01. For which of the following significance levels would the null hypothesis have been rejected in a two-tailed test?
   a. 5%.
   b. 1%.
   c. .1%.

17. If the p value of a test is .035, what is the smallest significance level at which the null hypothesis will be rejected for:
   a. A single-tailed test?
   b. A two-tailed test?

18. A study of the number of hours worked monthly by full-time nurses led to a test of H_0: $\mu = 192$ versus H_1: $\mu \neq 192$. A sample of 112 nurses produced $\bar{x} = 201$ hours with $s = 38$. Approximating σ by s, find the p value of the test. Is $\bar{x} = 201$ significant? Highly significant?

9.6   LARGE-SAMPLE TESTS CONCERNING THE MEAN

Most of the central ideas of simple hypothesis testing have been discussed in the preceding sections. One variation that has been omitted up to now is the approximation of the population standard deviation σ by the sample standard deviation s. This presents no problem. We have already had experience with this in constructing confidence intervals and know that when large samples are used, the normal distribution may be used with

$$Z = \frac{\overline{X} - \mu}{s/\sqrt{n}}$$

FIGURE 9.16   CRITICAL REGIONS

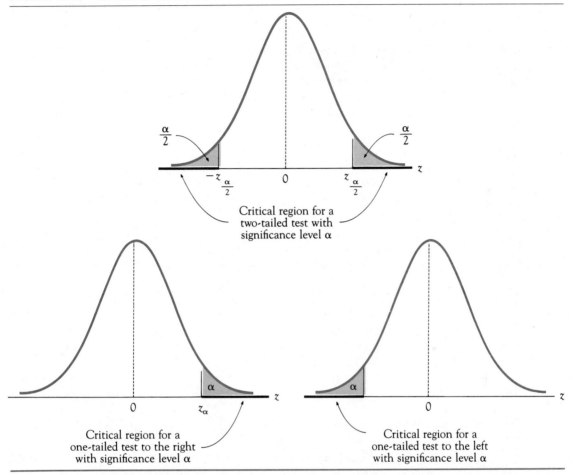

We now summarize the testing process.

PROCEDURE FOR TESTING THE NULL HYPOTHESIS
H_0: $\mu = \mu_0$ **AGAINST AN ALTERNATE HYPOTHESIS H_1**

Given: The alternate hypothesis H_1, the size n, the mean $\bar{x}$, and the standard deviation s of a random sample from the population being tested.

Assumption: The sample size n is at least 30.

1. Choose a one- or two-tailed test, based on the form of H_1.
2. Select a significance level α, such as $\alpha = .05$ or $\alpha = .01$.
3. Find the critical value(s) of the test. For a one-tailed test, this will be either z_α or $-z_\alpha$. For a two-tailed test, the values are $\pm z_{\alpha/2}$.

4. Compute

$$z = \frac{\bar{x} - \mu_0}{s/\sqrt{n}}$$

5. Compare the value of the test statistic to the critical value(s).
   - Case I—H_1: $\mu \neq \mu_0$. Reject H_0 if $z > z_{\alpha/2}$ or $z < -z_{\alpha/2}$.
   - Case II—H_1: $\mu > \mu_0$. Reject H_0 if $z > z_\alpha$.
   - Case III—H_1: $\mu < \mu_0$. Reject H_0 if $z < -z_\alpha$.

Even though it is not a part of the test, the p value should be computed for reporting purposes. Figure 9.16 shows the form of the critical regions for these three types of tests.

Example 9.12   A detergent manufacturer has its machine set to load 16.30 ounces of laundry soap into nominal 1 pound boxes. Samples of the production are routinely examined to ascertain that the boxes are not being appreciably overloaded or underloaded. Test the hypothesis that the loading process is in control if a sample of 36 boxes had a mean of 16.39 and a standard deviation of .21 ounces.

Let μ be the mean weight of the lot sampled. The loading process will be in control if $\mu = 16.30$. Thus we will test

$$H_0: \quad \mu = 16.30$$
$$H_1: \quad \mu \neq 16.30$$

A two-tailed test is required, since there is no direction specified in the alternate hypothesis.

Since no significance level is given we shall choose $\alpha = .05$. The critical values of Z are $\pm z_{.025} = \pm 1.96$. Using $\bar{x} = 16.39$ and $s = .21$, the value of the test statistic is

$$z = \frac{16.39 - 16.30}{.21/\sqrt{36}} = 2.57$$

Since $z = 2.57 > 1.96$ we reject the null hypothesis and conclude that the loading process is not in control. The p value is

$$p = P(Z > 2.57) = .0051$$

In Figure 9.17 we have shown the relation of the sample mean $\bar{x} = 16.39$ to the asserted $\mu = 16.30$. Note that it is more than 2.5 standard errors above the expected value—a most unlikely occurrence. ∎

In examples and problems it is necessary to both describe the situation and supply the sample data necessary to complete the test. The reader should ignore the sample data until the alternate hypothesis has been decided upon. Otherwise a bias may be injected into the testing procedure through choosing what appears to be the more favorable alternate hypothesis.

FIGURE 9.17   CRITICAL REGIONS

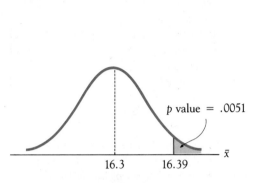 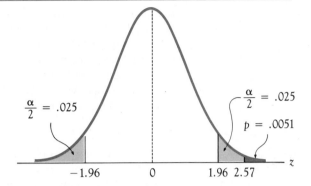

The probability under H_0
of a sample mean as large
as 16.39 is .0051.

The critical region of
the 5% significance
level two-tailed test.

Example 9.13   Rats that are raised in a laboratory environment have a mean life span of around 24 months. A sample of 31 rats reared to adulthood in a germ-free environment had life spans with a mean of 27.3 and a standard deviation of 5.9 months. Does this type of rearing have an effect on the life span of the laboratory rat? If there is an effect, we would expect it to increase the life span. Thus, if μ is the mean life of such rats, we wish to test:

$$H_0: \quad \mu = 24$$
$$H_1: \quad \mu > 24$$

A single-tailed test to the right is required. Choosing $\alpha = .05$, the critical value of Z is $z_{.05} = 1.65$. Using $\bar{x} = 27.3$ and $s = 5.3$ we next compute

$$z = \frac{27.3 - 24}{5.9/\sqrt{31}} = 3.11$$

FIGURE 9.18 $z = 3.11$ EXCEEDS THE CRITICAL VALUE $z_{.05}$

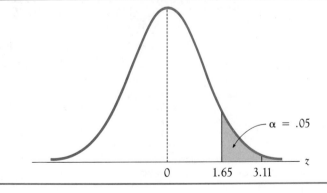

Since $z = 3.11$ greatly exceeds the critical value (see Figure 9.18), we reject the null hypothesis that the mean life span is 24 months. In the absence of confounding factors we attribute an effect to the rearing environment. The p value of the test is $p = P(Z > 3.11) = .0009$.

As a matter of interest we mention that laboratory rats that are reared and live in a specific pathogen-free (SPF) environment have an expected life span of about three years. These SPF environments are quite valuable to the researcher who needs to carry out long-term studies on animals without having to control for viral and bacterial factors. ■

Example 9.14
A Case Study

In several studies it has been reported that the natural age at menopause of non-smoking women is around 49.5 years. In a study entitled "Cigarette Smoking and Natural Health," (David Kaufman et al., *American Journal of Public Health*, vol. 70, 1980), a sample of 47 heavy smokers had a mean age at menopause of 47.6 years. Assuming a standard deviation of 4.6 years, test the hypothesis that cigarette smoking is associated with the early onset of menopause.

Let μ be the mean age at the onset of menopause for the heavy smokers. We wish to test that this mean is less than for the nonsmokers. Thus we will test

$$H_0: \quad \mu = 49.5$$
$$H_1: \quad \mu < 49.5$$

A single-tailed test to the left is dictated by the alternate hypothesis. Using $\alpha = .05$, the critical value of Z is $-z_{.05} = -1.64$. The value of the test statistic is

$$z = \frac{47.6 - 49.5}{4.6/\sqrt{47}} = -2.83$$

Since $z = -2.83 < -1.64$ the null hypothesis is rejected. The p value of the test is $p = P(Z < -2.83) = .0023$.

One brief footnote: In the actual study, ex-smokers and light, moderate, and heavy smokers were compared to nonsmokers by using tests on the difference of the sample means (next chapter). We have adapted the data to this simple hypothesis test. The conclusion is the same in both cases: the mean age at menopause for smoking women is below that of non-smoking women and the difference becomes more pronounced with increased smoking activity (see Table 9.2). ■

TABLE 9.2   MEAN AGE AT MENOPAUSE AMONG 656 NATURALLY POSTMENOPAUSAL WOMEN AGES 60–69, ACCORDING TO NUMBER OF CIGARETTES SMOKED

Number of cigarettes*	Number of women	Mean age at menopause (years)	Mean difference from never smokers (years)	Standard error of difference
Never-smoker	434	49.4	—	—
Ex-smoker	10	49.2	0.2	1.5
1–14/day	66	48.0	1.4	0.6
15–24/day	99	47.6	1.8	0.5
≥25/day	47	47.6	1.8	0.7

*Refers to smoking status at time of interview. All smokers began smoking before age 35; all ex-smokers stopped before age 35.

It will be recalled that the significance level of a test is the probability of a type I error. Thus, by using a small α, we seek to avoid rejecting the null hypothesis when it is actually true. The possibility of a type II error always looms when the decision is not to reject H_0. It is very easy to construct plausible situations where the probability β of this type error is as large as .5. The fact that a type II error is a distinct possibility should always be kept in mind. The decision to not reject a null hypothesis H_0 does not in any way establish it as the true state of affairs. We should feel comfortable with this decision only if β is small. In the last section of this chapter we explore in more detail how to keep both α and β small through the use of larger samples.

EXERCISE SET 9.3

Part A

1. **a.** Test H_0: $\mu = 146$ against H_1: $\mu \neq 146$ using $\bar{x} = 150$, $s = 14$, $n = 64$, and $\alpha = .05$. Find the p value.
   b. Test H_0: $\mu = 218$ against H_1: $\mu < 218$ using $\bar{x} = 210$, $s = 15$, $n = 30$, and $\alpha = .05$. Find the p value.
   c. Test H_0: $\mu = 24$ against H_1: $\mu > 24$ using $\bar{x} = 29.5$, $s = 15$, $n = 36$, and $\alpha = .01$. Find the p value.

2. Given the following sets of conditions, test the indicated hypothesis and find the p value:
   a. H_0: $\mu = 118$, H_1: $\mu \neq 188$; $\bar{x} = 115$, $s = 20$, $n = 100$, and $\alpha = .05$.
   b. H_0: $\mu = 56$, H_1: $\mu > 56$; $\bar{x} = 61$, $s = 30$, $n = 144$, and $\alpha = .01$.

3. A random sample of 100 building blocks has a mean weight of 15.6 pounds and a standard deviation of $s = .8$ pounds. Test the hypothesis that all blocks of this manufacturer have a mean weight of 16 pounds. Use $\alpha = .02$. What is the p value?

4. A certain grade of hamburger is supposed to contain at most 30% fat. A consumer group purchases a random sample of 30 packages from a market chain and found an average fat content of 31.4% with a standard deviation of 2.8%. Can they conclude that the fat content of this chain's hamburger exceeds 30% Use $\alpha = .02$. Give the p value of the test.

5. A survey of faculty workloads at a large university in 1979 showed an average of 44.50 hours per week of intructional and academic-related work. Recently the university switched to the quarter system. Test the hypothesis that the workload is unchanged and did not increase as predicted if a random sample of 30 faculty members showed workloads with a mean of 45.75 hours and a standard deviation of 7.4 hours per week. Use $\alpha = .05$.

6. Suppose the I.Q.s of graduating high school seniors have a mean of 106. A metropolitan area selects 55 students at random and will use them to demonstrate (hopefully) that students from this area have I.Q.s exceeding the national average of 105. Of the 55 students only 43 actually consented to undergo the extensive evaluation procedure and for these the mean and standard deviation were $\bar{x} = 110$ and $s = 15$. What is the conclusion if:
   a. $\alpha = .01$?
   b. $\alpha = .05$?
   Which significance level do you think the school district would use in making their reports to the parents?

7. The discharge temperature of coolant water from a nuclear reactor is stipulated to have a mean of 45°C. An environmental group believes it to be higher and on 39 different occasions obtains temperature readings that have a mean of 46.4°

and a standard deviation of 3.6°. Formulate the null and alternate hypotheses, state the assumptions for the test, and carry it out. Use $\alpha = .025$.

8. The production level of an assembly line before modernization was 30 units per hour. The modernization was supposed to increase output by 25% but the increase may in fact be less. Test the hypothesis that the full 25% increase did occur if a random sample of 40 production rates had a mean of 36.2 and a standard deviation of 5.2 units per hour.

9. The mean length of time for a company's computer to perform a daily general ledger update has been found to be 2 hours and 14 minutes. A new accounting package is being considered and is given a 45-day trial. For these 45 days the general ledger updating times had a mean of 2 hours and 3 minutes and a standard deviation of 36 minutes. If these times can be considered a random sample, would the company be justified in concluding that the general ledger updating will be done more efficiently with this new accounting package? Use $\alpha = .01$.

Part B

10. The mean assembly time for a certain item on an assembly line has been established at 23.5 minutes. A new procedure in which employees are allowed to rotate from job to job to avoid boredom is implemented. After the employees have had an opportunity to adjust to the procedure, a sample of 50 assembly times is taken. These times have a mean of 24.6 minutes and a standard deviation of 6 minutes. In terms of production, does the assembly line appear to be operating as efficiently as before?

11. In testing H_0: $\mu = 45$ against H_1: $\mu > 45$, a random sample of size n produced $\bar{x} = 46.6$ and $s = 6.0$. If a 5% significance level is used, what is the smallest sample size for which the null hypothesis would be rejected?

12. Before the implementation of a new scanning procedure, the average time to check out a supermarket customer was 5.6 minutes. When a random sample of 120 customers was observed with the new procedure, the checkout times had an average of 5.0 minutes with a standard deviation of $s = 2.3$ minutes. Can we conclude that the new procedure requires less time than the previous method? Use $\alpha = .02$.

13. In computing total payload weight, an airline has been using the figure of 45 pounds for the average weight of the carry-on luggage of passengers. In a recent sample of 600 passengers, the carry-on luggage weights were found to have a mean of 46.3 pounds and a standard deviation of 20 pounds. Can the airline conclude that the mean carry-on weight is no longer 45 pounds? Use $\alpha = .05$. Compute the p value of the test.

14. A random sample of size $n = 40$ is drawn from a normal population and found to have a mean of 70 and a standard deviation of 20. What is the smallest significance level for which H_0: $\mu = 75$ will be rejected in favor of H_1: $\mu \neq 75$?

15. In testing H_0: $\mu = 150$ against H_1: $\mu < 150$, a sample of size 225 produced $\bar{x} = 144$ and $s = 38.5$. What is the smallest significance level for which the null hypothesis would be rejected?

16. It is rumored that some journals do not publish research results if they are not significant (at the 5% level). If this were true, would one- or two-tailed tests give researchers an edge in having their results published? Before jumping to too severe a conclusion about researchers, consider whether or not the researcher in most instances will know the direction of the change in the mean if indeed there has been one.

17. Do students cope well with stress in learning situations? A standardized learning test based on word and phrase recognition has a mean of 300 and a standard

deviation of 100. Fifty students were given this test under conditions designed to increase stress; their mean score was 278. At the $\alpha = .001$ significance level, can we conclude that the students' learning is impaired by stress?

18. The mean of a population is asserted to be at least 40. Test this assertion, given a random sample of $n = 60$ for which $\bar{x} = 38.8$ and $s = 8$. Use $\alpha = .05$.

9.7   SMALL-SAMPLE TESTS CONCERNING THE MEAN

When the sample size is less than 30, the t distribution must be used in place of the normal. This of course requires the additional assumption that the population is normal or at least approximately so, since the central limit theorem no longer applies. With the exception of using t in place of z, there is no change in the testing procedure.

PROCEDURE FOR TESTING THE NULL HYPOTHESIS H_0: $\mu = \mu_0$ AGAINST AN ALTERNATE HYPOTHESIS H_1

Given: The alternate hypothesis H_1, the mean $\bar{x}$, the standard deviation s, and the size n of a random sample from the population being tested.

Assumption: The population is normal.

1. Choose a one- or two-tailed test, based on the form of H_1.
2. Select a significance level α, such as $\alpha = .05$ or $\alpha = .01$.
3. Using $\nu = n - 1$ degrees of freedom, find the critical values. For a one-tailed test this will be either t_α or $-t_\alpha$. For a two-tailed test the values are $\pm t_{\alpha/2}$.
4. Compute

$$t = \frac{\bar{x} - \mu_0}{s/\sqrt{n}}$$

5. Compare the value of the test statistic to the critical value(s).
   - Case I—H_1: $\mu \neq \mu_0$. Reject H_0 if $t > t_{\alpha/2}$ or $t < -t_{\alpha/2}$.
   - Case II—H_1: $\mu > \mu_0$. Reject H_0 if $t > t_\alpha$.
   - Case III—H_1: $\mu < \mu_0$. Reject H_0 if $t < -t_\alpha$.

Thus, loosely speaking, we reject H_0 if the computed t value is more extreme than a critical value.

Example 9.15

A time-and-motion study concluded that the mean number of worker-hours to assemble a certain light truck is 136.0. Test this assertion if the actual assembly times for 14 randomly selected trucks were found to have a mean of 140.2 hours and a standard deviation of 8.7 hours.

Since we have no way of knowing the direction of the error if the time-and-motion study is incorrect, we shall test:

$$H_0: \quad \mu = 136$$
$$H_1: \quad \mu \neq 136$$

This alternate hypothesis requires a two-tailed test. With $\alpha = .05$ and $\nu = 14 - 1 = 13$ degrees of freedom, the critical values are $\pm t_{.025} = \pm 2.160$ (see Figure 9.19). Next we compute t using $\bar{x} = 140.2$ and $s = 8.7$:

$$t = \frac{140.2 - 136.0}{8.7/\sqrt{14}} = 1.81$$

Since $t = 1.81$ does not exceed the critical value $t_{.025} = 2.160$, we do not reject the null hypothesis.

The p value of the test is $p = P(t > 1.81)$. For $\nu = 13$ the nearest table entry is $P(t > 1.771) = .05$. Thus we can report that $p < .05$. ∎

FIGURE 9.19   THE t DISTRIBUTION FOR $\nu = 13$ DEGREES OF FREEDOM

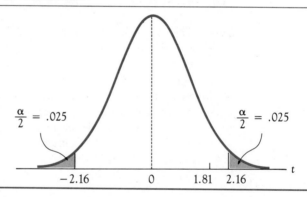

Example 9.15 should be considered in terms of the discussion at the end of the last section. We have not found reason to reject the null hypothesis. Does this mean that we have established $\mu = 136.0$ hours conclusively as the assembly time? Of course not. We would have been led to the same decision with a null hypothesis of, say, $\mu = 138.0$ or $\mu = 139.0$. If there is additional substantial evidence to support $\mu = 136.0$ then this result simply adds to it.

Example 9.16

The average spring snow depth on a certain plateau in the Sequoia National Forest has averaged 50 feet since it was first measured some 100 years ago. Test at a 5% significance level the assertion that there is a greater mean depth this year if the snow depths at 16 randomly selected sites had a mean of 57.0 feet and a standard deviation of 14.1 feet.

We let μ be the mean depth this year and test

$$H_0: \quad \mu = 50.0$$
$$H_1: \quad \mu > 50.0$$

With $\alpha = .05$ and $\nu = 15$, the critical value of t is $t_{.05} = 1.753$. Since

$$t = \frac{57.0 - 50.0}{14.1/\sqrt{16}} = 1.99 > 1.753$$

the null hypothesis is rejected and we conclude that the mean depth this year exceeds 50 feet. The p value is $p = P(t > 1.99) < .05$. See Figure 9.20. ∎

FIGURE 9.20 THE PROBABILITY OF A TEST VALUE t EXCEEDING 1.99 IS LESS THAN .05

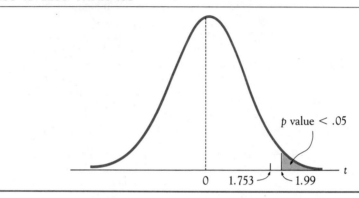

As we pointed out with confidence intervals, the normality condition imposed on the use of the t distribution with small samples may be dropped entirely for large samples. As a result, large-sample hypothesis tests of the mean are made with either the standard normal or the t distribution.

EXERCISE SET 9.4

Part A

1. Assuming a normally distributed population, test H_0: $\mu = 10$ against H_1: $\mu > 10$, given a sample of size 12 for which $\bar{x} = 11.2$ and $s = 2.3$. Use the significance level $\alpha = .01$. Also, estimate the p value.
2. Test the null hypothesis $\mu = 145$ against $\mu > 145$ using the significance level $\alpha = .05$, given that $n = 16$, $\bar{x} = 156$, and $s = 30$. Assume the population is normally distributed.
3. A random sample of size $n = 20$ from a normal population yielded $\bar{x} = 91.4$ and $s = 12.5$. Using the significance level $\alpha = .01$, test:

$$H_0: \quad \mu = 100$$
$$H_1: \quad \mu \neq 100$$

Assume the population has a normal distribution.

4. A random sample of eight observations from a normally distributed population yielded $\bar{x} = 19.4$ and $s = 6.2$.
   a. Test H_0: $\mu = 15$ against H_1: $\mu \neq 15$, using $\alpha = .05$.
   b. Test H_0: $\mu = 15$ against H_1: $\mu > 15$, using $\alpha = .05$.
5. A test that measures self-image is administered to 20 science students; their scores have a mean of 88 and a standard deviation of 24. If, in general, student scores

on this test are approximately normal with a mean of 80, can we conclude that science students have a better self-image than students in general? Use $\alpha = .05$.

6. A company packs sheet metal screws in 10 pound boxes. On sampling 10 of these boxes, a quality control inspector found weights with a mean of 9.8 and a standard deviation of .19 pounds. Is she justified in concluding that boxes are not being filled to 10 pounds on the average? Use $\alpha = .05$.

7. The hydrocarbon emissions of a certain type of truck engine are reported to be 29.2 parts per million (ppm). When 12 new trucks with this type of engine were tested by a state emissions control agent, the emissions had a mean of 40.6 ppm with a standard deviation of 8.1 ppm. At the $\alpha = .05$ significance level, can the agent conclude that truck engines have a mean emission level exceeding 29.2 ppm? You may assume the normal distribution for the hydrocarbon emissions.

8. Twenty motorcycles are selected at random and tested for noise. The resulting mean and standard deviations are $\bar{x} = 84$ decibels and $s = 18$. Can we conclude that the mean noise level of the motorcycles exceeds the suggested standard of $\mu = 75$ decibels? Use $\alpha = .05$ and assume the normal distribution. Also, estimate the p value.

Part B

9. A random sample of five observations from a normal population produced 38.2, 40.5, 49.1, 41.2, and 30.6. Test the null hypothesis that $\mu = 45$ against $\mu < 45$ using the significance level $\alpha = .01$.

10. The consumer protection agency of a large western state randomly sampled premium gasoline rated 91 octane sold by a small chain of service stations. Over several months, the sampling produced the following octane measurements:

$$88.2, \ 88.6, \ 88.3, \ 88, \ 89.4, \ 87.1, \ 89.1$$

   a. Were the stations selling a lower octane rated gasoline as premium?
   b. Test the hypothesis that regular gasoline (88 octane rating) was being sold as premium.

11. A report on the migratory habits of California black deer stated that these deer spend an average of four weeks in traveling from their summer to their winter ranges. In a study of these deer, 13 were captured and fitted with radio transmitting collars. One of these transmitters failed and three of the deer were killed by mountain lions. The migration times of the remaining deer had a mean of 31.4 days and a standard deviation of 4.0 days. Is this sufficient evidence to warrant rejecting the claim of the report? Use $\alpha = .05$ and assume the normal distribution.

12. The skeletal weight of the mature blue whale is reported to be around 6500 pounds. Five such whales died from natural causes and their skeletons were subsequently cleaned and reassembled for museum displays. The skeletal weights had a mean of 5830 pounds and a standard deviation of 1280 pounds. If these five whales can be treated as a random sample, is this sufficient evidence to reject the 6500 pound claim for the mean skeletal weight of the blue whale? Use $\alpha = .01$ and assume the normal distribution for the skeletal weights.

13. Many researchers elect to use the t distribution for hypothesis testing even when the sample size is large. Suppose in a large sample test of the mean, the null hypothesis is rejected when the t distribution is used. Would this also be the case if the normal distribution were used (assuming the same significance level α)?

9.8   TESTS CONCERNING A PROPORTION

Tests of hypotheses involving the proportion of a population having some characteristic or property are similar to those of the mean. In essence we examine a random sample; if its proportion $\hat{p}$ differs markedly from that asserted for the population in the null hypothesis, then the null hypothesis is rejected.

In our study of sampling distributions we saw that if the population proportion has the value $p = p_0$, then the distribution of the sample proportion $\hat{P}$ is approximately normal, provided both $np_0 \geq 5$ and $nq_0 \geq 5$. (Here $q_0 = 1 - p_0$.) As a consequence, in carrying out a two-sided test of H_0: $p = p_0$ against H_1: $p \neq p_0$ using, say, a 5% significance level, two options are available:

1. Compute the critical values

$$p_0 \pm 1.96 \sqrt{\frac{p_0 q_0}{n}}$$

of $\hat{P}$ and reject H_0 if the proportion $\hat{p}$ from the sample is more extreme than either of these.

2. Using the proportion $\hat{p}$ from the sample, compute

$$z = \frac{\hat{p} - p_0}{\sqrt{\dfrac{p_0 q_0}{n}}}$$

and reject H_0 if z lies outside the interval from -1.96 to 1.96. The second approach is preferred; see Figure 9.21.

FIGURE 9.21   CORRESPONDING CRITICAL REGIONS FOR $\hat{P}$ and Z

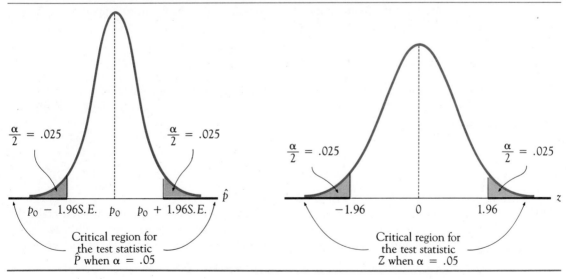

Similar approaches hold for single-tailed tests. The testing procedure, which we now summarize for completeness, is simply a restatement of that for the mean, with the obvious modification of computing the test statistic Z.

PROCEDURE FOR TESTING THE NULL HYPOTHESIS H_0: $p = p_0$ AGAINST AN ALTERNATE HYPOTHESIS H_1

Given: The alternate hypothesis H_1, the proportion $\hat{p}$, and the size n of a random sample from the population being tested.

Assumptions: $np_0 \geq 5$ and $nq_0 \geq 5$ where $q_0 = 1 - p_0$

1. Choose a one- or two-tailed test from the directional nature of H_1.
2. Select a significance level α such as $\alpha = .05$ or $\alpha = .01$.
3. Find the critical value(s) for the test. For a one-tailed test this will be either z_α or $-z_\alpha$. For a two-tailed test these are $\pm z_{\alpha/2}$.
4. Compute

$$z = \frac{\hat{p} - p_0}{\sqrt{\dfrac{p_0 q_0}{n}}}$$

5. Compare the value of the test statistic to the critical value(s):
   - Case I—H_1: $p \neq p_0$. Reject H_0 if $z > z_{\alpha/2}$ or $z < -z_{\alpha/2}$.
   - Case II—H_1: $p > p_0$. Reject H_0 if $z > z_\alpha$.
   - Case III—H_1: $p < p_0$. Reject H_0 if $z < -z_\alpha$.

Example 9.17

Suppose a nationwide survey showed that 70% of the adult population favored the concept of euthanasia for the terminally ill. Test whether this also holds for the over-55 age group if 684 of a sample of 900 are found to support the concept. We will test

$$H_0: \quad p = .70$$
$$H_1: \quad p \neq .70$$

Choosing $\alpha = .05$, the critical values for this two-tailed test are $\pm z_{.025} = \pm 1.96$. The sample proportion is $\hat{p} = 684/900 = .76$. The value of the test statistic is

$$z = \frac{.76 - .70}{\sqrt{\dfrac{(.70)\,(.30)}{900}}} = 3.93$$

Since $3.93 > 1.96$, the null hypothesis is rejected. It would seem safe to take the analysis one step farther and assert that a greater percentage of this age group favor euthanasia than does the general population. The p value of the test is

$$p = P(Z > 3.93) < .0001 \quad \blacksquare$$

Be careful of hasty conclusions about proportions when small samples are involved. As we have pointed out on previous occasions, the standard error of $\hat{P}$ tends to be large with moderate size samples. As a result, what appears to be a relatively large difference between the population and sample proportion may not prove significant.

Example 9.18

Before the introduction of a yearly emission inspection program, it was determined that 44% of a state's autos failed to meet emission standards. One year after the test was instituted, it was found that in a random sample of 150 autos, only 38% failed to meet the standards. Test whether the program has been successful in improving compliance with emission standards, using a 1% significance level.

We let p be the proportion of autos that presently fail to meet the emission standards and test

$$H_0: \quad p = .44$$
$$H_1: \quad p < .44$$

For a single-tailed test to the left with $\alpha = .01$, the critical value is $-z_{.01} = -2.33$. We note that $np_0 = 150(.44) = 66$ and $nq_0 = 150(.56) = 84$ both far exceed the lower limit of 5 and thus the use of the normal distribution is appropriate.

Using $\hat{p} = .38$, we find

$$z = \frac{.38 - .44}{\sqrt{\dfrac{(.44)\,(.56)}{150}}} = -1.48$$

Since $z = -1.48$ is not less than the critical value, the null hypothesis is not rejected (see Figure 9.22). In short, we cannot verify an effect from the testing program in terms of reducing the number of autos failing to comply with emission standards. The p value of the test is $P(Z < -1.48) = .0694$, so the conclusion

FIGURE 9.22 $z = -1.48$ **IS NOT IN THE CRITICAL REGION OF THE TEST**

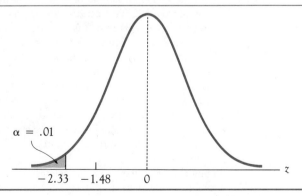

$\alpha = .01$

$-2.33 \quad -1.48 \quad \quad 0$

would have been the same with a 5% significance level test. In this case the standard error is

$$S.E. = \sqrt{\frac{(.44) \quad (.56)}{150}} = .041 = 4.1\%$$

To be highly significant, the difference between $\hat{p}$ and p must be numerically greater than $z_{.01}(.041) = 2.33(.041) = .096 = 9.6\%$. ■

Following the development we used for the mean, it would now be natural to treat the testing of proportions by using small samples. This can be done by using the binomial distribution and working with the exact count X, as we did in Chapter 6. Unfortunately the small-sample complications we discussed above do not disappear when we change from the normal to the binomial distribution. For this reason we do not cover this type of testing here.

EXERCISE SET 9.5

Part A

1. Given the following conditions, test the indicated hypothesis and find the p value:
   a. H_0: $p = .4$, H_1: $p \neq .4$, $n = 144$, $\hat{p} = .43$, and $\alpha = .05$.
   b. H_0: $p = .7$, H_1: $p > .7$, $n = 900$, $\hat{p} = .74$, and $\alpha = .01$.
2. Of 144 students surveyed on a certain campus, 72 were registered to vote. Can we conclude from this that fewer than 60% of the students on this campus are registered? Use $\alpha = .05$.
3. About four of every nine people reportedly favor the having holidays scheduled over long weekends. For college students it is believed that the proportion would be higher. What conclusion can be drawn if a sample of 625 students contained 296 who favored this scheduling approach? Use $\alpha = .01$.
4. Suppose 47% of the population of a certain region suffer at least one cold per year. A sample of 225 of those residents who were heavy users of vitamin C contained only 102 who had a cold during a year-long study. Can we conclude that people who take vitamin C in large doses have fewer colds? Use $\alpha = .05$.
5. According to the 1980 census, 64% of people live in the state in which they were born. The Chamber of Commerce for a certain popular state believes this percentage is too high for their state. Test this hypothesis using a 1% significance level if in a random sample of 1940 residents, 1145 were native-born.
6. According to the 1980 census, about 16% of all adults have had at least four years of college. Test that this percentage is too low for California if in a random sample of 1225 residents, 245 had completed at least four years of college. (Note: The percentage in the sample is correct for California; we have adjusted the actual figures to the problem.)
7. In a metropolitan area, the proportion of homeowners who use a certain lawn care product has previously been .16. Following an extensive advertising campaign, 400 homeowners are surveyed and 82 are found to be using the lawn care product. Can we conclude that the campaign increased the use of the product? Use $\alpha = .10$. Also, find the p value.
8. Suppose that 62% of the families in a town own their homes. A survey is made among the subscribers to cable television services in the city. Of 1500 subscribers surveyed, 68% were found to be home owners. Test the hypothesis that the

proportion of cable television subscribers who own their homes is greater than the proportion in the town. Use $\alpha = .10$.

9. Suppose that 300 tosses of a pair of dice result in 42 sevens. Can we conclude from this that the dice are biased against sevens? Use $\alpha = .01$.

Part B

10. Over the years a business education department of a university has found that its management science course has a failure rate of 25%. In an attempt to reduce this rate, the department starts enforcing the lower division mathematics prerequisites for the course. Of the 1280 students enrolled in the management science course during the next four semesters, 289 fail to complete it. Can we conclude that enforcing the prerequisites has improved the failure rate? Use $\alpha = .05$.

11. A coin is tossed 250 times, producing only 100 heads. Can we conclude with a reasonable certainty that the coin is biased in favor of tails?

12. In an ESP experiment, a deck of 20 cards is prepared that contains five cards with red faces, five with blue, five with white, and five with green. A card is selected at random and after concentrating on it the subject states what color he believes it to be. The card is returned to the deck, the deck is thoroughly shuffled, and the test repeated, 100 times in all. Using a two-tailed test and $\alpha = .05$, test the hypothesis that the subject has no ESP if the number of correctly identified cards is

    a. 32.

    b. 10.

    c. Think about the conclusion in part b. How can this predicament be avoided?

13. Suppose it has been established through long-term studies that 45% of all women having a radical mastectomy have a five-year relapse-free survival period. The effectiveness of CMF (Cyclophosphamide, Methotrexate, and Fluorouracil) in postoperative adjuvant chemotherapy was studied at two dosage levels. For level 1, the recommended dosage, it was found that 77% of a sample of 78 women had a relapse-free five-year period. For level 2 (about half the recommended dosage of CMF), 50% of a sample of 222 women had a relapse-free five-year period.

    a. Using $\alpha = .05$, test whether the relapse free percentage for the level 1 population exceeds $p = .45$.

    b. Proceed as in part a for the level 2 population.

    (Source: "The Effectiveness of CMF in Post Adjuvant Chemotherapy," Gianni Bonadonna, M.D., and P. Valagussa, *New England Journal of Medicine*, January 1981. Note: The 45% figure that is used here actually resulted from a control group in this study.)

14. A marketing research company conducts a test of a client's soft drink mix by having consumers compare it with the dominant mix on the market. When 400 consumers are asked to express a preference, 234 choose the new mix. Can the research company conclude that given a choice, consumers prefer their client's product? Use $\alpha = .05$.

9.9　THE POWER OF A TEST

In our preliminary remarks on hypothesis testing, the probabilities α and β of type I and type II errors were defined as follows:

$$\alpha = P(H_0 \text{ is rejected} | H_0 \text{ is true})$$
$$\beta = P(H_0 \text{ is not rejected} | H_1 \text{ is true})$$

The rationale for selecting a small significance level such as $\alpha = .05$ is, of course, to minimize the possibility of a type I error. A more positive way of thinking is to note that $1 - \alpha = 1 - .05 = .95$ is the probability of the complementary event, namely accepting H_0 when H_0 is true. A small α always assures us that this probability is near 1.

By the same token, $1 - \beta$ is the probability of accepting H_1 when H_1 is true. We would also like for this to be near 1, but as we have seen β may vary considerably and need not be small. For example, if $\beta = .25$, then $1 - \beta = .75$. Certainly the closer $1 - \beta$ is to 1, the more powerful the test must be. Thus we make the following definition.

DEFINITION 9.9

The *power* of a test is given by

$$\text{power} = 1 - \beta$$

where β is the probability of a type II error.

Figure 9.23 shows the area represented by the power and its relation to α and β for the case $H_0: \mu = \mu_0$ and $H_1: \mu = \mu_1$.

FIGURE 9.23   POWER $= 1 - \beta = P$ (H_1 IS ACCEPTED$|H_1$ IS TRUE)

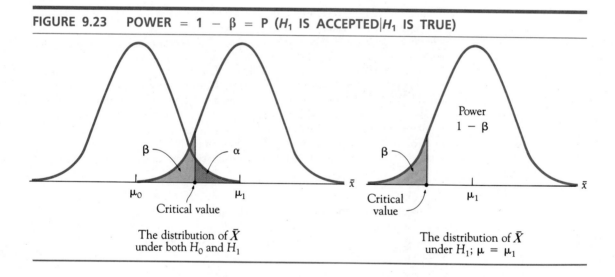

The distribution of $\bar{X}$ under both H_0 and H_1

The distribution of $\bar{X}$ under H_1; $\mu = \mu_1$

To compute the power of a test we need an alternate hypothesis that contains a specific value of the parameter μ or p. (Admittedly this is rarely the case in practice.) Having this, the power is computed directly by power $= P(H_1$ is accepted$|H_1$ is true).

Example 9.19   A 5% significance level test based on the mean of a random sample of size $n = 25$ is planned for

$$H_0: \quad \mu = 68$$
$$H_1: \quad \mu = 71$$

Assuming the population is normal with a standard deviation $\sigma = 8$, find the power of the test. The two competing distributions are shown in Figure 9.24.

FIGURE 9.24   THE DISTRIBUTION OF $\overline{X}$ UNDER $\mu = 68$ AND $\mu = 71$

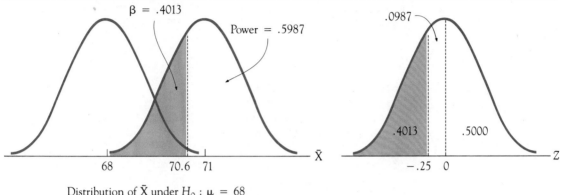

Distribution of $\overline{X}$ under $H_0 : \mu = 68$
and under $H_1 : \mu = 71$

$P(Z > -.25) = .5987$

Using $\alpha = .05$, the critical value of Z is $z_{.05} = 1.64$; for $\overline{X}$ it is $\bar{x} = 68 + 1.64\ (8/\sqrt{25}) = 70.6$. Since H_1 is accepted if $\overline{X} > 70.6$,

$$\text{power} = P(\overline{X} > 70.6|\mu = 71)$$

Note that the Z score will now be computed by using $\mu = 71$ as the true state of affairs. Thus

$$z = \frac{70.6 - 71.0}{8/\sqrt{25}} = -.25$$

and

$$\text{power} = P(Z > -.25) = .5987$$

The probability of a type II error is $\beta = 1 - \text{power} = .4013$. ■

Obviously we would not describe the above test as powerful. We note that the closer the critical value of $\overline{X}$ is to $\mu = 68$, the smaller β will be and the more powerful the test. This state of affairs can be brought about by decreasing the standard error. One way to do this is to increase the sample size.

Example 9.20

Compute the power of the previous test if the sample size is $n = 100$ rather than 25.

With everything else the same, the critical value of $\overline{X}$ is now $\bar{x}_1 = 68 + 1.64\ (8/\sqrt{100}) = 69.3$. From Figure 9.25 we see:

$$\text{power} = P(\overline{X} > 69.3 | \mu = 71) = P(Z > -2.12) = .9821$$

and

$$\beta = 1 - \text{power} = .0179 \quad \blacksquare$$

FIGURE 9.25   POSSIBLE DISTRIBUTIONS OF $\overline{X}$

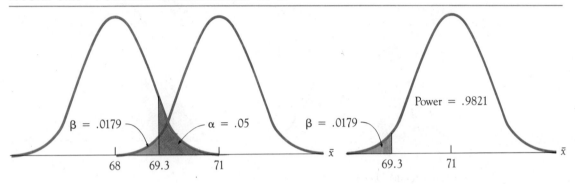

The distribution of $\overline{X}$ under H_0 and H_1

The power increases as the critical
value approaches $\mu = 68$

This example illustrates very clearly that the power increases with the sample size. It can also be shown that if we have an estimate of the alternate value of μ and a common variance σ^2, then it is possible to determine the sample size that will produce a power as close to 1 as desired.

In practice, the option of using a large sample size to reduce the standard error and to increase the power is rarely exercised. In most instances where one is testing for some effect, large samples are just not practical. An alternate approach preferred by researchers is to reduce the standard error by reducing the variance. This is done by controlling for variables not critical to the study and thereby working with a more homogeneous population. For example, in testing whether or not high noise levels affect classroom performance, one might limit the students to one grade level, restrict the subject matter to one particular subject such as reading, and work with only two noise levels rather than a full range.

EXERCISE SET 9.6

Part A

1. If the probability of a type II error is .65, what is the power of the test?
2. If the power of a test is .87, what is the probability of a type II error?
3. Is the power of a test given by

$$P(H_0 \text{ is rejected} | H_0 \text{ is false})$$

Explain your answer.

4. In a test of H_0: $\mu = 154$ against H_1: $\mu = 160$ with $\alpha = .01$ and $n = 64$, what is the power of the test if there is a common variance $\sigma^2 = 364$?

5. Suppose we are testing H_0: $p = .5$ against H_1: $p = .44$, using a sample of size 900 with $\alpha = .05$. What is the power of the test?

6. The mean diameter of a shipment of bolts is either .24 or .27 inches. To determine which, a sample will be selected and its mean will be used to test H_0: $\mu = 24$ against H_1: $\mu = .27$ using a 1% level of significance. Assuming $\sigma = .06$, what is the power of the test if the sample size is:
   a. 40?
   b. 100?
   c. 1000?

7. In testing H_0: $\mu = 68$ against H_1: $\mu > 68$ using a significance level α, what happens to the power as:
   a. α is decreased?
   b. α is increased?

8. In testing H_0: $\mu = \mu_0$ against H_1: $\mu = \mu_1$ at a significance level α, what happens to the power if μ_1 is decreased? Show graphically. (Assume $\mu_1 > \mu_0$)

Part B

9. Using the significance level $\alpha = .05$, a test is planned of

$$H_0: \quad p = .55$$
$$H_1: \quad p = .40$$

What size sample will be needed if the power of the test is to be .95?

10. We plan to test

$$H_0: \quad \mu = 45$$
$$H_1: \quad \mu = 48$$

If the variance in either case is 36, what size sample will ensure a power of at least .90 when the significance level is $\alpha = .05$?

11. A garage owner believes the average charge quoted for a four-cylinder tune-up by independent garages to be $39.00. A random sample of 36 garages produces quotes with a mean of $42.50 and a standard deviation of $10.00. In a test with $\alpha = .05$, what is the probability that H_0: $\mu = 39.00$ will be rejected if the true mean is $45.00? (Assume the garage owner uses a one tail test to the right.)

12. Let μ be the mean distance that students commute to a university. Using a significance level of .05 and the mean of a random sample of size $n = 81$, we plan to test

$$H_0: \quad \mu = 15$$
$$H_1: \quad \mu = 16.5$$

If the standard deviation is $\sigma = 5$, find and interpret α, β, $1 - \alpha$, and $1 - \beta$.

13. The significance level and power of a test are .10 and .83, respectively.
   a. What is the probability of rejecting H_0 if H_0 is true?
   b. What is the probability of not rejecting H_0 if H_0 is true?
   c. What is the probability of not rejecting H_0 if H_1 is true?
   d. What is the probability of rejecting H_0 when H_1 is true?
   e. What is the probability of accepting H_1 if H_0 is true?
   f. What is the probability of accepting H_1 when H_1 is true?

COMPUTER PRINTOUT

Figure 9.26 shows Minitab commands and a printout for two tests. In the first test,

$$H_0: \quad \mu = 60$$
$$H_1: \quad \mu \neq 60$$

In the second,

$$H_0: \quad \mu = 55$$
$$H_1: \quad \mu \neq 55$$

The same data were used with both tests. All features of the testing, including the computation of $\bar{x}$ and s, are done by the program. It should be noted that Minitab requires the use of the t-distribution when the population standard deviation σ is unknown.

FIGURE 9.26 COMPUTER PRINTOUT FOR TWO SIMPLE HYPOTHESIS TESTS

```
MTB >SET THE FOLLOWING SAMPLE DATA INTO COLUMN C1
DATA>52.3, 67.8, 64.4, 51.2, 59.7, 63.0
DATA>67.3, 58.8, 74.5, 67.3, 65.6, 71.8
DATA>END
MTB >TTEST OF MU = 60.0, USE THE DATA IN C1
   C1         N =   12      MEAN
   =       63.642     ST.DEV. =       7.11


   TEST OF MU =      60.0000 VS. MU N.E.     60.0000
   T =   1.775
   THE TEST IS SIGNIFICANT AT  0.1036
   CANNOT REJECT AT ALPHA = 0.05

MTB >TTEST OF MU = 55.0, USE THE DATA IN C1
   C1         N =   12      MEAN
   =       63.642     ST.DEV. =       7.11


   TEST OF MU =      55.0000 VS. MU N.E.     55.0000
   T =   4.212
   THE TEST IS SIGNIFICANT AT   0.0015

MTB >STOP
```

CHAPTER CHECKLIST

Vocabulary

Hypothesis
Null hypothesis
Alternate hypothesis
Type I error

Critical value of a test
Two-tailed test
One-tailed test
Decision rule

Type II error

Significance level of a test

p value of a test

Power of a test

Techniques

Identifing type I and type II errors in a given test of hypotheses

Finding the probability of a type I error when the decision rule is specified

Finding the probability of a type II error when both the decision rule and a concise alternate hypothesis are specified

Performing a large-sample test of hypotheses that concern the mean using the normal distribution; computing the p value

Performing a small-sample test of hypotheses that concern the mean using the t distribution, assuming the sampled population is normal; estimating the p value

Performing a large-sample test of hypotheses concerning the population proportion, using the normal distribution; computing the p value

Finding the power of a test when the decision rule and a concise alternate hypothesis are given

Notation

H_0	The null hypothesis
H_1	The alternate hypothesis
α	The probability of a type I error; also, the significance level of the test
β	The probability of a type II error
p value	The p value of a hypothesis test
$1 - \beta$	The power of a test

CHAPTER TEST

1. Explain each of the following:
   a. The difference between a type I and a type II error.
   b. The significance level of a test.
   c. The p value of a test.
   d. The power of a test.
2. In a test of

$$H_0: \quad \mu = \mu_0$$
$$H_1: \quad \mu \neq \mu_0$$

   with $\alpha = .04$ and $\beta = .15$:
   a. What is the probability that H_0 will be rejected if in fact H_0 is true?
   b. What is the probability that H_0 will not be rejected if in fact H_0 is true?
   c. Suppose H_1 is true. What is the probability that H_1 will be accepted?
3. In a large sample test of

$$H_0: \quad p_0 = p_0$$
$$H_1: \quad p > p_0$$

   with a significance level $\alpha = .02$, the data lead to a rejection of H_0.
   a. Would H_0 have been rejected if the significance level was $\alpha = .10$?

    b. Would H_0 have been rejected if the significance level was $\alpha = .01$?

    c. If a two-tailed test had been used with $\alpha = .05$, would H_0 have been rejected?

4. The p value of a test is .04. What is the smallest significance level for which H_0 would be rejected if the test is:

    a. A two-tailed test?

    b. A single-tailed test?

5. On a random sample of 185 airline flights, 130 contained at least one screaming infant. At the .05 level of significance, is this evidence sufficient to reject a public interest's group claim that 75% of all airline flights have a crying child as a passenger? Also compute the p value and interpret.

6. A health care organization claims that 60% of all patients admitted to hospitals are women. Test this hypothesis using $\alpha = .05$ if a random sample of 1200 hospital admission records showed 62.2% of the patients to be women. Compute the p value of the test.

7. A prominent investment counselor believes that the peak productivity of businesspeople is reached at the age of 55. A random sample of 16 retired businesspeople showed their peak productivity ages to have a mean of 52.8 years and a standard deviation of 8 years. Assuming the normal distribution for these ages, do these data provide evidence to refute the counselor's claim? Use $\alpha = .05$.

8. Forty water specimens from a large lake had dissolved oxygen contents with a mean of 14.3 parts per million (ppm) and a standard deviation of 3.4 ppm. Do these data present sufficient evidence to indicate that the oxygen concentration is below 16 ppm? Use $\alpha = .05$. Compute the p value.

9. We plan to test

$$H_0: \quad \mu = 80$$
$$H_1: \quad \mu > 80$$

using the mean of a random sample of size $n = 36$. Suppose the decision rule is to reject H_0 if $\bar{x} > 85$. What is the significance level of the test if $\alpha = 10$?

10. We plan to test

$$H_0: \quad \mu = 400$$
$$H_1: \quad \mu = 450$$

and will reject H_0 if the mean of a random sample of size $n = 50$ exceeds 425. Assuming a standard deviation of 100, find the probability of a type II error and compute the power of the test.

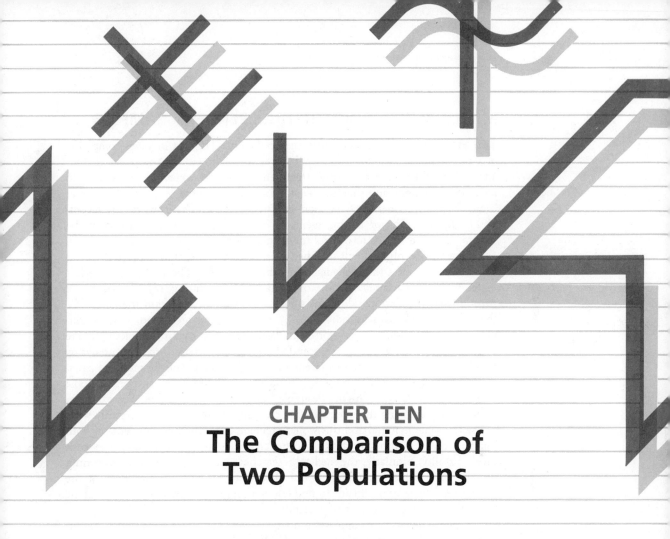

CHAPTER TEN
The Comparison of
Two Populations

Possibly the most common type of hypothesis testing involves the comparison of two populations. Reduced to its simplest terms, the question is whether or not two populations have the same mean or the same proportion exhibiting some property. One source of such problems is before-and-after comparisons of the means, as might be made with some new medical therapy or a new training technique for workers. Another source is the comparison of quantities such as incomes, education, or intelligence between different stratas of the general populations.

Additionally, there are problems involving percentages or proportions. Do two different ethnic groups have the same percentage of unemployed? Is a political candidate equally supported by both men and women? Does an advertising campaign appeal to the same percentage of people in the West as in the East? The possibilities are almost endless.

There are few really new ideas in this chapter. Apart from new sampling distributions, we are mostly extending the concepts that were developed for single-population hypothesis testing. In spite of this, the reader will quickly recognize that the potential for meaningful applications is increased enormously by this material.

351

10.1   MATCHED PAIRS VERSUS INDEPENDENT SAMPLES

Two approaches are used in comparing populations. The first involves comparing matched samples from the two populations, as might be done in a before-and-after study of the effect of some treatment. Here the experimenter wishes to reduce the variance and thereby the standard error, so that any change or effect caused by the treatment may be more easily recognized. This reduction is accomplished by matching or pairing members of the first sample with those of the second in such a way as to control for as many extraneous variables as possible.

One way to obtain such matched pairs is to use the same subject twice, as when the effect of a single medication is being studied. The patient is a perfect match for her- or himself. On the other hand, in comparing, say, the salaries of two different populations, different subjects must be used to form the pairs. We then try to match pairs of individuals by race, sex, age, profession, years of employment, education, and so forth. In both cases we obtain samples that are highly dependent on one another and the testing procedure involves analyzing the differences between the measurements on the members of the matched pairs.

A second approach is to use independent samples from the two populations. Matched samples, while highly desirable, are not always possible. For example, an instructor studying the effects of two teaching strategies would certainly want to use the same material and this would preclude using the same students in both samples. But it is also very difficult to match students so as to produce the same level of ability. Similarly, the medical researcher comparing two therapies cannot use the same subjects for both if the effects of one will obviate those of the second. And it is very difficult to find enough subjects for two samples, let alone to be able to match them physiologically.

When independent samples are used, the analysis is concerned with the difference between the two samples as a whole, not between pairs of observations. The approach then is to compare the two sample means or sample proportions and determine if their difference is due to the treatment or technique being investigated rather than simply to a randomness for which we have not been able to control.

Beginning with the next section we develop the distributions needed for testing the equality of two population means by using independent samples. Once this is complete, we will examine the relatively simple testing procedure when matched samples are used. Finally, the test for the equality of two population proportions using independent samples is considered.

10.2   THE DISTRIBUTION OF $\overline{X}_1 - \overline{X}_2$

In discussing two populations some additional notation is needed. The two populations we shall call 1 and 2. Their respective means are μ_1 and μ_2, their variances σ_1^2 and σ_2^2. Subscripts are used throughout to identify parameters, samples, and variables associated with these two populations. Thus the sample from population 1 is of size n_1, its mean is $\overline{x}_1$, and its variance is s_1^2. The sample mean $\overline{x}_1$ is one value of the sample statistic $\overline{X}_1$. A complete summary is given in Table 10.1.

TABLE 10.1 NOTATION USED WITH TWO POPULATIONS

Parameter	Population 1	Population 2
Population mean	μ_1	μ_2
Population variance	σ_1^2	σ_2^2
Sample size	n_1	n_2
Sample mean	$\bar{x}_1$	$\bar{x}_2$
Sample variance	s_1^2	s_2^2
Mean (variable)	$\overline{X}_1$	$\overline{X}_2$

In testing for the equality of two population means μ_1 and μ_2, we actually consider their difference $\mu_1 - \mu_2$. If these means are equal, then this difference is zero: $\mu_1 - \mu_2 = 0$. This gives us a specific number with which to compare the difference $\bar{x}_1 - \bar{x}_2$ of the corresponding sample means. If this difference is small we attribute it to randomness and do not reject the null hypothesis of no difference. If, however, the difference in the sample means is large, then this is evidence that the population means are not equal. Determinations of what are small or large differences require knowledge of the distribution of the variable $\overline{X}_1 - \overline{X}_2$ whose values are these differences.

For large sample sizes n_1 and n_2, the central limit theorem tells us that both $\overline{X}_1$ and $\overline{X}_2$ are approximately normal. Also, as we have pointed out previously, the sum and difference of two independent normal variables are also normal. The use of large independent random samples thus assures us that $\overline{X}_1 - \overline{X}_2$ is approximately normal.

As would be expected, the difference of two independent random variables has a mean that is the difference of their respective means:

$$\mu_{\overline{X}_1 - \overline{X}_2} = \mu_1 - \mu_2$$

Thus under the null hypothesis H_0: $\mu_1 = \mu_2$, the difference $\overline{X}_1 - \overline{X}_2$ has a zero mean.

The variance of the difference is a little more complex. Subtracting two variables does not produce a new variable with a smaller variance. Rather, it is larger: the variances add. Using the fact that these variances are the squares of the standard errors, we find the variance of $\overline{X}_1 - \overline{X}_2$ to be

$$\sigma^2_{\overline{X}_1 - \overline{X}_2} = \sigma^2_{\overline{X}_1} + \sigma^2_{\overline{X}_2} = \frac{\sigma_1^2}{n_1} + \frac{\sigma_2^2}{n_2}$$

The square root of this expression is the standard error of the difference $\overline{X}_1 - \overline{X}_2$. Since there will be no opportunity for confusion it will also be denoted simply by $S.E.$ Thus

$$S.E. = \sqrt{\frac{\sigma_1^2}{n_1} + \frac{\sigma_2^2}{n_2}}$$

Theorem 10.1 summarizes this discussion:

THEOREM 10.1

Let $\overline{X}_1$ and $\overline{X}_2$ be variables associated with the means of independent random samples of size n_1 and n_2 from two populations having means μ_1 and μ_2 and variances σ_1^2 and σ_2^2. When both sample sizes are large, the difference $\overline{X}_1 - \overline{X}_2$ has a distribution that is approximately normal with mean

$$\mu_{\overline{X}_1 - \overline{X}_2} = \mu_1 - \mu_2$$

and standard error

$$S.E. = \sqrt{\frac{\sigma_1^2}{n_1} + \frac{\sigma_2^2}{n_2}}$$

Equivalently,

$$Z = \frac{(\overline{X}_1 - \overline{X}_2) - (\mu_1 - \mu_2)}{\sqrt{\sigma_1^2/n_1 + \sigma_2^2/n_2}}$$

may be approximated by the standard normal distribution when both n_1 and n_2 are 30 or more. In most applications, of course, we must approximate the variances σ_1^2 and σ_2^2 by those of the samples. Then we use

$$Z = \frac{(\overline{X}_1 - \overline{X}_2) - (\mu_1 - \mu_2)}{\sqrt{s_1^2/n_1 + s_2^2/n_2}}$$

Finally, if we assume that the population means are equal (i.e., $\mu_1 - \mu_2 = 0$), this last expression simplifies to

$$Z = \frac{\overline{X}_1 - \overline{X}_2}{\sqrt{s_1^2/n_1 + s_2^2/n_2}}$$

Example 10.1

The hourly starting rates for a sample of 40 government secretarial positions had a mean of 7.45 and a standard deviation of 1.15 (dollars). For a sample of 36 positions in the private sector there was a mean of 6.83 and a standard deviation of 1.10. Estimate the probability of a difference this large (and in this direction) if the mean hourly rates for secretaries are the same in both the government and private sectors.

Let population 1 be the salaries of the government secretaries. Their sample has the larger mean; by this choice the difference $\overline{x}_1 - \overline{x}_2$ will be a positive number and easier to work with. The information from the two samples is given in Table 10.2.

TABLE 10.2 SUMMARY OF THE TWO SAMPLES OF SECRETARIAL SALARIES

Population	Sample Size	Mean	Variance
1	$n_1 = 40$	$\overline{x}_1 = 7.45$	$s_1^2 = (1.15)^2 = 1.3225$
2	$n_2 = 36$	$\overline{x}_2 = 6.83$	$s_2^2 = (1.10)^2 = 1.210$

The observed difference in the sample means is

$$\bar{x}_1 - \bar{x}_2 = 7.45 - 6.83 = .62$$

The distribution of $\overline{X}_1 - \overline{X}_2$ is shown in Figure 10.1. Note that the distribution is centered at 0 since we are assuming $\mu_1 = \mu_2$ in H_0. The desired probability $P(\overline{X}_1 - \overline{X}_2 > .62)$ is shaded. To compute the corresponding

FIGURE 10.1 THE DISTRIBUTION OF $\overline{X}_1 - \overline{X}_2$ WHEN $\mu_1 = \mu_2$

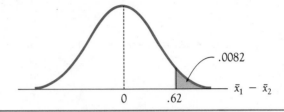

Z score we need the standard error of the difference of the means, which is

$$S.E. = \sqrt{\frac{1.3225}{40} + \frac{1.210}{36}} = .258$$

Thus

$$z = \frac{\bar{x}_1 - \bar{x}_2}{S.E.} = \frac{.62}{.258} = 2.40$$

The observed difference is thus 2.4 standard errors above the expected value. The probability of this occurring is

$$P(Z > 2.40) = .0082$$

Thus it is extremely unlikely that a difference this size would occur if the mean salaries for the two sectors were the same. ■

Bear in mind that two types of approximations were made in the above computations. First, we approximated the sampling distribution of $\overline{X}_1 - \overline{X}_2$ by the normal distribution. Second, we needed to approximate the population variances σ_1^2 and σ_2^2 by those of the sample. The central limit theorem, of course, assures us that there should be no serious error in the computed probability based on these approximations.

If the sample sizes are small, the central limit theorem no longer applies. If it should happen that both populations are approximately normal with equal variances, then the t distribution may be used for the distribution of the difference $\overline{X}_1 - \overline{X}_2$.

The only new problem is how best to obtain an estimate of the common variance of the two populations. This is solved by "averaging" or *pooling* the sum of the squares of the deviations which make up the individual sample variances. Notice that if s^2 is the variance of a sample of size n, then the sum of the squares of the individual deviations is $(n - 1)s^2$:

$$(n - 1)s^2 = (n - 1) \frac{\sum_{i=1}^{n} (x_1 - \bar{x})^2}{n - 1} = \sum_{i=1}^{n} (x_i - \bar{x})^2$$

Thus $(n_1 - 1)s_1^2 + (n_2 - 1)s_2^2$ is the pooled sum of these squares. Now the two samples have $(n_1 - 1)$ and $(n_2 - 1)$ degrees of freedom and the pooled sum of squares must have $(n_1 - 1) + (n_2 - 1) = n_1 + n_2 - 2$ degrees of freedom. Consequently the pooled approximation to the population variance σ^2 is given by the "average"

$$s_p^2 = \frac{(n_1 - 1)s_1^2 + (n_2 - 1)s_2^2}{n_1 + n_2 - 2}$$

Now, when $\sigma_1 = \sigma_2 = \sigma$, the standard error of $\bar{X}_1 - \bar{X}_2$ is

$$S.E. = \sqrt{\frac{\sigma^2}{n_1} + \frac{\sigma^2}{n_2}} = \sqrt{\sigma^2 \left(\frac{1}{n_1} + \frac{1}{n_2} \right)}$$

Approximating σ^2 by the pooled variance s_p^2 yields the approximation

$$S.E. = \sqrt{s_p^2 \left(\frac{1}{n_1} + \frac{1}{n_2} \right)}$$

The essential details of the foregoing discussion are contained in Theorem 10.2.

THEOREM 10.2

Suppose two independent samples are drawn from normal populations having means μ_1 and μ_2 and a common variance σ^2. If the samples have means $\bar{x}_1$ and $\bar{x}_2$, variances s_1^2 and s_2^2, and are of size n_1 and n_2, then

$$t = \frac{(\bar{x}_1 - \bar{x}_2) - (\mu_1 - \mu_2)}{\sqrt{s_p^2 (1/n_1 + 1/n_2)}}$$

where

$$s_p = \sqrt{\frac{(n_1 - 1)s_1^2 + (n_2 - 1)s_2^2}{n_1 + n_2 - 2}}$$

is the value of a variable having the t distribution with $\nu = n_1 + n_2 - 2$ degrees of freedom.

Again, in those applications where $\mu_1 = \mu_2$, the computation of the t value reduces to

$$t = \frac{\bar{x}_1 - \bar{x}_2}{\sqrt{s_p^2 (1/n_1 + 1/n_2)}}$$

Since the normal and t distributions are almost identical when the number of degrees of freedom is large, one finds the t distribution used almost exclusively in the literature with the difference $\overline{X}_1 - \overline{X}_2$. Keep in mind however, that when sample sizes drop below 30 or so, it is necessary to have a common variance and some semblance of normality for both populations.

EXERCISE SET 10.1

Part A

1. Give the graph, the mean, and the standard error of the distribution of $\overline{X}_1 - \overline{X}_2$ corresponding to the following information:
   a. $\mu_1 = 200$, $\sigma_1 = 30$, $n_1 = 100$;
      $\mu_2 = 200$, $\sigma_2 = 45$, $n_2 = 150$.
   b. $\mu_1 = 36.2$, $\sigma_1 = 8.95$, $n_1 = 50$;
      $\mu_2 = 32.0$, $\sigma_2 = 6.53$, $n_2 = 64$.
2. The following information is for independent samples from two normal populations with equal variances:

   $$\text{Sample 1: } n_1 = 64, \ \overline{x}_1 = 134, \ s_1 = 15$$
   $$\text{Sample 2: } n_2 = 100, \ \overline{x}_2 = 123, \ s_2 = 28$$

   a. Find an approximation to the standard error of $\overline{X}_1 - \overline{X}_2$.
   b. Is such a large observed difference in sample means very likely if $\mu_1 = \mu_2$?
3. Two populations have means $\mu_1 = \mu_2 = 500$ and standard deviations $\sigma_1 = 100$ and $\sigma_2 = 200$. Suppose independent random samples of sizes $n_1 = 64$ and $n_2 = 80$ are to be drawn.
   a. Discuss the distribution of $\overline{X}_1 - \overline{X}_2$ (draw a graph and give the mean and standard error).
   b. Find the probability that the mean of a sample from population 1 will exceed that of a sample from population 2 by at least 40.
4. Samples of student achievement test scores from two areas of a large city, A and B, produced the results below. What is the probability of a difference in sample means at least as large as that observed (and in the same direction) if achievement scores from both areas have the same mean?

Area A	Area B
$n = 40$	$n = 60$
$\overline{x} = 107$	$\overline{x} = 112$
$s^2 = 81$	$s^2 = 72$

5. Independent random samples from two normal populations with a common variance σ^2 are given below. Find the pooled estiamte to σ^2 and use this value to estimate the standard error of $\overline{X}_1 - \overline{X}_2$.

   $$\text{Sample 1: } 8, 12, 15, 19$$
   $$\text{Sample 2: } 6, 9, 12$$

6. In each of the following, compute the pooled estimate to the population variance and the estimate of the standard error of the difference $\overline{X}_1 - \overline{X}_2$.
   a. $n_1 = 10$, $s_1 = 23$; $n_2 = 12$, $s_2 = 27$
   b. $n_1 = 4$, $s_1 = 37$, $n_2 = 7$, $s_2 = 31$

Part B

7. We use the notation x_{1i} to denote the ith observation of sample 1. The members of sample 1 are thus $x_{11}, x_{12}, x_{13}, x_{14}, \ldots, x_{1n_1}$. Similarly the members of sample

2 are $x_{21}, x_{22}, x_{23}, x_{24}, \ldots, x_{2n_2}$. In this notation the first subscript selects the sample, the second the observation within the sample. Thus x_{25} is the fifth observation of sample 2. Write out and interpret what is represented by each of the following:

a. $\displaystyle\sum_{i=1}^{3} (x_{1i} - \bar{x}_1)^2$.

b. $\displaystyle\sum_{i=1}^{4} (x_{2i} - \bar{x}_2)^2$.

c. $\dfrac{\displaystyle\sum_{i=1}^{3} (x_{1i} - \bar{x}_1)^2 + \sum_{i=1}^{4} (x_{2i} - \bar{x}_2)}{3 + 4 - 2}$.

8. Independent samples from normal populations with a common variance produced the following results (we continue with the notation introduced in problem 7):

$$n_1 = 10, \; \bar{x}_1 = 74, \; \sum_{i=1}^{10} (x_{1i} - \bar{x}_2)^2 = 1440$$

$$n_2 = 8, \; \bar{x}_2 = 68, \; \sum_{i=1}^{8} (x_{2i} - \bar{x}_2)^2 = 1800$$

a. Find the pooled estimate to the common population variance.
b. Find the estimate to the standard error of the difference $\bar{X}_1 - \bar{X}_2$.
c. By how many standard errors does $\bar{x}_1$ exceed $\bar{x}_2$?

9. Give some considerations in trying to obtain matched samples to study the following:
   a. The effect of an improved fuel injection system on the miles per gallon obtained by automobiles.
   b. The difference in the cost of fringe benefits for workers in the public and private sectors.
   c. The durability of two grades of exterior house paint.

10. How might independent samples be obtained in the studies of problem 9?

11. The following results were obtained from an experiment that compared the absorption times (in minutes) of two orally administered antihistamines into patients' bloodstreams:

$$\text{Antihistamine 1: } n_1 = 12, \; \bar{x}_1 = 25.3, \; s_1 = 8.2$$
$$\text{Antihistamine 2: } n_2 = 10, \; \bar{x}_2 = 22.0, \; s_2 = 7.2$$

Assuming the absorption times for both antihistamines have the same normal distribution, what is the probability of obtaining sample means that differ by at least the amount observed; that is, find

$$P(|\bar{X}_1 - \bar{X}_2| > 3.3)$$

12. An eight-week study of two diet plans with independent samples of obese college students produced the following weight losses (in pounds):

$$\text{Sample 1: } 5.0, \; 0.6, \; 3.4, \; 1.2, \; 1.1, \; 2.2, \; 4.0$$
$$\text{Sample 2: } 2.3, \; 5.2, \; 8.7, \; 4.1, \; 3.5, \; 6.3$$

Is a difference in sample means as large as actually observed (and in the direction found) very likely if the distribution of weight losses is the same for both plans? What assumptions are necessary?

10.3 TESTING THE EQUALITY OF TWO POPULATION MEANS

Now that we have the distribution of the difference $\overline{X}_1 - \overline{X}_2$ we can proceed with the tests for H_0: $\mu_1 = \mu_2$ against such alternatives as H_1: $\mu_1 \neq \mu_2$, H_1: $\mu_1 > \mu_2$, and H_1: $\mu_1 < \mu_2$. In the event of a directional alternative, we can avoid the need to consider $\mu_1 < \mu_2$ by choosing population 1 as the one with the larger assumed mean.

The testing procedures are very similar to those for a single population. Under the assumption $\mu_1 = \mu_2$ of the null hypothesis, the difference $\overline{X}_1 - \overline{X}_2$ has a zero mean. If the difference $\overline{x}_1 - \overline{x}_2$ of the two sample means is on the order of 2 or 3 standard errors from 0, then we reject the null hypothesis. Of course, the actual critical values used are determined by the significance level.

PROCEDURE FOR TESTING THE NULL HYPOTHESIS H_0: $\mu_1 = \mu_2$ AGAINST AN ALTERNATE HYPOTHESIS H_1 USING LARGE SAMPLES

Given: The alternate hypothesis H_1, the means $\overline{x}_1$ and $\overline{x}_2$, the variances s_1^2 and s_2^2, and the sizes n_1 and n_2 of random samples from two populations with means μ_1 and μ_2.

Assumptions: The samples are independent.

The sample sizes are at least 30.

1. Decide on a one- or two-tailed test, based on the form of H_1.
2. Choose a significance level α such as $\alpha = .05$ or $\alpha = .01$.
3. Find the critical value(s) of Z. For a two-tailed test these are $\pm z_{\alpha/2}$. For a one-tailed test this will be either z_α or $-z_\alpha$.
4. Compute the standard error of $\overline{X}_1 - \overline{X}_2$:

$$S.E. = \sqrt{\frac{s_1^2}{n_1} + \frac{s_2^2}{n_2}}$$

5. Compute the value of the test statistic Z:

$$z = \frac{\overline{x}_1 - \overline{x}_2}{S.E.}$$

6. Compare the value of the test statistic with the critical value.
   - Case I–H_1: $\mu_1 \neq \mu_2$. Reject H_0 if $z < -z_{\alpha/2}$ or $z > z_{\alpha/2}$.
   - Case II—H_1: $\mu_1 > \mu_2$. Reject H_0 if $z > z_\alpha$.
   - Case III—H_1: $\mu_1 < \mu_2$. Reject H_0 if $z < -z_\alpha$.

Example 10.2

It is believed that students in the sciences tend to be somewhat younger than in the humanities. Test this assertion using the hypothetical data of Table 10.3.

TABLE 10.3 SUMMARY OF STUDENT AGES FROM TWO SAMPLES

College of major	Sample size	Mean	Variance
Science	125	20.8	5.29
Humanities	100	23.2	13.69

Let μ_1 be the mean age of the humanity majors, μ_2 that of those in the sciences. We wish to test:

$$H_0: \quad \mu_1 = \mu_2$$
$$H_1: \quad \mu_1 > \mu_2$$

A single-tailed test to the right is called for by H_1.

Choosing $\alpha = .05$, the critical value of Z is $z_{.05} = 1.64$. Using the sample sizes $n_1 = 100$, $n_2 = 125$ and the variances $s_1^2 = 13.69$, $s_2^2 = 5.29$, the standard error is

$$S.E. = \sqrt{\frac{s_1^2}{n_1} + \frac{s_2^2}{n_2}} = \sqrt{\frac{13.69}{100} + \frac{5.29}{125}} = .423$$

The value of the test statistic is

$$z = \frac{\bar{x}_1 - \bar{x}_2}{S.E.} = \frac{23.2 - 20.8}{.423} = 5.67$$

The difference is highly significant and we reject the hypothesis of equal mean ages for these two student populations. The p value is

$$p = P(Z > 5.67) = \; <.0000007 \quad \blacksquare$$

Example 10.3
A Case Study

Samples of springs from the Colorado Plateau and the Basin and Range Provinces of Arizona were examined in connection with an assessment of geothermal resources. The distributions of their air-adjusted temperatures are given in Figure 10.2. Treating these as independent samples, test the hypothesis that the mean temperature of the thermal springs in the Basin and Range Province exceeds that of the Plateau.

We let μ_1 be the mean temperature of springs in the Basin and Range Province, μ_2 the mean for the Plateau. We will test

$$H_0: \quad \mu_1 = \mu_2$$
$$H_1: \quad \mu_1 > \mu_2$$

We proceed directly to the computation of the value of the test statistic. For this we need the standard error. Since n_1 and n_2 are large,

$$S.E. = \sqrt{\frac{(3.37)^2}{71} + \frac{(2.88)^2}{121}} = .478$$

FIGURE 10.2 A COMPARISON OF TEMPERATURE DISTRIBUTION OF SPRINGS IN THE COLORADO PLATEAU AND BASIN AND RANGE PROVINCES

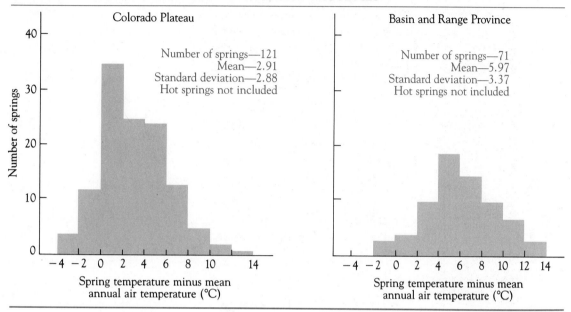

Then

$$z = \frac{\bar{x}_1 - \bar{x}_2}{S.E.} = \frac{5.97 - 2.91}{.478} = 6.40$$

Obviously the null hypothesis will be rejected for any reasonable significance level α. Thus we conclude that the mean temperature of the thermal springs of the Basin and Range Province exceeds the mean of those on the Colorado Plateau. (Source: "Thermal Springs of Arizona," James Witcher, *Earth Sciences and Mineral Resources in Arizona*, vol. 11, no. 2, June 1981.) ■

Now, suppose the samples are not large. If the populations have a common variance and are approximately normal, then a test based on the t distribution can be used. Apart from computing the standard error by using a pooled estimate of the population variance, the steps are almost identical to those used with large samples. Nevertheless we detail them for completeness.

PROCEDURE FOR TESTING THE NULL HYPOTHESIS H_0: $\mu_1 = \mu_2$ AGAINST AN ALTERNATE HYPOTHESIS H_1 USING SMALL SAMPLES

Given: The alternate hypothesis H_1, the means $\bar{x}_1$ and $\bar{x}_2$, the variances s_1^2 and s_2^2, and the sizes n_1 and n_2 of random samples from the two populations with means μ_1 and μ_2.

Assumptions. The samples are independent.

The populations are normal and have a common variance.

1. Decide on a one- or two-tailed test, based on the form of H_1.
2. Choose a significance level α, such as $\alpha = .05$ or $\alpha = .01$.
3. Using $\nu = n_1 + n_2 - 2$ degrees of freedom, find the critical value(s) of the test: For a two-tailed test these are $\pm t_{\alpha/2}$. For a one-tailed test this will be either t_α or $-t_\alpha$.
4. Compute the pooled estimate s^2 to the common population variance:

$$s_p^2 = \frac{(n_1 - 1)s_1^2 + (n_2 - 1)s_2^2}{n_1 + n_2 - 2}$$

5. Compute the standard error of the difference $\overline{X}_1 - \overline{X}_2$, using

$$S.E. = \sqrt{s_p^2\left(\frac{1}{n_1} + \frac{1}{n_2}\right)}$$

or $s_p\sqrt{1/n_1 + 1/n_2}$

6. Compute the value of the test statistic T:

$$t = \frac{\overline{x}_1 - \overline{x}_2}{S.E.}$$

7. Compare the value of the test statistic and the critical value(s):
   - Case I—H_1: $\mu_1 \neq \mu_2$. Reject H_0 if $t > t_{\alpha/2}$ or $t < -t_{\alpha/2}$.
   - Case II—H_1: $\mu_1 > \mu_2$. Reject H_0 if $t > t_\alpha$.
   - Case III—H_1: $\mu_1 < \mu_2$. Reject H_0 if $t < -t_\alpha$.

Example 10.4

A farmer tests two diets, say 1 and 2, for 12 weeks on samples of young turkeys to see if either offers any weight gain advantage. The results are summarized in Table 10.4. Assuming the weight gains by each diet are approximately normal and have a common variance, test the hypothesis that neither diet offers a weight gain advantage.

TABLE 10.4 SUMMARY OF WEIGHT GAINS IN OUNCES FOR SAMPLES OF TURKEYS ON TWO DIETS

Diet	Sample size	Mean	Standard deviation
1	15	40.2	7.5
2	15	36.4	7.1

We shall test the hypothesis that there is no difference in mean weight gains for turkeys maintained on either diet. Thus the hypotheses are

$$H_1: \quad \mu_1 = \mu_2$$
$$H_1: \quad \mu_1 \neq \mu_2$$

A two tailed test is dictated by this alternate hypothesis. Since the significance level is unspecified, choose $\alpha = .05$. The number of degrees of freedom is $\nu = n_1 + n_2 - 2 = 15 + 15 - 2 = 28$. The critical values are $\pm t_{\alpha/2} = \pm t_{.025} = \pm 2.048$. For the standard error, we first must compute the pooled estimate to the population variance.

$$s_p^2 = \frac{(n_1 - 1)s_1^2 + (n_2 - 1)s_2^2}{n_1 + n_2 - 2} = \frac{(15 - 1)(7.5)^2 + (15 - 1)(7.1)^2}{15 + 15 - 2} = 53.33$$

$$S.E. = \sqrt{s_p^2 \left(\frac{1}{n_1} + \frac{1}{n_2}\right)} = \sqrt{53.33 \left(\frac{1}{15} + \frac{1}{15}\right)} = 2.67$$

Finally, the t score is computed:

$$t = \frac{\bar{x}_1 - \bar{x}_2}{S.E.} = \frac{40.2 - 36.4}{2.67} = 1.42$$

Clearly $t = 1.42$ does not fall in the critical region, and so the null hypothesis is not rejected. While there is a difference in the sample means it is not large enough to allow us to conclude that either diet offers any advantage as to weight gain. The p value is $p = P(t > 1.42) < .10$. ■

Example 10.5

The arsenic concentrations in the water supplies of two communities, call them 1 and 2, are reported as 120 micrograms per liter and 85 micrograms per liter. Test whether there is also a difference in the arsenic concentrations of the blood of residents of these two communities, given the survey summary of Table 10.5.

TABLE 10.5 ARSENIC CONCENTRATIONS IN MICROGRAMS PER LITER IN THE BLOOD OF TWO COMMUNITIES

Community	Sample size	Mean	Standard deviation
1	19	.40	.15
2	35	.28	.17

Let μ_1 be the mean arsenic level in the blood of residents of the first community, μ_2 that of the second. We wish to test

$$H_0: \quad \mu_1 = \mu_2$$
$$H_1: \quad \mu_1 > \mu_2$$

This directional alternate hypothesis is used since we expect to see the higher arsenic concentration in the drinking water to carry over to the blood. Thus a single-tailed test to the right should be used.

Choosing $\alpha = .05$ and using $\nu = 19 + 35 - 2 = 52$ degrees of freedom, the critical value of t is $t_{.05} = 1.684$. Note that the nearest entries in Table VI are 1.684 ($\nu = 40$) and 1.671 ($\nu = 60$). The larger of these t values is used.

The concentration of chemicals in the blood of long-term residents of communities are generally distributed approximately normally, so there should be no problem with the normality assumption required by the test. And the close agreement between the sample standard deviations indicates that the equal variance assumption is reasonable.

The pooled estimate of the variance is

$$s_p^2 = \frac{(n_1 - 1)s_1^2 + (n_2 - 1)s_2^2}{n_1 + n_2 - 2} = \frac{(19 - 1)(.15)^2 + (35 - 1)(.17)^2}{19 + 35 - 2} = .02668$$

and the standard error is

$$S.E. = \sqrt{s_p^2 \left(\frac{1}{n_1} + \frac{1}{n_2} \right)} = \sqrt{.02668 \left(\frac{1}{19} + \frac{1}{35} \right)} = .0465$$

Then

$$t = \frac{\bar{x}_1 - \bar{x}_2}{S.E.} = \frac{.40 - .28}{.0465} = 2.58$$

Since $t = 2.58$ far exceeds the critical value $t_{.05} = 1.684$, the null hypothesis is rejected and we conclude that the mean blood arsenic concentration of the first community exceeds that of the second. The p value is $p = P(t > 2.58) < .01$. ■

There remains the question of how to proceed with a small-sample test when the normality or equal variance assumption does not apply. Several nonparametric tests are available for this. We recommend the Mann-Whitney test, which is discussed in Chapter 13.

As we have pointed out previously, one finds in the literature that the t distribution is used even when the sample sizes are large enough to justify the normal distribution. In particular, the large-sample test of the equality of two population means with unequal variances is carried out with the t distribution, using $\nu = n_1 + n_2 - 2$ degrees of freedom. We see this in the next example.

Example 10.6
A Case Study

Are teachers more likely to refer boys than girls for counseling when both exhibit identical behavorial or learning difficulties? This question is investigated in the report "Sex Bias in School Referrals" by Mark K. Gregory of the University of Wisconsin (*Journal of School Psychology*, vol. 15, no. 1, 1977). In this study teachers were given descriptive folders of children who had learning or behavorial problems in five categories. All descriptions for a given category were exactly the same. However, roughly half were labeled with a girl's name, the others with a boy's. As part of the evaluation process the teachers were asked to assess on a scale of 1 to 4 the likelihood of their referring the child for counseling. Here a rating of 4 indicated "very likely to refer" and a 1 "very rarely would refer."

The learning and behavorial categories, along with a summary of the analysis of the ratings given by the women teachers, are given in Table 10.6. There is a significant difference between the referral scores for the boys and the girls in the categories of arithmetic learning and withdrawn and aggressive behavior.

Further, when the scores were compiled for all categories and all teachers, both male and female, the t score obtained from the difference of the means was

$t = 2.22$ (with $\nu = 138$). Thus the author found sound evidence for rejecting the hypothesis that the referral prospects for girls is the same as that for boys when each have the same problem. As the author points out, many other studies support these findings. One study cited by the author reported a referral of ratio three girls to 10 boys!

TABLE 10.6 SEX BIAS IN SCHOOL REFERRALS: MEAN LIKELIHOOD OF REFERRAL AND t TESTS FOR FEMALE TEACHERS

Category	Child	N	Mean	SD	t value	p
Reading	Jenny	51	3.372	.631	.14	.444
	Jimmy	62	3.354	.704		
Withdrawn	Ben	51	3.000	.748	3.08	.0015
	Beth	62	2.532	.863		
Gifted	Mary	51	2.372	.871	1.60	.056
	Martin	62	2.629	.814		
Arithmetic	John	51	2.902	.855	2.77	.0035
	Joan	62	2.435	.934		
Aggressive	Tammy	51	3.000	.872	2.12	.018
	Tommy	62	3.338	.809		

To complete the discussion, let's examine the computation of one of the reported t scores, say from the arithmetic category. First we need the standard error of the difference. Assuming different variances,

$$S.E. \sqrt{\frac{(.855)^2}{51} + \frac{(.934)^2}{62}} = .1685$$

Using the scores for the boys as population 1, the t score is

$$t = \frac{2.902 - 2.435}{.1685} = 2.77$$

as reported. From our limited t tables with $\nu = 62 + 51 - 2 = 111$, we can only determine that $p < .005$. If we use the normal approximation to the t distribution, then the p value is approximately $p = P(Z > 2.77) = .0028$, which is quite close to the reported value. ∎

10.4 ESTIMATING $\mu_1 - \mu_2$

When the hypothesis of equal means is rejected, we may wish to estimate their difference. The difference $\bar{x}_1 - \bar{x}_2$ of their sample means is of course a point estimate of $\mu_1 - \mu_2$. If a confidence interval is needed, the boundaries are simply a certain number of standard errors above and below this point estimate. This, of course, depends on whether the distribution of

$$\frac{(\overline{X}_1 - \overline{X}_2) - (\mu_1 - \mu_2)}{S.E.}$$

can be approximated by the normal or t distribution and the computation of the standard error. And this we determine by the type of hypothesis test that would be used under the given conditions.

For example, if a 95% confidence interval is needed for $\mu_1 - \mu_2$ and large samples are used, the boundaries are located $z_{.025} = 1.96$ standard errors above and below the observed difference $\bar{x}_1 - \bar{x}_2$. The standard error is computed by using

$$S.E. = \sqrt{\frac{s_1^2}{n_1} + \frac{s_2^2}{n_2}}$$

In the vocabulary of Chapter 8, the upper and lower limits are

$$(\bar{x}_1 - \bar{x}_2) \pm E$$

where the maximum error of the estimate is given by $E = z_{.025}(S.E.)$.

PROCEDURE FOR CONSTRUCTING A LARGE-SAMPLE CONFIDENCE INTERVAL FOR THE DIFFERENCE $\mu_1 - \mu_2$ OF THE MEANS OF TWO POPULATIONS

Given: The level of confidence $1 - \alpha$, the means $\bar{x}_1$ and $\bar{x}_2$, the variances s_1^2 and s_2^2, and the sizes n_1 and n_2 of random samples from the two populations.

Assumptions: The samples are independent.
                   The samples sizes are at least 30.

1. Find $z_{\alpha/2}$.
2. Compute the standard error by:

$$S.E. = \sqrt{\frac{s_1^2}{n_1} + \frac{s_2^2}{n_2}}$$

3. Compute the maximum error of the estimate:

$$E = z_{\alpha/2}(S.E.)$$

4. Compute the upper and lower limts $(\bar{x}_1 - \bar{x}_2) \pm E$ and write the confidence interval.

Example 10.7

In the study of thermal spring temperatures (Example 10.3), the hypothesis of equal mean temperatures was rejected. The data summary is repeated below. Find a 99% confidence interval for the difference $\mu_1 - \mu_2$.

Area	Sample size	Mean	Standard deviation
Basin and Range (1)	71	5.97	3.37
Colorado Plateau (2)	121	2.91	2.88

Since $\alpha = 1 - .99 = .01$, $\alpha/2 = .005$, and $z_{.005,} = 2.58$. Repeating the previous computation, the standard error is

$$S.E. = \sqrt{\frac{(3.37)^2}{71} + \frac{(2.88)^2}{121}} = .478$$

The maximum error of the estimate is $E = z_{.005}(S.E.) = (2.58)(.478) = 1.23$. Finally, the upper and lower boundaries are

$$(\bar{x}_1 - \bar{x}_2) + E = (5.97 - 2.91) + 1.23 = 4.29$$
$$(\bar{x}_1 - \bar{x}_2) - E = (5.97 - 2.91) - 1.23 = 1.83$$

The confidence interval is thus

$$1.83 < \mu_1 - \mu_2 < 4.29$$

Briefly, the difference of the mean temperatures is between 1.83 and 4.29° C (with a 99% confidence).

For the case of equal variances, the t distribution is used when the populations are normal.

PROCEDURE FOR CONSTRUCTING A SMALL-SAMPLE CONFIDENCE INTERVAL FOR THE DIFFERENCE $\mu_1 - \mu_2$ OF THE MEANS OF TWO POPULATIONS

Given: The level of confidence $1 - \alpha$, the means $\bar{x}_1$ and $\bar{x}_2$, the variances s_1^2 and s_2^2, and the sizes n_1 and n_2 of random samples from the two populations.

Assumptions: The samples are independent.

The populations are normal and have a common variance.

1. Find $t_{\alpha/2}$, using $\nu = n_1 + n_2 - 2$ degrees of freedom.
2. Compute the standard error, using

$$S.E. = \sqrt{s_p^2 \left(\frac{1}{n_1} + \frac{1}{n_2} \right)}$$

or $s_p \sqrt{1/n_1 + 1/n_2}$ where

$$s_p^2 = \frac{(n_1 - 1)s_1^2 + (n_2 - 1)s_2^2}{n_1 + n_2 - 2}$$

3. Compute the maximum error of the estimate:

$$E = t_{\alpha/2}(S.E.)$$

4. Compute the upper and lower limits $(\bar{x}_1 - \bar{x}_2) \pm E$ and write the confidence interval.

Example 10.8

Find a 95% confidence interval for the difference in the means of the blood arsenic concentrations of Example 10.5. The data are repeated below.

Community	Sample size	Mean	Standard deviation
1	19	.40	.15
2	35	.28	.17

Both the pooled estimate of the variance and the standard error have previously been computed. The results were:

$$s_p^2 = .02668 \qquad s_p = .1633 \qquad S.E. = .0465$$

For $\alpha = .05$, $\alpha/2 = .025$. With $\nu = 19 + 35 - 2 = 52$ degrees of freedom, $t_{.025} = 2.021$, approximately. Since $\nu = 52$ is not available we have used the next smallest entry, namely $\nu = 40$ (which has a somewhat larger t value).

The maximum error of the estimate is

$$E = t_{.025}(S.E.) = (2.021)(.0465) = .094$$

The upper and lower limits $(\bar{x}_1 - \bar{x}_2) \pm E$ are .214 and .026. Thus to two decimal places, the confidence interval is

$$.03 < \mu_1 - \mu_2 < .21$$

Of course, a point estimate for the difference in the means of the blood arsenic concentrations is

$$\bar{x}_1 - \bar{x}_2 = .12$$

$$\text{(micrograms per liter)} \quad \blacksquare$$

EXERCISE SET 10.2

Part A

1. Using each set of results below, performing a 5% significance level test of

$$H_0: \quad \mu_1 = \mu_2$$
$$H_1: \quad \mu_1 \neq \mu_2$$

	Population	Sample size	Mean	Standard deviation
a.	1	30	118	14
	2	40	113	9
b.	1	10	164	25
	2	12	150	23

*Assume the populations are normal with equal variances

2. Independent random samples from two normal populations with a common variance are summarized below. Using $\alpha = .05$, test $H_0: \mu_1 = \mu_2$ against H_1: $\mu_1 > \mu_2$.

Sample 1	Sample 2
$n_1 = 15$	$n_2 = 18$
$\bar{x}_1 = 70$	$\bar{x}_2 = 62$
$s_1 = 14$	$s_2 = 18$

3. Given the following summary of data from two independent random samples, find a 99% confidence interval for the difference $\mu_1 - \mu_2$ of population means.

Population	Sample size	Mean	Standard deviation
1	50	145.2	40.5
2	30	132.4	34.2

4. A summary of the mercury content in micrograms per liter in samples of water from two lakes is given. Test the hypothesis that the mercury contamination is the same for both lakes.

Lake	Sample size	Mean	Standard deviation
1	36	42.4	18.3
2	36	50.0	27.6

5. Two makes of quartz watches are tested for durability by placing samples of each in a high humidity chamber that is subjected to continuous vibration. A summary of the number of days until failure for the watches of each sample is given below. Test the hypothesis that the watches are equally durable under the test conditions. Assume the failure times are normally distributed and use a pooled variance estimate and significance level $\alpha = .05$.

Sample 1	Sample 2
101.4	118.7
91.3	102.5
108.0	99.4
104.8	129.1
102.2	110.1

6. Two models of furnaces each have output ratings of 85000 British thermal units per hour. Given the data summary below, show that the hypothesis of equal outputs is rejected for $\alpha = .01$ and then find a 95% confidence interval for the difference of the mean outputs of these two models. Use a pooled variance estimate and assume the populations are normal.

Watch type	Sample size	Mean	Standard deviation
1	12	89,500	4700
2	12	83,600	4400

7. The blood pressures of 25 men and 25 women, each suffering some degree of hypertension, are measured while the subjects are in a relaxed condition. They

are then taught the nonoccult features of transcendental meditation: a relaxed position, a passive attitude, and controlled breathing. After mastering this, their blood pressures are measured while they are practicing the meditation. The reductions in their systolic pressures are summarized below.

Type	Sample size	Mean	Standard deviation
Systolic/Men	25	8.9	6.2
Systolic/Women	25	5.0	6.0

Test the hypothesis that men and women derive equal reductions in blood pressures through the use of this meditation. Use a pooled variance estimate and $\alpha = .05$. (The data are fictitious but quite similar to those obtained in several studies. See "Historical and Clinical Considerations of the Relaxation Response," Robert Benson et al., *American Scientist*, vol. 65, July 1977.)

8. Find a confidence interval for the difference in the means of the systolic blood pressures of problem 7.

9. Seventy-two students were randomly selected from the sixth grade classes of three schools and divided into experimental and control groups. They were then both given the same test on creative thinking. However, the conditions for taking the test were quite different. For the control group, the room was quiet and comfortable. For the experimental group a series of distractions were arranged: an alarm bell was set off, two kittens were released into the room and then removed, the lights went off and then came on again, a book was dropped, and so on. The question was whether or not the distractions would reduce the students ability to think creatively. The results are summarized below. Conduct a test, using $\alpha = .025$, of the hypothesis that creative thinking is diminished by distractions.

Group	Sample size	Mean	Standard deviation
Control	36	106.31	13.4
Experimental	36	95.48	15.6

(The data here are fictitious; the problem was suggested by a study entitled "The Effects of Distraction on Sixth Grade Students in Testing Situations," Landa Trentham, *Journal of Educational Measurement*, vol. 12, no. 1, 1975.)

10. Find a 95% confidence interval for the difference $\mu_1 - \mu_2$ in problem 9. Specifically what are the two populations and what does the difference $\mu_1 - \mu_2$ represent in this context?

11. A company is considering the relocation of its business office to one of two possible areas and is interested in the difference in rental housing costs for the two areas. Random samples of rents for two-bedroom apartments in these two areas are summarized below. Find a 90% confidence interval for the difference in mean costs for two-bedroom apartments in these areas.

Area	Sample size	Mean	Standard deviation
1	40	$630.00	$105.50
2	32	$548.40	$ 84.60

12. The owners of two makes of personal computers are sampled as to their total dollar outlay for equipment and software. The results are given below. Using $\alpha = .01$, test the hypothesis that on the average the owners of make B made a greater dollar outlay than the owners of make A. What are the necessary assumptions?

Make	Sample size	Mean	Standard deviation
A	21	$3456.20	$740.40
B	16	$3185.60	$987.00

Part B

13. Individuals with anorexia nervosa generally exist in a semifasting state. Among their numbers, however, are patients who exhibit a condition called bulimia, characterized by "rapid consumption of large amounts of food in a short time" between fasting bouts. The results below are part of a study that attempted to characterize individuals within these two subgroups. Other findings from this study will be us used in the next exercise set on proportons. Entries are $\bar{x} \pm s$.

Characteristic	Bulimia group ($n = 49$)	Fasting group ($n = 56$)	p values
Depression scores	30.0 ± 5.9	27.0 ± 5.4	
Obsession scores related to food	18.0 ± 3.3	17.2 ± 3.9	
Somatization scores	15.7 ± 5.2	14.0 ± 3.6	
Anxiety scores	11.5 ± 3.4	10.4 ± 3.4	

   a. Compute the missing p value.
   b. Compute 95% confidence intervals for the difference of the means where justified by a p value less than, say, .05.
(Source: "Bulimia," Regina Casper, M.D., et al., *Archives of General Psychiatry*, vol. 37, 1980.)

14. Two video-taped lectures are prepared by an instructor and shown to two classes of 110 and 125 students. Both lectures contained the same subject matter. However, the first version is delivered in a very serious fashion, the second with a great deal of humor relating to the concepts. The students were immediately tested on the material and then again after six weeks with no forewarning. The results are summarized below. All scores are mean ±1 standard deviation.

Type of lecture	Sample size	First test	Second test
Serious	110	17.0 ± 5.2	14.5 ± 4.0
Humorous	125	18.1 ± 4.8	16.9 ± 5.7

   a. Test the hypothesis that humor has no effect on immediate recall. (Use the results of the first test only.)

   b. Test for a long-range effect of humor. (Use the results of the second test only.)

15. In a study of plasma glucose response in adults to dietary fructose, 12 men with a high insulin response to sucrose and 12 normal men were tested with diets containing various levels of fructose. The physical characteristics of the two samples are summarized below.

   a. Test the hypothesis that the mean fasting plasma insulin level for normal men is less than that for men who are hyperinsulinemic.

   b. Formulate and test a hypothesis about the mean heights of the populations from which the two samples are drawn.

PRETEST VALUES FOR HYPERINSULINEMIC AND NORMAL MEN

Parameter	Control	Hyperinsulinemic
Age, years	39.8 ± 2.4	39.5 ± 2.1
Height, centimeters	176 ± 2.4	174 ± 1.7
Weight, kilograms	80.5 ± 3.2	81.4 ± 2.3
Insulinogenic score	0 ± 0	10.75 ± 1.12
Fasting plasma insulin, microunits per milliliter	20.9 ± 0.80	23.7 ± 0.79

All values are means ± S.E. of 12 observations.

(Source: "Effects of Dietary Fructose on Plasma Glucose and Hormone Responses in Normal and Hyperinsulinemic Men," Judity Hallfrisch et al., *Journal of Nutrition*, vol. 13, no. 9, September 1983.)

16. The results from the mathematical survey of the National Assessment of Educational Progress (NAEP) for the 1967–1969 and the 1975–1977 periods are summarized on page 373.

   a. Using the total test scores, test the hypothesis that preservice teachers in the 1975–1977 period were better prepared in mathematics than their counterparts during the first survey period. Use $\alpha = .05$.

   b. Using total test scores, test the hypothesis that for the 1975–1977 period, the total test scores of the preservice teachers had a mean exceeding that of the in-service teachers. Use $\alpha = .05$.

   c. Using the operations subtest results, test the hypothesis that there was no difference between the means of inservice teachers' scores for these two periods. Use $\alpha = .05$.

Library

Project

The following articles make use of the t test to compare two population means. Locate one of these articles in the periodical section of the library and report on it. Pay particular attention to the purpose of the study, the selection of the experimental subjects, the procedures used in obtaining the data, the reporting of the statistical analysis, and the conclusions.

a. "Long-Term Effects of Activity and of Calcium and Phosphorous Intake on Bones and Kidneys of Female Rats," K. Bauer and P. Griminger, *Journal of Nutrition*, vol. 113, no. 10, Oct. 1983.

b. "Suppression of Postoperative Pain," R. Dionne et al., *Journal of Clinical Pharmacology*, vol. 23, no. 1, Jan. 1983.

STATISTICAL DATA FOR 1967–69 AND 1975–77 STUDIES; 65 TEST QUESTIONS

| | 1967–69 Data | | | | | 1975–77 Data | | | | | | |
| | Preservice Teachers (N = 887) | | Inservice teachers (N = 177) | | Comparison: pre 67 vs. in 67 | | Preservice teachers (N = 737) | | Inservice teachers (N = 241) | | Comparison: pre 75 vs. in 75 | Comparison: pre 67 vs. pre 75 | Comparison: in 67 vs. in 75 |
Subtest	Mean	Standard deviation	Mean	Standard deviation			Mean	Standard deviation	Mean	Standard deviation			
Geometry	5.14	1.59	4.63	1.60	*		6.05	2.00	4.99	2.17	*	*	*
Number theory	5.37	2.01	4.75	2.20	*		6.15	2.09	5.62	2.42	*	*	*
Numeration system	4.65	1.69	4.42	1.64			5.52	1.89	4.93	2.02		*	*
Fractional numbers	6.15	2.26	5.83	2.22			5.66	2.45	5.35	2.51		*	*
Structural properties	4.41	2.19	3.69	2.23	*		5.77	2.74	4.33	2.85	*	*	*
Sets	3.29	1.24	3.04	1.33	*		3.88	1.37	3.15	1.58	*	*	*
Operations	4.36	1.72	4.27	1.77			5.04	1.90	4.77	2.04			*
Test source	33.37	8.88	30.70	8.90	*		38.07	10.97	33.13	12.16	*	*	*

*Differences that are statistically significant

(Source: "Today's Elementary School Teachers Are Better prepared in Mathematics," Fred L. Pigge, T. C. Gibney, and J. L. Ginther, Arithmetic Teacher, March 1979.)

c. "The Effect of Equal Status Contact on Ethnic Attitude," J. Spangenberg and E. Nel, *Journal of Social Psychology*, vol. 121 (2nd half), Dec. 1983.

d. "Transfer Effect of Word Recognition Strategies," J. McNeil and L. Donant, *Journal of Reading Behavior*, vol. 12, no. 2, Summer 1980.

e. "Principals' Attitude and Knowledge about Handicapped Children," R. Cline, *Exceptional Children*, vol. 48, no. 2, Oct. 1981.

f. "The Effect of Distraction on Sixth Grade Students in a Testing Situation," L. Trentham, *Journal of Educational Measurement*, vol. 12, no. 1, Spring 1975.

g. "Coronary Artery Bypass Surgery," C. Jenkins, *Journal of the American Medical Association*, vol. 250, no. 6, August 1983.

h. "Need Satisfaction of Older Persons Living in Communities and Institutions," L. Tickle, *American Journal of Occupational Therapy*, vol. 35, no. 19, Oct. 1981.

i. "Television Nonviewers: An Endangered Species?" M. Beeck and J. Robinson, *Journal of Consumer Research*, vol. 7, no. 4, March 1981.

j. "Core 12: A Controlled Study of the Impact of 12 Hour Scheduling," M. Mills et al., *Nursing Research*, vol. 33, no. 6, Nov.-Dec., 1983.

k. "Plasma Vitamin C Concentration in Patients in a Psychiatric Hospital," C. Schordah et al., *Clinical Nutrition*, vol. 37, no. 6, Dec. 1983.

10.5   PAIRED DIFFERENCE COMPARISON OF TWO MEANS

We now consider the design of a paired difference experiment for testing the equality of two population means. To place this subject in context, we note that if the two populations in question have large variances, then the standard error of the difference $\overline{X}_1 - \overline{X}_2$ will also be large for moderately sized samples. Thus any attempt to establish that the population means are different by applying, say, a two-standard-error test criterion to the difference of two sample means is almost certainly doomed to failure.

As an example, suppose we are interested in whether a proposed food additive for a bread has an effect on the amount of bread a family consumes. The approach until now would have been to obtain two independent random samples of families and regularly supply one with the treated bread (the experimental group) and the other with untreated bread (the control group). We would then carefully monitor the amounts each consumes over, say, a four-week period and compare their means.

Unfortunately, because of different family sizes, physical activities, and eating habits we can easily predict that there will be considerable variation in the amount of bread consumed by the families of the populations corresponding to the control and experimental groups. Thus the population variances will be large and it will be difficult to detect an effect of the additive (if indeed there is one) by using the sample means.

A totally different approach is to use only one sample of families as both the experimental and control group. We could do this by alternating in some random fashion the type of bread that is supplied over, say, an eight-week period. We would then consider the difference between the amounts of each type of bread that a family consumes. If the additive has no effect then these differences should all be close to zero.

This approach produces the ideal matched pair experiment: ideal in the sense that the same subjects are used in both the experimental and control groups. If it was not possible to use the same families for both ends of this experiment, because of, say, the time factor or monitoring consideratons, it might still be possible to imitate the design to a high degree by matching families according to characteristics such as those mentioned above that determine the amount of bread a family eats. If good matches are obtained, it would still seem reasonable to expect the differences in the amounts of bread consumed by the matched pairs to be close to zero provided there is no effect due to the additive.

Paired differences may also be used for testing two competing items under matched conditions. To illustrate, suppose we wish to test whether two makes of tires have the same mean mileage life. If five autos are equipped with tires of the first make and five others with tires of the second make, there will be considerable variation in the observed mileages because of differences in the characteristics of the autos, driving habits of the drivers, and road conditions.

Such variation can be minimized by matching up the conditions under which both makes of tires are tested, for example by mounting two tires of each make on each of the test autos. Since there is a difference in wear between the front and rear as well as from side to side, we would also randomly assign one make of each tire to the front as well as to the rear. The pair on the front gives us one matched pair, the pair on the rear a second. If there is no difference in the performances of these two makes of tires, the paired differences of the mileages of the matched pairs should all be close to zero.

Obviously, the samples we obtain under matched conditions are not independent, since the performances of the two items are closely related. As a result, using the t test with the difference of the sample means is not appropriate. The paired difference test that we will now develop uses instead the mean of the paired differences. If there is no difference in the performance of two items (such as the two makes of tires) or if there is no treatment effect (as in the case of the bread additive), then the expected value of these differences is zero.

Before launching into the development of the paired difference test, we introduce some notation that will be useful not ony here but throughout the text whenever two or more samples are involved. Here we have two population variables, X_1 and X_2. To indicate observations of these, two subscripts are needed. The first identifies the variable, the second the observation. Thus the two matched samples which we use are indicated by:

$$\text{Sample 1: } x_{11}, x_{12}, \ldots, x_{1n}$$
$$\text{Sample 2: } x_{21}, x_{22}, \ldots, x_{2n}$$

The design of the experiment now involves matched conditions or subjects and corresponding pairs of observations. We assume that it is x_{11} which is matched or paired with x_{21}, x_{12} with x_{22}, and in general x_{1i} with x_{2i}. Since we are interested in the difference between the observations of each pair, we let

$$d_i = x_{1i} - x_{2i}$$

The observations and the differences are generally arranged as shown in Table 10.7.

TABLE 10.7   DISPLAY OF MATCHED PAIR DATA

Pair	Sample 1	Sample 2	Difference
1	x_{11}	x_{21}	$d_1 = x_{11} - x_{21}$
2	x_{12}	x_{22}	$d_2 = x_{12} - x_{22}$
$\vdots$	$\vdots$	$\vdots$	$\vdots$
n	x_{1n}	x_{2n}	$d_n = x_{1n} - x_{2n}$

In testing the equality of the two population means or the equivalent hypothesis of no treatment effect, we have seen that it is natural to examine how close the mean of the differences is to 0. We already have the background for this if we treat the differences $d_1, d_2, \ldots, d_n$ as a random sample of a population variable D. This sample has a mean $\bar{d}$ and standard deviation s. The sample mean $\bar{d}$ is a value of the sample statistic $\bar{D}$. There is nothing new here. Only the symbols have been changed. Think of the corresponding situation with $x_1, x_2, \ldots, x_n, X$, and $\bar{X}$.

Now what can we say about the distribution of $\bar{D}$? For the moment let's focus on the most common case: the sample size n is small and the populations are both normal. Since $D = X_1 - X_2$, it follows that D is also normal and thus $\bar{D}$ is normal. Also $\mu_{\bar{D}} = \mu_D = \mu_1 - \mu_2$. The standard deviation of $\bar{D}$ is not so simple. But this is of little consequence, since it will have to be approximated by the standard deviation s of the differences. As a result the standard error of $\bar{D}$ is estimated by

$$S.E. = \frac{s}{\sqrt{n}}$$

We now see that

$$t = \frac{\bar{d} - (\mu_1 - \mu_2)}{s/\sqrt{n}}$$

is the value of a variable having the t distribution with $\nu = n - 1$ degrees of freedom. Finally, under the null hypothesis $H_0: \mu_1 = \mu_2$, this simplifies to

$$t = \frac{\bar{d}}{s/\sqrt{n}}$$

PROCEDURE FOR TESTING THE NULL HYPOTHESIS $H_0: \mu_1 = \mu_2$ AGAINST AN ALTERNATE HYPOTHESIS H_1 USING MATCHED PAIRS

Given: The alternate hypothesis H_1 and the mean $\bar{d}$ and standard deviation s of the differences $d_1, d_2, \ldots, d_n$ of paired observations from the two populations.

Assumptions: The populations are normal with a common variance.

1. Decide on a one- or two-tailed test.
2. Choose a significance level α.
3. Using $\nu = n - 1$, find the critical value(s). For a two-tailed test these are $\pm t_{\alpha/2}$. For a one-tailed test this will be either t_α or $-t_\alpha$.

4. Compute the value of the test statistic:

$$t = \frac{\bar{d}}{s/\sqrt{n}}$$

5. Compare the value of the test statistic to the critical value(s).
- Case I—H_1: $\mu_1 \neq \mu_2$. Reject H_0 if $t > t_{\alpha/2}$ or $t < -t_{\alpha/2}$.
- Case II—H_1: $\mu_1 > \mu_2$. Reject H_0 if $t > t_\alpha$.
- Case III—H_1: $\mu_1 < \mu_2$. Reject H_0 if $t < -t_\alpha$.

It should be noted that it is not necessary that both populations be normal for the paired t test, only that the differences of the means be normally distributed. Further, as the sample size increases, the central limit theorem comes into play and the normality conditions can be relaxed. In fact, for large samples the normal distribution may be used in place of the t distribution. In practice, however, large samples are rarely used in a paired difference experiment.

Example 10.9

In a study of the test-retest reliability of an adult intelligence rating scale, the test-retest I.Q.s of 12 psychiatric patients were compared at six-week intervals. The results, along with the computation of the mean and standard deviation of the differences, are given in Table 10.8.

TABLE 10.8   TEST-RETEST DATA

Subject	Test	Retest	Difference d	d^2
1	93	91	+2	4
2	89	89	0	0
3	84	78	+6	36
4	85	82	+3	9
5	90	97	−7	49
6	99	98	+1	1
7	97	95	+2	4
8	96	96	0	0
9	91	88	+3	9
10	88	90	−2	4
11	79	81	−2	4
12	85	81	+4	16

From the table, we see that

$$\Sigma d_i = 10 \qquad \Sigma d_i^2 = 136$$

$$\bar{d} = \frac{10}{12} = .83$$

$$s = \sqrt{\frac{136 - (10)^2/12}{12 - 1}} = 3.41$$

We shall test the hypothesis that there is no difference in the test and retest scores of psychiatric patients on this scale. Thus if μ_1 and μ_2 are the means of patients on the test and retest phases, respectively, we test H_0: $\mu_1 = \mu_2$ against H_1: $\mu_1 \neq \mu_2$. Choosing $\alpha = .05$, the critical values of t for a two-tailed test with $\nu = 12 - 1 = 11$ are $\pm t_{.025} = \pm 2.201$. The value of the test statistic is

$$t = \frac{\bar{d}}{s/\sqrt{n}} = \frac{.83}{3.41/\sqrt{12}} = .84$$

This value is not within the critical region (see Figure 10.3). Thus the null hypothesis is not rejected. ■

FIGURE 10.3
CRITICAL REGION FOR THE TEST H_0: $\mu_1 = \mu_2$, H_1: $\mu_1 \neq \mu_2$

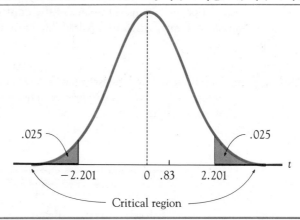

We note in passing that measures of test-retest reliabilities are generally obtained from correlation studies, which is the subject of the next chapter.

A $(1 - \alpha) \times 100\%$ confidence interval for $\mu_1 - \mu_2$ is easily derived as

$$\bar{d} - E < \mu_1 - \mu_2 < \bar{d} + E$$

where

$$E = t_{\alpha/2} \frac{s}{\sqrt{n}}$$

We shall not formally present this procedure since it is the same as for a single variable.

Example 10.10

In an attempt to improve productivity, the work environment for electronic assemblers at a factory was redesigned. The difference of the after and before productivity rates for a sample of nine assemblers had a mean of 3.2 and a standard deviation of 2.5 units per hour. Test the hypothesis that the mean productivity is greater in this new environment. If so, find a 95% confidence interval for the difference in the mean rates.

We begin by computing the p value. First,

$$t = \frac{\bar{d}}{s/\sqrt{n}} = \frac{3.2}{2.5/\sqrt{9}} = 3.84$$

Then using $v = 9 - 1 = 8$, the p value is

$$p = P(t > 3.84) < .005$$

Thus the hypothesis of no change in the mean productivity will be rejected for $\alpha = .05$, $\alpha = .01$, and even $\alpha = .005$.

For a 95% confidence interval, $\alpha = .05$. With $v = 8$, $t_{\alpha/2} = t_{.025} = 2.306$. The maximum error of the estimate is

$$E = 2.306 \left(\frac{2.5}{\sqrt{9}}\right) = 1.9$$

The upper and lower limits are 3.2 ± 1.9 and the confidence interval is

$$1.3 < \mu_1 - \mu_2 < 5.1$$

This interval is somewhat large because of the small sample size. The point estimate to the improvement in productivity is of course $\bar{d} = 3.2$ units per hour. ∎

EXERCISE SET 10.3

Part A

1. Suppose the paired differences $d_i = x_{1i} - x_{2i}$ resulting from a test of two items under matched conditions are as follows:

$$d_i = 12, d_2 = 8, d_3 = -6, d_4 = 18, d_5 = 0, \text{ and } d_6 = -2.$$

Assuming these differences are from a normal distribution,
   a. Test H_0: $\mu_1 = \mu_2$ against H_1: $\mu_1 \neq \mu_2$, using $\alpha = .05$.
   b. Find a 95% confidence interval for the difference $\mu_1 - \mu_2$.
2. Suppose in a test with paired differences we conclude that the null hypothesis H_0: $\mu_1 = \mu_2$ should not be rejected. What is the equivalent conclusion about the mean μ_D of the paired differences?
3. To test two water repellants for use on rainware, five garments were cut in half (by weight). One half of each garment was then treated with repellant A, the other with repellant B. The garments were then placed in extremely moist atmospheres for an hour, removed, and weighed again. The results, in the form of ounces of water absorbed during testing, are summarized below. Test the hypothesis of no difference in the protection from moisture by these repellants. Use $\alpha = .05$. What assumptions are necessary?

			Garment		
Treatment	1	2	3	4	5
A	1.4	4.3	14.2	5.0	2.2
B	1.7	3.9	14.6	4.2	2.0

4. Suppose eight matched pairs from normal populations with equal variances have differences such that

$$\sum_{i=1}^{8} d_i = 24 \qquad \text{and} \qquad \sum_{i=1}^{8} d_i^2 = 97$$

Test the hypothesis H_0: $\mu_1 = \mu_2$ against H_1: $\mu_1 \neq \mu_2$ and find a 95% confidence interval for the difference $\mu_1 - \mu_2$.

5. Two appraisal services routinely perform appraisals of jewelry and diamonds for the courts of a certain community in connection with the settlement of estates. Their independent appraisals in dollars of six randomly selected jewelry items are given below. Test the hypothesis that there is no difference in the appraisals offered by these two services. Use $\alpha = .05$ and assume the appraisals are normal distributed with equal variances.

Item	Appraiser 1	Appraiser 2
1	1650	1340
2	410	430
3	820	830
4	1010	970
5	300	300
6	120	110

6. A discount chain is considering a new design for its customer checkout areas. One of the newly designed areas is installed in each of eight randomly selected stores of the chain. To compare the customer processing rate of the new area with that currently in use, eight clerks were selected at random and monitored in both checkout areas. The average number of customers processed per hour in both environments are given below. Using $\alpha = .025$, test the hypothesis that the new checkout area improves the customer processing rate. Assume the differences are from a normal distribution.

Clerk	Rate (Present)	Rate (New)
1	23	28
2	18	19
3	25	23
4	22	22
5	17	22
6	24	23
7	19	25
8	20	24

7. A small business consulting firm specializing in data processing systems is interested in whether either of two computer systems (A or B) has a lower mean monthly operation cost. Five small businesses using system A are matched according to size, volume of business, number of employees, and the like with five businesses using system B. The differences between the operating costs (system A − system B) for these five pairs had a mean $\bar{d} = \$840$ and standard

deviation s = $450. Assume the normal distribution for the differences of the operating costs.
  a. At the 5% significance level, test the hypothesis that system B has a mean monthly operating cost lower that that of system A.
  b. Find a 95% confidence interval for the difference $\mu_A - \mu_B$ in the mean monthly operating costs of these two systems.
8. A mathematics department gives a screening test for its calculus sequence. Those students who fail the test must take a refresher precalculus course. Students who completed the course during a spring semester are selected at random at the beginning of the fall semester and again given the screening test. The scores are given below.

Test	Student					
	1	2	3	4	5	6
First test	65	58	62	43	54	66
Second test	73	65	58	40	60	67

Assuming the test scores are normally distributed with a common variance.
  a. Test the hypothesis that the mean score in the screening test is not affected by the precalculus course.
  b. Find a 95% confidence interval for the difference between the means of the posttest and pretest scores.

Part B

9. Many educators believe that low test scores on nationwide tests are partly due to students' unfamiliarity with the test form itself. A random sample of 20 third grade students were given an arithmetic achievement test which made use of a multiple choice answer form. Both their worksheets and the multiple choice answer forms were then graded. The results are given below. Test the hypothesis that on the average, arithmetic achievement scores of students at this grade level are lowered through the use of the multiple choice answer form. Use $\alpha = .05$ and assume the grades are from normal distributions with equal variances.

Student	Work Sheet Grade	Multiple Choice Form Grade	Student	Work Sheet Grade	Multiple Choice Form Grade
1	45	41	11	40	40
2	37	32	12	25	22
3	32	32	13	50	46
4	17	22	14	38	37
5	29	26	15	13	14
6	30	29	16	36	36
7	39	31	17	44	42
8	43	42	18	25	22
9	40	36	19	18	21
10	49	47	20	42	40

10. Suppose two independent random samples with five observations each have been selected to test H_0: $\mu_1 = \mu_2$ against H_1: $\mu_1 > \mu_2$. After recording the observa-

tions in a table similar to 10.7, we decide to pair up the adjacent entries and also carry out a test using paired differences.

a. For, say, $\alpha = .025$, which test will have the greatest critical value? Assume t is the test statistic.

b. In what sense is it more difficult to detect a difference in population means with the paired difference design when there is no reduction in variances because of the pairing?

11. Without doubt, many families waste electricity. A simple but ingenious monitor developed by an electrical engineer is designed so that once it is set to an average hourly rate (based on a 16 hour day), it activates a flashing red light whenever this rate is exceeded. A study is conducted of 10 households to see if this monitor has an effect on electrical usage. Their monthly usage (in kilowatt-hours) before and after installation of the monitor is given below. Can we safely conclude that the monitor is effective in reducing electrical consumption? What assumptions are necessary?

Test	Home									
	1	2	3	4	5	6	7	8	9	10
Before	940	1370	1030	2030	1540	2300	1800	910	640	1200
After	900	1230	1060	2100	1250	2200	1820	900	630	1110

12. List some variables for which one should try to control in a paired difference experiment to compare:

a. The weather resistance of two grades of house paint.

b. The effects of two birth control pills on the blood pressure of women.

c. The mathematical achievement of high school students who have used two different textbook series.

d. The blood lead levels of people working in a battery factory and in nonrelated industries.

Library

Project

The following articles make use of the paired t test to compare two population means. Locate one of these articles in the periodical section of the library and report on it. Pay particular attention to the purpose of the study, the selection of experimental subjects, the procedures used in obtaining the data, the reporting of the statistical analysis, and the conclusions.

a. "Metal Content of Roadside Dust," W. Lau and M. Wong, *Environmental Research*, vol. 31, no. 1, June 1983.

b. "Sampling for Farm Studies in Geography," G. Clark and D. Gordon, *Geography*, vol. 65, part 2, no. 287, April 1980.

c. "Coronary Artery Bypass Surgery," C. Jenkins et al., *Journal of the American Medical Association*, vol. 250, no. 6, August 1983.

d. "The Effect of Weight Training and Running Exercise Intervention Programs on the Self-Esteem of College Women" C. Trujillo, *International Journal of Sports Psychology*, vol. 14, no. 3, 1983.

e. "Lateral Eye Movement and Dream Recall in Males," A. LeBoeuf, *Imagination, Cognition, and Personality*, vol. 3, no. 1, 1983–1984.

f. "Impact of Heart Disease Risk Factor Screening Survey on an Upper Class Community," M. Austin et al., *Journal of Community Health*, vol. 8, no. 1, Fall 1982.

g. "The Effect of Vitamin C on Maximum Grip Strength and Muscular Endurance," R. Keith and E. Merrill, *Journal of Sports Medicine*, vol. 23, no. 3, Sept. 1983.

10.6 THE DISTRIBUTION OF $\hat{P}_1 - \hat{P}_2$

It will be recalled that for a large sample size n, the distribution of the sample proportion $\hat{P}$ is approximately normal with mean and standard error

$$\mu_{\hat{P}} = p$$

$$S.E. = \sqrt{\frac{pq}{n}}$$

Here $q = 1 - p$ and p is the proportion of the sampled population having the property of interest. Or equivalently, viewed as a sequence of n binomial trials, p is the probability of success on each trial.

Now for the present case of two populations and two proportions $\hat{P}_1$ and $\hat{P}_2$ from large independent samples, the distribution of the difference $\hat{P}_1 - \hat{P}_2$ exactly parallels that of $\overline{X}_1 - \overline{X}_2$: the mean of $\hat{P}_1 - \hat{P}_2$ is $p_1 - p_2$, the variances may be summed, and the distribution may be approximated by the normal.

THEOREM 10.3

For large independent samples of size n_1 and n_2 from populations with binomial parameters p_1 and p_2, the distribution of the difference $\hat{P}_1 - \hat{P}_2$ is approximately normal with mean and standard error

$$\mu_{\hat{P}_1 - \hat{P}_2} = p_1 - p_2$$

$$S.E. = \sqrt{\frac{p_1 q_1}{n_1} + \frac{p_2 q_2}{n_2}}$$

In most inference problems we must approximate $p_1 q_1$ and $p_2 q_2$ by using results from the samples. If, for example, it has been established that the population proportions are not equal, then we use the point estimates from the samples. The standard error is then estimated by

$$S.E. = \sqrt{\frac{\hat{p}_1 \hat{q}_1}{n_1} + \frac{\hat{p}_2 \hat{q}_2}{n_2}}$$

(Estimate of the standard error of the difference when $p_1 \neq p_2$).

If, on the other hand, the population proportions are equal, a situation similar to theat with equal variances for the mean arises in which we must obtain a pooled estimate of the common proportion p. There are two equivalent ways to obtain this value. First, if we use the observed proportions $\hat{p}_1$ and $\hat{p}_2$, the pooled estimate is

$$\hat{p} = \frac{n_1 \hat{p}_1 + n_2 \hat{p}_2}{n_1 + n_2}$$

Or, if x_1 and x_2 are the observed number of successes in the two samples, then this same estimate is given by

$$\hat{p} = \frac{x_1 + x_2}{n_1 + n_2}$$

We set $\hat{q} = 1 - \hat{p}$ and the standard error is then estimated by

$$S.E. = \sqrt{\hat{p}\hat{q}\left(\frac{1}{n_1} + \frac{1}{n_2}\right)}$$

(Estimate of the standard error when $p_1 = p_2$).

Example 10.11

In a study of the difference in a candidate's support among men and women, independent samples of 200 men and 250 women are polled. Eighty-five of the men and 120 of the women were found to support the candidate. Find estimates of the standard error for both the cases $p_1 \neq p_2$ and $p_1 = p_2$.

Suppose first $p_1 \neq p_2$. The two sample proportions are

$$\hat{p}_1 = \frac{120}{250} = .48 \quad \text{and} \quad \hat{p}_2 = \frac{85}{200} = .425$$

As usual, we have chosen population 1 as the one whose sample proportion is largest, to avoid negative differences.

With $\hat{q}_1 = 1 - .48 = .52$, $n_1 = 250$, $\hat{q}_2 = 1 - .425 = .575$ and $n_2 = 200$, the estimate of the standard error is

$$S.E. = \sqrt{\frac{(.48)(.52)}{250} + \frac{(.425)(.575)}{200}} = .0471 \ (4.71\%)$$

If $p_1 = p_2 = p$, the pooled estimate for p using $x_1 = 120$ and $x_2 = 85$ is found as

$$\hat{p} = \frac{120 + 85}{250 + 200} = .456$$

Or, if we use $\hat{p}_1 = .48$ and $\hat{p}_2 = .425$ from above, the pooled estimate may also be obtained by

$$\hat{p} = \frac{250(.48) + 200(.425)}{250 + 200} = \frac{120 + 85}{250 + 200} = .456$$

The estimate of the standard error is now

$$S.E. \ \sqrt{(.456)(.544)\left(\frac{1}{250} + \frac{1}{200}\right)} = .0473 \ (4.73\%)$$

The two estimates are quite close because $\hat{p}_1$ and $\hat{p}_2$ are not very different. ∎

To assure a normal distribution for the difference $\hat{P}_1 - \hat{P}_2$, large samples are needed. Precisely how large is difficult to state. In most instances samples of size

30 or larger for which np and nq are at least 5 will suffice. One troublesome exception occurs when either of these proportions is near 0 or 1. Apart from this, these conditions give us quite a bit of latitude in our testing and no attempt will be made to deal with the remaining case.

10.7 TESTING THE EQUALITY OF TWO POPULATION PROPORTIONS

For the hypothesis test we consider here, the sample sizes are always large and the null hypothesis is the assertion of equal proportions: $p_1 = p_2$ or, equivalently, $p_1 - p_2 = 0$. Substituting this into

$$Z = \frac{(\hat{P}_1 - \hat{P}_2) - (p_1 - p_2)}{S.E.}$$

we obtain the test statistic

$$Z = \frac{\hat{P}_1 - \hat{P}_2}{S.E.}$$

Here the standard error is found by using the pooled estimate

$$S.E. = \sqrt{\hat{p}\hat{q}\left(\frac{1}{n_1} + \frac{1}{n_2}\right)}$$

This test statistic is very similar to the one used in testing the equality of two means. And this similarity continues with the decision rule and the testing procedure.

PROCEDURE FOR TESTING THE NULL HYPOTHESIS H_0: $p_1 = p_2$ AGAINST AN ALTERNATE HYPOTHESIS H_1.

Given: The proportions $\hat{p}_1$ and $\hat{p}_2$ and the sizes n_1 and n_2 of two samples from two populations with binomial parameters p_1 and p_2.

Assumptions: The samples are independent.

Both samples are large.

1. Decide on a one- or two-tailed test, based on the directional nature of H_1.
2. Choose a significance level α.
3. Find the critical values. For a two-tailed test, these are $\pm z_{\alpha/2}$. For a one tailed test, this will be either z_α or $-z_\alpha$.
4. Compute the pooled estimate to the assumed common proportion:

$$\hat{p} = \frac{n_1\hat{p} + n_2\hat{p}_2}{n_1 + n_2}$$

5. Compute the estimate of the standard error:

$$S.E. = \sqrt{\hat{p}\hat{q}\left(\frac{1}{n_1} + \frac{1}{n_2}\right)}$$

where $\hat{q} = 1 - \hat{p}$.

6. Compute the value of the test statistic:

$$z = \frac{\hat{p}_1 - \hat{p}_2}{S.E.}$$

7. Compare the test and critical value(s).
   - Case I—$H_1: p_1 \neq p_2$. Reject H_0 if $z > z_{\alpha/2}$ or $z < -z_{\alpha/2}$.
   - Case II—$H_1: p_1 > p_2$. Reject H_0 if $z > z_{\alpha}$.
   - Case III—$H_1: p_1 < p_2$. Reject H_0 if $z < -z_{\alpha}$.

Example 10.12   Of 300 case records of cancer fatalities from the year 1968, 50 were found to involve women. A similar sample of 500 records from 1980 showed 125 of the victims to be women. Is there evidence to reject the assertion that the proportion of cancer fatalities is the same for women for these two years?

We let p_1 be the proportion of women among the cancer victims in 1980 and p_2 the proportion for 1968. The corresponding estimates from the samples are

$$\hat{p}_1 = \frac{125}{500} = .250 \quad \text{and} \quad \hat{p}_2 = \frac{50}{300} = .167$$

We will test:

$$H_0: p_1 = p_2$$
$$H_1: p_1 \neq p_2$$

Choosing $\alpha = .05$, the critical values for a two-tailed test are $\pm z_{.025} = \pm 1.96$.

We shall compute the pooled estimate to the common proportion p by using the actual counts:

$$\hat{p} = \frac{x_1 + x_2}{n_1 + n_2} = \frac{125 + 50}{500 + 300} = .219$$

The estimate of the standard error is thus

$$S.E. = \sqrt{\hat{p}\hat{q}\left(\frac{1}{n_1} + \frac{1}{n_2}\right)} = \sqrt{(.219)(.781)\left(\frac{1}{500} + \frac{1}{300}\right)} = .0302$$

The value of the test statistic is

$$z = \frac{\hat{p}_1 - \hat{p}_2}{S.E.} = \frac{.250 - .167}{.0302} = 2.75$$

Since $z = 2.75 > z_{.025}$ the null hypothesis is rejected and we conclude that the fatality rate in 1980 is different from that for 1968. The p value is $p = P(Z > 2.75) = .003$. ■

This last example was constructed from data in the 1980 Surgeon General's Report, which showed that in 1968, women accounted for one in six fatal cases of cancer, while in 1980 the number was one in four. Most of the increase was in lung cancer and is probably due to the increase in cigarette smoking among women.

Example 10.13

A study is made of two types of federally funded college admission programs. In a sample of 450 students admitted into the first program, 60% successfully completed their first year. In an independent sample of 200 from the second program, only 55% managed to survive. Using a 2% significance level, determine if this is sufficient evidence to support the conclusion that the first program is more successful than the second.

We test H_0: $p_1 = p_2$ against H_1: $p_1 > p_2$ where p_1 and p_2 are the proportions successfully completing the first year in the first and second programs, respectively.

The critical value for a one-sided test to the right with $\alpha = .02$ is

$$z_{.02} = 2.05$$

The pooled estimate of the common proportion p is

$$\hat{p} = \frac{n_1\hat{p}_1 + n_2\hat{p}_2}{n_1 + n_2} = \frac{450(.60) + 200(.55)}{450 + 200} = .585$$

The standard error is estimated by

$$S.E. = \sqrt{(.585)(.415)\left(\frac{1}{450} + \frac{1}{200}\right)} = .0419$$

The value of the test statistic is thus

$$z = \frac{.60 - .55}{.0419} = 1.19$$

Since $z < z_{.02}$, the null hypothesis of equality is not rejected. The p value is $p = P(Z > 1.19) = .1170$. ■

The last example demonstrates again the necessity of large samples for inferences about proportions. With the sample sizes in the example, the standard error is almost 4.2%. Even if the samples were doubled, the standard error would only be reduced to around 3%. A good rule of thumb with a two-standard error decision rule is not to become too excited with a 4.5 to 5% difference unless the sample sizes are around 800 or better.

When we construct a confidence interval for the difference $p_1 - p_2$, the pooled estimate for a common proportion is no longer used. The construction procedure is summarized below.

PROCEDURE FOR CONSTRUCTION A $(1 - \alpha) \times 100\%$ CONFIDENCE INTERVAL FOR THE DIFFERENCE $p_1 - p_2$ OF TWO BINOMIAL PARAMETERS

Given: The proportions $\hat{p}_1$ and $\hat{p}_2$ of samples of size n_1 and n_2 from the two populations having parameters p_1 and p_2.

Assumptions: The samples are independent.

The sample sizes are large.

1. Compute the estimate of the standard error:

$$S.E. = \sqrt{\frac{\hat{p}_1\hat{q}_1}{n_1} + \frac{\hat{p}_2\hat{q}_2}{n_2}}$$

where $\hat{q}_1 = 1 - \hat{p}_1$ and $\hat{q}_2 = 1 - \hat{p}_2$.

2. Compute the maximum error of the estimate:

$$E = z_{\alpha/2}(S.E.)$$

3. Compute the upper and lower limits $(\hat{p}_1 - \hat{p}_2) \pm E$ and indicate the confidence interval.

Example 10.14

The difference $\hat{p}_1 - \hat{p}_2$ in the cancer fatality proportions for women between 1968 and 1980 was found to be highly significant in Example 10.12. Find a 95% confidence interval for the difference $p_1 - p_2$. The essential data are as follows:

$$n_1 = 500, \hat{p}_1 = .250; n_2 = 300, \hat{p}_2 = .167$$

The standard error is now estimated by

$$S.E. = \sqrt{\frac{\hat{p}_1\hat{q}_1}{n_1} + \frac{\hat{p}_2\hat{q}_2}{n_2}}$$

$$= \sqrt{\frac{(.250)(.750)}{500} + \frac{(.167)(.833)}{300}} = .0290$$

Using $z_{.025} = 1.96$, the error of the estimate is

$$E = 1.96(.0290) = .0568$$

The upper and lower limits are $(.250 - .167) \pm .0568$ and the confidence interval is

$$.0262 < p_1 - p_2 < .1398$$

Thus the difference in the rate between these two years is somewhere between 2.62% and 13.98%, with 95% confidence. Again, the small samples contribute to this rather large interval. ■

EXERCISE SET 10.4

Part A

1. Test H_0: $p_1 = p_2$ against H_1: $p_1 \neq p_2$:
   a. Using $\alpha = .05$, $n_1 = 450$, $\hat{p}_1 = .40$, $n_2 = 600$, and $\hat{p}_2 = .35$.

    b. Using $\alpha = .01$, $n_1 = 300$, $x_1 = 64$, $n_2 = 450$, and $x_2 = 80$.

2. Residents of two communities are polled to determine their support for a proposal to construct a nuclear plant within their community. Given the results below, test the hypothesis that the proposal is equally supported in both communities.

Community	Sample size	Support (X)
A	1200	664
B	900	542

3. To test a new medication for the treatment of arthritis, 600 arthritis sufferers were divided randomly into two samples. The first sample was given the new medication; the second served as the control group and received a placebo. Given the results below, test the hypothesis that the new medication is no better than the placebo; that is, it is ineffective. Use $\alpha = .01$.

Sample	Sample size	Positive report
Treatment	320	42.0%
Control	280	35.5%

4. Fifteen hundred people from each of two income groups were surveyed as to their voter registration. It was found that 62.4% of the high-income and 58.2% of the low-income group were registered. Is this sufficient evidence to conclude at a 1% significance level that the percentage of registered voters in the high-income group exceeds that in the low-income group?

5. Two makes of automobiles, A and B, are crashed into a barrier at 35 miles per hour as part of a safety test. Suppose that samples of size 40 are used and that studies indicate that 64.5% of the drivers of model A would have survived while only 51% of those of model B would have made it through safely. Can we conclude at a 5% significance level that model A is safer than model B in protecting the driver when the crash conditions are as in the test?

6. Seven hundred people received an advertisement, 400 by first class and 300 were by second class mail. Careful interviews were then conducted to determine whether the advertisement was actually read. It was found that of those who had read the advertisement, 220 had received the first class mailing and 128 had received the 2nd class. Estimate the true difference in the proportions of people who respond to these two types of mailing strategies.

7. A cereal company sampled 300 individuals with its presweetened version of a certain grain cereal. At the same time, another 400 people tried the unsweetened version of the cereal. At the end of a two-week period, the participants were asked whether they would purchase the cereal. Positive responses were made by 184 of these using the presweetened version and 205 of those with the unsweetened. Test the hypothesis that the percentage who would buy either version of this cereal is the same. Use $\alpha = .05$.

8. Three hundred autos were stopped at random and tested for compliance with federal emission standards; 112 failed the test. Then 110 autos were selected at random, tuned, and driven for approximately 5000 miles. When these were

tested, 30 failed to meet the emission standards. Is there evidence here that mandatory-tune ups would significantly reduce the compliance rate for autos?

9. A merchandise mail order house has developed two fliers (A and B) for several specially selected items. To test their effectiveness, 1000 of each are mailed to independent random samples of past customers. Within three weeks, 18% of those receiving flier A and 14.5% of those receiving flier B responded with an order. Using $\alpha = .01$, test the hypothesis that advertising flier A is the more effective of the two.

10. In a test of a vaccine against Asian flu, 300 chickens are selected and divided equally into experimental and control groups, the experimental group receiving the vaccine. Both groups were then infected with the influenza virus. Sixty-three of the 150 experimental chickens and 92 of the controls subsequently died from the virus. Find a 99% confidence interval for the *increase* in the proportion surviving when the vaccine is used.

11. In a study, random samples of 300 male and 400 female middle level business executives were rated as to their objectivity in the evaluation of personnel merit raise recommendations. Two hundred and forty-five of the women and 176 of the men were rated as being totally objective in their evaluations. Find a 95% confidence interval for the difference in the proportions of female and male executives who are objective in their personnel evaluations.

Part B

12. Six hundred and fifty individuals are randomly selected and asked to rank two perfume fragrances, say A and B. Forty-four percent liked fragrance A and 56% preferred fragrance B. We want to test whether the same proportions like fragrance A and B. Is the hypothesis test of this section directly applicable?

13. Show that if the sample proportions are equal ($\hat{p}_1 = \hat{p}_2$), then the pooled estimate of the common proportion p is the common sample value.

14. Show that if the sample sizes are equal, then the pooled estimate of the common proportion p is just the average of the sample proportions $\hat{p}_1$ and $\hat{p}_2$.

15. Suppose it is estimated that p_1 and p_2 are around .40. If equally large samples are used, for what size would a difference of $.03 = 3\%$ in the sample proportions be significant in a two tailed test with $\alpha = .05$?

16. Here are some additional clinical features of the patients in the bulimia–anorexia nervosa study (see problem 13 of Exercise Set 10.2).

Characteristic/feature	Bulimia group N = 49	Fasting group N = 56
a. Admires a thin person	19	12
b. Hyperactive childhood	14	9
c. Outgoing personality	42	32
d. Disinterest in sex	39	34
e. Anorexia nervosa started with voluntary dieting	42 (yes)	50 (yes)

Test for a significant difference in proportions for each characteristics a, b, c, and d.

COMPUTER PRINTOUT

The independent sample t test for the equality of two population means is easily carried out with the aid of Minitab. The printout in Figure 10.4 covers the test

$$H_0: \mu_1 = \mu_2$$
$$H_1: \mu_1 \neq \mu_2$$

In this example, the weight gains (in ounces) of hatchery trout on two different feeding plans are compared. Note that a confidence interval for the difference $\mu_1 - \mu_2$ is included as part of the analysis.

FIGURE 10.4 MINITAB TWO-SAMPLE t TEST

```
MTB >SET THE WEIGHT GAINS FROM PLAN A INTO C1
DATA>32, 36, 34, 40, 18, 25, 21, 27, 42, 45, 36, 34
DATA>22, 38, 39
DATA>END
MTB>SET THE WEIGHT GAINS FROM PLAN B INTO C2
DATA>32, 31, 45, 18, 30, 31, 28, 23, 40, 21, 32, 35
DATA>26, 16
DATA>END
MTB >POOLED T TEST FIRST SAMPLE IN C1, SECOND SAMPLE
IN C2
    C1       N =  15    MEAN =     32.600    ST.DEV. =
8.21
    C2       N =  14    MEAN =     29.143    ST.DEV. =
8.04
    DEGREES OF FREEDOM = 27
    A 95.00 PERCENT C.I. FOR MU1-MU2 IS (    -2.7411,
9.6553)
    TEST OF MU1 = MU2 VS. MU1 N.E. MU2
    T =  1.145
    THE TEST IS SIGNIFICANT AT 0.2624
    CANNOT REJECT AT ALPHA = 0.05
MTB>STOP
```

CHAPTER CHECKLIST

Vocabulary
Independent samples
Matched pairs
Pooled estimate of the variance
Paired differences
Dependent samples

Techniques
Finding the distribution of $\overline{X}_1 - \overline{X}_2$ for large samples; (including the mean and standard error)

Testing for the equality of two populations means using large independent samples

Testing for the equality of two population means using independent samples from normal populations and a pooled variance estimate

Performing a paired difference test of the equality of two population means using matched pairs from normal populations

Constructing a confidence interval for the difference $\mu_1 - \mu_2$ using the means and variances of two large independent samples

Constructing a confidence interval for the difference $\mu_1 - \mu_2$ using the means and variances of two samples from normal populations and a pooled variance estimate

Finding the distribution of $\hat{P}_1 - \hat{P}_2$ for large samples; (including the mean and standard error)

Performing a test of the equality of two populations proportions using the proportions from two large independent samples

Constructing a confidence interval for the difference $p_1 - p_2$ of two population proportions using the proportions $\hat{p}_1$ and $\hat{p}_2$ of large samples.

Notation

x_{1i}	The ith observation of sample 1
x_{2i}	The ith observation of sample 2
$\mu_{\bar{x}_1 - \bar{x}_2}$	The mean of $\bar{X}_1 - \bar{X}_2$
$\sigma_{\bar{x}_1 - \bar{x}_2}$	The standard error of $\bar{X}_1 - \bar{X}_2$
d_i	The ith paired difference $(x_{1i} - x_{2i})$
$\bar{d}$	The mean of the paired differences
D	The paired difference variable

CHAPTER TEST

1. Using a 5% significance level and the data below from two independent samples, test

$$H_0: \mu_1 = \mu_2$$
$$H_1: \mu_1 \neq \mu_2$$

Population I	Population II
$n_1 = 80$	$n_2 = 75$
$\bar{x}_1 = 1930$	$\bar{x}_2 = 2150$
$s_1 = 500$	$s_2 = 550$

2. Using a 5% significance level and the data below, test

$$H_0: p_1 = p_2$$
$$H_1: p_1 < p_2$$

You may assume the samples are independent.

Population I	Population II
$n_1 = 400$	$n_2 = 500$
$\hat{p}_1 = .600$	$\hat{p}_2 = .660$

3. Independent random samples from two normal populations with a common variance produced the data below. Perform a 1% significance level test of

$$H_0: \mu_1 = \mu_2$$
$$H_1: \mu_1 < \mu_2$$

Population I	Population II
$n_1 = 10$	$n_2 = 8$
$\bar{x}_1 = 16$	$\bar{x}_2 = 21$
$s_1 = 7$	$s_2 = 8$

4. Using the data in problem 1, construct a 95% confidence interval for $\mu_2 - \mu_1$.
5. Using the data of problem 2, construct a 90% confidence interval for $p_2 - p_1$.
6. A remedial algebra program at a university consists of a self-paced course built around modules that the student completes with the help of peer tutors. No formal lecture or problem-solving discussions are involved. To measure the program's effectiveness, five students are given a test at the beginning of the course and then again at the end. The test scores are shown below. Can we conclude at a 5% significance level that the student's algebraic skills (as measured by this test) are improved by this course? What assumptions are required?

Student	Pretest	Posttest
1	60	75
2	52	54
3	66	80
4	58	56
5	61	68

7. A paired difference experiment was conducted with 10 matched pairs of experimental subjects. The ten differences of their respective scores had a mean $\bar{d} = 400$ and a standard deviation $s = 420$. Using $\alpha = .01$, test for a difference in the means of the two populations. Assume the differences are from a normal distribution.

8. A researcher is interested in whether there is a difference in the salaries of registered nurses in two metropolitan areas, say I and II. The results below were obtained from two independent random samples. Do these data provide sufficient evidence at the .05 level of significance to indicate that the salaries of nurses in area I are, on the average, less than those of area II? Find the p value of the test.

Area I	Area II
$n_1 = 60$	$n_2 = 70$
$\bar{x}_1 = \$24{,}690$	$\bar{x}_2 = \$24{,}960$
$s_1 = \$730$	$s_2 = \$670$

9. In a study of secondary mathematics teachers, an agency of a state's department of education found that 42% of a sample of 525 teachers did not believe that the use of a computer would significantly improve their classroom effectiveness. During a similar study five years ago, 46% of a sample of 400 teachers held a similar view. Test the hypothesis that there has been no change in the teachers' opinions. Use $\alpha = .05$ and assume the samples are independent.

10. Two installation procedures are being evaluated for the ignition system of a jet engine. The question of interest is whether the procedure has an effect on the in-flight restart time of the engine. The results of tests on two independent samples of such engines are summarized below; the data given are restart times in seconds. Test for a difference in the restart times due to the installation procedure, using $\alpha = .05$. What assumptions are necessary?

Procedure 1	Procedure 2
$n_1 = 10$	$n_2 = 8$
$\bar{x}_1 = 26.8$	$\bar{x}_2 = 18.3$
$s_1 = 12.0$	$s_2 = 10.7$

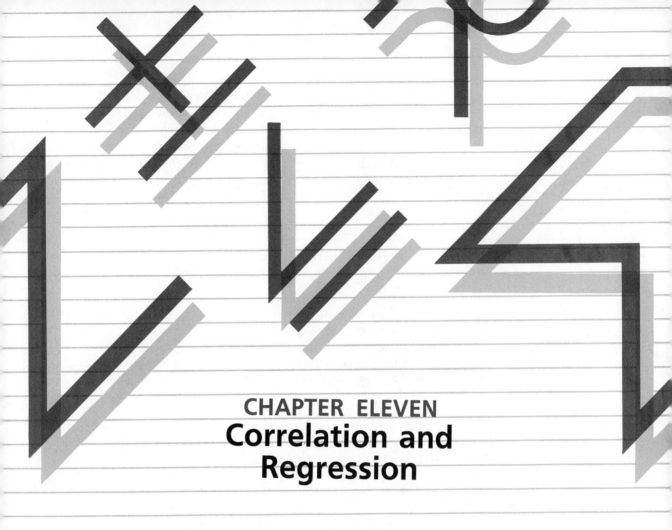

CHAPTER ELEVEN
Correlation and Regression

Several variables are often studied simultaneously on the same set of experimental subjects: I.Q.'s and SAT scores for high school seniors; yields and planting densities of trees in avocado orchards; law school grades and scores on the Law School Admission Test. Frequently the important question is whether or not there is an association or correlation between the values of the variables. Establishing whether such an association exists and assessing its strength is the subject of *correlation analysis*.

Regression analysis is concerned with finding a mathematical model or formula that relates the values of one variable to those of the other. For example, is there a formula that will statistically predict the blood pressure of an adult male from his age? Of course, perfect results are not expected from a regression formula. But if we input an age such as 40, it should produce a number such as 137, about which the systolic pressures of men of this age will cluster. Ideally, we would like this predicted value of 137 to be mean systolic pressure for all men of age 40.

In most instances it is not reasonable to expect to relate one variable exclusively to a second. As we have observed on previous occasions, the variability present in a quantity such as blood pressure is not due to a single variable such as age. Such other factors as weight, cholesterol level, or occupational stress all contribute. Thus beyond two variables are multivariable correlation and regression problems. In this chapter we confine our attention for the most part to two-variable problems, where it is assumed that the regression model is a straight line. In the last two sections we briefly indicate some extensions of the theory to the general multivariable problem.

11.1   CAUSALITY AND CORRELATION

When a correlation is found between two variables there is a temptation to make a great leap and infer a cause-and-effect relationship between the variables. One must be extremely cautious with such pronouncements. In general they are not justified unless there is other experimental evidence present.

For example, if we examine annual military appropriations and annual paper clip production in the United States since 1948, we find that both have increased and that there is a strong correlation between these variables. But this finding does not suggest that one of these trends is responsible for the other. In this case there is a third variable, time, to which both are related.

Or consider the study in which a correlation was established between achievement scores of elementary school children and the amounts by which they are overweight. Does this result suggest a causal relationship? Or are both related to common hereditary and behavorial factors? In any event, while the variables may be correlated, we should be very skeptical of any causal relationship without additional evidence.

Basically, an inference of causality is not justified in a correlational study because of the nature of the data: since they consist of random pairs of observations we cannot assume that the value of one variable is responsible for the second. In an experimental study, by contrast, the values of one variable may be varied in some specific manner and we observe corresponding values of the other. Here it may be possible to discern a cause-and-effect relationship. Regardless of how one chooses to view the matter, the message is clear: do not assume that correlated variables are causally related.

11.2   THE COVARIANCE

Most of the development in this chapter is concerned with the existence of a relationship between two variables. Since graphs and plots of data are often used, it is better to use the variable names X and Y rather than X_1 and X_2 as in the last chapter. Associated with these variables are means and standard deviations μ_X, μ_Y, σ_X, and σ_Y and a joint probability distribution that assigns probabilities to events that now consist of pairs of values (x, y).

There are many ways in which two variables may be related mathematically. Among these, it is the *linear* or *straight-line* relationship that proves most useful in practical problems. Throughout this chapter we shall assume the existence of such a relationship. In Figure 11.1 are given the plots or scatter diagrams of

FIGURE 11.1   A LINEAR TENDENCY IS INDICATED IN THE SCATTER DIAGRAMS (A) AND (B)

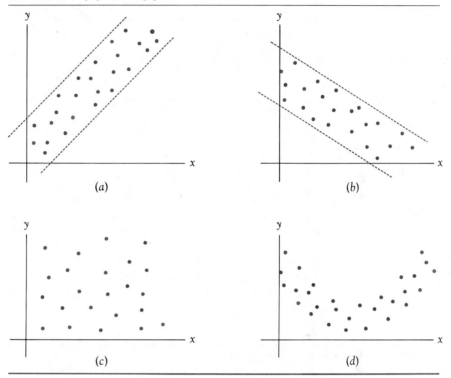

samples of data from four distributions. In (a) and (b) the points (x, y) do not fall on a straight line. They do, however, fall within a narrow band defined by two straight lines, and this suggests a linear trend or relationship between the variables. In (c) there is no discernable pattern. In (d) there is a relationship but it is not linear; it is *quadratic* or *parabolic*.

When we study the joint distribution of two variables, we need a new parameter, the *covariance*, which measures how X and Y jointly vary or cluster about their respective means. The covariance is a sort of two-dimensional version of the variance of a single variable. If we recall that the variance of X was defined by

$$\sigma_X^2 = E[(X - \mu_X)^2] = E[(X - \mu_X)(X - \mu_X)]$$

then we arrive at Definition 11.1.

DEFINITION 11.1

Let X and Y have a joint probability distribution with means and standard deviations μ_X, μ_Y, σ_X, and σ_Y. The *covariance* σ_{XY} of X and Y is

$$\sigma_{XY} = E[(X - \mu_X)(Y - \mu_Y)]$$

The counterpart of the covariance for a sample is the sample covariance.

DEFINITION 11.2

The covariance s_{XY} of the sample (x_1, y_1), (x_2, y_2), . . ., (x_n, y_n) is

$$s_{XY} = \frac{\sum_{i=1}^{n} (x_i - \bar{x})(y_i - \bar{y})}{n - 1}$$

In Figure 11.2 the scatter diagrams for three sets of data are given. The computation of their corresponding covariances are given immediately below.

A positive covariance indicates that the factors of the products $(x_i - \bar{x})(y_i - \bar{y})$ tend to both be positive or negative. This situation can occur, for example, if there is a tendency for one of the variables to increase as the other increases (see Figure 11.2b). A negative covariance results if the factors tend to be of opposite signs; this can occur if one variable decreases as the other increases (see Figure 11.2a). If the covariance is 0, as in 11.2c, there can be no linear

---

FIGURE 11.2   SCATTER DIAGRAMS

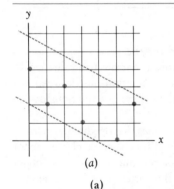

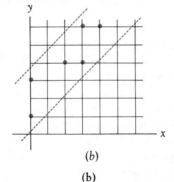

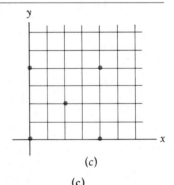

| (a) | (b) | (c) |

(a)

x	y	$x - \bar{x}$	$y - \bar{y}$	$(x - \bar{x})(y - \bar{y})$
0	4	−3	2	−6
1	2	−2	0	0
2	3	−1	1	−1
3	1	0	−1	0
4	2	1	0	0
5	0	2	−2	−4
6	2	3	0	0
21	14			−11

$\bar{x} = 21/7 = 3$ $s_{xy} = -11/6$
$\bar{y} = 14/7 = 2$ $= -1.83$

(b)

x	y	$x - \bar{x}$	$y - \bar{y}$	$(x - \bar{x})(y - \bar{y})$
0	1	−2	−3	6
0	3	−2	−1	2
2	4	0	0	0
3	4	1	0	0
3	6	1	2	2
4	6	2	2	4
12	24			14

$\bar{x} = 12/6 = 2$ $s_{xy} = 14/5$
$\bar{y} = 24/6 = 4$ $= 2.80$

(c)

x	y	$x - \bar{x}$	$y - \bar{y}$	$(x - \bar{x})(y - \bar{y})$
0	0	−2	−2	4
0	4	−2	2	−4
2	2	0	0	0
4	4	2	2	4
4	0	2	−2	−4
10	10			0

$\bar{x} = 10/5 = 2$ $s_{xy} = 0/4$
$\bar{y} = 10/5 = 2$ $= 0.00$

relationship between X and Y. Additionally, we obtain the following important result.

THEOREM 11.1

If X and Y are independent random variables, then their covariance is zero, that is, $\sigma_{XY} = 0$.

The converse of this theorem is not true: two variables can have a covariance of zero and not be independent.

11.3   THE CORRELATION COEFFICIENT

The preceding results suggest that the covariance is a natural starting point for investigating the strength of a linear relationship between two variables. Unfortunately it is not a dimensionless quantity; its value depends on the actual size of the coordinates. If, for example, we take two variables and divide each by two, the new variables should have the same degree of dependence as the original, but the covariance will be reduced by a factor of four. Such problems can be eliminated, however, by using the covariance of the standardized variables:

$$Z_X = \frac{X - \mu_X}{\sigma_X} \qquad Z_Y = \frac{Y - \mu_Y}{\sigma_Y}$$

This covariance is called the linear correlation coefficient of X and Y.

DEFINITION 11.3

Let X and Y have a joint distribution with means and standard deviations μ_X, μ_Y, σ_X, and σ_Y. The *linear correlation coefficient* ρ (rho) is the number

$$\rho = E[Z_x \cdot Z_y] = E\left[\left(\frac{X - \mu_X}{\sigma_X}\right)\left(\frac{Y - \mu_Y}{\sigma_Y}\right)\right]$$

The correlation coefficient ρ measures the strength of the linear relationship between X and Y in the following sense: ρ is some number between $+1$ and -1.

DEFINITION 11.4

The linear correlation coefficient r for the sample (x_1, y_1), (x_2, y_2), . . ., (x_n, y_n) is given by

$$r = \sum_{i=1}^{n} \frac{z_x z_y}{n - 1} = \frac{\displaystyle\sum_{i=1}^{n} \left(\frac{x_i - \bar{x}}{s_X}\right)\left(\frac{y_i - \bar{y}}{s_Y}\right)}{n - 1}$$

If $\rho = 0$ there is no linear relationship. If $\rho = \pm 1$ there is a perfect linear relationship between X and Y.

The correlation coefficient for a sample is defined in an analogous manner. The statements we made for ρ apply equally well to the sample correlation coefficient r. It has values between $+1$ and -1; either of the limits indicates that all of the data points fall on a single straight line. A value of $r = 0$ indicates the absence of any linear relation.

When evidence of a linear relation is exhibited, say, by a scatter diagram, intermediate values of r measure the strength of this relation. The sign of r is also informative. If r is positive, the tendency is for y to increase with x. Likewise, if r is negative, the tendency is for y to decrease as x increases. It is natural to describe these situations by saying that there is a *positive* or *negative linear correlation*. When $r = 0$, we say the variables are *uncorrelated*. Remember, however, that this statement applies only in a linear sense.

The scatter diagrams for several sets of data are given in Figure 11.3. Study these carefully. Note that in *(f)* there is an obvious quadratic relation between x and y that is not reflected in the value of r. We reemphasize that r should be used only when there is evidence of a linear relation.

FIGURE 11.3   SCATTER DIAGRAMS

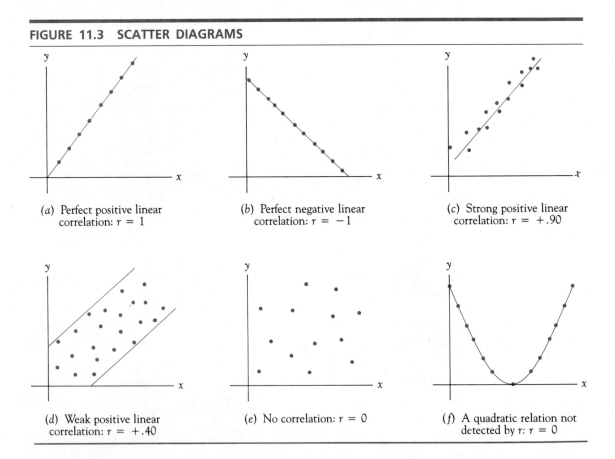

(a) Perfect positive linear correlation: $r = 1$

(b) Perfect negative linear correlation: $r = -1$

(c) Strong positive linear correlation: $r = +.90$

(d) Weak positive linear correlation: $r = +.40$

(e) No correlation: $r = 0$

(f) A quadratic relation not detected by r: $r = 0$

11.4 COMPUTING r

A simplified formula for computing the correlation coefficient

$$r = \frac{\sum_{i=1}^{n} \left(\dfrac{x_i - \bar{x}}{s_X}\right)\left(\dfrac{y_i - \bar{y}}{s_Y}\right)}{n - 1}$$

may be obtained by substituting for s_X and s_Y and performing some algebra along the lines of that used with the sample variance in Chapter 3.

THEOREM 11.2

An alternate formula for the correlation coefficient is as follows:

$$r = \frac{n \cdot \sum_{i=1}^{n} x_i y_i - \left(\sum_{i=1}^{n} x_i\right)\left(\sum_{i=1}^{n} y_i\right)}{\sqrt{n \cdot \sum_{i=1}^{n} x_i^2 - \left(\sum_{i=1}^{n} x_i\right)^2} \sqrt{n \cdot \sum_{i=1}^{n} y_i^2 - \left(\sum_{i=1}^{n} y_i\right)^2}}$$

To use this formula in hand computation, a table with columns consisting of the values x, y, x^2, y^2, and xy is prepared. The sums needed are the column totals.

Example 11.1

The correlation between the age X and weight Y of men with the same type of bone structure is under study. The data from a sample of eight men are given in

---

FIGURE 11.4 SCATTER DIAGRAM FOR AGE VERSUS WEIGHT DATA OF TABLE 11.1. SUBJECT NUMBER INDICATED ABOVE EACH POINT.

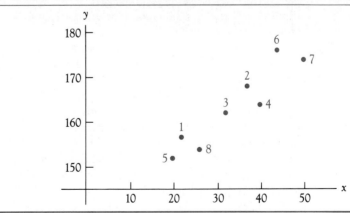

the first two columns of Table 11.1. The first step is to examine the scatter diagram (Figure 11.4). A linear correlation appears to exist. The value of r is

$$r = \frac{8(44,902) - 271(1307)}{\sqrt{8(9989) - (271)^2} \ \sqrt{8(214,085) - (1307)^2}} = .94$$

This indicates an extremely high linear correlation. ∎

TABLE 11.1   AGE, VERSUS WEIGHT DATA FOR COMPUTING r IN EXAMPLE 11.1

Subject	x	y	x^2	y^2	xy
1	22	157	484	24,649	3454
2	37	168	1369	28,224	6216
3	32	162	1024	26,244	5184
4	40	164	1600	26,896	6560
5	20	152	400	23,104	3040
6	44	176	1936	30,976	7744
7	50	174	2500	30,276	8700
8	26	154	676	23,716	4004
Totals	271	1307	9989	214,085	44,902

A correlation coefficient as high as the $r = .94$ obtained in the last example is rare except with contrived data. The next sample is more typical of what is obtained in actual studies.

Example 11.2
A Case Study

Is I.Q. related to the home environment of the individual? This is the subject of a study entitled "Race, I.Q., and the Middle Class" by Frances Trotman (Columbia University, *Journal of Educational Psychology*, vol. 69, no. 3). One topic of this study is the possibility of a linear correlation between the I.Q. scores of children and the intellectual environment ratings of their homes. In all, 50 white and 50 black ninth grade girls from a middle class suburban school district were involved in the study. From the scatter diagram (Figure 11.5), a positive correlation between these ratings is apparent. This is confirmed by a correlation coefficient $r = .63$. The correlation $r = .68$ for the black girls was higher than that for the white ones ($r = .37$). The author suggests several possible explanations for this, including a lack of heterogenity in the sample of white families. In the final analysis the author concluded that "there was a direct relationship between the intellectual home environment and the child's I.Q." Further, from other data within the report she went on to conclude that among black families, the "intellectual home environment was at least as good a predictor of the child's academic achievement as was the child's intelligence test score."

FIGURE 11.5   CHILDREN'S I.Q.'S AND INTELLECTUAL HOME ENVIRONMENT SCORES FOR BLACK AND WHITE FAMILIES

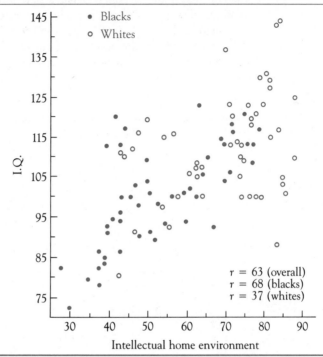

As we saw in the age-weight study (Example 11.1), large sums can easily arise in computing r by the alternate formula. These may be reduced, however, through division or subtraction by some constant. As long as all the observations of a variable are treated the same, the correlation coefficient is unchanged. Thus in the age-weight study we could subtract 20 from each age and 150 from each weight, or we could divide each age by 10 and each weight by 100. Combinations of both these transformations may also be used. This technique is illustrated in the next example. We admit, however, that it is of limited utility when a computer or hand-held calculator is available.

The next example is concerned with the test-retest reliability of a process. If an examination or any type of measuring instrument is to be useful, it should have capability of reproducing approximately the same results when observations are made under similar conditions or on the same subjects. In the case of individuals taking a test for, say, intelligence or anxiety, we have to assume that no learning or change results from taking the test. One proposed measure of the test-retest reliability is the correlation coefficient between the two sets of observations. Typically, one sees r values of .70 to .90 cited in the literature for the reliability of popular intelligence, anxiety, and depression scales.

Example 11.3   Twelve students in a program for educable mentally retarded (EMR) youth were randomly selected and administered a new form of an intelligence test on two occasions. The scores X and Y from the two testings are given in the first two columns of Table 11.2. In the third and fourth columns are modified scores x^* and y^* obtained by subtracting 75 from each of the original scores. These are used to complete the remainder of the table and to compute r. The scatter diagram is given in Figure 11.6.

FIGURE 11.6 SCATTER DIAGRAM

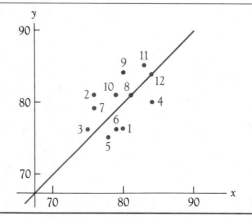

TABLE 11.2 TEST SCORES

Subject	x	y	x^*	y^*	$(x^*)^2$	$(y^*)^2$	x^*y^*
1	80	76	5	1	25	1	5
2	76	81	1	6	1	36	6
3	75	76	0	1	0	1	0
4	84	80	9	5	81	25	45
5	78	75	3	0	9	0	0
6	79	76	4	1	16	1	4
7	76	79	1	4	1	16	4
8	81	81	6	6	36	36	36
9	80	84	5	9	25	81	45
10	79	81	4	6	16	36	24
11	83	85	8	10	64	100	80
12	84	84	9	9	81	81	81
Sums			55	58	355	414	330

From these data, we find that

$$r = \frac{(12)(330) - (55)(58)}{\sqrt{(12)(355) - (55)^2} \; \sqrt{(12)(414) - (58)^2}} = .55$$

Such a low correlation indicates a poor test-retest reliability. It does, however, show a positive linear trend, as would be expected. If the test had perfect reliability, all observations would have fallen on the line $y = x$ shown in the scatter diagram. ■

11.5   TESTING FOR A CORRELATION

If X and Y are uncorrelated, then $\rho = 0$. However, because of the sampling process the coefficient r will rarely if ever be 0. Thus the question arises of just what values of r may be regarded as significant. The answer to this question is found in the sampling distribution of r.

Unfortunately, for general values of ρ, this distribution requires more mathematics than we care to introduce. For the special case of $\rho = 0$ where both X and Y are independent normal variables, the distribution is somewhat simpler and depends only on the sample size. Its form is sketched in Figure 11.7; we omit further details.

Actually, for testing the null hypothesis of no correlation ($\rho = 0$) we need only the critical values of r under the assumption $\rho = 0$. These are given in Table VIII and as usual are single-tailed values (see Figure 11.7).

FIGURE 11.7   THE DISTRIBUTION OF r WHEN $\rho = 0$

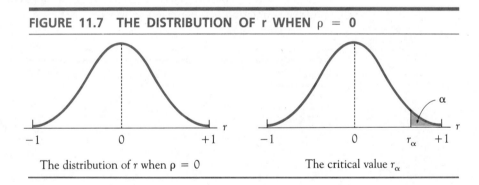

The distribution of r when $\rho = 0$ The critical value r_α

A brief description of the test for no correlation between two variables is as follows:

PROCEDURE FOR TESTING THE NULL HYPOTHESIS OF NO CORRELATION H_0: $\rho = 0$ AGAINST AN ALTERNATE HYPOTHESIS H_1.

Given: A random sample of n observations from the joint distribution of X and Y: (x_1, y_1), (x_2, y_2), . . ., (x_n, y_n).
Assumptions: X and Y are independent normal variables.
1. Select a significance level α.

2. Compute the sample correlation coefficient

$$r = \frac{n\sum\limits_{i=1}^{n} x_i y_i - \left(\sum\limits_{i=1}^{n} x_i\right)\left(\sum\limits_{i=1}^{n} y_i\right)}{\sqrt{n\sum\limits_{i=1}^{n} x_i^2 - \left(\sum\limits_{i=1}^{n} x_i\right)^2} \sqrt{n\sum\limits_{i=1}^{n} y_i^2 - \left(\sum\limits_{i=1}^{n} y_i\right)^2}}$$

3. Compare r to the critical value(s).
   - Case I—H_1: $\rho \neq 0$. Reject H_0 if $r > r_{\alpha/2}$ or $r < -r_{\alpha/2}$.
   - Case II—H_1: $\rho > 0$. Reject H_0 if $r > r_\alpha$.
   - Case III—H_1: $\rho < 0$. Reject H_0 if $r < -r_\alpha$.

Example 11.4

In the age-weight study (Example 11.1) involving $n = 8$ men, a sample correlation coefficient $r = .94$ resulted. Using a 5% significance level, test

$$H_0: \rho = 0$$
$$H_1: \rho > 0$$

Since this is a one-tailed test, the critical value is $r_{.05}$. Entering Table VIII in the row $n = 8$ we find $r_{.05} = .621$. The computed $r = .94$ far exceeds this value (see Figure 11.8). Thus we reject the null hypothesis and conclude that there is a positive correlation between age and weight. ■

FIGURE 11.8 $r = .94$ FAR EXCEEDS $r_{.05}$. REJECT H_0

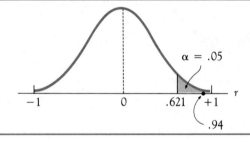

Example 11.5

Ten randomly selected individuals at an outpatient psychiatric clinic were given two tests, each of which supposedly measures depression. The correlation coefficient for the resulting scores was $r = .61$. Using $\alpha = .01$, test

$$H_0: \rho = 0$$
$$H_0: \rho \neq 0$$

For $n = 10$, the critical values are $\pm r_{.005} = \pm.765$. Since $r = .61$ we do not reject the null hypothesis of no correlation. In short, there does not appear to be a close linear association between the scores on the two tests. ■

It should be noted (as a rule of thumb) that only when the sample size is at least 16 does a sample correlation coefficient of $r = .5$ prove significant in a two-tailed test.

EXERCISE SET 11.1

Part A

1. Determine whether the covariance of each of the following should be positive or negative.

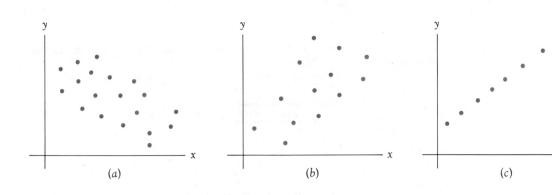

(a)   (b)   (c)

2. Suppose each of the following pairs of variables are linearly related. Would you expect the covariance to be positive or negative? Explain.
   a. Heights of mothers and daughters.
   b. The weight and height of individuals.
   c. The age and price of a certain make of auto (no antiques, please).
   d. Natural gas usage and the prevailing temperature.
   e. I.Q. and SAT scores.

3. Plot the scatter diagram and find the covariance for each of the following:

a. x	y	b. x	y	c. x	y
0	2	0	4	0	0
1	2	1	4	0	4
2	3	2	4	4	0
3	3	3	4	4	4
4	6				
5	5				

4. Suppose each of the following pairs of variables are linearly related. Would you expect a positive or negative correlation? Explain.
   a. The temperature Y at a height X above ground level.
   b. The yearly income Y and the number of years of education X of individuals who are not self-employed.
   c. The concentration Y of smog at an altitude X feet directly above Los Angeles City Hall.
   d. The pulse rate Y of an individual after X minutes of exercise.

5. A study of married men was made to determine if a correlation exists between

their annual income and the amount of life insurance they carry. Using the results below,

a. Construct the scatter diagram.
b. Compute the correlation coefficient r.
c. Test for a positive correlation using $\alpha = .05$.

Income x (thousands of dollars)	44	22	30	110	70	50	20	80
Insurance y (thousands of dollars)	95	10	50	200	130	80	0	190

6. The height y and base diameter x of five trees of a certain variety produced the following data.

x (inches)	y (inches)
1.0	31
1.0	36
2.0	70
3.0	94
5.0	127

a. Prepare a scatter diagram.
b. Compute the correlation coefficient r using the alternate formula.

7. Unemployment and interest rates are given below for randomly selected times over the last 35 years.
a. Prepare a scatter diagram.
b. Compute r and test for a correlation using $\alpha = .05$.

Unemployment rate	3.8	3.2	5.5	4.1	5.5	5.5	4.5	3.8	3.6	5.9	5.6	7.7	6.0
Interest rate	3.2	3.2	3.7	3.5	4.7	5.1	4.8	5.6	6.2	9.1	8.2	10.6	9.0

8. Eleven instructors were asked to evaluate themselves, using the same questionnaire used by their students. These self-evaluations and the average of the instructors' student evaluations are tabulated below. Test for a positive correlation.

Student	4.0	6.2	10.0	5.4	7.1	2.5	7.0	2.1	3.4	8.3	7.6
Instructor	6.2	7.0	10.0	5.0	7.5	5.4	9.3	8.6	5.2	9.0	7.0

9. What is the correlation coefficient of a set of pairs obtained by matching numbers with themselves, that is, $(x_1, x_1), (x_2, x_2), \ldots, (x_n, x_n)$?

10. The following is a random sample of right- and left-hand strengths of 13 individuals. Is a test for correlation justified on these data?

| Left hand | 2.0 | 3.2 | 2.2 | 2.2 | 4.5 | 5.8 | 7.4 | 1.5 | 2.9 | 5.1 | 9.6 | 5.9 | 3.3 |
| Right hand | 5.9 | 4.1 | 5.5 | 4.5 | 3.6 | 4.9 | 6.0 | 5.7 | 3.8 | 5.9 | 8.5 | 7.0 | 3.8 |

Part B

11. The means $\bar{x}$ and $\bar{y}$ of a set of data locate a point in the x-y plane. Through this point we draw vertical and horizontal lines (as shown) and label the four quadrants thus created as I, II, III, and IV. Complete the following table, giving the signs of the indicated quantities in these four quadrants.

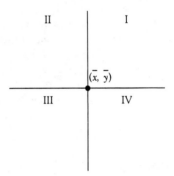

Quadrant	$(x - \bar{x})$	$(y - \bar{y})$	$(x - \bar{x})(y - \bar{y})$	$\dfrac{(x - \bar{x})(y - \bar{y})}{s_x \cdot s_y}$
I				
II				
III				
IV				

12. Suppose 10 points are selected from the lines below and r is computed. What value of r would result?

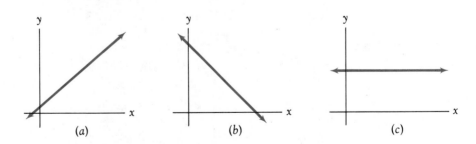

(a)   (b)   (c)

13. In a 1977 study by Martin and Kidwell (*Educational and Psychological Measurements*, 1977, 37), the newly revised Wechsler Intelligence Scale for Children (WISCR) was compared with the Slossen Intelligence Test (SIT), an abbreviated form of the Stanford Binet Test. In tests on 33 children the following correlation coefficients were obtained between the SIT scores and the full WISCR, as well as the verbal and performance portions.

	WISCR (Full)	WISCR (Verbal)	WISCR (Performance)
SIT	.79	.82	.50

   a. Test for a positive correlation between the WISCR verbal scores and the SIT scores. Use $\alpha = .025$.

   b. Test for a positive correlation between the WISCR performance score and the SIT scores. Use $\alpha = .025$.

14. Sixteen sets of twins who for various reasons were reared apart are brought together and compared. (Surprisingly enough, when this is actually done there are many behaviorial similarities). When their I.Q.'s are tested, the correlation coefficient is $r = .78$. Test for a correlation, using $\alpha = .025$.

15. In a one-sided test of H_0: $\rho = 0$ against H_1: $\rho > 0$, a sample $r = .42$ was obtained. For what sample sizes is this not significant? (Use $\alpha = .05$.)

11.6   THE LEAST SQUARES LINE

We now turn to the problem of finding the line that best represents the linear trend of a set of data. Important to the development is the assumption that the data have been obtained under controlled conditions in which particular values of one of the variables, X, are specified and the corresponding values of the second variable, Y, are obtained as observations. We then refer to X as the *independent* or *predictor variable* and Y as the *dependent* or *criterion variable*. For example, in testing a medication we might administer a specific amount x of the medication and observe the duration y of the effect. It is natural to think of y as being determined by the value x of the predictor variable.

In the scatter diagram of Figure 11.9 several lines are drawn, each of which

FIGURE 11.9 WHICH LINE BEST REPRESENTS THE LINEAR TREND OF THE DATA?

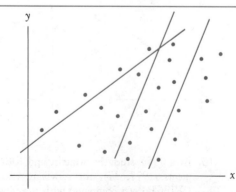

appears to represent to some degree the linear tendency of the data. The question is, which line does this best? More generally, of all possible lines we might draw, how do we determine the one that best fits the data? To obtain a unique answer which is not dependent on what is most appealing to the eye of the beholder, we follow the *least squares* approach.

To illustrate the least squares approach, we begin with a simple scatter diagram having only five points (Figure 11.10). We assume that the equation of the best-fitting line is written in slope intercept form, $y = b_0 + b_1x$. Now, for each specified value x_i of the predictor variable, there is a corresponding predicted value y_i' of the criterion variable where

$$y_i' = b_0 + b_1x_i$$

Since each x value is specified, the only reason an observed point (x_i, y_i) does not fall on this line is due to an error or difference e_i, sometimes called *noise*, in the corresponding value of the criterion variable. These differences are computed as follows:

$$e_1 = y_1 - y_1' = y_1 - (b_0 + b_1x_1)$$
$$e_2 = y_2 - y_2' = y_2 - (b_0 + b_1x_2)$$
$$\cdots \cdots \cdots \cdots \cdots \cdots \cdots$$
$$e_5 = y_5 - y_5' = y_5 - (b_0 + b_1x_5)$$

The five errors are shown in Figure 11.10; note that e_1 and e_3 are positive, e_2 and e_5 are negative, and e_4 is 0, since the point (x_4, y_4) is on the line. It would be fruitless to try to use the sum of the errors as a measure of how closely the points cluster about the line, since the positive and negative errors would cancel each other. Instead, as with the variance, we consider their squares and look for the line that minimizes the sum of these squares. Herein lies the origin of the name "the least squares line."

FIGURE 11.10 SAMPLE POINTS AND THE LEAST SQUARES REGRESSION LINE

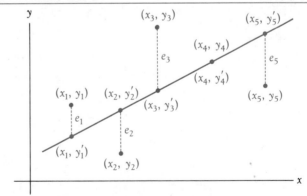

A way to think of this concept is as follows. For each line we might draw through the given set of data, we can compute a different sum of squares E. Some

---

DEFINITION 11.4

The *least squares line* for a set of data (x_1, y_1), (x_2, y_2), . . ., (x_n, y_n) is that line $y = b_0 + b_1 x$ which minimizes the sum of the squares of the errors

$$E = e_1^2 + e_2^2 + \ldots + e_n^2$$

where $e_i = y_i - (b_0 + b_1 x_i)$.

---

of these values of E will be large, others small. The line for which E is the smallest is the least squares line. It is also called the *least squares regression line* or simply the *regression line* for the sample.

When the mathematics is performed to obtain the minimal sum E, we find that the coefficients b_0 and b_1 are the solutions of the following system of equations:

$$nb_0 + \left(\sum_{i=1}^{n} x_i \right) b_1 = \sum_{i=1}^{n} y_i$$

$$\left(\sum_{i=1}^{n} x_i \right) b_0 + \left(\sum_{i=1}^{n} x_i^2 \right) b_1 = \sum_{i=1}^{n} x_i y_i$$

These are called the *normal equations*.

Example 11.6

Find the least squares line for the data in Table 11.3. We have extended the table to obtain the sums $\sum x_1^2$ and $\sum x_i y_i$ needed for the normal equations:

$$5b_0 + 12b_1 = 21$$
$$12b_0 + 46b_1 = 69$$

TABLE 11.3

x	y	x^2	xy
0	1	0	0
1	3	1	3
2	4	4	8
4	7	16	28
5	6	25	30
12	21	46	69

If we multiply the first equation by 12 and the second by 5, we obtain the same coefficient (60) for b_0 in both equations:

$$60b_0 + 144b_1 = 252$$
$$60b_0 + 230b_1 = 345$$

Subtracting the second equation from the first yields

$$-86b_1 = -93$$

Then $b_1 = 93/86 = 1.0814$. Substituting this value back into the original first equation, we obtain

$$5b_0 + 12(1.0814) = 21$$

Solving for b_0, we obtain $b_0 = 1.6046$. Thus the equation of the regression line is

$$y = 1.6046 + 1.0814x$$

The relationship of the given points to this line is shown in Figure 11.11. ■

FIGURE 11.11   THE GRAPH OF THE LINE
$y = 1.6046 + 1.0814x$

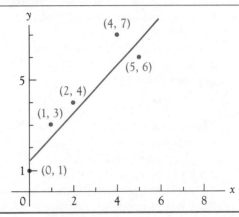

We can avoid the algebra of the last problem by solving the system of normal equations once and for all. The results are as given in Theorem 11.3.

THEOREM 11.3

The coefficients b_0 and b_1 of the least squares regression line are given by

$$b_1 = \frac{n \sum_{i=1}^{n} x_i y_i - \left(\sum_{i=1}^{n} x_i \right)\left(\sum_{i=1}^{n} y_i \right)}{n \sum_{i=1}^{n} x_i^2 - \left(\sum_{i=1}^{n} x_i \right)^2}$$

$$b_0 = \bar{y} - b_1 \bar{x}$$

Example 11.7   Using the data of the previous example, we compute b_0 and b_1 directly as

$$b_1 = \frac{5(69) - 12(21)}{5(46) - (12)^2} = 1.0814$$

$$b_0 = \frac{21}{5} - (1.0814)\frac{12}{5} = 1.6046 \quad \blacksquare$$

The reader is cautioned that these formulas are quite sensitive to rounding. To avoid problems, carry as many decimal places as practical until both coefficients have been obtained.

Before we look at another example, a few observations about the coefficients b_0 and b_1 of the least squares line are appropriate. The coefficient b_0 is the y-intercept or y-coordinate of the point where the line intersects the y-axis. The coefficient b_1 is the slope of the line, which is simply the change in y, denoted Δy, for a unit change in x, that is for $\Delta x = 1$. We can also express this by $b_1 = \Delta y/\Delta x$.

The sign of the slope is much like the sign of the covariance. If it is positive, then both x and y increase or decrease together. If the slope is negative, the behavior of y is opposite to that of x: if x increases, then y decreases, and if x decreases, then y increases.

For example, suppose the regression line relating the grade point average y of students to the number of hours x that they work per week on part time jobs has been found to be

$$y = 3.20 - .05x$$

The y-intercept is $b_0 = 3.20$, and this is the predicted grade point average for students working $x = 0$ hours per week.

The slope of this line, $b_1 = -.05$, is negative, indicating that the tendency is for y to decrease as x increases. Specifically, for each increase of 1 hour of work per week, the predicted grade point average will decrease by .05 points. Thus if there is an increase of, say, $\Delta x = 20$ in the number of hours worked, then the predicted grade point average y will change by

$$\Delta y = b_1 \cdot \Delta x = (-.05)(20) = -1.0$$

That is, it will decrease by 1 point.

Example 11.8   Eight students in a mathematics anxiety workshop are given a questionnaire on math anxiety and an inventory test on basic arithmetic skills. The scores for both are given in Table 11.4. Find the equation of the least square regression line of y on x for these data and interpret the results. Once the table has been extended as shown, the coefficients are readily found:

$$b_1 = \frac{8(21087) - (492)(379)}{8(32894) - (492)^2} = -.842754$$

$$\bar{x} = \frac{492}{8} = 61.5$$

$$\bar{y} = \frac{379}{8} = 47.375$$

$$b_0 = 47.375 - (-.842754)(61.5) = 99.204$$

TABLE 11.4 ANXIETY AND ARITHMETIC SKILL SCORES

Student	x (Skills)	y (Anxiety)	x^2	y^2	xy
1	70	35	4900	1225	2450
2	80	30	6400	900	2400
3	62	41	3844	1681	2542
4	90	28	8100	784	2520
5	45	60	2025	3600	2700
6	30	80	900	6400	2400
7	65	55	4225	3025	3575
8	50	50	2500	2500	2500
Sums	492	379	32,894	20,115	21,087

After we round b_0 and b_1 to $b_0 = 99.20$ and $b_1 = -.84$, we obtain the equation of the regression line, $y = 99.20 - .84x$. The y-intercept is $b_0 = 99.20$. This is the predicted anxiety level for a score of $x = 0$ on the arithmetic skills test.

The slope is $b_1 = -.84$; this indicates a tendency for the anxiety scores to decrease as the skills test scores increase. That is, a decrease in mathematics anxiety is predicted with increasing arithmetic skills.

The scatter diagram of the data and the graph of the regression lines are given in Figure 11.12. The line was graphed by using the intercept (0, 99.20)

FIGURE 11.12 SCATTER DIAGRAM OF ANXIETY AND ARITHMETIC SKILLS SCORES AND THE LEAST SQUARES LINE $y = 99.20 - .84x$

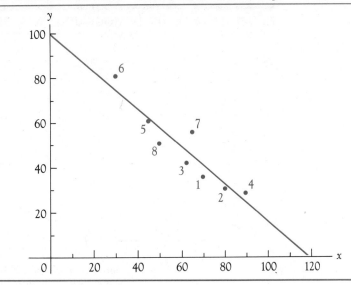

and the point (50, 47.20). The coordinate $y = 47.20$ was obtained by substituting $x = 50$ into the equation $y = 99.20 - .84x$. ∎

If into the equation of the least squares line we substitute $b_0 = \bar{y} - b_1\bar{x}$, we obtain

$$y = b_0 + b_1 x = \bar{y} - b_1\bar{x} + b_1 x$$
$$= \bar{y} + b_1(x - \bar{x})$$

This equation, $y = \bar{y} + b_1(x - \bar{x})$, is called the *point slope form* of the equation of the least squares line. From this we readily find that if $x = \bar{x}$, then $y = \bar{y}$. In other words, the least squares line contains the point whose coordinates are the means of the x and y values composing the data.

Example 11.9

In the anxiety–arithmetic skills study of Example 11.8, we found $b_1 = .84$, $\bar{x} = 61.5$, and $\bar{y} = 47.4$ (rounded). The point slope form of the equation of the least squares line is thus

$$y = 47.4 - .84(x - 61.5)$$

Clearly this line contains the point (61.5, 47.4). ∎

While the use of regression with experimental studies has been emphasized, it is by no means restricted to this application. It is quite common to find regression being used with the purely random observations that characterize correlational studies. When this is done, care should be exercised to eliminate as much error as possible in gathering the data.

Example 11.10
A Case Study

As part of a study by University of Wisconsin wildlife ecologists, an index was developed to predict the density of rabbit populations in different areas, using

FIGURE 11.13 LYNX DEMOGRAPHY IN ALBERTA

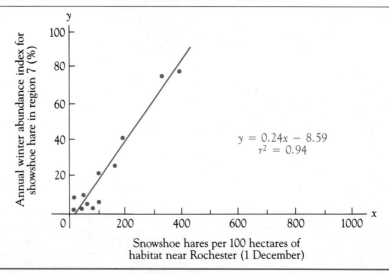

Annual winter abundance index for showshoe hare in region 7 (%)

$y = 0.24x - 8.59$
$r^2 = 0.94$

Snowshoe hares per 100 hectares of habitat near Rochester (1 December)

reports from trappers and hunters. This index (0—poor to 100—abundant) was then tested against actual count data gained from intensive studies over several winters of the snowshoe hare population in a test area near Rochester, Alberta. The results are shown in Figure 11.13. (Source: "Lynx Demography During a Snowshoe Hare Decline in Alberta," C. J., Brand and L. B. Keith, *Journal of Wildlife Management*, vol. 43, no. 4, 1979.) ■

11.7 THE STANDARD ERROR OF ESTIMATE

Once the equation of the regression line has been found it can be used to predict values of y corresponding to specific values of x. These predicted values are denoted by y′ and are computed by

$$y' = b_0 + b_1 x$$

Thus if the regression formula relating the production cost y (in dollars) of a book to the number of pages x is $y = 8.43 + .025x$, then the predicted cost for a book with $x = 400$ pages is

$$y' = 8.43 + .025(400) = 18.43$$

For certain values of x, namely $x_1, x_2, \ldots, x_n$, we have observed y values $y_1, y_2, \ldots, y_n$, and we can also compute predicted values $y'_1, y'_2, \ldots, y'_n$. The closer each y_i is to y'_i, the closer the point (x_i, y_i) is to the regression line. An important measure of the overall fit of the line to the given data is the standard error of estimate.

DEFINITION 11.5
The *standard error of estimate* is defined by

$$S_e = \sqrt{\frac{\sum_{i=1}^{n} (y_i - y'_i)^2}{n - 2}}$$

The standard error of estimate is constructed much in the same spirit as the standard deviation. The quantity under the radical is an "average" of the squares of the deviations between the observed and predicted values. The divisor of $n - 2$ in place of n arises in connection with the need for an unbiased estimator of a certain variance needed in later work. And the square root simply gives a measure in terms of the original units of y rather than their squares.

Example 11.11

A summary of the growths y of young wheat sprouts over a seven day period at various controlled humidity levels x is given in Table 11.5. Find the equation of the regression line and compute the standard error of estimate.

TABLE 11.5 HUMIDITY LEVEL—GROWTH DATA

x (%)	y (inches)	x^2	xy	y'	$(y - y')$	$(y - y')^2$
40	1.0	1600	40.0	1.20	$-.20$	.0400
45	1.5	2025	67.5	1.66	$-.16$	.0256
50	1.8	2500	90.0	2.12	$-.32$	.1024
50	2.2	2500	110.0	2.12	$+.08$	.0064
55	2.8	3025	154.0	2.57	$+.23$	.0529
55	3.0	3025	165.0	2.57	$+.43$	.1849
60	3.2	3600	192.0	3.03	$+.17$	.0289
65	3.7	4225	240.5	3.49	$+.21$	.0441
70	3.9	4900	273.0	3.94	$-.04$	.0016
75	4.0	5625	300.0	4.40	$-.40$	.1600
565	27.1	33025	1632.0			.6468

First we complete the x^2 and xy columns of the table and use these to compute the slope and intercept of the regression line:

$$b_1 = \frac{10(1632.0) - (565)(27.1)}{10(33025) - (565)^2} = .09147$$

$$b_0 = \frac{27.1}{10} - .09147 \frac{(565)}{10} = -2.4581$$

Rounding these coefficients to two decimal places, the equation of the least squares line is $y = -2.46 + .09x$. This predicts an approximate growth of .09 inches for each 1% increase ($\triangle x = 1$) in the humidity level. Note that this is a case where the y-intercept is meaningless: $x = 0$, $y = -2.46$ defies interpretation in terms of growth.

Next we compute the predicted values of y and complete the table. For $x_1 = 40$, $y_1' = -2.4581 + .09147(40) = 1.2$ (rounded). Then $y_1 - y_1' = 1.0 - 1.2 = -.2$ and $(y_1 - y_1')^2 = .04$. The remaining entries are computed in a similar manner. From the total of the last column we obtain

$$S_e = \sqrt{\frac{.6468}{10 - 2}} = .28$$

This extremely small value indicates a good fit of the line to the data (see Figure 11.14). ∎

If we substitute $y'_i = b_0 + b_1 x_i$ into the expression for S_e and then expand the squares and collect terms, we obtain a simpler computation formula for the standard error of estimate.

THEOREM 11.4

$$S_e = \sqrt{\frac{\sum y_i^2 - b_0 \left(\sum y_i \right) - b_1 \sum x_i y_i}{n - 2}}$$

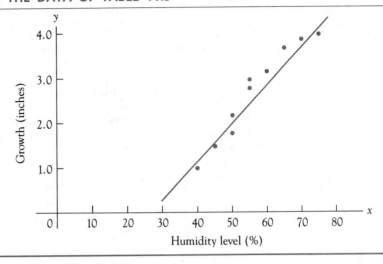

FIGURE 11.14 THE SCATTER DIAGRAM AND THE REGRESSION LINE
FOR THE DATA OF TABLE 11.5

Example 11.12

In the wheat growth example above, we found the coefficients $b_0 = -2.4581$ and $b_1 = .09147$. Also we had

$$\sum_{i=1}^{10} y_i = 27.1$$

and

$$\sum_{i=1}^{10} x_i y_i = 1632.0$$

If we now extend Table 11.4 to produce a column of y_i^2, we find $\sum y_i^2 = 83.31$. Then

$$S_e = \sqrt{\frac{83.31 - (-2.4581)(27.1) - (.09147)(1632.0)}{10 - 2}}$$

$$= \sqrt{\frac{83.31 + 66.6145 - 149.2790}{8}} = .28 \quad \blacksquare$$

**EXERCISE SET
11.2**

Part A

1. For the data in the table below,
   a. Draw the scatter diagram.
   b. Find the equation of the regression line of y on x.
   c. Draw the regression line on the scatter diagram.
   d. Compute the standard error of estimate.

x	y
0	0
1	0
2	2
1	2

2. For the data in the table below,
  a. Draw the scatter diagram.
  b. Find the equation of the least squares regression line $y = b_0 + b_1 x$.
  c. Draw the least squares line on the scatter diagram.
  d. Compute the standard error of estimate.

x	y
0	4
3	0
1	2
2	3

3. The average occupancy rate and the number of full-time equivalent employees of a random sample of six small community hospitals (60–100 beds) are given below.

Hospital	Occupancy rate (%)	Full-time equivalents
1	62	253
2	70	260
3	80	300
4	50	225
5	75	280
6	65	255

  a. Find the equation of the least squares regression line relating the number of full-time equivalent employees y to the occupancy rate x.
  b. What is the predicted number of full-time equivalent employees for a 50% occupancy rate?
  c. What is the predicted number of full-time equivalent employees for a 100 percent occupancy rate?
  d. Compute the standard error of estimate S_e.

4. A mathematics department has developed a screening test whose scores x are between 0 and 100 inclusively. Along with this they have developed a linear model $y = .7x + 28$ that predicts a student's percentage of the total points in a college algebra course.
  a. What percentage is predicted for a screening score of 100? Of 0?
  b. What change in the predicted percentage will accompany a one point change in the screening score? A 10 point change?

5. A random sample of 12 cities produced the following data for population and number of municipal employees:

	City											
	1	2	3	4	5	6	7	8	9	10	11	12
Population x (thousands)	102	90	140	99	51	120	110	67	61	68	54	170
Employees y	1290	780	1370	925	495	1040	1530	610	560	590	360	2200

    a. Construct a scatter diagram.

    b. Explain what the scatter diagram shows about the relationship between the number of employees and the population of the city.

    c. Compute the regression equation for the above data and graph the line on the scatter diagram.

    d. Compute the standard error of estimate for the regression equation.

    e. Estimate the number of employees in the government of a city whose population is 110,000.

6. As part of meterological study, trained observers were asked to estimate the speed y of moving storm systems, which were compared with the speed x obtained by radar trackings. The results are given below. (Speeds are in miles per hour.)

x	8	9	10	10	12	18	20	30	36	37
y	11	12	15	18	16	15	21	28	35	33

    a. Construct the scatter diagram and draw what appears to be a good fitting line.

    b. Find the equation of the regression line and graph it on the scatter diagram.

Part B

7. A mail-order firm is interested in estimating the number of orders that need to be processed on a given day from the weight of the mail received. A close monitoring of the mail on eight randomly selected business days produced the results below. Find the equation of the least squares regression line relating the number of orders to the weight of the mail.

Mail x (pounds)	Orders y
40	140
55	205
20	65
28	104
45	150
15	30
70	305
60	265

8. Using $(\bar{x}, \bar{y})$ as the point, find the point slope form of the least squares line in problem 3.

9. A study by a large discount chain of the relationship between inventory shrinkage (loss due to theft, shoplifting, etc.) and the number of clerks on duty produced the following data.

Number of clerks x (daily average)	45	40	35	30	25
Weekly inventory shrinkage y (hundreds of dollars)	12	13	16	18	22

    a. Find the regression line of y on x.
     b. What shrinkage is predicted for $x = 38$ clerks?
     c. Find the standard error of estimate.
10. The college entrance examination scores of five students and their grade point averages after the completion of their freshman year are given below.

Score x	Grade point average y
1050	3.1
910	2.6
1180	3.8
840	2.0
1240	3.7

    a. Find the equation of the least squares regression line.
     b. What is the increase in the predicted grade point average for each 100 point increase in the entrance examination score?
11. For a given set of data (x_1, y_1), (x_2, y_2), . . ., (x_n, y_n) the regression equation of x on y is of the form $x = b_0' + b_1'y$. The formulas for these coefficients are obtained from those for b_0 and b_1 by interchanging x's and y's throughout. Find the regression equation of x on y for the data of problem 9. Draw this line on the scatter diagrams along with the regression line of y on x obtained in number 9.
12. Suppose the data for each of the following studies indicate a linear trend. Would the slope of the least square lines be positive or negative?
    a. The number of items produced (x) and the production cost per item (y).
    b. Annual rainfall in Iowa (x) and the yield per bushel of the state's corn crop (y).
    c. The price of gold (x) and the price of oil (y).
    d. The advertising budget (x) of a company and its sales volume (y).
13. A line has slope $b_1 = 4.25$. Find the change Δy in y for the given change in x.
    a. $\Delta x = 1$.
    b. $\Delta x = 3$.
    c. $\Delta x = -4$.
14. Data on the effect of the hole size (in square millimeters) of salt shakers and the amount of salt (in grams) used on meals was obtained for 586 adults patronizing the cafeteria of a large department store in Sydney, Australia. The resulting regression line $y = .17 + .07x$ is shown in the figure below. Test results from other sites are also plotted to show the utility of this model for predicting average salt usage.
    a. At the restaurants (C and D) the holes of the salt shakers had an area of 8 mm^2. What is the average weight of salt usage predicted for this size of hole?

b. The authors feel that a hole size of 3 mm^2 is optimal in reducing the salt intake of adults. (Apparently, when the holes are smaller, the salt users attempt to enlarge them with their forks, etc.) What average usage is predicted for this optimal size? (Source: "Salting of Food," H. Greenfield, J. Maples, and R. Wills, *Nature*, vol. 301, Jan. 27, 1983.)

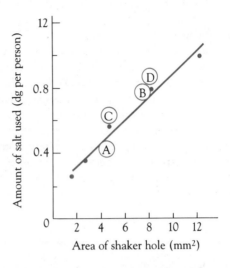

Library

Project

The following articles contain applications of simple correlation and regression. Locate one of them and report on it.

a. "Social Epidemiology of Overweight;" C. Ross and J. Mirowsky, *Journal of Health and Social Behavior*, vol. 24, no. 1, Jan. 1983.

b. "The Effect of Food Concentration of Swimming Patterns, Feeding Behavior, Ingestion, Assimilation, and Respiration by Daphnia;" K. Porter et al., *Limnology and Oceanography*, vol. 27, no. 5, Sept. 1982.

c. "Attitudes Toward Authority and Authoritarian Personality Characteristics;" K. Rigby and E. Rump, *Journal of Social Psychology*, vol. 116, 1982.

d. "Car Dive and Collision Cost;" D. Vaillancourt and N. Pulling, *Accident Analysis and Prevention*, vol. 15, no. 3, June 1983.

e. "Time Use and Activities in Junior High Classes;" J. Sanford and C. Evertson, *Journal of Educational Research*, vol. 73, no. 3, Jan. 1983.

f. "Self-Esteem as a Function of Masculinity in Both Sexes;" J. Antill and J. Cunningham, *Journal of Consulting and Clinical Psychology*, vol. 47, no. 4, 1979.

g. "Lactate Dehydrogenase Isoenzyme (A Rapid Assay of Myocardial Disease);" M. McCoy, *Archives of Internal Medicine*, vol. 143, March 1982.

h. "Hope Index Scale (An Instrument for the Objective Assessment of Hope);" A. Obayuwana, *Journal of the National Medical Association*, vol. 74, no. 8, Aug. 1982.

i. "Duplication of Diet Study on Mercury Intake by Fish Consumers in the United Kingdom;" J. Sherlock, et al., *Archives of Environmental Health*, vol. 37, no. 5, Sept.-Oct. 1982.

j. "Vitamin A, D, and E Status in a Finnish Population: A Multivitamin Study;" M. Parvianen and T. Koskinen, *Clinical Nutrition*, vol. 37c, no. 6, Dec. 1983.

11.8   REGRESSION ANALYSIS

We now wish to examine some inferences that can be made about the variable Y by using the predicted y' values from the least squares regression equation. To do so, certain assumptions about the distribution of Y are necessary. To see how these arise, consider again the wheat growth–humidity level study discussed above. For a particular humidity level x there were only one or two observations. Suppose, however, that a flat of 5000 sprigs had been maintained at a humidity level x. Then 5000 observed growths would have resulted; in all likelihood these would have a distribution strongly resembling the normal.

From our experience we know the normality condition is often met in many practical situations. Thus the first assumption we make is that the distribution of Y for each specific x is normal.

For each humidity level x at which the wheat sprigs are maintained, we would expect to obtain a different normal distribution. But there should be some similarity among them. It is not unreasonable to assume that when the humidity level is changed, the distribution of the growths is simply shifted by some amount

FIGURE 11.15   THE DISTRIBUTION OF Y FOR $x = 50$, $x = 60$, AND $x = 70$ SUPERIMPOSED ON THE REGRESSION LINE OF THE MEAN $\mu_{Y|x}$

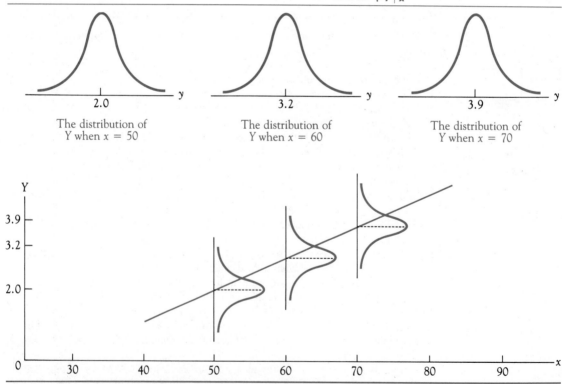

and that the variance is unchanged. Thus the second assumption is that these normal distributions have a common variance σ^2 and differ only in their means. We adopt the notation $\mu_{Y|x}$ for the mean of the distribution of Y for the specified x.

Finally, when we specify a humidity level x and compute the predicted growth $y' = b_0 + b_1 x$, we interpret this result as a sort of average or expected growth for this humidity level. Thus y' should be a point estimate of the mean $\mu_{Y|x}$. Since these estimates fall on a straight line, there may be situations where the same is true of the means $\mu_{Y|x}$. Theoretical considerations confirm this conjecture.

Thus the third and final assumption is that the means $\mu_{Y|x}$ of the different normal distributions fall on a straight line. We use the Greek leters β_0 and β_1 for the intercept and slope of this theoretical regression line, which has the equation

$$\mu_{Y|x} = \beta_0 + \beta_1 x$$

The importance of this formula is that it gives the mean of the distribution of Y for each x.

In Figure 11.15 we show the distributions of Y for three different humidity levels superimposed on the regression line of their means.

As we have seen, the least squares line contains the point $(\bar{x}, \bar{y})$ and can be written in the form $y' = b_1(x - \bar{x}) + \bar{y}$. From this it follows that the slope is critical to inferences based on the predicted value y'. By using the assumptions discussed above, it is possible to derive the distribution of the variable B_1 whose values are the slopes of the least square lines fitted to samples of size n using a specific set of values of x.

THEOREM 11.5

Suppose the distribution of Y for each x is normal and that these distributions have a common variance σ^2 and means given by $\mu_{Y|x} = \beta_0 + \beta_1 x$. Then the slope B_1 associated with the least squares regression lines from samples of size n has a normal distribution with mean $\mu_{B_1} = \beta_1$ and standard error

$$S.E. = \frac{\sigma}{\sqrt{n-1}\, s_X} = \frac{\sigma}{\sqrt{\dfrac{n \displaystyle\sum_{i=1}^{n} x_i^2 - \left(\sum x_i\right)^2}{n}}}$$

Since the standard derivation σ is not known, it must be approximated by the standard error of estimate S_e. For small samples this leads to the use of the t rather than then the normal distribution. Also, the number of degrees of freedom is $v = n - 2$ rather then $n - 1$.

THEOREM 11.6

Under the assumptions of Theorem 11.5

$$t = \frac{b_1 - \beta_1}{\dfrac{S_e}{\sqrt{\dfrac{n \sum x_i^2 - \left(\sum x_i\right)^2}{n}}}}$$

is the value of a variable having the t distribution with $\nu = n - 2$ degrees of freedom.

As usual with large samples, we may replace t by z and use the standard normal distribution.

Finally, the usual algebra leads to the form of a confidence interval, whose construction is described below.

PROCEDURE FOR CONSTRUCTING A $(1 - \alpha) \times 100\%$ CONFIDENCE INTERVAL FOR THE SLOPE β_1 OF THE THEORETICAL REGRESSION LINE

Given: n pairs of observations (x_1, y_1), (x_2, y_2), . . ., (x_n, y_n).

Assumptions: The distribution of Y for each x is normal. These distributions have a common variance and their means satisfy an equation of the form:

$$\mu_{Y|x} = \beta_0 + \beta_1 x$$

1. Find $t_{\alpha/2}$ using $\nu = n - 2$ degrees of freedom.
2. Compute the standard error using

$$S.E. = \frac{S_e}{\sqrt{\dfrac{n \sum_{i=1}^{n} x_i^2 - \left(\sum_{i=1}^{n} x_i\right)^2}{n}}}$$

3. Compute the slope b_1 of the regression line.
4. Compute the upper and lower limits of the confidence interval:

$$b_1 \pm t_{\alpha/2}(S.E.)$$

Example 11.13   Using the data of Example 11.11, find a 95% confidence interval for the slope β_1 of the theoretical regression line relating the mean growth of the wheat sprouts to the humidity level. From the previous computations we have

$$n = 10$$

$$\sum_{i=1}^{10} x_i^2 = 33025$$

$$\sum_{i=1}^{10} x_i = 565$$

$$b_1 = .091$$

$$S_e = .28$$

With $\nu = 10 - 2 = 8$, we find $t_{.025} = 2.31$.
The standard error for β is next found by

$$S.E. = \frac{S_e}{\sqrt{\dfrac{n\sum_{i=1}^{n} x_i^2 - \left(\sum_{i=1}^{n} x_i\right)^2}{n}}} = \frac{.28}{\sqrt{\dfrac{10(33025) - (565)^2}{10}}} = .0084$$

The limits of the confidence interval are $.091 \pm 2.31(.0084) = .091 \pm .019$. Thus

$$.072 < \beta_1 < .110$$

A moment's consideration will convince the reader that it is the mean $\mu_{Y|x}$ that is most valuable to us in most cases. As would be expected, the best point estimate of this is

$$y' = b_0 + b_1 x$$

Further, using this equation and the confidence interval for the slope, we can show that the upper and lower limits of a $(1 - \alpha) \times 100\%$ confidence interval for $\mu_{Y|x}$ are

$$y' \pm t_{\alpha/2} S_e \sqrt{\frac{1}{n} + \frac{n(x - \bar{x})^2}{n\sum_{i=1}^{n} x_i^2 - \left(\sum_{i=1}^{n} x_i\right)^2}} \qquad (\nu = n - 2)$$

The assumptions are exactly as with the confidence interval for the slope. We omit the formal statement of the construction procedure.

Example 11.14 A study was conducted by a city fire and rescue service of the time y needed for a rescue unit to reach a residence located x miles from the rescue station. Using their results, shown in Table 11.6, find a 95% confidence interval for the mean time $\mu_{Y|3.0}$ needed to reach a residence located 3.0 miles from the station.

TABLE 11.6 TIME AND DISTANCE STUDY DATA

x	y	x^2	y^2	xy
2.5	11.0	6.25	121.00	27.50
3.4	13.5	11.56	182.25	45.90
1.1	9.4	1.21	88.36	10.34
1.0	10.2	1.00	104.04	10.20
1.3	10.0	1.69	100.00	13.00
2.1	11.1	4.41	123.21	23.31
4.1	14.3	16.81	204.49	58.63
3.5	12.6	12.25	158.76	44.10
.7	8.6	.49	73.96	6.02
19.7	100.7	55.67	1156.07	239.00

First we find the equation of the regression line.

$$b_1 = \frac{9(239.00) - (19.7)(100.7)}{9(55.67) - (19.7)^2}$$

$$= 1.4805$$

$$\bar{x} = \frac{19.7}{9} = 2.1889$$

$$\bar{y} = \frac{100.7}{9} = 11.1889$$

$$b_0 = 11.1889 - 1.4805(2.1889) = 7.9482$$

Thus the regression line has the equation $y' = 7.95 + 1.48x$ when the coefficients are rounded to two places.

Next we compute S_e, using the alternate formula

$$S_e = \sqrt{\frac{1156.07 - 7.95(100.7) - 1.48(239.00)}{9 - 2}} = .50$$

Next, we compute the predicted value for $x = 3.0$ and obtain $y' = 7.95 + 1.48(3.0) = 12.39$. This is the point estimate (in minutes) of the mean response time to a location 3 miles from the station.

Using $v = 9 - 2 = 7$ and $t_{.025} = 2.365$, the 95% confidence interval for $\mu_{Y|3.0}$ is

$$12.39 \pm 2.365(.50) \sqrt{\frac{1}{9} + \frac{9(3.0 - 2.1889)^2}{9(55.67) - (19.7)^2}}$$

$$= 12.39 \pm .48$$

Thus

$$11.91 < \mu_{Y|3.0} < 12.87$$

The mean response time for a rescue unit to a residence 3 miles from the station is somewhere between 11.9 and 12.9 minutes. ∎

One might expect that the confidence intervals for $\mu_{Y|x}$ would produce a confidence "band" of constant width about the least squares line. A close examination of the formula

$$y' \pm t_{\alpha/2}\, S_e \sqrt{\frac{1}{n} + \frac{n(x - \bar{x})^2}{n \sum x_i^2 - (\sum x_i)^2}}$$

shows, however, that as x moves away from the mean $\bar{x}$ the term $(x - \bar{x})^2$ increases and the confidence intervals grow wider. The result looks like the set of confidence intervals in Figure 11.16.

FIGURE 11.16 95% CONFIDENCE INTERVALS FOR THE MEANS $\mu_{Y|x}$. THREE CONFIDENCE INTERVALS ARE DRAWN.

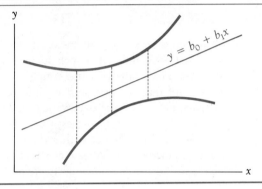

Example 11.15
A Case Study

The scatter diagram of Figure 11.17 is from the report "Air and Blood Lead Levels in a Battery Factory," (Elihu Richter, et al., Hebrew University, *Environmental Research*, no. 20, 87–98, 1979). Using the 95% confidence band, we can read off confidence limits for blood lead levels of approximately 40 and 50 micrograms per

FIGURE 11.17   RELATIONSHIP BETWEEN LEAD IN AIR (µg/m³) AND IN BLOOD (µg/100 ml) OF WORKERS IN BATTERY FACTORY ($n = 59$).

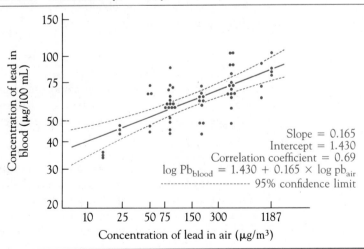

100 milliliters when the airborne concentration of lead is $x = 25$ micrograms per cubic meter. Logarithmic scales account for the uneven spacing of the units along the axes.

From the graph in Figure 11.1 it is easy to see that the closer x is to the mean $\bar{x}$ the better we can expect y' to approximate $\mu_{Y|x}$. And as x moves away from $\bar{x}$, the accuracy of the estimate should decrease. Extrapolation, the process of making estimates or predictions when x is outside the range of values used in constructing the regression line should, as a result, be avoided.

11.9   REGRESSION AND THE CORRELATION COEFFICIENT

Most regression studies are based on the assumption of a linear relation between the variables X and Y. For such studies the correlation coefficient r simply serves as a measure of how well the sample data fit the least squares line. As a matter of fact, inferences based on r are not justified in a controlled experiment, since the pairs (x_i, y_i) are not determined randomly.

We do not wish to suggest that r is not related to the least squares line. As we have seen, the sign of both r and the slope b_1 indicates whether the tendency is for y to increase or decrease as x increases. The reason that both have the same sign is shown in the following theorem.

THEOREM 11.7

The slope of the regression line is related to the correlation coefficient and the sample standard deviations by:

$$b_1 = \frac{r \cdot s_Y}{s_X}$$

Thus the slope depends on the strength of the association between x and y and the sample standard deviations of x and y.

An interesting result is obtained if we transform the data pairs (x_i, y_i) to pairs of standardized scores (z_{x_i}, z_{y_i}) by using

$$z_x = \frac{x - \bar{x}}{s_X} \qquad z_y = \frac{y - \bar{y}}{s_Y}$$

The correlation coefficient for these standardized pairs is exactly the same as for the original data. However, both sets of standardized scores have means of 0 and standard deviations of 1. Thus the least squares regression line relating z_y to z_x is

$$z_y = r z_x$$

In other words, when we transform all the data using standardized variables, the slope of the resulting regression line is simply the correlation coefficient.

One further result concerns how well the predicted y' values correlate with the observed values of y. Using the fact that y' is related to x by a linear equation $y' = b_0 + b_1 x$, we can easily show that the correlation between the observed and predicted values is exactly the same as that of the original data. From this it follows that no other line can predict values that will correlate better.

11.10   THE COEFFICIENT OF DETERMINATION

While r is of limited importance in regression studies, the same cannot be said of its square.

DEFINITION 11.6

The *coefficient of determination* is the square, r^2, of the correlation coefficient.

This coefficient is important in singling out what proportion of the observed variation in the criterion variable is due to the regression on the predictor variable and what part is due to random errors. We sketch the development briefly.

We begin to omitting the divisor $(n - 1)$ in the denominator of the variance s_y^2 and obtain the sum of the squares of the individual deviations. This is called the total variation.

DEFINITION 11.7

The *total variation* is the sum

$$\sum_{i=1}^{n} (y_i - \bar{y})^2$$

Now, there are two factors that can contribute to this variation. The first is the variation due to or explained by the regression of y on x. This involves the predicted y_i' values and their mean $\bar{y}'$. However, it can be shown that this mean is that of the original y values.

DEFINITION 11.8

The *explained variation* is the sum

$$\sum_{i=1}^{n} (y_i' - \bar{y})^2$$

The second contributor to the total variation is the sum of the random errors $\sum e_i^2$ where $e_i = y_i - y_i'$.

DEFINITION 11.9

The *error variation* is

$$\sum_{i=1}^{n} (y_i - y_i')^2$$

Now, with considerable algebra one can show the not-too-surprising result that the error and explained variations add to give the total variation.

THEOREM 11.8

$$\sum_{i=1}^{n} (y_i - \bar{y})^2 = \sum_{i=1}^{n} (y_i - y_i')^2 + \sum_{i=1}^{n} (y_i' - \bar{y})^2$$

This theorem is perhaps best remembered as follows:

$$\text{total variation} = \text{error variation} + \text{explained variation}$$

If we divide this equation through by the total variation we obtain

$$1 = \frac{\text{error variation}}{\text{total variation}} + \frac{\text{explained variation}}{\text{total variation}}$$

The second term on the right is that proportion of the total variation explained by the regression equation relating y to x. We omit the details of showing that it is equal to the coefficient of determination.

THEOREM 11.9

$$r^2 = \frac{\sum_{i=1}^{n} (y_i' - \bar{y})^2}{\sum_{i=1}^{n} (y_i - \bar{y})^2} = \frac{\text{explained variation}}{\text{total variation}}$$

And of course $1 - r^2$ is that proportion due to the random errors.

Example 11.16   In an insurance company study relating the scores y by students on a written driver's training test after x hours of class instruction, the correlation coefficient was $r = .83$. Since $r^2 = .69$, about 69% of the variation in the test scores is explained by the linear regression that exists with the number of hours of instruction. ■

When considerable variation is observed in a variable, its source is always of interest, which leads to correlation and regression studies. And although a large correlation coefficient may be obtained, its square, which explains the contribution to the total variation by the regression, will be considerably smaller. For example, if $r = .6$, then $r^2 = .36$ and only 36% of the total variation is explained by the regression. This small value would suggest the existence of other variables of possibly greater importance.

EXERCISE SET 11.3

Part A

1. Find the 95% confidence interval for the slope β_1 of the theoretical regression line, using the following information: $n = 20$, $b_1 = .83$, $S_e = 4.283$, $\Sigma x^2 = 22500$, and $\Sigma x = 660$.

2. Using the following results, find a 95% confidence interval for the slope β_1 of the theoretical regression line of y on x: $n = 5$, $\Sigma x = 15$, $\Sigma y = 45$, $\Sigma x^2 = 55$, $\Sigma y^2 = 451$, $\Sigma xy = 156$, and $S_e = 3.48$.

3. Given: $n = 25$, $y = -1.60 + .06x$, $\Sigma x = 1730$, and $\Sigma x^2 = 124,200$. Find 95% confidence intervals for $\mu_{Y|68}$.

4. A study of the relationship between the weekly sales and operating costs of a large company produced the following results:

Cost (thousands) x	Sales (ten of thousands) y	x^2	y^2	xy
8	22	64	484	176
8	20	64	400	160
8	25	64	625	200
9	25	81	625	225
10	26	100	676	260
10	22	100	484	220
10	29	100	841	290
12	30	144	900	360
14	35	196	1225	490
14	31	196	961	434
15	38	225	1444	570
15	36	225	1296	540
18	40	324	1600	720
22	41	484	1681	902
22	44	484	1936	968
22	39	484	1521	858
24	46	576	2116	1104
25	48	625	2304	1200
28	50	784	2500	1400
30	56	900	3136	1680
Σ: 324	703	6220	26,755	12,757

a. Plot the scatter diagram.
b. Find the equation of the regression line and graph it on the scatter diagram.
c. Find 95% confidence intervals for $\mu_{Y|10}$, $\mu_{Y|20}$, and $\mu_{Y|30}$ and show these on the scatter diagram.

5. For the data below, $r = .7898$ and the least squares line is $y = .267 + 1.114x$.
    a. Complete the table and find the total variation, the explained variation, and the error variation.
    b. How are the quantities in part a related?
    c. Compare the ratio of the explained variation to the total variation with the coefficient of determination.

x	y	y'	$(y - \bar{y})^2$	$(y' - \bar{y})^2$	$(y - y')^2$
1	0				
2	5				
3	4				
4	3				
5	5				
6	8				

6. The data of a regression study relating family food cost y and disposable income x had a correlation coefficient $r = .55$. What percent of the total variation in the observed food costs of the families is accounted for by the disposable family income?

7. A study of the effect on mileage obtained with different sizes of metering jets in a particular model of a compact auto produced the following data.

Size of jet x	.32	.35	.38	.40	.45
Miles per gallon y	28	30	25	24	19

    a. Find the equation of the regression line.
    b. Compute the standard error of estimate.
    c. Find a 95% confidence interval for the slope β_1 and interpret these bounds.

8. For the data in problem 7, Find a 95% confidence interval for the mean $\mu_{Y|x}$ for $x = .38$.

9. A sampling of the average prices of leaded and unleaded gasoline prices in the Los Angeles area during the period 1973 to 1982 is given below.

Regular leaded x	36.7	59.4	71.6	118.2	124.2
Unleaded y	39.1	62.1	77.1	123.6	130.0

What percent of the variation in y is explained by the regression with x?

10. Suppose a regression line has been obtained relating the suicide rate y to the unemployment rate x in the United States. Further, the sample correlation coefficient has been found to be $r = .84$. Interpret r^2 in terms of these variables.

11. The life expectancy of an infant who survives the first few weeks of birth is given below for a random selection of times over the last 40 years.

Year of birth x	1950	1960	1970	1977	1978
Life expectancy y	71.1	73.1	74.8	77.1	77.2

Find the equation of the least squares line relating the standardized variable z_y to z_x.

Part B

12. If a group of observations (x_i, y_i) fall on a straight line, what percent of the total variation in y is due to the regression with x when the slope of the line is $b_1 = \frac{1}{2}$? $b_1 = 0$?

13. From the expression relating the coefficient of determination to the explained and total variation, we find, for the explained variation,

$$\Sigma \, (y' - \bar{y})^2 = r^2 \, \Sigma \, (y - \bar{y})^2$$

Using this (and the relation between the total, explained, and error variations) show that the error variation will be very small when r is near 1 or -1—that is, nearly all the variation in y is then explained by the regression with x.

14. Using the information in problem 13, show that if r is near 0, then the explained variation is very close to 0 and the error variation is very nearly the total variation; that is, the regression of y on x is of little help in predicting more accurate values of y.

15. Theorem 11.6 provides the basis for testing simple hypotheses about the slope β_1 of the theoretical regression line $y = \beta_0 + \beta_1 x$.
   a. Using $\alpha = .05$ and the data of problem 1, test

$$H_0: \beta_1 = .75$$
$$H_1: \beta_1 > .75$$

   b. Using $\alpha = .01$ and the data of problem 3, test

$$H_0: \beta_1 = .15$$
$$H_1: \beta_1 \neq .15$$

16. In a study involving 576 military contracts, the regression equation relating the cost overrun Y (expressed as a percent of the bid price) to the bid price X (millions of dollars) was found to be $y = 300 - .25x$. Further, for the data, $r = .78$. What percent of the cost overrun variability is not explained by the correlation with bid price?

17. An auto manufacturer made a study in eight sales areas of the relation of sales Y to the area's advertising budget X ($r = .62$) and to the average family income Z for the area ($r = .84$). Which variable is the better linear predictor of sales and how much more so?

18. In a regression study involving $n = 15$ pairs of observations, the regression line was found to be $y' = 3.8 + 1.9x$. Further, $S_e = 2.4$, $\bar{x} = 10.8$ and $s_X = 1.4$. Find a 95% confidence interval for the mean $\mu_{Y|6.0}$. Hint:

$$s_x^2 = \frac{n \, \Sigma \, x^2 - (\Sigma x)^2}{n(n - 1)}$$

19. When an area is shaved on the back of laboratory rats and is painted repeatedly with a certain chlorine-bearing compound, many of the rats develop cancer. A scientist observed the following relationship between the percentage y of rats acquiring cancer and the percentage x of the recommended amount of a certain vitamin in their diet.

Vitamin content x (%)	40	60	75	90	100
Cancer y (%) rate	82	71	69	67	63

What percent of the total variation in the cancer rate can be explained by the regression with the content of the vitamin in the diet?

20. The following is a letter to the editor of the *New England Journal of Medicine* from June 14, 1970. Read it carefully. It uses medical insight coupled with correlation and regression analysis to arrive at an extremely interesting hypothesis. (References that appeared with the original letter have been omitted.)

Salt Intake and Mortality from Stroke

To the Editor: The declining incidence of stroke is a puzzling phenomenon. Garraway et al. found a decrease in Rochester, Minnesota, beginning in 1945, before any effective treatment was available. A similar decrease was observed in Baltimore, beginning in 1940. In the United States as a whole, a striking decline has been reported since 1925, but it has been attributed to classification errors. However, according to our experience over the past 25 years, improved classification in Belgium and other countries has resulted in an increase of reported stroke mortality and not in a decrease. A possible explanation for this problem is offered by a different cause of mortality, stomach cancer. The incidence of stomach cancer has also been decreasing in the United States since 1930. In Japan the incidence of both stroke and stomach cancer began to decline after 1960.

A similar pattern of stroke and stomach cancer has been observed between different countries (plotting age-adjusted death rates from strokes against those from stomach cancer in 24 countries in 1974 gives a regression line in which the slope of cerebrovascular accident = 1.87 + 6.53 vs. stomach cancer [$r = 0.81$, $p < 0.002$]); within the United States over time (as shown in Figure 1, there is a similar regression line when annual death rates from cerebrovascular accident are plotted against those from stomach cancer for 1950 to 1974); between social classes (both levels being higher in the lowest social classes); and between different regions in England and Wales and in Japan.

The association between the decreases in deaths from stroke and stomach cancer could be spurious. However, the slope and the intercept of the stomach cancer–stroke regression line are of the same order of magnitude between different countries and within the United States over time. The probability of this occurring by chance is indeed low.

The hypothesis was therefore presented that salt intake could be the linking factor between stroke and stomach cancer. Salt intake has declined in Western countries, perhaps because the introduction of refrigerators and freezers has made salt preservation of food superfluous. In Belgium the amount of sodium chloride in 24-hour urine samples, standardized to 1.77 g of creatinine, was followed in 5265 samples from 1960 on. The average decline from 15.3 g in 1966 to 8.2 g from 1978 to 1979. A relation between salt intake and stroke could be due to the possible influence of sodium chloride on blood pressure. On the other hand, a relation between salt and stomach cancer has been reported in Japan since 1954. Since antihypertensive drugs probably have no effect on stomach cancer, an improvement in antihypertensive treatment could lead to an additional decrease in stroke mortality, thus explaining the gradual departure from linearity in the stomach cancer–stroke regression line obtained during the past few years in the United States (Fig. 1).

If this hypothesis is confirmed, lowering salt intake in the population could reduce mortality from stroke and stomach cancer.

J. V. Joossens
H. Kesteloot
A. Amery
St. Rafaël University Hospital
B-3000 Leuven, Belgium

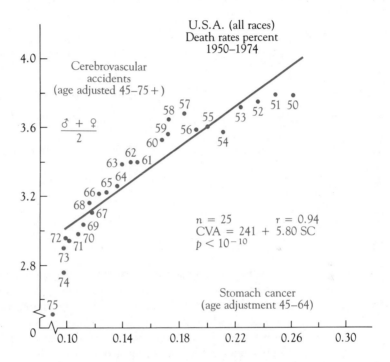

Answer the following questions:

a. What was the cerebrovascular accident (CVA) rate in 1950? In 1970?

b. Has there been an increase or decrease in the stomach cancer (SC) rate since 1950?

c. What is the significance of the positive slope of the regression line?

d. What do the authors believe is the reason for the relationship between these two rates? Why do they believe that this is not due to chance?

e. What reason is suggested for the poor fit of the data to the line from 1971 to 1975?

11.11   MULTIVARIATE CORRELATION

Often a variable Y is jointly related to several variables $X_1, X_2, \ldots, X_k$. In this section we briefly sketch some extensions of correlation to the case where more than two variables are involved.

Using double subscripts, a sample of n observations is now of the form: $(x_{11}, x_{21}, \ldots, x_{k1}, y_1)$ $(x_{12}, x_{22}, \ldots, x_{k2}, y_2), \ldots, (x_{1n}, x_{2n}, \ldots, x_{kn}, y_n)$. As usual, the first subscript identifies the variable, the second the observation.

For each pair of variables a correlation coefficient may be computed. For two predictor variables such as X_i and X_j we use the notation r_{ij}. Note that r_{ji} would be equally appropriate, since the order in which the variables are specified has no effect on the correlation coefficient. The correlation coefficient between the criterion variable Y and a predictor variable X_i is denoted both by r_{Yi} and r_{iY}.

In practice the correlation coefficients are summarized in an array called the *matrix of correlation coefficients* or simply the *correlation matrix*. The form of this matrix for the case of four variables is given in Figure 11.18.

FIGURE 11.18   CORRELATION MATRIX FOR FOUR VARIABLES

	X_1	X_2	X_3	Y
X_1	1	r_{12}	r_{13}	r_{1y}
X_2	r_{21}	1	r_{23}	r_{2y}
X_3	r_{31}	r_{32}	1	r_{3y}
Y	r_{y1}	r_{y2}	r_{y3}	1

Note that the main diagonal entries of the correlation matrix are all 1, since the correlation coefficient for any variable with itself is 1. Also, the matrix is symmetrical about the main diagonal, since the coefficients are independent of the order in which the variables are chosen. One can easily gain a sense of the relative strength of the correlation between any one variable and the others by simply scanning across the row or down the column headed by that variable.

Example 11.17

A study of marketing operations in several areas by a major retailer produced the information shown in Table 11.7 on sales volume Y, population X_1, advertising budget X_2, and number of outlets X_3.

TABLE 11.7
SALES, POPULATION, AND ADVERTISING BUDGET DATA

Sales area	x_1 (Millions)	x_2 (Thousands)	x_3	y (Thousands)
1	1.4	65	16	2160
2	3.2	80	28	6010
3	2 .5	40	19	3980
4	1.1	60	20	2900
5	6.0	150	108	8260
6	1.8	62	24	3570
7	2.0	58	25	3000

With the aid of a preprogrammed pocket calculator we obtain the correlation coefficients and construct the correlation matrix in Figure 11.19.

FIGURE 11.19
CORRELATION MATRIX

	x_1	x_2	x_3	y
x_1	1.00	.89	.93	.96
x_2	.89	1.00	.96	.85
x_3	.93	.96	1.00	.87
y	.96	.85	.87	1.00

Thus, for example, we see that $r_{13} = r_{31} = .9343$. Also, by looking across the last row or down the last column we see that Y has a higher correlation with X_1 than with any other predictor variable. ■

Suppose we temporarily limit the discussion to the case where there are just three variables: X_1, X_2, and Y. The correlation coefficient r_{1Y} does not measure the relationship that exists solely between X_1 and Y. Included is an effect due to the correlation between each of these variables and the unmentioned third variable X_2. A new correlation coefficient can be computed in which the effect of the third variable is removed. This is called a *partial correlation coefficient* and is denoted by $r_{1Y \cdot 2}$. The variable following the dot is the one whose effect is being removed. This coefficient is computed as follows:

$$r_{1Y \cdot 2} = \frac{r_{1Y} - r_{12} \cdot r_{Y2}}{\sqrt{(1 - r_{12}^2)(1 - r_{Y2}^2)}}$$

Similarly, the partial correlation coefficient between X_2 and Y when the effect of X_1 is removed is found by

$$r_{2Y \cdot 1} = \frac{r_{2Y} - r_{21} \cdot r_{Y1}}{\sqrt{(1 - r_{21}^2)(1 - r_{Y1}^2)}}$$

Finally,

$$r_{12 \cdot Y} = \frac{r_{12} - r_{1Y} \cdot r_{2Y}}{\sqrt{(1 - r_{1Y}^2)(1 - r_{2Y}^2)}}$$

Example 11.18

A correlation study of systolic blood pressure Y, age X_1, and weight X_2 of adult males produced the data in Table 11.8 and the correlation matrix in Figure 11.20. Compute the partial correlation coefficients $r_{1Y \cdot 2}$ and $r_{2Y \cdot 1}$ and interpret.

TABLE 11.8   BLOOD PRESSURE, AGE, AND WEIGHT DATA

x_1	x_2	y
20	140	120
25	150	124
30	158	130
40	170	136
50	175	142
50	180	150

FIGURE 11.20   CORRELATION MATRIX

	X_1	X_2	Y
X_1	1.0000	.9812	.9716
X_2	.9812	1.0000	.9758
Y	.9716	.9758	1.0000

$$r_{1Y \cdot 2} = \frac{.9716 - (.9812)(.9758)}{\sqrt{(1 - [.9812]^2)(1 - [.9758]^2)}} = .3352$$

Thus when the effect of the correlation with weight is removed, the correlation between age and blood pressure is .3352. Equivalently, this is the correlation between age and blood pressure when the weight is held constant.

$$r_{2Y \cdot 1} = \frac{.9758 - (.9812)(.9716)}{\sqrt{(1 - [.9812]^2)(1 - [.9716]^2)}} = .4919$$

We interpret this to mean that when age is held constant, the correlation between systolic blood pressure and weight is .4919. ■

The notion of a partial correlation coefficient may be extended to the case where several variables are simultaneously considered. For example, the notation $r_{Y1 \cdot 234}$ is the partial correlation coefficient for Y and X_1 when X_2, X_3, and X_4 are held constant or adjusted for. Such coefficients are best computed by using matrix algebra, a subject which we will not introduce here.

Finally, with regard to correlation we mention that it is also possible to derive a multiple correlation coefficient R. Among its many applications, R^2 is used in regression to estimate the proportion of the total variation in Y that is due to the regression with the predictor variables X_1, X_2, . . ., X_k. In this usage R^2 has the same role as the coefficient of determination for two variables. Again, the mathematics is beyond what we consider here.

11.12   MULTIPLE LINEAR REGRESSION

We now briefly examine the subject of linearly relating a criterion variable Y to several predictor variables $X_1, X_2, \ldots, X_k$. Recall that the linear equation for one predictor variable is $y = \beta_0 + \beta_1 x$. For k such variables the natural extension is

$$y = \beta_0 + \beta_1 x_1 + \beta_2 x_2 + \ldots + \beta_k x_k$$

Variables satisfy such an equation are said to be *linearly related.*

Variables whose values approximately satisfy a linear equation are said to have a *linear tendency.* We shall not attempt to define this notion in greater depth. We note, however, that for the case of two independent variables,

$$y = \beta_0 + \beta_1 x_1 + \beta_2 x_2$$

is the graph of a flat surface called a *plane* (see Figure 11.21). Thus a three-dimensional sample $(x_{11}, x_{21}, y_1), (x_{12}, x_{22}, y_2), \ldots, (x_{1n}, x_{2n}, y_n)$ has a linear tendency if these points fall on or near some plane.

FIGURE 11.21   A PLANE

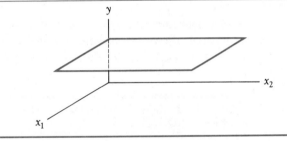

The least squares regression equation is now of the form:

$$y' = b_0 + b_1 x_1 + b_2 x_2 + \ldots + b_k x_k$$

Again this equation best fits a sample of n observations, in the sense that the sum of the squares

$$\sum_{i=1}^{n} (y_i - y_i')^2$$

is smaller if we use the predicted values from the least squares equation than if we use those from any other linear equation in these variables.

The least squares regression equation is interpreted as an approximation to the true regression equation:

$$\mu_Y = \beta_0 + \beta_1 x_1 + \beta_2 x_2 + \ldots \beta_k x_k$$

Here we use the notation μ_Y rather than $\mu_{Y|x_1 x_2 \ldots x_k}$ for the mean of the distribution of Y when the predictor variables have the specific values $x_1, x_2, \ldots, x_k$.

In the case of three variables Y, X_1, and X_2, the least squares regression equation is

$$y' = b_0 + b_1 x_1 + b_2 x_2$$

The system of normal equations that determine these coefficients is:

$$nb_0 + \left(\sum_{i=1}^{n} x_{1i}\right)b_1 + \left(\sum_{i=1}^{n} x_{2i}\right)b_2 = \sum_{i=1}^{n} y_i$$

$$\left(\sum_{i=1}^{n} x_{1i}\right)b_0 + \left(\sum_{i=1}^{n} x_{1i}^2\right)b_1 + \left(\sum_{i=1}^{n} x_{1i}x_{2i}\right)b_2 = \sum_{i=1}^{n} x_{1i}y_i$$

$$\left(\sum_{i=1}^{n} x_{2i}\right)b_0 + \left(\sum_{i=1}^{n} x_{2i}x_{1i}\right)b_1 + \left(\sum_{i=1}^{n} x_{2i}^2\right)b_2 = \sum_{i=1}^{n} x_{2i}y_i$$

A similar system of $k + 1$ equations exists for the coefficients $b_0, b_1, b_2, \ldots, b_k$ of the least squares equation

$$y' = b_0 + b_1x_1 + b_2x_2 + \ldots b_kx_k$$

The effort needed to solve these normal systems is tremendous. Additionally, they tend to be quite sensitive to round off and a great number of decimal places must be carried in the calculations. In all but the simplest of cases they are generally solved with the aid of a computer. The next example illustrates the three-variable case.

Example 11.19   A state auto club believes the cost Y of restoring an auto's emission control system to its original level of efficiency is linearly related to the mileage X_1 and age X_2 of the auto. Find the least squares equation for the data of Table 11.9.

TABLE 11.9 COMPUTATION OF THE COEFFICIENTS OF THE NORMAL EQUATIONS

x_1 (Thousands)	x_2 (Months)	y (Dollars)	x_1^2	x_2^2	x_1x_2	x_1y	x_2y
4	2	34	16	4	8	136	68
10	8	60	100	64	80	600	480
16	20	116	256	400	320	1856	2320
30	36	172	900	1296	1080	5160	6192
42	50	228	1764	2500	2100	9576	11400
46	51	240	2116	2601	2346	11040	12240
50	49	265	2500	2401	2450	13250	12985
198	216	1115	7652	9266	8384	41618	45685

The system of normal equations is:

$$7b_0 + 198b_1 + 216b_2 = 1115$$
$$198b_0 + 7652b_1 + 8384b_2 = 41618$$
$$216b_0 + 8384b_1 + 9266b_2 = 45685$$

We solve these equations by eliminating b_0 from the second and third equations using the first equation and then using the new second equation to eliminate b_1 from this new third equation. We find that

$$b_2 = 1.3191438 \qquad b_1 = 3.4509215 \qquad b_0 = 20.968925$$

Rounding to two decimal places, the least squares equation is

$$y' = 20.97 + 3.45x_1 + 1.32x_2$$

The coefficient $b_1 = 3.45$ indicates that for each increase of one thousand miles in the mileage x_1, the cost increases by \$3.45 (assuming there is no change in age). The coefficient $b_2 = 1.32$ predicts a \$1.32 increase in repair cost for each month of age, again assuming the mileage is held constant. We can interpret $b_0 = 20.97$ as a fixed cost that is added on regardless of the age or mileage. ■

In the two-variable case the standard error of estimate S_e proved to be a measure of the agreement between the predicted and observed y values. The same is true here. It is, however, necessary to modify the definition of the standard error as follows:

$$S_e = \sqrt{\frac{\sum\limits_{i=1}^{n} (y_i - y_i')^2}{n - (k + 1)}}$$

Here k is the number of independent variables.

Example 11.20   Using the cost, mileage, and age data and the least squares equation from Example 11.19, find the standard error of estimate. Table 11.10 contains the necessary information. The entries of the last column have been rounded. Then

$$S_e = \sqrt{\frac{313.05}{7 - 3}}$$
$$= 8.85 \quad ■$$

TABLE 11.10
COST, MILEAGE, AND AGE DATA

x_1	x_2	y	y'	$(y - y')^2$
4	2	34	37.41	11.63
10	8	60	56.79	10.30
16	20	116	102.57	180.36
30	36	172	171.99	00.00
42	50	228	231.87	14.98
46	51	240	246.99	48.86
50	49	265	258.15	46.92
				313.05

When a large sample is involved and we make the usual normality and common variance assumptions about the distributions of Y, then $y' \pm 2S_e$ is roughly a 95% confidence interval for the mean μ_Y. Unfortunately large samples are not that common, and so this relation is of limited usefulness. Still, an estimate of two to three standard errors often turns out to be in the right ballpark even for small samples.

Because of the different variances of the predictor variables, it is often difficult to evaluate the relative importance of their contribution to the total variation of the criterion variable Y. One way to avoid this problem is to standardize the variables. Recall that we mentioned the possibility of such standardization for the regression of Y on X, indicating that the least squares equation then turned out to be $z_y = rz_x$. And it was r^2, the square of the coefficient of z_x, which answered the question of the relative contribution to the total variation due to the regression with x. When we extend this idea to k independent variables we obtain a least squares equation of the form:

$$z_y = \beta'_1 z_{x1} + \beta'_2 z_{x2} + \ldots + \beta'_k z_{xk}$$

where

$$z_{xi} = \frac{x_i - \bar{x}_i}{s_{xi}}$$

is the standardized equivalent of x_i.

These coefficients $\beta'_1, \beta'_2, \ldots, \beta'_k$ are termed the *beta coefficients* or *beta weights* and should not be confused with the betas of the theoretical regression equation. Actually, there are theoretical betas and estimates of them, but the distinction is rarely observed.

The most direct way to find these beta weights is to fit a least squares equation to the standardized data. Obviously this involves a great deal of work. One must first standardize the data, which requires the sample mean and standard deviation of each variable. Then the normal equation be found and solved. All of this effort is best left to a computer.

Once we have these beta weights we can use them and the ratios of their squares to decide on the relative importance of the different independent variables in the total variation.

Example 11.21

In the cost-mileage-age study in Example 11.19, the least squares regression equation was found to be

$$y = 20.97 + 3.45x_1 + 1.319x_2$$

For the standardized variables z_{x1}, z_{x2}, and z_y, the least squares equation is

$$z_y = .62z_{x1} + .40z_{x2}$$

Since $\beta'_1 = .62$ is the largest in magnitude of the beta weights, z_{x1} (and thus x_1) is the most important of the independent variables. The ratio

$$\frac{(\beta'_1)^2}{(\beta'_2)^2} = \frac{(.62)^2}{(.40)^2} = 2.4$$

tells us that z_{x1} is about 2.4 times as important as z_{x2} in accounting for the total variation in the variable z_y. ■

This use of the squares of the beta weights suggest that they are related to the partial correlation coefficients. This is indeed the case, but the mathematics is too complex to pursue here.

Example 11.22

A Research Study

Can sociological explanations of suicide be adapted to predict suicide rates for nations? This is the subject of a report entitled "Suicide: A Comparative Analysis," by Steven Stack (*Social Forces*, vol. 57, no. 2, Dec. 1978). In this report the author proposes a model for predicting a nation's suicide rate S using its level of industrialization I, rate of economic growth E, and the degree F of female participation in its labor force. The author outlines in detail why these variables have been singled out. In particular, female participation is viewed as an index of social or status integration. Research indicates that the greater the female labor force participation, the less the status of integration and the greater the suicide rate. Official United Nation data from 45 nations were used in the study. The correlation matrix for the variables is reproduced in Table 11.11. From this it is seen that the female participation F correlates more highly with the suicide rate S ($r_{FS} = .5866$) than with any other variable. Table 11.12 summarizes the coefficients of the regression equation relating S to the variables F, I, and E. Thus

$$S = -6.377 + .364F + .002I + .958E$$

The beta coefficients are given in the final column of Table 11.12.

In the author's view, "the degree of female participation in the labor force is the most important determinant of suicide (beta = .399). The level of industrialization (beta = .308) is second and the rate of economic growth is third (beta = .257)." Finally, after adjusting the square of the multiple correlation coefficient R for the sample size, the author concludes that "taken together, these variables account for 43% of the variance in suicide rates."

TABLE 11.11 MATRIX OF PEARSON PRODUCT MOMENT CORRELATION COEFFICIENTS[a]

	I	S	F	E
I	1.0000	0.5042[b]	0.4922[b]	0.0004
S		1.0000	0.5866[b]	0.3130[c]
F			1.0000	0.1406
E				1.0000

[a]I = index of industrialization, GNP/capita, 1966; S = suicide rate, 1970; F = female participation in the labor force, 1960; E = the rate of economic growth, 1960–72.
[b]Statistically significant at the .005 level ($p < .005$), one-tailed test.
[c]Statistically significant at the .025 level ($p < .025$), one-tailed test.

TABLE 11.12 THE EFFECTS OF FEMALE PARTICIPATION IN THE LABOR FORCE, LEVEL OF INDUSTRIALIZATION, AND THE RATE OF ECONOMIC GROWTH ON THE SUICIDE RATE ($N = 45$)[a]

Variable	Regression coefficient	Standard error of coefficient	Computed value of t	Beta coefficient
F	0.364	0.121	3.014[b]	0.399
I	0.002	0.001	2.348[c]	0.308
E	0.958	0.430	2.227[c]	0.257

[a]Intercept $= -6.37792$, $R = .69$, $R^2 = .47$, and adjusted $R^2 = .43$.
[b,c]See footnotes to Table 11.11.

EXERCISE SET 11.4

Part A

1. A correlation study of the high school grade point average x_1, the SAT score x_2, and college grade point average y after one semester resulted in the following correlation matrix:

	x_1	x_2	y
x_1	1.00	.51	.45
x_2	.51	1.00	.97
y	.45	.97	1.00

  a. Interpret the entries of the matrix.
  b. Compute the partial correlation coefficients $r_{1Y\cdot2}$ and $r_{2Y\cdot1}$.
  c. Interpret the partial correlation coefficients found in part b.

2. Fill in the missing entries in the following correlation matrix:

	x_1	x_2	x_3	y
x_1	1.0	.3		
x_2			.5	.7
x_3	.2			
y	.4		.8	

3. For the data below:
  a. Find the normal equations for the coefficients of the regression equation $y = b_0 + b_1x_1 + b_2x_2$.
  b. Solve the system.
  c. Compute the standard error S_e of the estimate.

x_1	x_2	y
0	0	2
1	1	6
2	1	8
1	2	9
0	4	6
2	0	8

4. In an agricultural study, the following data were obtained on the height y of a plant maintained for six weeks in soil with a pH x_1, nitrogen content x_2, and humidity level x_3. Find the system of normal equations for the coefficient of the regression equation

$$y = b_0 + b_1 x_1 + b_2 x_2 + b_3 x_3$$

x_1	x_2	x_3	y
5	60	3	5
7	60	4	3
6	70	3	4
6	70	4	6
5	80	4	7

5. The equation $y = 14.20 + .02x_1 + .04x_2 + .13x_3$ is the multiple regression equation derived from a small business study relating:
   - y Computer usage (hours weekly)
   - x_1 Number of employees
   - x_2 Sales volume (thousands of dollars per week)
   - x_3 Number of inventory categories
   a. How many hours of usage is predicted for a company with 100 employees, a sales volume of \$250,000, and 450 inventory categories?
   b. If other factors are held constant, what is the change in predicted usage for an additional 50 inventory categories?
6. A regression study relating disposable family incomes X_1, interest rates X_2, and per capita increase in savings had a multiple correlation coefficient $R = .62$. What proportion of the variation in the savings increases is explained by the regression with disposable incomes and interest rates?
7. The equation $z_y = .34z_{x1} + .23z_{x2} - .47z_{x3} + .12x_{x4}$ is the regression equation in terms of standardized variables for a study involving factors that contribute to anxiety in working women. Assuming

$$y = \text{the measure of the level of anxiety}$$
$$x_1 = \text{the measure of job stress}$$
$$x_2 = \text{the measure of social stress}$$
$$x_3 = \text{the age of the woman}$$
$$x_4 = \text{the measure of job security}$$

   a. Which variable has the greatest effect on the anxiety level?
   b. Rank the variables in terms of their importance to anxiety.

Part B

8. Since

$$R^2 = \frac{\text{explained variation}}{\text{total variation}} = 1 - \frac{\text{error variation}}{\text{total variation}}$$

it follows that once the regression equation is known and the predicted y' values are known, then we can find R^2 by

$$R^2 = 1 - \frac{\Sigma (y_i - y'_i)^2}{\Sigma (y_i - \bar{y})^2}$$

Find the coefficient of multiple determination for the study of problem 3.

9. The regression equation for a set of data relating wheat yield y (bushels per acre) to fertilizer x_1 (pounds per acre) and seasonal rainfall x_2 (inches) was found to be $y = 34 + .030x_1 + .75x_2$. If we hold one of the independent variables constant, then a two-dimensional graph relating y to the other variable can be obtained. Such graphs are sometimes useful in analyzing the behavior of y.

   a. Holding x_2 fixed at $x_2 = 2$ inches, graph the resulting equation using y and x_1 axes. Repeat the process for $x_2 = 10$ and $x_2 = 20$.

   b. Holding x_1 fixed at $x_1 = 100$ pounds, graph the resulting equation using a y-x_2 axis system. Repeat the process for $x_1 = 400$ and $x_1 = 600$.

   c. What is the role of b_1 when x_2 is held constant? What is the role of b_2 when x_1 is held constant?

10. A study of the relation between attitudes toward a nuclear power plant in a small community and the likelihood of various outcomes as perceived by the residents produced the following table of regression coefficients between these attitudes and outcomes:

MULTIPLE REGRESSION ANALYSIS OF GENERAL ATTITUDES TOWARD THE NUCLEAR POWER PLANT

Predictor Variable	Simple r	Multiple r	Multiple r^2	Change in multiple r^2	F^a
Hazards	$-.59^c$	.59	.35	.35	49.75^c
Economic growth	$.51^c$	.68	.46	.11	54.99^c
Social disruption	$-.46^c$	.71	.50	.04	17.03^c
Lower costs	$.43^c$	.72	.52	.02	10.90^c
Community visibility[b]	.12	(.72)	(.52)	(0)	

[a]$df = 1316$ for a total of 321 cases.
[b]Not in the equation.
[c]$p < .001$.
[d]$p < .01$.

The first column contains the simple correlation coefficients between the general attitude scores and the subject's view that the plant would likely produce hazards, increase economic growth, and so on. The correlation of community visibility with attitude was found not to be significant and was dropped. The second column gives the multiple correlation coefficients as the variables are added in sequence to the model. Thus the .68 entry indicates that the multiple correlation between attitude and the first two variables is $R = .68$. When the first three variables are considered, R increases to .71, and so on. The next column is simply the squares of the multiple correlation coefficients and the final column are the changes in these coefficients of multiple determination as the additional variables are added to the model. This is an example of a *stepwise* selection process for adding new variables to a model.

The authors also gave the regression equation as follows:

$$\text{attitude} = 2.06 - .50(\text{hazards}) + .70(\text{economic growth})$$
$$- 36(\text{social disruption}) + .26(\text{lower costs})$$

a. About what percent of the variance in attitudes toward the proposed nuclear plant is accounted for by this regression equation?

b. What is the significance of the coefficients of the regression equation? Interpret their meaning.

c. If an individual believed there was a high likelihood of increased economic growth and lower power costs for the community and little chance of hazards or social disruption, what type of score (positive or negative) would the equation predict?

d. If an individual saw a high likelihood of hazards and social disruption and little chance of community economic gains or reduction in power costs, what kind of an attitude score is predicted (positive or negative)? (Source: "Community Attitudes Toward a Proposed Nuclear Power Generating Facility as a Function of Expected Outcomes," Eric Sundstrom et al., *Journal of Community Psychology,* 1977, 5.)

Library

Project

The following research articles make use of multivariate correlation and regression. Select one and report on it. Note that some of these papers may involve procedures yet to be discussed. If so, limit your discussion to the correlation and regression aspects of the study.

a. "A Pedagogic Application of Multiple Regression-Precipation in California," P. Taylor, *Geography,* vol. 65, part 3, no. 288, July 1980.

b. "The Public and Care by Non-Physicians: Health Policy Considerations," B. Lavin, *Journal of Sociology and Social Welfare,* vol. 7, no. 3, May 1980.

c. "The Interrelated Behavior of Exchange Rates," L. Soenen, *Business Economics,* vol. 17, no. 4, Sept. 1982.

d. "Advertising and Sales Relationship for Tooth Paste," R. Carlson, *Business Economics,* vol. 16, no. 4, Sept. 1981. (Also read the following report.)

e. "Forecasting Sales on the Basis of Advertising Budget: The Case of Crest Tooth Paste," M. Schreiber, *Business Economics,* vol. 17, no. 3, May 1982. (Read in conjunction with the preceding report.)

f. "Allocation of Time to Mass Media," J. Harnik and M. Schlinger, *Journal of Consumer Research,* vol. 7, no. 4, March 1981.

COMPUTER PRINTOUT

Regression and correlation studies are generally performed with the aid of a computer. The Minitab instructions and printout for an analysis of English and Mathematics ACT scores are given in Figure 11.22. The analysis of variance summary pertains to topics that we take up in Chapter 12.

FIGURE 11.22 MINITAB COMPUTER PRINTOUT: REGRESSION STUDY—Cont.

```
MTB >SET ACT ENGLISH SCORES INTO C1
DATA>33 32 24 25 22 20 11 10 17 22 19 25 28 34 27 18
DATA 10
DATA>END
MTB >SET ACT MATH SCORES INTO C2
DATA>32 30 25 23 21 18 15 12 20 20 20 23 28 30 19 22
DATA 12
DATA>END
MTB >PLOT C2 VS C1
  C2
35.0+
    -
    -
    -                                                        *
    -
30.0+                                                      *    *
    -
    -
    -                                              *
    -
25.0+                                      *
    -
    -                                        2
    -                          *
    -                                     *
20.0+                         *  *     *
    -
    -                                           *
    -                            *
    -
    -
15.0+          *
    -
    -
    -         2
    -
10.0+
       +---------+---------+---------+---------+---------+C1
      6.0      12.0      18.0      24.0      30.0      36.0
MTB >REGRESS C2 ON 1 PREDICTOR IN C1

THE REGRESSION EQUATION IS
Y =     5.67 + 0.726 X1

                                   ST. DEV.    T-RATIO =
      COLUMN    COEFFICIENT        OF COEF.     COEF/S.D.
        --           5.6718         1.8051          3.14
X1      C1           0.7257         0.0773          9.39

THE ST. DEV. OF Y ABOUT REGRESSION LINE IS
S =      2.332
WITH (  17- 2) = 15 DEGREES OF FREEDOM

R-SQUARED = 85.5 PERCENT
R-SQUARED = 84.5 PERCENT, ADJUSTED FOR D.F.
```

Continued.

```
FIGURE 11.22   MINITAB COMPUTER PRINTOUT: REGRESSION
STUDY—cont

ANALYSIS OF VARIANCE

    DUE TO        DF        SS      MS = SS/DF
    REGRESSION     1    479.458      479.458
    RESIDUAL      15     81.601        5.440
    TOTAL         16    561.059


            X1          Y     PRED. Y    ST.DEV.
  ROW      C1         C2      VALUE     PRED. Y    RESIDUAL ST.RES.
   15     27.0    19.000     25.265      0.678      -6.265  -2.81R


R DENOTES AN OBS. WITH A LARGE ST. RES.
```

CHAPTER CHECKLIST

Vocabulary

Linearly related
Scatter diagram
Covariance
Correlation coefficient
Predictor variable
Criterion variable
Least squares regression line
Theoretical regression line
Slope
Intercept

Normal equations
Standard error of estimate
Coefficient of determination
Error variation
Explained variation
Total variation
Correlation matrix
Partial correlation coefficient
Multiple linear regression
Beta coefficients, beta weights

Techniques

Preparing a scatter diagram and checking for a linear trend
Computing and interpreting the covariance
Computing and interpreting the sample correlation coefficient
Performing a test for linear correlation using the sample correlation coefficient
Finding the equation of the least squares line
Interpreting the slope and intercept of the least squares line
Computing and interpreting the standard error of estimate
Interpreting the predicted values of the theoretical regression equation
Computing and interpreting the coefficient of variation
Computing and interpreting the correlation matrix
Computing and interpreting partial correlation coefficients
Finding the normal equations for the coefficients of the multilinear least squares equation with two predictor variables
Interpreting the coefficients of the predictor variables of a multilinear least squares equation
Interpreting the beta coefficients (beta weights) of a multilinear equation.

Notation

σ_{XY} Theoretical covariance of X and Y
s_{XY} Sample covariance of X and Y
ρ Theoretical correlation coefficient for X and Y
r Sample correlation coefficient for X and Y
Δy A change in y
Δx A change in x
b_0 The y-intercept of the least squares line
b_1 The slope of the least squares line
y' The value of y corresponding to x as predicted by the least squares line
β_0 The intercept of the theoretical regression line
β_1 The slope of the theoretical regression line
$\mu_{Y|x}$ The mean of Y when x is specified.
S_e The standard error of estimate
r^2 The coefficient of determination
r_{ij} The sample correlation coefficient for the variables X_i and X_j
r_{iY} The sample correlation coefficient for the variables X_i and Y
z_x The standardized z score corresponding to x
β_i' The beta weight or beta coefficient of the standardized variable Z_{X_i}

CHAPTER TEST

1. Three scatter diagrams are given below. For which of the corresponding sets of data is:
   a. The covariance positive? Negative?
   b. The coefficient of correlation r positive? Negative?
   c. The slope of the best fitting least squares line positive? Negative?

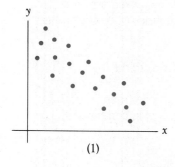

(1)

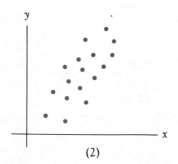

(2)

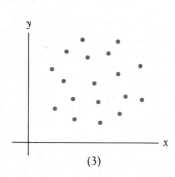

(3)

2. The daily high and low temperatures in Los Angeles on seven randomly selected spring days are given below.
   a. Draw a scatter diagram
   b. Test for a positive correlation between these temperature extremes. Use $\alpha = .025$.

Day	High temperature	Low temperature
1	66	48
2	70	49
3	66	46
4	50	34
5	43	28
6	54	39
7	82	65

3. Aptitude test scores and corresponding grade point averages of a random sample of 11 students are given below.

Aptitude score x	G.P.A. y
68	2.4
75	2.7
40	0.9
80	3.1
86	3.5
55	1.6
90	3.7
68	2.1
70	2.6
60	2.9
96	3.8

  a. Find the regression line of y on x.
  b. What grade point average is predicted for an aptitude score of 70?
  c. Find the standard error of estimate.
4. A researcher is interested in whether a correlation exists between educational achievement of elementary school children (as measured by a standardized test) and the incomes of their families. In a study involving a random sample of $n = 20$ children, the researcher found a correlation coefficient of $r = .65$ between these variables. Does this finding support the hypothesis of a positive correlation between achievement and family income? Use $\alpha = .005$.
5. In a regression study relating the daily sucrose intake x (in grams) to antisocial behavior (measured by an antisocial behavior index y) in young prison inmates, the following results were obtained: $n = 10$, $\bar{x} = 14.5$, $\bar{y} = 21.2$, $\Sigma x_i = 145$, $\Sigma x_i^2 = 2785$, and $S_e = 4.1$. The equation for the regression line was $y = 6.7 + 1.0x$. Find the 95% confidence interval for the mean antisocial index score corresponding to a daily sucrose intake of 40 grams. What assumptions are necessary?

6. Use the results of problem 5 to obtain a 95% confidence interval for the slope β of the theoretical regression line relating the antisocial index score to sucrose intake.

7. Suppose the data of problem 2 is standardized by using

$$z_x = \frac{x - \bar{x}}{s_x} \qquad z_y = \frac{y - \bar{y}}{s_y}$$

a. What is the correlation coefficient for the standardized data?
b. What is the equation of the least squares line for the standardized data?

8. A tabulation of the pulse rate increases of experimental subjects due to varying dosages of a stimulant is given below. Compute the coefficient of determination and interpret the results.

Dosage x (grains)	Pulse rate increase y (beats/minute)
10	4
20	5
30	8
40	10
50	15

9. A study of the weight Y, height X_1, and age X_2 of adult males produced the regression equation

$$y = -180 + 5x_1 - .4x_2$$

a. What weight is predicted for an individual who is 65 inches tall and 30 years of age?
b. What is the significance of the coefficient $b_1 = 5$? $b_2 = -.4$?

10. A study correlating employee job satisfaction with number of years of employment and number of promotions produced the following results:

Job satisfaction score (y)	Years of employment (x_1)	Number of promotions (x_2)
60	6	2
70	7	3
80	9	2
50	5	1
70	3	2
90	7	4
30	2	0

a. Find the matrix of correlation coefficients.
b. Compute the partial correlation coefficient $r_{1y \cdot 2}$ and interpret.
c. Compute the partial correlation coefficient $r_{2y \cdot 1}$ and interpret.

11. The equation $z_y = .10z_{x1} + .45z_{x2} + .22z_{x3}$ is the regression equation relating the standardized variable z_y to z_{x1}, z_{x2}, and z_{x3}. Compare the importance of z_{x2} to z_{x1} in accounting for the total variation in z_y.

CHAPTER TWELVE
The Analysis of Variance

$\mathbf{W}$hat factors are responsible for the variability that produces a distribution? Certainly not just random errors. In most instances, variables for which we have not controlled play a major role.

To achieve some perspective on this idea, consider a study of blood pressure in men. If we simply select men at random there will be considerable variability in the observed pressures. If we were to limit the study to men between the ages of 35 and 40, the variability would be reduced. If we controlled for weight by considering only those men who are at most 5% overweight, there would be a further reduction. An additional reduction would result if we controlled for occupation in terms of stress. Many other factors no doubt come to mind. It is clear that the identification of such factors is of great importance; such is the impetus for many statistical investigations.

The *analysis of variance*, commonly referred to as ANOVA, is a statistical method for detecting factors that produce variability in responses or observations. The approach is to control for a factor by specifying different values or treatment levels for it to see if there is an effect. Each time this is done we can think of having sampled a potentially different population (different in the sense of having different means). Thus from a statistical point of view, we can test the hypothesis of no effect by testing for equality of several population means. As should be apparent, factors that have an effect produce a change in the variation of the sample means; this change is what we hope to detect. Our approach is to compare the means of sums of squares that are in fact estimators of a common population variance. Hence the name, the analysis of variance.

The analysis of variance was originated by Sir Ronald Fischer (1890–1962) and found early application in agricultural research. Today its use is commonplace throughout all disciplines that use statistical research.

12.1   FACTORS AND TREATMENTS

As the reader may have inferred from the above discussion, factors and variables are closely related.

DEFINITION 12.1

A *factor* of an experiment is a controlled independent variable.

Factors, as used in the analysis of variance, are generally categories and classifications, such as teaching methods, surgical procedures, or age groups. These factors may affect such variables as grades, survival rates, and blood pressure. These are called response variables; their values are called responses.

Note that the idea of a factor does not exclude a quantitative variable such as the level of fluoridation in drinking water. In most cases, however, it will be better to think of fluoridation as a factor and to consider the various fluoridation levels as treatment levels or simply treatments.

DEFINITION 12.2

Specific controlled values of a factor are called *treatment levels*.

Treatment levels need not be identified by numerical values. In fact, they are often subcategories or subclasses that taken together determine the factor.

Example 12.1

In Table 12.1 several factors and possible treatment levels are given that might be considered in studying the learning process in the elementary school classroom.

As a response variable one might use achievement levels on some standardized test or possibly the mean rating of a team of independent observers. ∎

TABLE 12.1 FACTORS AND TREATMENT LEVELS IN THE LEARNING PROCESS

Factor	Treatment levels
Textbook	different textbook series
Teaching method	different teaching strategies
Class size	different class sizes
Teacher preparation	types of degrees
Classroom assistants	number of assistants
Classroom decor	colors of walls

As should be obvious, its ability to deal with nonnumeric factors makes the analysis of variance a powerful tool.

12.2 ONE-WAY ANALYSIS OF VARIANCE

The simplest form of the analysis of variance occurs when there is but one factor and the question is whether the treatment levels produce different responses. If so, a *treatment effect* is said to exist. For example, is the textbook a factor of importance in the learning process? Or in the case of surgery for breast cancer, is there any difference in the long-term survival rate of those women having a radical mastectomy and those having only a partial one?

First let us look at our assumptions. We shall assume that each factor has only a fixed number of treatment levels and that each is tested. To each treatment corresponds a population. Thus if there are k treatments, there are k populations; these we assume to be normally distributed with means $\mu_1, \mu_2, \ldots, \mu_k$. Further, we assume that any treatment effect is reflected in the means and that these treatments do not alter the occurrence of uncontrolled factors that produce the random errors. Thus the final assumption is that these populations have a common variance σ^2.

To test the hypothesis of no treatment effect, we test the equality of the means $\mu_1, \mu_2, \ldots, \mu_k$, using random samples from the k populations. For the case where the populations are existing entities, such as the grades of business majors at four universities, these samples are obtained by selecting the requisite number of subjects from each population and making the required observation. This is simply an extension of the process used with the two-independent-sample t test of Chapter 10.

In the case where specific treatments of, say, a medical nature are to be compared, the populations exist only hypothetically. The samples are then obtained by selecting a pool of experimental subjects and then randomly assigning each subject to receive one of the treatments. Those receiving the same treatment form a random sample which in ANOVA terminology is called a *treatment group*. These k treatment groups or the observations on them are the required independent samples. An experiment in which the treatment groups are obtained in this fashion is said to have a *completely randomized design*.

The k samples or treatment groups need not all be of the same size. While it is true that we generally start with equal-sized samples, subjects often drop out of the study and so we allow for this. These sample sizes we denote by n_1, n_2, . . ., n_k. The total number of observations in all the samples is denoted by n.

The double subscript notation first used with the observations of two samples readily extends to k samples:

$$\text{Sample 1: } x_{11}, x_{12}, \ldots, x_{1n_1}$$
$$\text{Sample 2: } x_{21}, x_{22}, \ldots, x_{2n_2}$$
$$\text{Sample 3: } x_{31}, x_{32}, \ldots, x_{3n_3}$$
$$\text{Sample } k: x_{k1}, x_{k2}, \ldots, x_{kn_k}$$

As usual, the first subscript locates the sample, the second the observation within that sample.

In attempting to analyze the sample data for evidence of a treatment effect, a logical quantity to consider is the sum of the squares of the deviations about the grand mean $\bar{x}$ of all the observations. The reader should recall that this sum is used in computing an estimate of the population variance.

DEFINITION 12.3

The *total sum of squares* SS is the sum

$$SS = \sum_i \sum_j (x_{ij} - \bar{x})^2$$

Here the double-sigma notation indicates a summation over all i and j; thus all observations are to be used.

It is clear that the total sum of squares SS should be large if there are in fact different population means because of a treatment effect. For then the observations of each sample, while continuing to cluster about their respective sample means, will not cluster closely about the grand mean.

When we attempt to identify precisely how the total sum of squares grows as the treatment effect becomes more pronounced, we discover that it splits into the sum of two components SSTr and SSE, the *treatment sum of squares* and the *error sum of squares*.

THEOREM 12.1

$$SS = SSTr + SSE$$

where

$$SSTr = \sum_i n_i(\bar{x}_i - \bar{x})^2$$

and

$$SSE = \sum_i \sum_j (x_{ij} - \bar{x}_i)^2$$

Here $\bar{x}_i$ denotes the mean of the ith sample.

Before continuing the development, we consider an example illustrating the notation we have just introduced.

Example 12.2

A company is interested in whether the day of the week is a significant factor in employee absenteeism. To keep matters simple, suppose only three workdays, Monday, Wednesday, and Friday are studied; thus there are three treatment levels. Suppose also that they have only the records of seven days, two Mondays, two Wednesdays, and three Fridays, for which other factors such as weather and proximity to payday are essentially the same. These three samples are detailed in Table 12.2.

TABLE 12.2

Sample 1 Monday	Sample 2 Wednesday	Sample 3 Friday
6	3	7
8	1	8
		9

Thus there are $n = 7$ observations and the sample sizes are $n_1 = 2$, $n_2 = 2$ and $n_3 = 3$. The samples are denoted by:

Sample 1: $x_{11} = 6$, $x_{12} = 8$
Sample 2: $x_{21} = 3$, $x_{22} = 1$
Sample 3: $x_{31} = 7$, $x_{32} = 8$, $x_{33} = 9$

The grand mean is

$$\bar{x} = \frac{(6 + 8) + (3 + 1) + (7 + 8 + 9)}{7} = \frac{42}{7} = 6$$

We use this in computing the total sum of squares:

$$SS = \sum_i \sum_j (x_{ij} - \bar{x})^2 = (6 - 6)^2 + (8 - 6)^2 + (3 - 6)^2 +$$
$$(1 - 6)^2 + (7 - 6)^2 + (8 - 6)^2 + (9 - 6)^2 = 52$$

To compute the treatment and error sums we need the individual sample means.

$$\bar{x}_1 = \frac{6 + 8}{2} = 7$$

$$\bar{x}_2 = \frac{3 + 1}{2} = 2$$

$$\bar{x}_3 = \frac{7 + 8 + 9}{3} = 8$$

Then

$$SSTr = \sum n_i(\bar{x}_i - \bar{x})^2 = 2(7 - 6)^2 + 2(2 - 6)^2 + 3(8 - 6)^2 = 46$$

$$SSE = \sum_i \sum_j (x_{ij} - \bar{x}_i)^2 = (6 - 7)^2 + (8 - 7)^2 + (3 - 2)^2 + (1 - 2)^2$$
$$+ (7 - 8)^2 + (8 - 8)^2 + (9 - 8)^2 = 6$$

As predicted,

$$SSTr + SSE = 46 + 6 = 52 = SS \quad \blacksquare$$

Now let us look more closely at these sums. Consider first the error sum SSE. Each term $(x_{ij} - \bar{x}_i)^2$ involves the difference between an observation and the mean of the sample to which it belongs. Thus the common treatment effect for the sample is subtracted out; as a consequence, the error sum does not depend on a possible treatment effect. The difference $(x_{ij} - \bar{x}_i)$ are due to random errors produced by uncontrolled factors. Hence the name "error sum."

Recalling that each

$$s_i^2 = \frac{\displaystyle\sum_{j=1}^{n_i-1} (x_{ij} - \bar{x}_i)^2}{n_i - 1} \qquad i = 1, 2, \ldots, k$$

is a point estimate of the population variance σ^2, it follows that we should be able to pool these to obtain an even better estimate of σ^2. Proceeding as in Chapter 10 with the case of two samples, the pooled estimate is

$$\sigma^2 \approx \frac{\displaystyle\sum_{i=1}^{k} 1\, (n_i - 1)s_i^2}{n - k} = \frac{\displaystyle\sum_{i=1}^{k} \sum_{j=1}^{n_i-1} (x_{ij} - \bar{x}_i)^2}{n - k} = \frac{SSE}{n - k}$$

where $n - k = (n_1 - 1) + (n_2 - 1) + \ldots + (n_k - 1)$. This "average" yields an unbiased estimate of σ^2 and, we note, does not depend on a possible treatment effect.

DEFINITION 12.4

The *error mean square* MSSE is defined to be

$$MSSE = \frac{SSE}{n - k} = \frac{\displaystyle\sum_i \sum_j (x_{ij} - \bar{x}_i)^2}{n - k}$$

The importance of the error mean square is twofold: It is independent of any treatment effect and it is an approximation to the common variance.

Now consider the treatment sum $SSTr$. As the name implies, it varies with the treatment effect. In fact it is quite sensitive to it. For when an effect exists, not all the population means μ_i are equal; as a result some are not well approximated by the grand mean $\bar{x}$. Thus while we expect $(\bar{x}_i - \mu_i)^2$ to be reasonably small for all samples, the same cannot be said of $(\bar{x}_i - \bar{x})^2$. This effect is further amplified when we multiply by the sample size and sum over all the samples.

On the other hand, if there is no treatment effect, there is a common population mean that is approximated by the grand mean $\bar{x}$. The deviations $(\bar{x}_i - \bar{x})$ of the treatment means from the grand mean can be used to produce a second unbiased estimator of σ^2, which is called the *treatment mean square* and is denoted by MSSTr. Additionally, when there are treatment effects, MSSTr should have values that are larger than σ^2.

DEFINITION 12.5

The *treatment mean square* is defined to be

$$MSSTr = \frac{\sum_{i=1}^{k} n_i(\bar{x}_i - \bar{x})^2}{k - 1} = \frac{SSTr}{k - 1}$$

In summary, when there is no treatment effect, both the error mean square and treatment mean square are estimators of σ^2. If, however, there is an effect, then this will be true of only the error mean square MSSE. The treatment mean square MSSTr will be considerably larger. We test for this by considering the size of the ratio.

$$f = \frac{MSSTr}{MSSE}$$

This should be close to 1 when there is no treatment effect. Otherwise it is larger.

As we have seen, the F distribution applies to the ratio of two sample variances from the same population. Since the numerator MSSTr has $v_1 = k - 1$ degrees of freedom and the denominator, MSSE, has $v_2 = n - k$, the specific distribution needed for a test of no treatment effect is $F(k - 1, n - k)$.

PROCEDURE FOR TESTING FOR TREATMENT EFFECT

H_0: There is no treatment effect ($\mu_1 = \mu_2 = \ldots = \mu_k$)
H_1: There is a treatment effect ($\mu_i \neq \mu_j$ for at least one pair of means)

Given: k independent samples with a total of n observations
Assumptions: The populations being tested are normal with a common variance.

1. Select a significance level α such as $\alpha = .05$ or $\alpha = .01$.
2. Find the critical value $f_\alpha(k - 1, n - k)$.
3. Compute $f = MSSTr/MSSE$.
4. Compare the value of the test statistic with the critical value. Reject H_0 if $f > f_\alpha(k - 1, n - k)$.

Example 12.3   Consider again the absenteeism study of Example 12.2. Test the hypothesis that absenteeism is the same for each of the three days. The samples are reproduced below.

Sample 1	Sample 2	Sample 3
6	3	7
8	1	8
		9

We have already computed the treatment and error sums and found

$$SSTr = 46 \qquad SSE = 6$$

Using the number of samples $k = 3$ and the total number of observations $n = 7$ we obtain the error and treatment mean squares:

$$MSSTr = \frac{SSTr}{k-1} = \frac{46}{2} = 23$$

$$MSSE = \frac{SSE}{n-k} = \frac{6}{7-3} = 1.5$$

Thus

$$f = \frac{MSSTr}{MSSE} = \frac{23}{1.5} = 15.33 \text{ (rounded)}$$

Assuming a significance level of $\alpha = .05$ we find

$$f_{.05}(2, 4) = 6.94.$$

Since $f > f_{.05}(2, 4)$ we reject the hypothesis that there is no difference in the mean number of absentees for these three days. ■

We remind the reader of the assumptions of normal distributions and a common variance for the treatment populations. When these are greatly violated, then the nonparametric Kruskal-Wallis test of Chapter 13 is recommended.

Here is one additional example. Again we have used small samples in order to keep the computations within reason.

Example 12.4   A builder is interested in whether location has an effect on the selling price of a new three-bedroom home in the 2000 square foot range. To keep matters simple, let us suppose there are only three areas in which new homes are being built and thus only three treatment levels. The three samples of selling prices (in thousands of dollars) are listed below in Table 12.3.

TABLE 12.3 SELLING PRICES OF HOMES (THOUSANDS OF DOLLARS)		
Area 1	Area 2	Area 3
89	103	97
94	98	110
99	91	103
	92	104
		106

Using the significance level $\alpha = .05$ we are to test:

H_0: Location has no effect on the selling price
H_1: Location does have an effect on the selling price

The grand mean is $\bar{x} = 98.83$.

The sample means are easily found: $\bar{x}_1 = 94$, $\bar{x}_2 = 96$, and $\bar{x}_3 = 104$. The terms of the treatment mean square are found by subtracting each sample mean from the grand mean, squaring, and then multiplying by the corresponding sample size. Thus

$$MSSTr = \frac{3(94 - 98.83)^2 + 4(96 - 98.83)^2 + 5(104 - 98.83)^2}{3 - 1} = 117.83$$

The terms of the error mean square are obtained by subtracting each observation from the mean of the sample to which it belongs and squaring.

$$MSSE = \frac{(89 - 94)^2 + \ldots + (103 - 96)^2 + \ldots + (97 - 104)^2 + \ldots + (106 - 104)^2}{12 - 3}$$
$$= 26.00$$

The quotient of these terms is

$$f = \frac{MSSTr}{MSSE} = \frac{117.83}{26.00} = 4.53 \text{ (rounded)}$$

The numerator has $v_1 = 3 - 1 = 2$ degrees of freedom, the denominator $v_2 = 12 - 3 = 9$. From Table VII we find $f_{.05}(2, 9) = 4.26$.

Since $f = 4.53 > 4.26$, the null hypothesis of no treatment effect is rejected at a 5% level of significance. ■

Note that in these examples we are actually testing the equality of several population means. Thus we have an extension of the two-population t test of Chapter 10.

12.3   SIMPLIFYING THE COMPUTATIONS

In Chapter 3 an alternate formula for the variance was introduced that involved both the sum and the sum of the squares of the observation. An analogous simplification is possible with the treatment and error mean squares MSSTr and MSSE. To this end we introduce the notation T_i for the total or sum of the ith sample and T for the grand total. Thus

$$T_i = \sum_j x_{ij}$$

and

$$T = \sum_i \sum_j x_{ij}$$

Omitting the algebraic manipulations, we state that the treatment and error sums can be computed as follows:

THEOREM 12.2

$$SSTr = \sum_i \frac{T_i^2}{n_i} - \frac{T^2}{n}$$

$$= \frac{T_1^2}{n_1} + \frac{T_2^2}{n_2} + \ldots + \frac{T_k^2}{n_k} - \frac{T^2}{n}$$

$$SSE = \sum_i \sum_j x_{ij}^2 - \sum_i \frac{T_i^2}{n_i}$$

$$= (x_{11}^2 + x_{12}^2 + \ldots + x_{k,n_k}^2) - \left(\frac{T_1^2}{n_1} + \frac{T_2^2}{n_2} + \ldots + \frac{T_k^2}{n_k} \right)$$

Dividing each of these sums by its respective degrees of freedom then yields the means. It is customary to arrange the computation as in Figure 12.1.

FIGURE 12.1
ARRANGEMENT OF DATA FOR THE ANALYSIS OF VARIANCE

	Sample 1	Sample 2 ...	Sample k	Grand Total
	x_{11} x_{11}^2	x_{21} x_{21}^2	x_{k1} x_{k1}^2	
	x_{12} x_{12}^2	x_{22} x_{22}^2	x_{k2} x_{k2}^2	
	...	...	...	
Size	n_1	n_2	n_k	n
Sum	T_1	T_2	T_k	T
Sum of squares				
Quotients				

Example 12.5   Consider again the data of Example 12.4 on the selling price of homes. The analysis would be arranged as in Figure 12.2.

FIGURE 12.2 ARRANGEMENT OF ANALYSIS OF VARIANCE COMPUTATION FOR HOME PRICE–AREA STUDY

	Sample 1		Sample 2		Sample 3		Grand Total
	x	x^2	x	x^2	x	x^2	
	89	7921	103	10609	97	9409	
	94	8836	98	9604	110	12100	
	99	9801	91	8281	103	10609	
			92	8464	104	10816	
					106	11236	
Sample size	$n_1 = 3$		$n_2 = 4$		$n_3 = 5$		$n = 12$
Sum	$T_1 = 282$		$T_2 = 384$		$T_3 = 520$		$T = 1186$
Sum of squares	26,558		36,958		54,170		117,686
Quotients	$\dfrac{T_1^2}{n_1} = 26,508$		$\dfrac{T_2^2}{n_2} = 36,864$		$\dfrac{T_3^2}{n_3} = 54,080$		117,452

Since $SSTr = \sum \dfrac{T_i^2}{n_i} - \dfrac{T^2}{n}$ we use the grand totals from rows 1, 2, and 4. Thus

$$SSTr = 117,452 - \frac{(1186)^2}{12} = 235.67$$

and

$$MSSTr = \frac{235.67}{3 - 1} = 117.83 \text{ (rounded)}$$

The error sum involves the grand totals from rows 3 and 4.

$$SSE = \sum_{i=1} \sum x_{ij}^2 - \sum \frac{T_i^2}{n_i}$$
$$= 117,686 - 117,452 = 234$$

FIGURE 12.3 FORM OF AN ANOVA TABLE

Source of Variation	Sum of Squares (SS)	Degrees of Freedom (df)	Mean Square (MS)	f
Treatment	$SSTr = \sum_i \dfrac{T_i^2}{n_i} - \dfrac{T^2}{n}$	$k - 1$	$MSSTr = \dfrac{SSTr}{k - 1}$	$f = \dfrac{MSSTr}{MSSE}$
Error	$SSE = \sum_i \sum_j x_{ij}^2 - \sum_i \dfrac{T_i^2}{n_i}$	$n - k$	$MSSE = \dfrac{SSE}{n - k}$	
Totals	$SS = SSTr + SSE = \sum \sum x_{ij}^2 - \dfrac{T^2}{n}$	$n - 1$		

Thus $MSSE = \dfrac{234}{12 - 3} = 26.00$. And as in the previous calculations, we find

$$f = \frac{MSSTr}{MSSE} = 4.53 \quad \blacksquare$$

In published reports, the essential results from an analysis of variance study are summarized in what is called an ANOVA table. The form of this table as well as the formulas for calculating its entries is given in Figure 12.3.

Example 12.6

The ANOVA table for the study of home prices (Example 12.4) is given below as Table 12.4. ■

TABLE 12.4   ANOVA TABLE FOR AREA–HOME PRICE STUDY

Source	SS	df	MS	f
Area of location	235.67	2	117.83	4.53
Error	234	9	26.00	
Total	469.67	11		

Example 12.7

Interpret the following ANOVA table (Table 12.5)

TABLE 12.5

Source	SS	df	MS	f	p
Treatment	510	3	170	5.86	<.01
Error	464	16	29		
Totals	974	19			

For $f = 5.86$, $p < .01$. Thus for any $\alpha > .01$ (say $\alpha = .025$), we can conclude that there is a difference in population means; equivalently, there is a treatment effect.

Since treatment $df = 3$, there were $k = 3 + 1 = 4$ treatment levels or samples involved in the study. Since total $df = 19$, there were $n = 19 + 1 = 20$ observations or subjects in all. If the samples were of equal size, then each contained $20/4 = 5$ observations. Finally $SSTr = 510$, $MSSTr = 510/3 = 170$, $SSE = 464$, and $MSSE = 464/16 = 29$. ■

We note that in the literature the treatment mean square $MSSTr$ and the error mean square $MSSE$ are often referred to as the "between-group" and "within-group" estimates of the common variance σ^2. This is because the value of $MSSTr$ depends on the difference between the samples or treatment groups, as opposed to $MSSE$, which depends on differences within each sample or treatment group.

Example 12.8
A Case Study

In a study entitled "Effects of Expected and Obtained Grades on Teacher Evaluations and Attributes of Performance," by C. E. Snyder and Mark Clair (*Journal of Educational Psychology*, vol. 68, no. 1, 1976), a group of 72 students were manipulated to expect a grade of A, B, or C on a test. Then, after listening to a lecture on the subject material by an instructor, they were given a test where the grades were also manipulated.

On the basis of these grades the students were then divided into a positively disconfirmed group (grade gain over the expected), a negatively disconfirmed group (grade loss over the expected), and a confirmed group (same grade as expected). For example, the positively disconfirmed group consisted of those who had expected a C and received a B or an A, or those who had expected a B and received an A.

The students were then asked to complete an evaluation of the instructor using the University of Kansas Curriculum Instruction Feedback Survey (CIFS) and those forms of Holmes' Teacher Assessment Blank (TAB) covering (1) quality of presentation, (2) degree of stimulation and motivation, and (3) clarity of the test.

A one-way analysis of variance was then performed on the teacher evaluations. As shown in Table 12.6, there were significant effects. The positively disconfirmed group rated the instructor higher than the confirmed group. Similarly, the negatively disconfirmed group rated the instructor lower than the confirmed group. This result held across all rating forms.

TABLE 12.6 MEANS AND SUBSEQUENT ANALYSES FOR THE FOUR TEACHER EVALUATION MEASURES AS A FUNCTION OF GRADE OUTCOME

Dependent variable	Positively disconfirmed *M*	Confirmed *M*	Negatively disconfirmed *M*	*f*	*df*	*p*
CIFS	2.91	2.41	2.00	11.01	2,69	<.001
TAB 1	3.15	2.87	2.58	10.29	2,69	<.001
TAB 2	2.71	2.46	2.08	8.45	2,69	<.001
TAB 3	2.88	2.91	2.45	2.77	2,69	<.10

In short, the hypothesized expected grade and that which was actually received had a significant effect on the subsequent evaluation of the instructor. ∎

EXERCISE SET 12.1

Part A

1. List possible factors influencing the distribution of each of the following.
   a. The selling price of a home.
   b. The grade on a test such as the SAT.
   c. The gross sales of a member store of a chain.
   d. The number of auto accidents in a city within a given week.
   e. The effectiveness of a medication for curbing appetite.
   f. The achievement level of a child.

2. Samples from three treatment populations are given below:

   Sample 1: 35, 37, 23
   Sample 2: 40, 46, 31, 28
   Sample 3: 20, 30, 35, 38, 40

   Use subscript notation to name each of these observations.

3. For the data below, compute the treatment, error, and total sum of squares. (Work with the deviations from the mean.) Then verify that $SS = SSTr + SSE$.

Sample 1	Sample 2	Sample 3	Sample 4
16	15	12	11
23	21	14	15
19	23	18	20

4. Twenty-five experimental subjects are randomly divided into five equal groups and each group is administered a treatment. What can be concluded in an ANOVA test with $\alpha = .05$ if $f = 8.9$?

5. A company has three package designs for a new product. In a trial evaluation, 15 stores from similar marketing areas are selected and supplied the new item. Only one package design, however, is used in any one store. The number of sales during the first 45 days following the introduction of the product are given below. Can the company conclude that the package design has an effect on sales? Arrange the computations as in Figure 12.1 and summarize the results in an ANOVA table. Use $\alpha = .05$.

Design I	Design II	Design III
205	240	260
240	285	280
255	340	360
280	360	410
340	370	450

6. Perform an analysis of variance on the data below to test the equality of the corresponding population means. Arrange the computations as in Figure 12.1 and summarize the results in an ANOVA table.

Sample 1	Sample 2	Sample 3
10	7	15
14	16	18
12	9	20
	10	14
		18

7. A report contained the following ANOVA table:

Source	SS	df	MS	f
Treatment	1020	5	204	2.83
Error	3456	48	72	
Total	4476	53		

a. How many sample or treatment groups were involved in the experiment?
b. What is the total number of observations or experimental subjects?
c. If the samples (or experimental groups) are all the same size, what is the size of each?
d. What is the critical value of F if $\alpha = .01$? Should the null hypothesis of equal means be rejected?

8. An analysis of variance table is given below.
    a. How many treatment groups or independent samples are implied?
    b. What is the total number of observations or experimental subjects?
    c. If $\alpha = .05$, should the null hypothesis of equal means be rejected?

Source	SS	df	MS	f
Between groups	1750	4	437.50	2.87
Within groups	6860	45	152.44	
Totals	8610	49		

9. Three makes of batteries used as power backups for computers were tested for hours of continuous service. Using the data below, test the hypothesis that each make has the same average service life. Use $\alpha = .025$.

Make 1	Make 2	Make 3
15.3	16.0	14.3
15.8	16.3	15.2
16.4	16.8	15.9
16.9	18.2	16.1
17.3	19.1	16.4

10. In an agricultural study of the yields of three different strains of wheat, the same quantity of each is sown on individual plots that are controlled for size, fertility, moisture, and other factors. The yields in pounds per plot are as follows:

Type 1	Type 2	Type 3
5	6	8
3	11	16
3	14	4
4	13	13

a. Compute the means $MSSTr$ and $MSSE$ directly from their definitions.
b. What do these quantities represent?
c. Using $\alpha = .01$, test the hypothesis of no difference in the yields of these three strains of wheat.

Part B

11. Does the daily hospital rate for a private room differ in New York, Chicago, and Los Angeles? Assuming the rates for each city are normally distributed with a common (homogeneous) variance, analyze the following rates (dollars per day) using $\alpha = .05$. Arrange the computations as in Figure 12.1 and summarize the results in an ANOVA table.

Los Angeles	Chicago	New York
205	200	215
180	180	195
190	170	220
225	155	245
230	160	200

12. If the same number is subtracted from all observations of a study, the variance(s) are not affected. This provides a way of avoiding the large numbers that might ordinarily result. In problem 11, subtract (say) 150 from each observation, carry out the analysis, and compare the results with those obtained previously.

13. A gas company is interested in whether the design of the heating system has an effect on the cost of heating a house. Three types of heating systems are compared. A random sample of five similar homes was selected for each of the systems and the total heating bill for the two months of January and February were obtained. Unfortunately one of the heating systems failed and could not be replaced immediately. In another home, the family was away for a considerable time. The results for the remaining 13 homes were as follows.

System 1	System 2	System 3
$189	$210	$190
187	230	200
190	195	180
207	215	220
		205

a. What is the response variable?
b. Test for equality of means using $\alpha = .05$.
c. What are some of the other factors that should have been held constant across all the samples?

14. Typists trained by three different systems are tested for their typing speed (words per minute). Assuming the speeds from each system are approximately normal with the same variance, test the hypothesis that the systems are equally effective. Use $\alpha = .05$.

System 1	System 2	System 3
85	90	85
90	105	80
105	110	90
100	95	100

15. In an oxides emission study of four different types of truck engines, the following data were reported (emissions are in parts per million):
- Engine 1: $n_1 = 10$, $\Sigma\ x_{1j}^2 = 180.50$, and $\Sigma\ x_{1j} = 36.8$
- Engine 2: $n_2 = 10$, $\Sigma\ x_{2j}^2 = 554.25$, and $\Sigma\ x_{2j} = 65.5$
- Engine 3: $n_3 = 7$, $\Sigma\ x_{3j}^2 = 451.63$, and $\Sigma\ x_{3j} = 53.5$

Using $\alpha = .05$, test the hypothesis that the mean emission rates for these three types of engines are the same. Report the results in an ANOVA table.

16. At a microcomputer conference the participants were asked to rate the conference on a continuous scale from 0 to 10. The participants were also classed in terms of their experience with computers. A random sampling of the scores from each of these experience classes is summarized in the following ANOVA table.

Source	SS	df	MS	f
Treatment (between groups)	480	2	240	26.7
Error (within groups)	360	40	9	
Total	840	42		

a. How many classes were used?
b. How many participants were involved in the survey?
c. What conclusion would be drawn regarding the effect of experience on the replies of the participants?

17. An ANOVA table is given below for a study involving samples of equal size.

Source	SS	df	MS	f
Treatment	630	4		
Error		15		
Total	1150			

a. Complete the table and test the hypothesis of no treatment effect using $\alpha = .05$.
b. Find the number of treatment levels and the size of each sample.

18. A consumer protection group wants to know whether there is a difference in the service life of four brands of light bulbs. A random sample of three bulbs of each brand is tested. The following information is then calculated: $T_1 = 40$,

$T_2 = 60$, $T_3 = 60$, $T_4 = 50$, and $\Sigma \Sigma x_{ij}^2 = 3930$. Test the hypothesis that there is no difference in the mean service lives of these four brands against the alternative hypothesis that there is a difference. Use $\alpha = .05$. Display the results in an ANOVA table.

19. A testing service is interested in whether the design of a test will affect test grades. Five easy and five difficult algebra problems are prepared and arranged to produce four distinct test forms. On form I, the first five questions are the easy ones, the second five the difficult ones. On form II, there is just the opposite arrangement, with the five difficult problems coming first. On form III, the problems alternate easy, difficult, easy, difficult. And on form IV, they alternate difficult, easy, difficult, easy. A random sample of 16 students is then given the test, with each student taking only one form. The results are given below.

   a. Test the null hypothesis of no treatment effect caused by the test design using $\alpha = .05$. Arrange the computations as in Figure 12.1.

   b. Give the ANOVA table.

Form I	Form II	Form III	Form IV
78	72	71	73
84	55	69	61
65	73	77	79
89	68	82	68

20. In a study of three treatment levels, sample 1 has four observations, sample 2 has 3, and sample 3 has 5. What do the following terms represent?

   a. $\dfrac{\sum\limits_{j=1}^{4} x_{1j}}{4}$

   b. $\dfrac{1}{3} \sum\limits_{j=1}^{3} x_{2j}$

   c. $\dfrac{\Sigma \Sigma x_{ij}}{12}$

Library

Project

The following research articles make use of one-way analysis of variance. Select one, locate it in the periodical section of the library, and report on it.

a. Perception of the Internal and External Auditor as a Deterrent to Corporate Irregularities;" W. Uecker, *The Accounting Review*, vol. 56, no. 3, July 1981.

b. "The Effects of Selected Death Education Curriculum Models on Death Anxiety and Death Acceptance;" D. Combs, *Death Education*, vol. 5, no. 1, Spring 1981.

c. "The Effects of Competition Upon the Endurance Performance of College Women;" S. Higgs, *International Journal of Sports Psychology*, vol. 3, no. 2, 1972.

d. "The Pennsylvania Domicillary Care Experiment;" S. Sherwood, *American Journal of Public Health*, vol. 73, no. 4, April 1983.

e. "Conceptual Systems Functioning as a Mediate Factor in the Development of Counseling Skills;" N. Lulwok and J. Hennessy, *Journal of Counseling Psychology*, vol. 29, no. 3, May 1982.

f. "Effect of Inservice Training Intensity on Teachers' Attitude Toward Mainstreaming;" B. Larriivee, *Exceptional Children*, vol. 48, no. 1, Sept. 1981.

g. "Academic and Social Self-Concept of the Academically Gifted;" A. Ross and M. Parker, *Exceptional Children*, vol. 47, no. 2, Sept. 1980.

h. "Morphine-Reduced Ventilation Without Changing Metabolic Rates in Exercise;" B. Martin et al., *Medicine and Science in Sports and Exercise*, vol. 12, no. 4, 1981.

12.4 THE TWO-WAY ANALYSIS OF VARIANCE

When an ANOVA study involves several factors, the computations become so tedious that the assistance of a computer is a near necessity. The important special case of two factors occurs frequently and will be considered here. The calculations are within reason and the approach provides excellent insights as to how the preceding analysis extends to multiple factors.

Suppose then we have two factors, which we call R and C, that are to be studied for an effect on some response variable. For example, R might be comprised of different surgical procedures and C different postoperative regimens of medication. Or in a marketing study of a new item, one might wish to determine the effect on sales of the package design R and the advertising program C.

There are three questions to be considered.

1. Is there a treatment effect due to R?

2. Is there a treatment effect due to C?

3. Is there a treatment effect due to the interaction of the two factors?

For example, in the surgical and medication study just mentioned, it may be that the recovery time associated with a particular technique varies with the follow-up medication. If so, there is an interaction effect. Or in the study of advertising programs and package designs, it may be that certain advertising programs are more successful when coupled with certain package designs. For instance, we would expect the package size, color, and lettering to be of greater importance in a highly visual program (television, billboards, etc.) than with a coupon discount mailer program.

The notation needed is best understood in terms of the arrangement of the observations in the analysis table. Recall that the two factors are to be called R and C. These terms were chosen to symbolize the *rows* and *columns* of the table. Continuing in this vein, we let r and c be the corresponding number of treatment levels for the factors R and C.

For each treatment level of R there will be a row in the analysis table. These we denote by $R_1, R_2, \ldots, R_r$. Similarly, the columns $C_1, C_2, \ldots, C_c$ represent the treatment levels of the factor C.

At the intersection of each row and column there is a *cell*; In all there are $r \cdot c$ of these cells. When we speak of, say, the (2, 3) cell we are referring to the one located at the intersection of the second row and the third column. At this point it might be a good idea to look at Figure 12.4, which shows the form of the analysis table. It probably will not all make sense just yet but you can begin to get the idea.

The entries of the (i, j) cell are the observations from those experimental units that jointly received treatments R_i and C_j. It turns out that to test for an interaction effect there must be at least two observations per cell. These must be obtained independently of one another and, insofar as possible, with all other factors held constant. In short, the experiment must be *replicated* or carried out several times. Considerations about the usefulness of the test in the case of somewhat different variances show that more powerful tests result when the number of observations per cell are all the same. This number will be donoted by n.

Three subscripts are now needed to locate an observation in the analysis table. The first two locate the cell, the third the observation within that cell. For example, if cell (2,3) contains four observations, these would be labeled.

$$x_{231}, x_{232}, x_{233}, \text{ and } x_{234}$$

FIGURE 12.4 ARRANGEMENT OF DATA FOR THE TWO-WAY ANALYSIS OF VARIANCE

Factor R	Factor C				Total
	Treatment C_1	Treatment C_2	. . .	Treatment C_c	
Treatment R_1	x_{111} x_{112} $\vdots$ x_{11n}	x_{121} x_{122} $\vdots$ x_{12n}	. . .	x_{1c1} x_{1c2} $\vdots$ x_{1cn}	$T_{1\cdot}$
Treatment R_2	x_{211} x_{212} $\vdots$ x_{21n}	x_{221} x_{222} $\vdots$ x_{22n}	. . .	x_{2c1} x_{2c2} $\vdots$ x_{2cn}	$T_{2\cdot}$
Treatment R_3	x_{311} x_{312} $\vdots$ x_{31n}	x_{321} x_{322} $\vdots$ x_{32n}	. . .	x_{3c1} x_{3c2} $\vdots$ x_{3cn}	$T_{3\cdot}$
. . .	. . .	. . .	. . .	. . .	. . .
Treatment R_r	x_{r11} x_{r12} $\vdots$ x_{r1n}	x_{r21} x_{r22} $\vdots$ x_{r2n}	. . .	x_{rc1} x_{rc2} $\vdots$ x_{rcn}	$T_{r\cdot}$
Total	$T_{\cdot 1}$	$T_{\cdot 2}$	. . .	$T_{\cdot c}$	T

Now refer again to Figure 12.4. The symbols $T_{1\cdot}, T_{2\cdot}, \ldots, T_{r\cdot}$ are used for the row totals. Having a dot in the second subscript position indicates a row total. Similarly, a dot in the first position is used for the column totals $T_{\cdot 1}, T_{\cdot 2}, \ldots, T_{\cdot c}$. The grand total of all the observations is denoted simply by T (some authors use $T_{\cdot\cdot}$).

Similarly, $\bar{x}_{i\cdot}$ represents the mean of the ith row, $\bar{x}_{\cdot j}$ that of the jth column, and $\bar{x}$ the grand mean. Also we shall need the means of the various cells. While $\bar{x}_{ij\cdot}$ is a natural notation for the mean of the (i, j) cell, we shall shorten it to simply $\bar{x}_{ij}$, leaving off the dot.

Before turning to the analysis, let us illustrate this notation with an example.

Example 12.9   The emission of carbon monoxide by an auto engine is being evaluated in relation
to two particular types of fuel systems (I and II) and three types of driving (city,
freeway, and mountain). Since different demands on the engine are translated
through the fuel system, an interactive effect is a definite possibility. Other factors, such as type of fuel, make of auto, operating conditions, and the like are
kept constant across all the tests; only the course of driving and the fuel system
are varied.

To keep calculations to a minimum, suppose we use only two independent
observations of the emissions (in centigrams per mile) for each of the $2 \times 3 = 6$
combinations of fuel system and driving course. These are detailed in the analysis
Table 12.7 below. The factors are the fuel system (R), with two treatment levels
and the driving course (C), with three levels.

**TABLE 12.7   OXIDE EMISSIONS IN RELATION TO
FUEL SYSTEMS AND TYPES OF DRIVING**

Fuel system	Course			Total
	City	*Freeway*	*Mountain*	
I	140	128	152	
	156	134	158	868
II	126	124	146	
	138	130	156	820
Total	560	516	612	1688

The column totals are $T_{.1} = 560$, $T_{.2} = 516$ and $T_{.3} = 612$. The row totals are
$T_{1.} = 868$ and $T_{2.} = 820$. The grand total is $T = 1688$; this quantity may be
found by adding either the row totals or the column totals.

The column, row, and grand means are easily found:

$$\bar{x}_{.1} = \frac{560}{4} = 140, \ \bar{x}_{.2} = \frac{516}{4} = 129, \ \bar{x}_{.3} = \frac{612}{4} = 153$$

$$\bar{x}_{1.} = \frac{868}{6} = 144.67, \ \bar{x}_{2.} = \frac{820}{6} = 136.67$$

$$\bar{x} = \frac{1688}{12} = 140.67$$

The mean of the $(1,1)$ cell is

$$\bar{x}_{11} = \frac{140 + 156}{2} = 148$$

Similarly,

$$\bar{x}_{12} = \frac{262}{2} = 131, \ \bar{x}_{13} = \frac{310}{2} = 155$$

$$\bar{x}_{21} = \frac{264}{2} = 132, \ \bar{x}_{22} = \frac{254}{2} = 127, \ \bar{x}_{23} = \frac{302}{2} = 151.$$

Notice that the mean for the city driving is simply the mean of the first column, which is $\bar{x}_{\cdot 1} = \dfrac{560}{4} = 140$. The mean over those trials using fuel system II is the mean of the second row: $\bar{x}_{2\cdot} = \dfrac{820}{6} = 136.67$. The mean of, say, the (1,2) cell, $\bar{x}_{12}$, is of course the mean of those trials over the freeway course using fuel system 1. And so forth. ■

The analysis for a single factor led to the consideration of the total sum of squares SS, an error sum SSE, and a treatment sum $SSTr$. These are related by the equation

$$SS = SSTr + SSE$$

We now have two factors R and C, whose treatment sums of squares will be donoted by SSR and SSC. If there is no interaction effect, we should expect the equation:

$$SS = SSR + SSC + SSE$$

On the other hand, if there is an interaction effect, this equation will not hold. Thus it seems natural to define the interaction treatment sum $SSRC$ in terms of this difference:

$$SSRC = SS - SSR - SSC - SSE$$

This we shall do. Since these sums can also be expressed in terms of squares of observations and their totals, we incorporate all of these concepts into one (grand) definition.

DEFINITION 12.6

Consider an ANOVA computation table having r rows, c columns, n entries per cell, and $N = nrc$ total entries. The total sum of squares SS is defined by

$$SS = \sum_i \sum_j \sum_k (x_{ijk} - \bar{x})^2 = \sum_i \sum_j \sum_k x_{ijk}^2 - \frac{T^2}{N}$$

The sum of squares between rows or the treatment sum for the factor R is defined by

$$SSR = \sum_i n \cdot c(\bar{x}_{i\cdot} - \bar{x})^2 = \sum_i \frac{T_{i\cdot}^2}{n \cdot c} - \frac{T^2}{N}$$

The sum of square between columns or the treatment sum for the factor C is defined by

$$SSC = \sum_j n \cdot r(\bar{x}_{\cdot j} - \bar{x})^2 = \sum_j \frac{T_{\cdot j}^2}{n \cdot r} - \frac{T^2}{N}$$

The error sum of squares is defined by

$$SSE = \sum_i \sum_j \sum_k (x_{ijk} - \bar{x}_{ij})^2 = \sum_i \sum_j \sum_k x_{ijk}^2 - \sum_i \sum_j \frac{\left(\sum_k x_{ijk}\right)^2}{n}$$

The interaction treatment sum is defined by

$$SSRC = SS - SSR - SSC - SSE$$

$$= \frac{T^2}{N} - \sum_i \frac{T_{i\cdot}^2}{n \cdot c} - \sum_j \frac{T_{\cdot j}^2}{n \cdot r} + \sum_i \sum_j \frac{\left(\sum_k x_{ijk}\right)^2}{n}$$

Obviously, we have taken some liberties with what constitutes a definition but this keeps the presentation simple. And we have exactly the same result as obtained by starting with an analysis of the total sum SS, namely

$$SS = SSR + SSC + SSRC + SSE$$

As with the one-factor analysis, we need the "means" of the treatment and error sums.

DEFINITION 12.7

$$MSSR = \frac{SSR}{r - 1}$$

$$MSSC = \frac{SSC}{c - 1}$$

$$MSSRC = \frac{SSRC}{(r - 1)(c - 1)}$$

$$MSSE = \frac{SSE}{rc(n - 1)}$$

As usual, the denominators of these defining ratios are the degrees of freedom of the corresponding "means." Now for the test.

PROCEDURE FOR A TWO-WAY ANALYSIS OF VARIANCE OF A ROW EFFECT, A COLUMN EFFECT, OR AN INTERACTION EFFECT

Given: An $r \times c$ two-way analysis of variance table with $n > 1$ observations per cell.

Assumptions: The populations corresponding to the individual cells are normally distributed and have a common variance.

• To test H_0: No row effect (or no treatment effect due to R) at a significance level α, compute

$$f_R = \frac{MSSR}{MSSE}$$

and reject H_0 if $f_R > f_\alpha (r - 1, rc [n - 1])$.

- To test H_0: No column effect (or no treatment effect due to C) at a significance level α, compute

$$f_C = \frac{MSSC}{MSSE}$$

and reject H_0 if $f_C > f_\alpha$ $(c - 1, \, rc\,[n - 1])$.

- To test H_0: No interaction (or no interaction effect due to R and C) at a significance level α, compute

$$f_{RC} = \frac{MSSRC}{MSSE}$$

and reject H_0 if $f_{RC} > f_\alpha$ $([r - 1][c - 1], \, rc\,[n - 1])$.

In each case, the rejection of the null hypothesis is taken as an indication of an effect.

Example 12.10

Consider again the emission study involving fuel systems and road courses. The analysis table is reproduced below (Table 12.8). Only now, adjacent to each observation has been entered its square.

TABLE 12.8   OXIDE EMISSIONS IN RELATION TO FUEL SYSTEM AND TYPES OF DRIVING

Fuel system	Course						Total
	City		Freeway		Mountains		
System I	140	19,600	128	16,384	152	23,104	$T_{1\cdot} = 868$
	156	24,336	134	17,956	158	24,964	
System II	126	15,876	124	15,376	146	21,316	$T_{2\cdot} = 820$
	138	19,044	130	16,900	156	24,336	
Total	$T_{\cdot1} = 560$		$T_{\cdot2} = 516$		$T_{\cdot3} = 612$		$T = 1688$

First we compute $\Sigma\,\Sigma\,\Sigma x_{ijk}^2 = 239{,}192$. Next, the individual sums:

$$SS = \Sigma\,\Sigma\,\Sigma x_{ijk}^2 - \frac{T^2}{N} = 239{,}192 - \frac{(1688)^2}{12} = 1746.67$$

$$SSR = \Sigma\,\frac{T_{i\cdot}^2}{n \cdot c} - \frac{T^2}{N} = \frac{(868)^2}{2 \cdot 3} + \frac{(820)^2}{2 \cdot 3} - \frac{(1688)^2}{12} = 192.00$$

$$SSC = \Sigma\,\frac{T_{\cdot j}^2}{n \cdot r} - \frac{T^2}{N} = \frac{(560)^2}{2 \cdot 2} + \frac{(516)^2}{2 \cdot 2} + \frac{(612)^2}{2 \cdot 2} - \frac{(1688)^2}{12} = 1154.67$$

In computing SSE we need

$$\sum_i \sum_j \frac{\left(\sum_k x_{ijk}\right)^2}{n}$$

Note that

$$\left(\sum_k x_{ijk} \right)$$

is simply the sum of the observations of the (i, j) cell. Thus we need the squares of the sums of each cell divided by the number of observations of that cell.

$$SSE = \sum_i \sum_j \sum_k (x_{ijk}^2) - \sum_i \sum_j \frac{\left(\sum x_{ijk} \right)^2}{n}$$

$$= 239{,}192 - \left[\frac{(296)^2}{2} + \frac{(262)^2}{2} + \frac{(310)^2}{2} + \frac{(264)^2}{2} + \frac{(254)^2}{2} + \frac{(302)^2}{2} \right]$$

$$= 239{,}192 - 238{,}888 = 304.00$$

From these sums we find SSRC by

$$SSRC = 1746.67 - 192.00 - 1154.67 - 304.00$$
$$= 96.00.$$

Finally, we compute the "means."

$$MSSR = \frac{SSR}{r - 1} = \frac{192.00}{2 - 1} = 192.00$$

$$MSSC = \frac{SSC}{c - 1} = \frac{1154.67}{3 - 1} = 577.34$$

$$MSSRC = \frac{SSRC}{(r - 1)(c - 1)} = \frac{96.00}{(2 - 1)(3 - 1)} = 48.00$$

$$MSSE = \frac{SSE}{rc(n - 1)} = \frac{304.00}{2 \cdot 3(2 - 1)} = 50.67$$

We now test the hypotheses of no effect using $\alpha = .05$.

$$f_R = \frac{MSSR}{MSSE} = \frac{192.00}{50.67} = 3.79$$

The critical F value is $f_\alpha(r - 1, rc[n - 1]) = f_{.05}(2 - 1, 2 \cdot 3[2 - 1]) = f_{.05}(1, 6) = 5.99$. Since $f_R = 3.79 < 5.99$ we do not reject the null hypothesis of no row effect. Or, we can say that the evidence is not strong enough to indicate an effect in oxide emissions due to the choice of fuel system.

Similarly,

$$f_C = \frac{MSSC}{MSSE} = \frac{577.34}{50.67} = 11.39$$

The critical F value is now $f_\alpha(c - 1, rc[n - 1]) = f_{.05}(2, 6) = 5.14$. We do reject the null hypothesis of no effect in emissions due to the road course, since $f_C = 11.39 > 5.14$.

Finally, we test for an interaction effect.

$$f_{RC} = \frac{MSSRC}{MSSE} = \frac{48.00}{50.67} = .95$$

Obviously, we do not reject the null hypothesis here, since our result is less than 1. With f_{RC} so small there appears to be little if any interaction effect. For illustrative purposes, we note however that the critical F value here is

$$f_\alpha([r - 1][c - 1], rc[n - 1]) = f_{.05}(2, 6) = 5.14 \quad \blacksquare$$

The form of the ANOVA table for the two-way analysis of variance is given in Figure 12.5, along with a summary of the relevant formulas needed to carry out the analysis.

FIGURE 12.5   FORM OF TWO-WAY ANALYSIS OF VARIANCE TABLE

Source	SS	df	MS	f
Rows (R)	SSR	$r - 1$	$MSSR = \dfrac{SSR}{r - 1}$	$f_R = \dfrac{MSSR}{MSSE}$
Columns (C)	SSC	$c - 1$	$MSSC = \dfrac{SSC}{c - 1}$	$f_C = \dfrac{MSSC}{MSSE}$
Interaction (RC)	SSRC	$(r - 1)(c - 1)$	$MSSRC = \dfrac{SSRC}{(r - 1)(c - 1)}$	$f_{RC} = \dfrac{MSSRC}{MSSE}$
Error	SSE	$rc(n - 1)$	$MSSE = \dfrac{SSE}{rc(n - 1)}$	
Total	SS	$rcn - 1$		

In computing the various sums of squares we recommend that the interaction sum be computed from the others by using the formula

$$SSRC = SS - SSR - SSC - SSE$$

The ANOVA table (Table 12.9) is for the emission study of Example 12.10.

TABLE 12.9   ANOVA TABLE FOR OXIDES EMISSION STUDY

Source	SS	df	MS	f
Rows (fuel system)	192.00	1	192.00	3.79
Columns (road course)	1154.67	2	577.34	11.39
Interaction	96.01	2	48.00	.95
Error	304.00	6	50.67	
Total	1746.67	11		

Example 12.11
A Case Study

How do women fare under weak and strong fair employment policies? This is the subject of a report, "Influence of Strong Versus Weak Fair Employment Policies and Applicant's Sex on Selection Decisions and Salary Recommendations in Management Simulation," (American Psychological Association, 0021–9010, 1979).

In this study 68 municipal administrators (57 men and 11 women) participated in a decision making process in which each was asked to follow either a
strongly worded or weakly worded fair employment policy statement in reviewing
and recommending a male or female applicant for a supervisory position. In the
case of a recommendation for hiring, a starting salary was also to be chosen. The
mean selection ratings and salary recommendations are given in Table 12.10. The
analysis of variance summary is given in Table 12.11.

TABLE 12.10   MEAN RATINGS FOR SELECTION DECISIONS AND SALARY RECOMMENDATIONS

Managerial action	Strong fair employment policy		Weak fair employment policy	
	Male applicant	Female applicant	Male applicant	Female applicant
Selection decision[*]	1.78	1.70	2.17	2.05
n	20	17	20	20
Salary recommendation	$11,070	$10,555	$10,830	$11,205
n	7	10	12	12

[*]Scores represent a scale ranging from 1 (reject) to 4 (accept immediately).

TABLE 12.11   SUMMARY OF THE ANALYSIS OF VARIANCE

Source	MS	df	f
Selection decision			
Policy (A)	.257	1	2.85[a]
Sex (B)	.22	1	—
A × B	.01	1	—
Within	.89	72	—
Salary recommendation			
Policy (A)	.62	1	—
Sex (B)	.01	1	—
A × B	11.03	1	6.25[b]
Within	1.76	37	—

[a]$p < .10$.
[b]$p < .025$.

As can be seen, the administrators were more reluctant to hire under the strong
affirmative action policy than the weak policy, but the actual differences were
not significant. Also, no interaction effect (fair employment policy × sex of
applicant) was not found for the selection decisions.

In the salary recommendation analysis, only the interaction effect of
policy × sex of the applicant proved significant. The authors believe that this
evinces "managerial reactance" in recommending lower starting salaries for
women when under strong affirmative action pressure to hire them.

Additional analysis also showed that starting salary recommendations for men and women applicants did not differ statistically under the weak affirmative action policy.

The conclusion, then, is that under a strong affirmative action policy women applicants tend to fare more poorly than under a weak policy.   ■

12.5   TWO-WAY ANALYSIS WITH NO INTERACTION EFFECT

If the two-way analysis indicates no significant interaction between the factors R and C, a more powerful ANOVA test can be obtained by combining the interaction and error square sums into a pooled error term and computing a pooled error mean square term.

DEFINITION 12.8

The *pooled error mean square* is defined to be

$$PMSSE = \frac{SSRC + SSE}{(r - 1)(c - 1) + rc(n - 1)}$$

Note that the degrees of freedom of this pooled mean is just the sum of those of the interaction and error means. The ANOVA test is now carried out by using the pooled error mean square in place of the original error mean square.

PROCEDURE FOR A TWO-WAY ANALYSIS OF VARIANCE OF A ROW EFFECT OR A COLUMN EFFECT

Given: An $r \times c$ two-way analysis of variance table with n observations per cell.

Assumptions: The $r \times c$ populations corresponding to the individual cells are normally distributed with a common variance.
    There is no interaction effect.

• To test H_0: No row effect (no treatment effect caused by R) against
H_1: A row effect, at a significance level α, compute

$$f_R = \frac{MSSR}{PMSSE}$$

and reject H_0 if $f_R > f_\alpha (r - 1, [r - 1][c - 1] + rc[n - 1])$.

• To test H_0: No column effect (no treatment effect caused by C) against
H_1: A column effect, at a significance level α, compute

$$f_C = \frac{MSSC}{PMSSE}$$

and reject H_0 if $f_C > f_\alpha (c - 1, [r - 1][c - 1] + rc[n - 1])$.

Note that we no longer have the restriction of two observations per cell. If it is known that there is no interaction, this test can be applied with one observation per cell.

Example 12.12

In the emission study of Example 12.10, we found no evidence of an interaction between the fuel system R and the road course C. It would then be appropriate to use the pooled mean square error in testing for a row or column effect. Using $SSRC = 96.01$ and $SSE = 304$, we find

$$PMSSE = \frac{96.01 + 304}{(2 - 1)(3 - 1) + 2 \cdot 3(2 - 1)} = \frac{400.01}{8} = 50.00$$

Then

$$f_R = \frac{MSSR}{PMSSE} = \frac{191.99}{50.00} = 3.84$$

$$f_C = \frac{MSSC}{PMSSE} = \frac{577.34}{50.00} = 11.55$$

Since $PMSSE$ has $2 + 6 = 8$ degrees of freedom, the critical value for testing for a row effect is $f_{.05}(1, 8) = 5.32$. Again we do not reject the null hypothesis of no row effect.

Similarly, for the column effect, the critical value is $f_{.05}(2, 8) = 4.46$ and the null hypothesis is rejected. ■

In the above example the conclusions using the pooled error mean square are exactly the same as when the error mean square was used. In general, however, this need not be the case.

Example 12.13

A Research Study

Patients with aphasia and schizophrenic disorders often display similar communication problems. The question of whether it is possible to differentiate between these disorders by factors of language deviations is the subject of many investigations. In a study entitled "Schizophrenic and Aphasic Language: Discriminable or Not?" by Marie Rausch, et al. (American Psychological Association, 0022-006X, 1980), the question of a differentiation based on the patients' ability to apply linguistic rules to word ordering tasks is discussed. There are thus two factors: the type of task and the type of patient.

For this study the tasks were classified as card or graphic. The card task involved arranging several card with one word per card to form a sentence, while the graphic task involved rearranging several words on one card to form the sentence. Four groups of two patients—each corresponding to aphasic, process schizophrenic, reactive schizophrenic, and normal—were used. Think of the patient groups as representing the rows, the two tasks as representing the columns. The measurements were the times needed to complete a task.

Table 12.12 summarizes the ANOVA of the times. The within-group entry corresponds to what we have called the error mean square. A significant effect of grouping ($f = 58.00$) was found. There was no significant difference in the card versus graphic task times, nor was there a group × task interaction. Further, when the pooled error mean square was computed (MS = 1121), the difference between the task times ($f = 2343/1121 = 2.09$) still did not prove significant.

In summary, these findings "indicate that the groups took significantly different amounts of time to complete the task but that the type of task had no consistent overall effect and that the groups were not affected differently by the tasks."

TABLE 12.12   ANALYSIS OF VARIANCE RESULTS

Source	df	MS	f
Group (A)	3	117,155	58.00*
Within group	36	2,020	
Task (B)			
(card vs. graphic)	1	2,343	2.09
A × B	3	377	<1
Pooled measures/subjects	36	1.21	

*$p < .001$.

TABLE 12.13   TUKEY TESTS ON TIME TO COMPLETE THE TASK

Group	M	Gap between groups		
		Aphasic	Process	Reactive
Normal	24.65	164.16*	22.90	13.86
Aphasic	188.81		141.26*	150.30*
Process	47.55			9.04
Reactive	38.51			

*$p < .01$.

The difference between the completion times of the various groups were subjected to further analysis (Table 12.13). The aphasic group took significantly longer to complete the task than the others and in fact proved to be responsible for the effect uncovered by the ANOVA. Further, it turned out that the two schizophrenic groups were indistinguishable from the normal group in their completion times. This latter analysis was carried out using a Tukey test, which can be found in more advanced texts. ■

EXERCISE SET 12.2

Part A

1. A tabulation of the observations for a two-way analysis of variance is given.

Factor R	Factor C		
	C_1	C_2	C_3
R_1	10 15	15 30	30 40
R_2	20 30	20 50	70 30

    a. Compute $\bar{x}$
    b. Compute $\bar{x}_{11}$
    c. Compute $\bar{x}_{\cdot 2}$
    d. Compute $T_{\cdot 1}$
    e. Compute $T_{2 \cdot}$
    f. Compute T

2. Write out each of the following:

    a. $\displaystyle\sum_{j=1}^{2}\sum_{k=1}^{3} x_{1jk}$

    b. $\displaystyle\sum_{i=1}^{2}\sum_{j=1}^{2}\sum_{k=1}^{3} x_{ijk}^{2}$

3. For the study in problem 1, test the hypotheses of no treatment or interaction effects at a 5% significance level. Prepare an ANOVA table summarizing the results.

4. A career inventory satisfaction test was adminstered to graduates from four nursing schools with both two and four year training programs. All participants had at least five years work experience and the study was concerned with the role of training and the school of graduation in job satisfaction. The results are summarized below. Test whether these factors have an effect on job satisfaction of nurses. Use $\alpha = .05$.

Length of program	School of graduation			
	School 1	*School 2*	*School 3*	*School 4*
2 years	120	125	140	130
	110	140	135	110
	125	130	145	110
4 years	85	90	120	100
	80	100	115	90
	75	80	110	100

5. For the ANOVA table given below,
    a. How many treatment levels does the factor R have?
    b. How many treatment levels does the factor C have?
    c. How many replications are there per treatment (i.e., how many entries per cell)?
    d. Complete the ANOVA test using $\alpha = .05$.

Source	SS	df	MS	f
Rows	40	4	11	
Columns	54	2	27	
Interaction	16	8	2	
Error	40	15	2.8	
Totals	118	29		

6. Describe the number of rows, columns, and entries per cell of the analysis of variance computation table that produced the following results:

Source	SS	df	MS	f
R	300	2	150	6.25
C	1860	3	620	25.83
R × C	558	6	92	3.83
Within	288	12	24	
Totals	3006	23		

7. Are the manufacturer and the cost factors in the life of television picture tubes? Tubes are selected from the three major producers, classified according to price, and tested for months of continuous operation until failure. The results (months) are given below. Carry out a two-way analysis of variance using $\alpha = .05$ and summarize the results in an ANOVA table. What specific hypotheses are being tested?

| Manufacturer | Price | | |
	Under $75.00	$75.00 to $125.00	Over $125.00
Company A	60.0 65.8	68.4 77.2	84.2 86.0
Company B	78.0 75.0	80.0 76.0	82.0 77.0
Company C	74.0 70.0	78.0 76.0	84.0 86.0

8. Complete the following ANOVA table and test for a row, column, and interaction effects. You should find no evidence to support an interaction effect. Carry out the tests a second time using the pooled mean square error.

Source	SS	df	MS	f Ratio
Row R	780	4		
Column C	180	3		
Interaction RC	205	12		
Error	1190	20		
Total				

9. Carry out an ANOVA test on the data of problem 3 assuming there is no interaction effect. Use $\alpha = .05$.

Part B

10. Four strains of corn are subjected to three different fertilizers to produce the following yields in bushels per plot. Apart from the fertilization, the plots were

identical; two such plots were used for each of the possible combinations to obtain the two observations per cell. Analyze the yields using a two-way ANOVA with $\alpha = .05$.

Fertilizer	Corn			
	Strain 1	Strain 2	Strain 3	Strain 4
Brand 1	80	110	90	100
	90	120	80	90
Brand 2	105	125	100	120
	100	135	110	180
Brand 3	100	100	80	110
	90	110	90	120

11. A computer supply house is considering three package designs for the diskettes it carries. It is also considering two types of display racks for the packages. The six possible combinations of package design and display rack are tested in 24 similar market areas. The sales in dollars for a ten day period are summarized below.
   a. Does the type of display rack affect sales?
   b. Is there a difference in sales attributable to the package design?
   c. Is there an interaction effect?

Display Rack	Package design		
	I	II	III
1	140	380	260
	170	270	200
	210	310	240
	200	340	300
2	320	490	500
	400	580	490
	380	390	400
	440	500	480

12. The following problem illustrates a $2 \times 3 \times 2$ way analysis of variance with three factors: (A) the sex of the personnel interviewer (who is called the "subject"), (B) the degree of authoritarianism of the interviewer (high, medium, or low), and (C) the sex of the job applicant. In this study applicants were all presented as equally qualified and were given evaluative, potency, and activity scores by the interviewers. The purpose of the study was to determine what if any role the authoritarianism of the interviewer played in the selection of male and female applicants for a job. The ANOVA table for each of these sets of scores is given below.
   a. Why would the sex of the interviewer be included as a factor, given the stated purpose of the study?
   b. For which of the set of scores, if any, was the sex of the interviewer significant?
   c. For which of the set of scores was the applicant's sex significant?
   d. For which of the set of scores was authoritarianism a significant factor?

e. For any of the sets of scores, was there a significant interaction effect between the authoritarianism and sex of the interviewer?

f. For which of the sets of scores was there a significant interaction effect between the sex of the applicant and the authoritarianism of the interviewer?

(Source: "Impact of Recruiter Authoritarianism and Applicant Sex on Evaluation and Selection Decisions in a Recruitment Interview Analogue Study," K. Simas and M. McCarrey, *Journal of Applied Psychology*, vol. 64, no. 5, 1979.)

SUMMARY OF ANALYSIS OF VARIANCE: Effects of Sex and Authoritarianism of Subject and Sex of Job Applicant on Evaluative Scores

Variable	df	MS	F
Sex of subject (A)	1	.51	.76
Authoritarianism (B)	2	.67	1.00
A × B	2	.66	.99
R:AB	78	.67	
Sex of applicant (C)	1	.29	1.61
A × C	1	.48	2.67
B × C	2	1.32	7.33[a]
A × B × C	2	.52	2.89
CR:AB	78	.18	

Note. R:AB is between-subjects error term. CR:AB is within-subjects error term.
[a]$p < .001$.

SUMMARY OF ANALYSIS OF VARIANCE: Effects of Sex and Authoritarianism of Subject and Sex of Job Applicant on Activity Scores

Variable	df	MS	f
Sex of subject (A)	1	.45	.73
Authoritarianism (B)	2	.20	.32
A × B	2	2.05	3.31[a]
R:AB	78	.62	
Sex of applicant (C)	1	1.44	5.33[b]
A × C	1	.10	.37
B × C	2	1.34	4.96[b]
A × B × C	2	.21	.78
CR:AB	78	.27	

Note. R:AB is between-subjects error term, CR:AB is within-subjects error term.
[a]$p < .05$.
[b]$p < .01$.

SUMMARY OF ANALYSIS OF VARIANCE: Effects of Sex and Authoritarianism of Subject and Sex of Job Applicant on Potency Scores

Variable	df	MS	f
Sex of subject (A)	1	.43	.83
Authoritarianism (B)	2	.09	.17
A × B	2	2.00	3.85[a]
R:AB	78	.52	
Sex of applicant (C)	1	27.90	77.50[b]
A × C	1	.03	.08
A × C	2	1.68	4.67[c]
A × B × C	2	.13	.36
CR:AB	78	.36	

Note. R:AB is between-subjects error term, CR:AB is within-subjects error term.
[a]$p < .05$.
[b]$p < .001$.
[c]$p < .01$.

Library

Project

The following articles make use of multiple ANOVA. Select one and report on it. When a significant effect (main or interaction) is reported, give an interpretation in terms of the variables involved.

a. "Development of Personal Space in Pre-School Children as a Function of Age and Day Care Experience," E. Sarafino and H. Helmuth, *Journal of Social Psychology*, vol. 115, first half. Oct. 1981.

b. "Home Defense and the Police," J. Feagin, *American Behavioral Scientist*, vol. 13, no. 5 and 6, Aug. 1970.

c. "Loneliness, Cooperation, and Conformity among American Undergraduates," R. Hansson and W. Jones, *Journal of Social Psychology*, vol. 115, first half, Oct. 1981.

d. "Sex Differences in Opinion Conformity and Dissent," D. Tuthill and D. Forsyth, *Journal of Social Psychology*, vol. 116, second half, April 1982.

e. "The Selling of Career Counseling," B. Kerr, *The Vocational Guidance Quarterly*, vol. 30, no. 3, March 1982.

f. "The Effect of Different Reinforcement Systems on Cooperative Behavior Exhibited by Children in Classroom Context," D. Buckholdt and J. Wodarski, *Journal of Research and Development in Education*, vol. 12, no. 1, 1978.

g. "Use of Cognitive Capacity in Reading Easy and Difficult Text," B. Britton et al., *Journal of Reading Behavior*, vol. 12, no. 1, Spring 1980.

h. "Wait Time as a Variable in Sex-Related Differences During Mathematical Instruction," D. Gore and D. Roumagoux, *Journal of Educational Research*, vol. 76, no. 5, May 1983.

i. "The Effect of Goal Specification on Counseling Outcome," J. King and G. Voge, *Journal of College School Personnel*, vol. 23, no. 3, July 1983.

j. "The Attitudes of Elementary School Children Toward School and Subject Matter," T. Haladyna and G. Thomas, *Journal of Experimental Education*, vol. 48, Fall 1979.

k. "Social Interest, Running, and Life Adjustment," J. Zarski et al., *The Personnel and Guidance Journal*, vol. 61, no. 3, Nov. 1982.

l. "Inhalation of NO_2 and Blood-Borne Cancer Cell Spread to the Lungs," A. Richters and K. Kuraitis, *Archives of Environmental Health*, vol. 36, no. 1, Jan.-Feb. 1981.

m. "Prevalence and Correlates of Passive Smoking," G. Friedman, et al., *American Journal of Public Health,* vol. 73, no. 4, April 1983.

n. "The Effect of Exercise on Mood and Cognitive Functioning," S. Lichtman and E. Poser, *Journal of Psychosomatic Research,* vol. 27, no. 1, 1983.

o. "The Effect of Aerobic Conditioning and Induced Stress," G. McGlynn et al., *Journal of Sports Medicine,* vol. 23, no. 3, Sept. 1983.

p. "'The Impact of Birth Order and Sex on Social Interest," L. Schneider and D. Reuterfors, *Journal of Individual Psychology,* vol. 37, no. 2, May 1981.

COMPUTER PRINTOUT

Three teaching methods are compared by using scores of tests given to students at the end of the teaching unit. The Minitab commands, the scores, and the resulting one-way analysis are shown in Figure 12.6. The confidence intervals for the three means are a nice addition to the analysis.

FIGURE 12.6 COMPUTER PRINTOUT: MINITAB ONE-WAY ANALYSIS OF VARIANCE

```
MTB >SET GRADES FROM METHOD I INTO COLUMN C1
DATA>42 98 87 56 78 49 81 90 67 72 85
DATA>END
MTB >SET GRADES FROM METHOD II INTO COLUMN C2
DATA>82 71 76 54 34 46 51 68 42 50
DATA>END
MTB >SET GRADES FROM METHOD III INTO COLUMN C3
DATA>55 49 62 93 71 60 43 47 51 80 35 44
DATA>END
MTB >AOVONEWAY ON C1-C3

ANALYSIS OF VARIANCE

DUE TO      DF        SS        MS=SS/DF      F-RATIO
FACTOR      2        1814.         907.          3.17
ERROR      30        8581.         286.
TOTAL      32       10395.

LEVEL              N         MEAN          ST. DEV.
C1                11         73.2            17.9
C2                10         57.4            15.9
C3                12         57.5            16.8

POOLED ST. DEV. =              16.9

INDIVIDUAL 95 PERCENT C. I. FOR LEVEL MEANS
(BASED ON POOLED STANDARD DEVIATION)
```

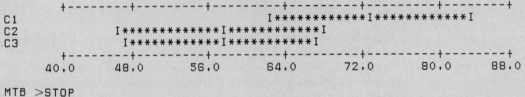

```
MTB >STOP
```

CHAPTER CHECKLIST

Vocabulary

Factor	Error mean square
Treatment	Treatment mean square
Treatment level	One-way ANOVA
Treatment effect	ANOVA table
Completely randomized design	Between-groups variance estimate
Total sum of squares	Within-groups variance estimate
Treatment sum of squares	Two-way ANOVA
Error sum of squares	Interaction effect
Treatment means	Replicated experiment
Grand mean	

Techniques

Performing a one-way ANOVA test and reporting the results in an ANOVA table

Interpreting an ANOVA table.

Performing a two-way ANOVA test where there is the possibility of an interaction and reporting the results in an ANOVA table

Performing a two-way ANOVA test where there is no interaction and reporting the results in an ANOVA table

Interpreting a two-way ANOVA table

Notation

SS	Total sum of squares
$SSTr$	Treatment sum of squares (one-way ANOVA)
SSE	Error sum of squares
$MSSTr$	Treatment mean square (one-way ANOVA)
$MSSE$	Error mean square
SSR	Treatment sum of squares for rows (two-way ANOVA)
SSC	Treatment sum of squares for columns (two-way ANOVA)
$SSRC$	Interaction sum of squares (two-way ANOVA)
$MSSR$	Treatment mean square for rows (two-way ANOVA)
$MSSC$	Treatment mean square for columns (two-way ANOVA)
$MSSRC$	Interaction mean square (two-way ANOVA)
$PMSSE$	Pooled error mean square (two-way ANOVA with no interaction effect)

CHAPTER TEST

1. In an ANOVA test, the F distribution is used with the ratio $MSSTr/MSSE$, since both $MSSTr$ and $MSSE$ are approximations to _____.

2. Why are the treatment mean square and error mean square also referred to as the between- and within-group estimates of the common variance σ^2?

3. In a one-way ANOVA test, the sum SS is partitioned into two other sums. What are these sums and what do they represent?

4. In a two-way ANOVA with replication on the treatment combinations, the sum SS is partitioned into four sums. What are these sums called? How are they defined?

5. The ANOVA table for a one-factor experiment is given below.
   a. How many treatments were involved?
   b. Assuming equal-sized treatment groups, how many observations were there in each?
   c. Using $\alpha = .05$, test for a treatment effect.

Source	SS	df	MS
Treatment	76	4	18.5
Error	39	15	2.6
Totals	115	19	

6. Complete the ANOVA table below.

Source	SS	df	MS
Treatment		5	20
Error	180		5
Totals			

7. An airport bordered by residental areas is planning to limit access to "quiet" jet airliners. Takeoff noise level readings in decibels for the four makes of jets operating from the airport were recorded for early morning flights on four randomly selected days. The results are given below. Using $\alpha = .05$, test for a difference in the mean noise levels of these four makes of airliners on takeoff.

Make of plane			
A	B	C	D
76	59	64	73
72	63	68	71
81	60	72	76
79	66	70	74

8. A poultry raiser wishes to determine the effects of three dietary programs on the weight gains of three varieties of chickens. An experiment is conducted in which two young chickens of each type are selected and maintained on each of the dietary programs for four weeks. The weight gains are summarized below.
   a. Test the hypothesis that the mean weight gains are equal across all three dietary programs. Use $\alpha = .05$.

b. Test the hypothesis that the mean weight gains are equal for the three poultry varieties. Use $\alpha = .05$.

c. Test for an interaction effect. Use $\alpha = .05$.

Variety of chicken	Dietary program		
	I	*II*	*III*
A	10 8	7 5	9 8
B	9 9	11 7	9 11
C	11 12	8 5	10 12

9. Rework number 8, assuming there is no interaction effect, using the pooled error mean square.

10. Using the data in problem 8, find:

a. $\displaystyle\sum_{i=1}^{3}\sum_{j=1}^{3}\sum_{k=1}^{2} x_{ijk}$

b. $\displaystyle\frac{\sum_{k=1}^{2} x_{23k}}{2}$

c. $\bar{x}_1.$

d. $\bar{x}_2.$

CHAPTER THIRTEEN
Nonparametric Statistics

The statistical tests developed in previous chapters all require that the distributions of the population variables be known or else the samples be of such size that the normality conclusion of the central limit theorem can be invoked. Often these conditions cannot be met. Additionally, we have yet to consider methods of analyzing ordinal data, which result from relative rankings rather than direct measurements.

Problems of unknown distributions, small samples, and ordinal data are quite common. Fortunately there are statistical tests that are appropriate when these conditions exist. Since these do not depend on the distributions of the variables or associated parameters such as the mean and variance, they should be called *distribution-free* or *nonparametric* tests. It is the practice, however, to refer to both types as nonparametric tests.

Nonparametric testing is an extensive subject. Textbooks and courses exist that are exclusively devoted to it. We make no pretense of covering the subject in its entirety; rather, we concentrate on what we believe to be the most powerful and popular of the tests available.

13.1   TESTS BASED ON THE CHI-SQUARE DISTRIBUTION

The chi-square distribution was introduced in Chapter 6 in connection with tests for the goodness of fit of a theoretical distribution to a particular population. A review of this test will show that it is nonparametric in nature. In this section we present two further important applications of the chi-square distribution. These are tests for the independence of two variables and for the quality of several population proportions, the latter sometimes being referred to as a test of *homogeneity*.

13.2   CONTINGENCY TABLES

Often we must work with qualitative variables whose values are descriptive rather than numerical. Using these descriptions we can divide the population into classes; we then speak of having *classified* the population. Thus for example we might classify students according to their class standing, major, or even marital status. For each of these classifications there is a qualitative variable whose values are the possible classifications. In the case of class standing, these might be freshman, sophomore, junior, senior, or graduate. With qualitative variables it is the count or relative frequency of each class that is used in the analysis.

When a population is classified according to two variables it is said to have been *cross-classified* or subjected to a *two-way classification*. While higher classifications are possible, they will not be considered here.

A *two-way contingency table* is a simple array that is used to present the results of a two-way classification of a sample. In such a table the classes of one of the variables are the row headings and those of the other are the column headings. In the cell at the intersection of a row and column is written the number of subjects observed jointly in the corresponding classes. The total of any row is called a *marginal row total*, that of any column a *marginal column total*. A marginal total is simply the number of subjects in the class that defines the row or column in question. The sum of either the marginal row or marginal column totals must equal the total number of subjects that have been cross-classified.

Example 13.1

The two-way classification of 75 students according to their grade point average and use of the library is given in Table 13.1. The "infrequent" class corresponds to fewer than two visits per week.

TABLE 13.1   CONTINGENCY TABLE: LIBRARY USE × GRADE POINT AVERAGE

Library use	g.p.a.			Total
	1.0–2.0	2.0–3.0	3.0–4.0	
Infrequent	7	11	9	27
Frequent	7	17	24	48
Total	14	28	33	75

The marginal row totals are 27 and 48. Thus of the 75 students, 27 were infrequent users and 48 were frequent users. Similarly from the marginal column totals, 14 of the students had a g.p.a. between 1.0 and 2.0, 28 between 2.0 and 3.0, and 33 between 3.0 and 4.0. The entry 7 in the cell at the intersection of the first row and first column indicates that of the 27 infrequent users, 7 had a g.p.a. between 1.0 and 2.0. Or equivalently, of the 14 students with a g.p.a. between 1.0 and 2.0, 7 were classified as being infrequent users of the library. If, for example, we want the number of students with a grade point average between 3.0 and 4.0 who were frequent users of the library, we look in the cell at the intersection of the third column and second row and find 24. ∎

13.3 THE CHI-SQUARE TEST OF INDEPENDENCE

Suppose we have two variables or classifications associated with the same population and we wish to test whether they are independent. A sample of subjects is randomly selected and cross-classified by the two variables to produce a contingency table. By assuming the variables are independent, we can also predict an expected frequency for each cell. Proceeding as with the goodness-of-fit test, we then compute

$$\sum \frac{(o - e)^2}{e}$$

where o and e are observed and expected frequencies, respectively, and the sum extends over all the cells of the table. If this value is too large, it indicates a poor agreement between the observed and expected frequencies and the null hypothesis of independence is rejected. As critical chi-square values we commonly use $\chi^2_{.05}$, $\chi^2_{.01}$, or $\chi^2_{.005}$. It turns out that for a contingency table with m rows and n columns, the number of degrees of freedom is $v = (m - 1)(n - 1)$.

To illustrate how the computations proceed, consider the contingency table for the classification of 100 adults as to their years of formal education and to the type of major investment program in which they participate (Table 13.2).

TABLE 13.2 CONTINGENCY TABLE: SAVING PLAN × YEARS OF EDUCATION

Investment	Education (years)				Total
	0–12	*12–14*	*14–16*	*Over 16*	**Total**
Banks, credit unions, etc.	17	10	11	7	45
Real estate	6	5	7	6	24
Stocks and bonds	9	8	6	8	31
Total	32	23	24	21	100

Now suppose the variables are independent. Then for any two events

$$P(A \text{ and } B) = P(A)P(B)$$

Thus for the joint occurrence of a subject in the classifications of row 1 and column 1:

P(row 1 and column 1 occurrence) = P(row 1 occurrence) $\times$ P(column 1 occurrence)

The best estimates of these two probabilities from the existing data are

$$P\text{(row 1 occurrence)} = \frac{45}{100} = \frac{\text{row 1 marginal total}}{\text{total number of subjects}}$$

$$P\text{(column 1 occurrence)} = \frac{32}{100} = \frac{\text{column 1 marginal total}}{\text{total number of subjects}}$$

Thus we use P(row 1 and column 1 occurrence) = (45/100)(32/100). Based on this the expected number of the 100 subjects in the (1, 1) or first row–first column cell is

$$e_{11} = \left(\frac{45}{100}\right)\left(\frac{32}{100}\right) \times 100 = \frac{45 \times 32}{100} = 14 \text{ (rounded)}$$

It is no accident that the number in the first row–first column cell is

$$e_{11} = \frac{\text{(row 1 marginal total)} \times \text{(column 1 marginal total)}}{\text{total number of subjects}}.$$

A similar computation for, say, the (2, 4) cell at the intersection of the second row and fourth column shows the expected value to be

$$e_{24} = \frac{\text{(row 2 marginal total)} \times \text{(column 4 marginal total)}}{\text{total number of subjects}}$$

In general, the expected number for a cell is found by multiplying the marginal totals of the row and column in which the cell occurs and dividing by the total number of subjects.

The computation of the expected number of adults in each of the 12 cells in Table 13.2 is given below. As usual, the first subscript locates the row, the second the column. The expected values are all rounded to two decimal places.

$$e_{11} = \frac{45 \times 32}{100} = 14.40 \qquad e_{12} = \frac{45 \times 23}{100} = 10.35 \qquad e_{13} = \frac{45 \times 24}{100} = 10.80 \qquad e_{14} = \frac{45 \times 21}{100} = 9.45$$

$$e_{21} = \frac{32 \times 24}{100} = 7.68 \qquad e_{22} = \frac{24 \times 23}{100} = 5.52 \qquad e_{23} = \frac{24 \times 24}{100} = 5.76 \qquad e_{24} = \frac{24 \times 21}{100} = 5.04$$

$$e_{31} = \frac{31 \times 32}{100} = 9.92 \qquad e_{32} = \frac{31 \times 23}{100} = 7.13 \qquad e_{33} = \frac{31 \times 24}{100} = 7.44 \qquad e_{34} = \frac{31 \times 21}{100} = 6.51$$

These results and the original observed frequencies are now summarized in Table 13.3. The observed frequencies are above the diagonals, the expected values based on the independence assumption are below them.

The discrepancy between the observed and expected values is now measured by

$$\chi^2 = \frac{(o - e)^2}{e}$$

Summing across the successive rows we find

$$\chi^2 = \frac{(17 - 14.40)^2}{14.40} + \frac{(10 - 10.35)^2}{10.35} + \frac{(11 - 10.80)^2}{10.80} + \frac{(7 - 9.45)^2}{9.45} +$$
$$\frac{(6 - 7.68)^2}{7.68} + \frac{(5 - 5.52)^2}{5.52} + \frac{(7 - 5.76)^2}{5.76} + \frac{(6 - 5.04)^2}{5.04} +$$
$$\frac{(9 - 9.92)^2}{9.92} + \frac{(8 - 7.13)^2}{7.13} + \frac{(6 - 7.44)^2}{7.44} + \frac{(8 - 6.51)^2}{6.51} = 2.79.$$

TABLE 13.3 CONTINGENCY TABLE WITH EXPECTED FREQUENCIES

Investment	Education 0–12	12–14	14–16	Over 16	Total
Banks, credit unions, etc.	17 / 14.40	10 / 10.35	11 / 10.80	7 / 9.45	45
Real estate	6 / 7.68	5 / 5.52	7 / 5.76	6 / 5.04	24
Stocks bonds	9 / 9.92	8 / 7.13	6 / 7.44	8 / 6.51	31
Total	32	23	24	21	100

Finally, to test at a 5% significance level the null hypothesis that these two classifications are independent against the alternate hypothesis that they are not, we compare $\chi^2 = 2.79$ with the critical value $\chi_{.05}$. Using $\nu = (3 - 1)(4 - 1) = 6$ degrees of freedom, we find $\chi_{.05}^2 = 12.59$. Since $\chi^2 = 2.79 < \chi_{.05}^2$ we do not reject the hypothesis that the type of investment plan which an individual follows is independent of the amount of education he or she may have.

PROCEDURE FOR PERFORMING THE CHI-SQUARE TEST OF INDEPENDENCE

H_0: Two variables of classification are independent.
H_1: The variables are dependent.
Given: The significance level α;
   A contingency table with m rows and n columns.
1. Using $\nu = (m - 1)(n - 1)$, find the critical value χ_α^2.
2. For each cell of the contingency table compute the expected frequency using

$$e_{ij} = \frac{(\text{row } i \text{ marginal total})(\text{column } j \text{ marginal total})}{\text{total number of subjects classified}}$$

3. Compute the value of the test statistic:

$$\chi^2 = \Sigma \frac{(o - e)^2}{e}$$

where the summation extends over all cells of the contingency table.
4. Compare the value of the test statistic and the critical value. Reject H_0 if $\chi^2 > \chi_\alpha^2$.

In this test the continuous χ^2 distribution is used to approximate a discrete distribution. For this approximation to be effective, it is necessary that the expected frequencies of the cells not be too small. It is recommended that not more than 20% of the cells have expected frequencies below 5. When this is not the case, it may still be possible to salvage part of the test by combining classes.

Example 13.2

The results of a cross-classification by grade and by major of a random sample of 130 students from a multisection statistics course are given in Table 13.4. Test the hypothesis that these classifications are independent.

TABLE 13.4 CONTINGENCY TABLE: GRADES × MAJOR

| Grade | Major | | | Total |
	Psychology	Sociology	Health	
A	3	7	2	12
B	4	18	11	33
C	9	33	17	59
D	5	9	12	26
Total	21	67	42	130

The frequencies of four cells (33%) are below 5 and this proves to be true of their expected values. One way to avoid this is to combine the A and B grade classes. Or we could combine the psychology and health classifications. The first choice would probably better serve the intent of the study. We obtain Table 13.5, in which all but one of the cells (11%) have expected frequencies greater than 5.

Using

$$\chi^2 = \Sigma \frac{(o - e)^2}{e}$$

we find $\chi^2 = 3.92$. Using $\nu = (3 - 1)(3 - 1) = 4$ degrees of freedom, we find the critical value is $\chi_{.05} = 9.50$. Thus $\chi^2 = 3.92$ is not significant and we do not reject the two classifications as being independent.

TABLE 13.5 CONTINGENCY TABLE WITH EXPECTED FREQUENCIES

Grade	Major Psychology	Major Sociology	Major Health	Total
A-B	7 / 7.27	25 / 23.19	13 / 14.54	
C	9 / 9.53	33 / 30.41	17 / 19.06	59
D	5 / 4.20	9 / 13.40	12 / 8.40	26
Total	21	67	42	130

13.4 THE CHI-SQUARE TEST OF HOMOGENEITY

In Chapter 10 a test for the equality of two population proportions was developed based on the normal distribution. On occasion it may happen that there are several proportions to be tested simultaneously. An even more complex situation arises when the several populations have all been classified according to the same variable. We generally do not expect an equality of proportions for all the classes of all the populations. We do, however, quite often need to test whether the proportions for a given class are equal across all populations and whether this is true for each class. If this proves to be the case, we say the populations are *homogeneous with respect to the variable of classification.*

The chi-square test of homogeneity is performed with the aid of contingency table, with the rows representing the populations and the columns the classes. The hypotheses are now:

H_0: The populations are homogeneous with respect to the variable of classification.
H_1: The populations are not homogeneous.

Independent samples from the populations are now required. And while the underlying rationale is considerably different, the test of homogeneity is computationally the same as for the test of independence. We omit the formal statement of the test.

Example 13.3

A political pollster is studying whether various age groups are homogeneous with respect to political affiliation. Three age groups are defined: 18 through 29, 30 through 49, and 50 or over. The classes are Democrat, Republican, Independent, and Not Registered. Independent samples of 200, 400, and 150 are selected from these populations and classified to produce Table 13.6 below. The expected values are also included.

TABLE 13.6   CONTINGENCY TABLE: AGE × PARTY AFFILIATION

Age group	Affiliation				Total
	Democratic	Republican	Independent	No registration	
10–29	60 / 80.53	50 / 60.00	34 / 30.93	56 / 28.53	200
30–49	190 / 161.07	130 / 120.00	52 / 61.87	28 / 57.07	400
50 and over	52 / 60.40	45 / 45.00	30 / 23.20	23 / 21.40	150
Total	302	225	116	107	750

For a 5% significance level test with $\nu = (4 - 1)(3 - 1) = 6$, the critical value is $\chi^2_{.05} = 12.6$. Against this we should compare:

$$\chi^2 = \frac{(60 - 80.53)^2}{80.53} + \frac{(50 - 60.00)^2}{60.00} + \frac{(34 - 30.93)^2}{30.93} +$$
$$\frac{(56 - 28.53)^2}{28.53} + \frac{(190 - 161.07)^2}{161.07} + \frac{(130 - 120.00)^2}{120.00} +$$
$$\frac{(52 - 61.87)^2}{61.87} + \frac{(28 - 57.07)^2}{57.07} + \frac{(52 - 60.40)^2}{60.40} +$$
$$\frac{(45 - 45.00)^2}{45.00} + \frac{(30 - 23.20)^2}{23.20} + \frac{(23 - 21.40)^2}{21.40}$$

However, for the fourth term we see

$$\frac{(56 - 28.53)^2}{28.53} = 26.45$$

This cell's contribution alone far exceeds the critical value, and so the hypothesis of homogeneity is rejected. The degree of party affiliation is not the same across these age groups.   ■

The next example illustrates the test of equality for several proportions. Note that the classification variable is now a binomial variable and the classes correspond to whatever represents "success" and "failure."

Example 13.4

Using the voter registration data of Example 13.3, test the hypothesis that the proportion of registered voters is the same for each age group. Let p_1, p_2, and p_3 be the proportion of registered voters aged 18 to 29, 29 to 49, and 50 or over, respectively. We wish to test:

H_0: $p_1 = p_2 = p_3$
H_1: Not all the proportions are equal.

By combining the data of each of the samples into the classes "registered" and "unregistered," we obtain Table 13.7 below.

TABLE 13.7 CONTINGENCY TABLE: AGE × REGISTRATION

Age group	Status Registered	Unregistered	Total
18–29	144 / 171.47	56 / 28.53	200
29–49	372 / 342.93	28 / 57.07	400
50 and over	127 / 128.60	23 / 21.40	150
Total	643	107	750

$$e_{11} = \frac{200 \times 643}{750} = 171.47 \qquad e_{12} = \frac{200 \times 107}{750} = 28.53$$

$$e_{21} = \frac{400 \times 643}{750} = 342.93 \qquad e_{22} = \frac{400 \times 107}{750} = 57.07$$

$$e_{31} = \frac{150 \times 643}{750} = 128.60 \qquad e_{32} = \frac{150 \times 107}{750} = 21.40$$

For a 5% significance level test with $\nu = (3 - 1)(2 - 1) = 2$ the critical value is $\chi^2_{.05} = 5.99$. The calculated chi-square value is

$$\chi^2 = \frac{(144 - 171.47)^2}{171.47} + \frac{(56 - 28.53)^2}{28.53} + \frac{(372 - 342.93)^2}{342.93} +$$
$$\frac{(28 - 57.07)^2}{57.07} + \frac{(127 - 128.60)^2}{128.60} + \frac{(23 - 21.40)^2}{21.40} = 48.26$$

This far exceeds the critical value and thus the null hypothesis is rejected: the proportions are not all equal. Note that the sample proportions are

$$\hat{p}_1 = \frac{144}{200} = .72 \qquad \hat{p}_2 = \frac{372}{400} = .93 \qquad \hat{p}_3 = \frac{127}{150} = .847$$

There appears to be substantial disagreement, and this is confirmed by the chi-square test. ∎

EXERCISE SET
13.1

Part A

1. A trade school classifies its electronic assemblers in terms of their speed and the quality of their finished products. The following contingency table resulted from a joint classification of a sample of these assemblers.

Speed	Quality			Total
	Poor	Satisfactory	Excellent	
Slow	8	14	18	40
Fast	15	12	8	35
Total	23	26	26	75

   a. How many assemblers were classified in all?
   b. How many assemblers were classified as slow?
   c. How many assemblers had their products classified as satisfactory?
   d. How many of the assemblers were fast and produced an excellent product?
   e. Complete the table to show the expected number in each cell when these two classifications are considered to be independent.
   f. Using $\alpha = .05$, test whether the variables of classification are independent.

2. Adults of varying ages were shown a new model of an auto in several different colors and then asked to complete a questionnaire that contained, among other information, their age and their preferred color. The joint classification of these responses is given below. Test for a relationship between age and color preference.

Age	Color			
	Gray	Maroon	Blue	White
Under 25	40	26	38	70
25–40	21	40	30	58
41–55	39	28	40	36
Over 55	28	21	14	11

3. Forty male and 50 female students at a university were asked to try three brands of soft drinks and select a favorite. Their responses are tabulated below. Is there a relationship between the sex of the student and the brand preferred?

Sex	Brand I	Brand II	Brand III
Male	9	12	19
Female	14	23	13

4. A sample of 300 seniors is classified with respect to their grade point average and their evaluation of their school. Test the hypothesis that the ratings and grade point averages are not related.

Ratings	Grade point average		
	2–2.75	2.75–3.5	3.5–4.0
Excellent	50	63	28
Satisfactory	28	35	23
Poor	14	20	39

5. Three hundred drivers (over 17 years of age) are sampled as to their age and the number of traffic tickets they have collected over the preceding 12 month period. The results are summarized below. Does the number of tickets received appear to be related to the age of the driver?

Number of tickets	Age		
	17–25	26–40	Over 40
0	23	38	45
1	18	31	40
2	45	25	9
3	10	4	5
Over 3	4	2	1

6. Samples of 300 workers from the eastern, central, and western United States were asked whether they would be interested in relocating if a better job were available. The results are given below. Using $\alpha = .01$, test for homogeneity with respect to area of residence.

Attitude	Eastern	Central	Western
Not interested	170	204	125
Possibly interested	84	69	90
Definitely interested	46	27	85

7. Undocumented workers of three nationalities were sampled to determine whether or not they planned to eventually return to their country of origin. The results are tabulated below. Test whether the percentage of undocumented workers who plan to return to their homeland is the same for each of these three nationalities.

Plans	Nationality I	Nationality II	Nationality III
Planned to return	215	137	251
Did not plan to return	85	63	184

8. The distribution of weights and pulse rates (at rest) of a sample of middle-aged women is given below. At a 5% significance level, test whether pulse rate is related to weight.

Pulse rate	Weight		
	Underweight	*Normal*	*Overweight*
Below 75	84	51	43
75–81	48	64	65
Above 81	28	50	67

9. A health network maintains four clinics throughout the area it serves and is interested in whether the rate of hospitalization is the same for each. Two hundred cases from each clinic are examined; the results are shown below. Does there appear to be a difference in the recommended hospitalization rates for these four clinics?

Recommendation	Clinic 1	Clinic 2	Clinic 3	Clinic 4
Hospitalization recommended	36	42	49	31
Hospitalization not recommended	164	158	151	169

Part B

10. The following chi-square tables are from a study of the differences in educational backgrounds of students who persist with an educational program. In all 612 students were tracked and evaluated over a two semester period. (Source: "Educational Background at Entry and Dropout From Part-Time Undergraduate Studies," Joti Bhatnagar, *Journal of Experimental Education*, 1976.)

TABLE 1 POSSESSION OF HIGH SCHOOL DIPLOMA

Group	High school diploma	
	Yes	*No*
Persisters	262 75.7%	84 24.3%
Drop-outs	173 65.0%	93 35.0%

$\chi^2 = 8.35; p < .005.$

TABLE 2 AVERAGE HIGH SCHOOL MATRICULATION GRADE

Group	0–60%	61–65%	66–70%	71–80%	80%+
Persisters	23 8.6%	79 29.5%	65 24.3%	72 26.9%	29 10.8%
Drop-outs	10 6.0%	51 30.5%	49 29.3%	44 26.3%	13 7.8%

TABLE 3   TYPE OF HIGH SCHOOL ATTENDED

| Group | Type of high school attended | | | | |
	Private	Public	Evening	Correspondence	Other
Persisters	39 11.3%	21 66.8%	63 18.2%	4 1.2%	9 2.6%
Drop-outs	27 10.2%	175 65.8%	52 19.5%	1 0.4%	11 4.1%

TABLE 4   DELAY BETWEEN LEAVING HIGH SCHOOL AND UNIVERSITY REGISTRATION

| Group | Delay | |
	Yes	No
Persisters	295 85.3%	51 14.7%
Drop-outs	207 77.8%	59 22.2%

a. According to the author, "students who did not possess a high school diploma showed a 10% higher rate of discontinuance. This difference was significant at a 1% level of confidence." Interpret this statement in light of Table 1.

b. Refer to Table 2. Test the hypothesis that the two student groups are homogeneous with respect to high school matriculation grade.

c. Refer to Table 3. Formulate a hypothesis and test it.

d. Refer to Table 4. Formulate a hypothesis and test it.

11. Samples of black and white home owners were surveyed as to their home defense orientation (personal role versus leaving it to the police). The respondents were also classified by several demographic variables. Two of the contingency tables are given below. (Source: "Home Defense and the Police," J. R. Feagin, *American Behavioral Scientist,* vol. 13, nos. 5 and 6.)

BLACK AND WHITE PERSPECTIVES ON HOME-DEFENSE (ORC SURVEY, 1968)

Question: "Do you think that people like yourself have to be prepared to defend their homes against crime and violence, or can the police take care of that?"

	Defend homes, %	Leave to police, %	No opinion/ other[a], %	Total percentage	Total n
Black sample	65	24	11	100	468
White sample	52	41	7	100	551

[a]About two-thirds of these answers were of the "don't know/no opinion" type.

HOME-DEFENSE ORIENTATION BY KEY DEMOGRAPHIC VARIABLES (ORC SURVEY, 1968)

	Black sample					White sample				
	Defend homes, %	Leave to police, %	No opinion/ other, %	Total percentage	n	Defend homes, %	Leave to police, %	No opinion/ other, %	Total percentage	n
Sex[a]										
Male	73	18	9	100	169	51	43	6	100	250
Female	60	28	13	101	299	53	40	8	101	301
Age[a]										
18–29	73	20	7	100	104	54	38	9	101	129
30–49	70	20	11	101	190	54	41	4	99	206
50 +	53	33	14	100	163	48	43	9	100	212
	$\chi^2 = 14.880, p < .006$					$\chi^2 = 4.621$, n.s.				
Region[a]										
Northeast	63	24	13	100	71	38	56	5	99	149
Northcentral	74	20	6	100	155	54	39	8	101	143
South	59	27	15	101	218	75	21	5	101	141
West	58	29	13	100	24	40	50	10	100	118
Urban[a]										
Under 2500 (rural)	64	15	21	100	75	67	27	6	100	192
2500– 100,000	51	42	8	101	65	52	41	8	101	106
100,000–1 million	65	23	12	100	128	44	49	7	100	167
Over 1 million	69	23	9	101	200	33	61	7	101	86
	$\chi^2 = 22.535, p < .001$					$\chi^2 = 36.804, p < .0001$				

[a]Percentages may not add to 100% because of rounding-off procedures. Numerical totals vary a bit from table to table because respondents with "no answer/no data" on a given demographic variable have been deleted.

a. Was there a significant difference in the proportion of blacks and whites who supported the concept of a personal role in home defense? NOTE: Some estimations will be necessary due to the rounding of the data.

b. Is there a relationship between the sex of the responder and the individual's home defense orientation for the blacks? For the whites? If the two samples are pooled, does this relationship prove significant?

c. Test for a relationship between region and orientation for both populations.

d. The relationship between the size of the city of residence (urban factor) and the home defense orientation proved significant for both populations. Examine the percentages for the white sample and interpret the obvious increase in support for the role of the police.

Library

Project

Scientific periodicals are filled with statistical investigations that in some way or another make use of a chi-square test. Locate one of the following articles and report the chi-square test used.

a. "Ethnicity and Child Rearing Practices in Uganda," J. Opolot, *Journal of Social Psychology*, vol. 116, 1982.

b. "Game Preferences of Delinquent and Nondelinquent Boys," A. Blum, *Journal of sociology and Social Welfare*, vol. 76, no. 2, May 1980.

c. "Female Career Preference and Androgyny," J. Clarey and A. Sanford, *The Vocational Guidance Quarterly*, vol. 30, no. 3, March 1983.

d. "Mentoring Among Teachers," M. Fagan and G. Walter, *Journal of Educational Research,* vol. 76, no. 2, Nov. 1982.

e. "The Effect of Residence Hall Living at College on Attainment of the Baccalaureate Degree," B. Levin and D. Clowes, *Journal of College Student Personnel,* vol. 23, no. 2, March 1982.

f. "Consumer Problems of Older Americans," K. Bernhardt, *Journal of Retailing,* vol. 57, no. 3, Fall 1981.

g. "A Longitudinal Comparison of Minority and Nonminority College Dropouts," E. Rugg, *The Personnel Guidance Counselor,* vol. 61, no. 4, Dec. 1982.

h. "Daily Variation and Other Factors Affecting the Occurrence of Cerebrovascular Accidents," C. Brackenridge, *Journal of Gerontology,* vol. 36, no. 2, March 1981.

13.5   THE MANN-WHITNEY TEST

The Mann-Whitney test, also known as the Wilcoxon two-sample test, is one of the most powerful of the nonparametric tests for comparing two populations. More specifically, it is used to test the null hypothesis that two populations have identical distributions against the alternate hypothesis that these distributions differ only with respect to location. In many applications it is used in place of the t test for the equality of two means when the normality assumption is questionable. A most useful feature of the test is that it can be applied when the observations are ranks (ordinal data) rather than measurements.

The first step in using the test is to combine the samples from the two populations and to rank the observations in ascending order by size. Ties are assigned the mean of ranks that would have been used had there been no ties. Next we compute the sums w_1 and w_2 of the ranks for each of the samples.

Example 13.5   Consider the two samples:

Sample 1: $x_1 = 10, x_2 = 14, x_3 = 14, x_4 = 17, x_5 = 20$
Sample 2: $y_1 = 11, y_2 = 15, y_3 = 17$

When these two samples are combined and ranked we obtain:

10	11	14	14	15	17	17	20	(observations)
x	y	x	x	y	x	y	x	(variables)
1	2	3.5	3.5	5	6.5	6.5	8	(ranks)

The rankings for each of the samples are as follows:

Sample 1		Sample 2	
x	Rank	y	Rank
10	1	11	2
14	3.5	15	5
14	3.5	17	6.5
17	6.5	$w_2 = 13.5$	
20	8		
$w_1 = 22.5$			

Now, if there is a difference in the location parameters of the two populations, most of the lower ranks will be assigned to the observations of one sample and the higher ones to those of the second sample. If, for example, all the lower ranks fall to sample 1, then its rank sum is

$$1 + 2 + 3 + \ldots + n_1 = \frac{n_1\,(n_1\,+\,1)}{2}$$

And if all the high ranks fall in sample 2, its rank sum is

$$(n_1\,+\,1)\,+\,(n_1\,+\,2)\,+\,\ldots\,(n_1\,+\,n_2) = n_1 \cdot n_2\,+\,\frac{n_2(n_2\,+\,1)}{2}$$

Regardless of how the ranks are assigned to the samples, we see that

$$u_1 = n_1 n_2 + \frac{n_1(n_1\,+\,1)}{2} - w_1 \qquad u_2 = n_1 n_2 + n_2(n_2\,+\,1) - w_2$$

are both nonnegative. Traditionally, the smaller of these two numbers has been used as the value of the test statistic U. If this value is too small, then the null hypothesis that both populations have the same location parameter is rejected.

The one-sided critical values of U for sample sizes through $n_1 = n_2 = 20$ are given in Table IX of the Appendix. The blanks indicate that there is no value of U for which the null hypothesis can be rejected at the stated significance level.

Example 13.6

Suppose $n_1 = 10$, $n_2 = 8$, and $u = 9$. With $\alpha = .05$, test

H_0: The populations have the same distributions.
H_1: The populations have different locations.

From Table IX with $n_1 = 10$ and $n_2 = 8$, we find $u_{.025} = 17$. Thus we should reject H_0 for $u < 13$. Since $u = 9$, the null hypothesis is rejected. This may be interpreted to mean that either the medians or the means of the two populations are different. ∎

The Mann-Whitney test may be summarized as below.

PROCEDURE FOR PERFORMING THE MANN-WHITNEY TEST

H_0: The two populations have identical distributions.
H_1: The populations differ with respect to location.

Given: Two independent samples of size n_1 and n_2 from the two populations in question. The significance level α.

Assumptions: The variable under study is continuous. (This does not preclude ordinal rankings as observations).

The distributions of the two populations are identical with the possible exception of location.

The sample sizes do not exceed 20.

1. Find the critical value $u_{\alpha/2}$ of U.
2. Combine the samples and rank the observations. Then compute the rank sums w_1 and w_2 of the two samples.
3. Choose the smaller of the following two numbers as the value u of the test statistic:

$$u_1 = n_1 n_2 + \frac{n_1(n_1 + 1)}{2} - w_1 \qquad u_2 = n_1 n_2 + \frac{n_2(n_2 + 1)}{2} - w_2$$

4. Compare u with $u_{\alpha/2}$. Reject H_0 if $u < u_{\alpha/2}$.

The limitation on sample sizes is due to the abbreviated table of critical values of U. We will show shortly how to avoid this by using the normal distribution. The test as stated is a two-tailed test. It can be easily modified to a single-tailed test by using u_α as the critical value and by choosing U to be the statistic that would be smaller in the event H_1 is true.

Example 13.7

The nitrous oxide emissions in parts per million for samples of two types of engines (I and II) are summarized below. Test the hypothesis that the nitrous oxide emissions of these two engines have the same distribution.

Type I (x)	1.41	1.35	1.30	1.20	1.05
Type II (y)	1.28	1.25	1.15	1.10	

Combining the observations we obtain the following rankings:

Observation:	1.05	1.10	1.15	1.20	1.25	1.28	1.30	1.35	1.41
Variable:	x	y	y	x	y	y	x	x	x
Rank:	1	2	3	4	5	6	7	8	9

Sample 1		Sample 2	
x	Rank	y	Rank
1.41	9	1.28	6
1.35	8	1.25	5
1.30	7	1.15	3
1.20	4	1.10	2
1.05	1	$w_2 = 16$	
$w_1 = 29$			

$$u_1 = 5 \cdot 4 + \frac{5(5 + 1)}{2} - 29 = 6 \qquad u_2 = 5 \cdot 4 + \frac{4(4 + 1)}{2} - 16 = 14$$

Thus the value of the test statistic is $u = 6$. Using this value, we plan to test:

H_0: The distribution of emission levels for the two engines are identical.
H_1: The two distributions differ as to location.

For a significance level $\alpha = .05$, the critical value is $u_{.025} = 1$. Thus we reject H_0 if $u < 1$. Since $u = 6$, the null hypothesis is not rejected. ∎

The following example covers the case where only rankings are available from the outset.

Example 13.8

Eight business executives and seven blue-collar workers are examined by a team of doctors and ranked as to their cardiovascular health. Letting X denote the executives and Y the blue collar workers the rankings given below were obtained. Test the hypothesis that there is no difference in the cardiovascular health of these two classes of workers.

Variable	x,	x,	$(y$	$=y$	$=x$),	x,	y,	y,	y,	$(y$	$=x$),	x,	y,	x,	x
Ranking	1	2	4	4	4	6	7	8	9	10.5	10.5	12	13	14	15

Sample 1 (executives)	Sample 2 (blue-collar workers)
Ranks	*Ranks*
1	4
2	4
4	7
6	8
10.5	9
12	10.5
14	13
15	$w_2 = 55.5$
$w_1 = 64.5$	

For a two-tailed 5% significance level test, the critical value is $u_{.025} = 10$. We reject H_0 if u is less than 10. From the above data,

$$u_1 = 8 \cdot 7 + \frac{8(8 + 1)}{2} - 64.5 = 27.5$$

$$u_2 = 8 \cdot 7 + \frac{7(7 + 1)}{2} - 55.5 = 28.5$$

Thus $u = 27.5$. The null hypothesis of no difference in the cardiovascular health of these two types of workers is not rejected.

The following interpretation may prove useful. Had actual scores or measurements of the health of the subjects been used and had they resulted in the above rankings, we could not conclude that there was a difference between the distributions of these two populations. ∎

Table IX provides critical values of U only for sample sizes up to 20. When both n_1 and n_2 are 20 or more, U is approximately normal and the standard normal distribution is used with

$$Z = \frac{U - \frac{n_1 n_2}{2}}{\sqrt{\frac{n_1 n_2 (n_1 + n_2 + 1)}{2}}}$$

In most instances with sample sizes this large one can generally use the t test for the difference between two means.

We mention in closing that the Mann-Whitney test is also used to test the null hypothesis that two independent samples come from the same population. This application should be clear since two samples from the same population should have nearly equal means or medians. If this is not the case, it can be taken as an indication that two distinct populations were involved.

EXERCISE SET 13.2

Part A

1. Two samples are as follows:

x	y
3	2
4	4
6	7
8	10
	12

   a. Combine the samples and rank the observations.
   b. Find the rank sum for each sample.
   c What is the minimum rank sum for two samples of size 4 and 5? The maximum rank sum?

2. Samples from two populations are given below. Using $\alpha = .05$ with the Mann-Whitney test, test the hypothesis that these two populations do not differ as to location.

Sample 1	28	34	40	40	45	48	48	60	66	70
Sample 2	36	42	48	56	58	62	67	69		

3. The double-occupancy daily hotel rates for random samples of hotels in two resort areas are given below. Does there appear to be a difference in the mean room cost between these two areas?

Area 1	Area 2
38	40
44	46
47	54
56	58
73	62
	65
	66

4. Physical therapists compared two postoperative therapies for patients with hip replacement surgery. The patients were ranked after 12 days as to their degree of full recovery as follows:

Treatment 1 (ranks)	1, 3, 5.5, 5.5, 8.5, 12
Treatment 2 (ranks)	2, 4, 7, 8.5, 10, 11

Test for a difference in the effectiveness of these two therapies.

5. Eight left-handed and 10 right-handed third grade students were selected and coached on several new mathematical concepts until all were mastered by each student. Eight days later they were given a mastery test. The results are given below. Using the Mann-Whitney test with $\alpha = .05$, test for a difference in the ability of left- and right-handed students to retain mathematical concepts.

Left-handed	64	78	82	84	88	90	91	91		
Right-handed	66	68	70	75	81	81	84	86	87	91

6. The declared value of purchases 20 foreign travelers from Europe and 30 from Asia are compared. When combined and ranked it is found that the rank sum for the European travleers is $w_1 = 230$; for Asians, it is $w_2 = 1045$. Test for a difference in the distributions of declared purchases of travelers from these two areas of the world.

Part B

7. According to *Forbes* magazine, women now purchase more than 40% of all new autos sold in this country. Suppose the cost of accessories is compared on the purchases of similar autos by 23 men and 22 women and the results below are obtained.

Men (dollars)	1430,1290,1150,2150,830,1580,1300,1900,1720,2650,1090, 590,860,1840,2780,2130,790,1050,1730,1650,2000,870,2560
Women (dollars)	1280,1340,1560,1430,1320,1610,1230,1500,1460,980,1570, 910,2030,1500,1440,1350,1410,1860,920,1000,1530,1690

Test whether the mean expenditure for accessories by men and women is equal. Computer assistance is recommended.

8. Ten adults having no stomach disorders (as evidenced by an endoscopic examination) were divided into two groups, I and II. Group I received heavy doses of plain aspirin, group II the same doses of buffered aspirin. After eight days their stomachs were again examined endoscopically. Varying degrees of aspirin-induced superficial ulcers, bleeding, and the like were present. The following ranks were obtained when the ratings for these two samples were pooled.

Aspirin	Buffered aspirin
1.5	1.5
3.5	2
7	3.5
7	7
9.5	9.5

Is there evidence to indicate that the buffered aspirin offers any protection to the stomach over the plain aspirin? (Source: "Endoscopic Evaluation of the Effects of Aspirin," Frank Lanza et al., *New England Journal of Medicine*, vol. 303, no. 3, July 1980.)

9. An attitude survey team is hired to compare employee attitudes toward local management at two factories (1 and 2). Random samples of $n_1 = 25$ and $n_2 = 20$ employees are obtained from the factories and asked to rate their management on a scale from 0 to 100. These ratings are then combined and ranked and the rank sums $w_1 = 335$ and $w_2 = 700$ are computed. Use the normal approximation to the distribution of U to test for a difference in attitude toward management at these two factories. use $\alpha = .01$.

10. Suppose w_1 and w_2 are the respective rank sums for samples of size n_1 and n_2.
    a. Show

    $$w_1 + w_2 = \frac{(n_1 + n_2)(n_1 + n_2 + 1)}{2}$$

    b. Verify this result using the samples in problem 1.

11. In a study to determine if adult female monkeys can distinguish the screams of their offsprings from those of others, the scream of a young monkey was played to a group consisting of the monkey's mother and two other female controls. This was repeated four times for each group and for four groups. The summary of the Mann-Whitney analysis of the latency (reaction) times and duration of interest time is given below. (Source: "Vocal Recognition in Free-Ranging Vervet Monkeys," D. L. Cheney and R. M. Seyfarth, *Animal Behavior*, 1980, vol. 28, no. 2.)
    a. Duplicate the calculations of w for the latency times of the mother D group. Note: Sample 1 is the mothers' times, sample 2 that of the two controls.
    b. Does the evidence support the view that the mean latency time of the mothers is less than that of the controls?
    c. Test the hypothesis that the duration of interest times for the mothers is greater than that for the controls.
    d. Do these results suggest that adult vervet monkeys are able to classify young monkeys into at least two categories (own offspring versus others) on the basis of voice alone?

LATENCY TO LOOK AND DURATION OF LOOKING AT THE SPEAKER FOLLOWING SCREAM PLAYBACK

	Trial 1	Trial 2	Trial 3	Trial 4	Overall mean	Mann-Whitney	
Latency							
Mother D	7	9	15	810+	210.3		
Control	11	10	30	810+	225.1	$u = 11.5$	
Control	810+	9	13	108		$p = 0.224$	
Mother B	8	29	5	10	13.0		
Control	5	41	35	810+	318.9	$u = 5$	
Control	30	810+	810+	10		$p = 0.036$	
Mother P	5	25	24	10	16.0		
Control	—	11	51	810+	364.4	$u = 4$	
Control	810+	44	810+	15		$p = 0.024$	
Mother TD	5	26	24	31	21.5		
Control	29	28	8	810+	127.9	$u = 9.5$	
Control	64	6	26	52		$p = 0.136$	
						Overall	
				Mothers	65.2	$u = 124$	
				Controls	255.7	$p = 0.003$	
Duration							
Mother D	93	494	0	146	183.3		
Control	0	182	111	422	211.8	$u = 19$	
Control	172	265	0	542		$p = 0.305$	
Mother B	676	457	736	613	620.5		
Control	320	0	29	283		$u = 2$	
Control	345	655	0	288		$p = 0.008$	
Mother P	702	140	87	139	267.0		
Control	39	46	0	103		$u = 1$	
Control	47	69	45	0		$p = 0.004$	
Mother TD	291	652	444	221	402.0		
Control	162	205	200	0		$u = 5$	
Control	19	541	396	163		$p = 0.036$	
						Overall	
				Mothers	368.2	$u = 140.5$	
				Controls	176.6	$p = 0.006$	

Units in each cell are number of frames, taken from 45 seconds of filming at 18 frames per second. A value of 810+ in the upper portion of the table indicates the female never looked at the speaker. These values treated as 810 in statistical tests. — = Control looking at speaker when trial began.

Library

Project

The Mann-Whitney U Test is used in the analysis of each of the following research reports. Locate one of these articles and report on it.

a. "Children's Social Interaction and Parental Attitudes among Hupa Indians and Anglo Americans," L. Bachtold, *Journal of Social Psychology*, vol. 116, 1982.

b. "Types of Life Events and Factors Influencing Their Seriousness Ratings," B. Chalmers, *Journal of Social Psychology*, vol. 121, second half, Dec. 1983.

c. "An Experiment in Computer-Assisted Accounting for Introductory Accounting," S. Groomer, *The Accounting Review*, vol. 56, no. 4, Oct. 1981.

d. "Cigarette Smoking among Peking High Schoolers," Y. Gong-shao and L. Wongsheng, *Hydie*, vol. 2, no. 2, July 1983.

e. "Behavioral Effects of Early Rearing Conditions and Neonatal Lesions on the Visual Cortex in Cats," P. Cornwell and W. Overman, *Journal of Comparative and Psychological Behavior*, vol. 95, no. 6, Dec. 1981.

f. "Visual Neglect in Parkinson's Disease," C. Villardita, et al., *Archives of Neurology*, vol. 40, no. 1, Nov. 1983.

13.6 THE WILCOXON MATCHED PAIR SIGNED RANK TEST

The t test for the equality of two population means using matched pairs relies on the distribution of the differences of the members of the pairs. The Wilcoxon test for identical distributions makes use of the ranks of these differences and in spirit is very similar to the Mann-Whitney test.

This test is very easy to perform. We first rank the differences $d_i = x_i - y_i$ of the n pairs (x_i, y_i) from smallest to largest, ignoring any signs. If a difference is 0, it is ignored and the sample size n is reduced by 1. If there is a tie between differences they are assigned the average of the ranks that would have been used.

Once the ranking is complete, we assign to each rank the sign of the difference from which the rank was derived. Then the sums T_+ and T_- of the positive and negative ranks are computed, ignoring the signs.

Under the null hypothesis of identical population distributions, the difference of these two sums should be near 0. In any event, an extremely small (or large) value of either of these sums will force a rejection of the null hypothesis. Historically, the smaller of these sums has been used and is denoted by T (no relation to the t distribution). For $n \leq 30$ the critical values T_α of the test statistic for both one- and two-tailed tests are given in Table X of the Appendix.

PROCEDURE FOR PERFORMING WILCOXON SIGNED RANK TEST

H_0: The population distributions are identical
H_1: The population distributions are different

Given: The significance level α.
   n matched pairs $(x_1, y_1), (x_2, y_2), \ldots, (x_n, y_n)$.
Assumptions: The variable under observation is continuous.
   The number of pairs n does not exceed 30.

1. Find the two-tailed critical value $T_{\alpha/2}$.
2. Compute the differences $d_i = x_i - y_i$, rank them, and assign to each the sign of the corresponding difference. Then compute the sums T_+ and T_- of these signed ranks but ignore the signs. The value T of the test statistic is the smaller of T_+ and T_-.
3. Compare the value of the test statistic and the critical value. Reject H_0 if $T < T_{\alpha/2}$.

Shortly we shall outline how to use the normal distribution with T when the number of pairs exceeds 30. Also we should note that single-tailed tests based on an alternate hypothesis as to the relative location of the two populations can be formulated if one is willing to make the additional assumption, as in the Mann-Whitney test, that the two populations differ at most as to location.

Example 13.9   Nine pairs of matched students are tracked through a one-semester course in mathematics. The first member of each pair is provided with free tutoring as needed. Otherwise all students have the same instructor, text, tests, and so on. Their class grades as a percentage of the total number of possible points are given in Table 13.8. Does the free tutoring significantly affect the classroom performance of students (assuming the matching was done on variables known to be highly correlated with classroom performance)?

TABLE 13.8   CLASS GRADES AND THEIR ASSIGNED RANKS

Pair	x	y	d	$|d|$	Rank	Signed rank
1	63	71	−8	8	5	−5
2	74	74	0	0	•	•
3	78	75	3	3	2	+2
4	98	90	8	8	5	+5
5	77	58	19	19	8	+8
6	85	89	−4	4	3	−3
7	62	70	−8	8	5	−5
8	83	69	14	14	7	+7
9	61	60	1	1	1	+1

$T_+ = 23$
$T_- = 13$

Note that no rank was assigned to the difference of 0 in the second pair. Once the remaining differences are arranged by increasing size they may easily be ranked:

Differences $|d|$	1	3	4	8	8	8	14	19
Ranks	1	2	3	5	5	5	7	8

The signed ranks are obtained from the ranks above by using the signs of the corresponding differences.

The sums of the signed ranks are found next:

$$T_- = 5 + 3 + 5 = 13$$
$$T_+ = 2 + 5 + 8 + 7 + 1 = 23$$

The value of the test statistic is thus $T = 13$.
Suppose we want to test at a significance level $\alpha = .05$

H_0: The populations of grades have the same distribution (no effect from tutoring).

H_1: The populations of grades have different distributions (an effect from tutoring).

In finding the critical value $T_{.025} = 4$ in Table X we use $n = 8$, since one of the nine pairs was not ranked. Since $T = 13$ is not less than this critical value the null hypothesis is not rejected. The tutoring has not significantly affected the class grades of the students. ■

When the sample size is larger than covered by Table X, the normal distribution may be used with

$$Z = \frac{T - \dfrac{n(n + 1)}{4}}{\sqrt{\dfrac{n(n + 1)(2n + 1)}{24}}}$$

In fact, the normal distribution can be used with sample sizes down to around $n = 10$ with good results.

EXERCISE SET 13.3

Part A

1. In a two-tailed Wilcoxon matched pair test, the following signed rank sums were obtained. At the given significance level, would the null hypothesis be rejected?
   a. $T_+ = 80$, $T_- = 40$, $n = 15$, $\alpha = .05$.
   b. $T_+ = 26$, $T_- = 52$, $n = 12$, $\alpha = .01$.
2. In a two-tailed Wilcoxon matched pair test, the following values of the test statistic were obtained. Using the given significance levels and sample sizes, would the null hypothesis be rejected?
   a. $T = 22$, $n = 11$, $\alpha = .05$.
   b. $T = 50$, $n = 20$, $\alpha = .01$.
3. In a before-and-after experiment involving a retraining procedure for stroke victims, the following dexterity measures were obtained:

Before training (x)	47	63	76	28	42	55	38	83	37	45	81
After training (y)	58	61	72	45	53	59	49	83	35	56	80

   a. Using $\alpha = .05$, test the hypothesis that the training procedure has no effect on the dexterity of stroke victims.
   b. Now examine the original data more closely, particularly that of the patients having low initial scores (say below 60). Formulate a new hypothesis and test it.
4. Eight male and eight female executives are matched as to age, education, duration of employment, type of job, and size of company with which they are employed. The salaries of these eight matched pairs are given below. Test the hypothesis that the sex of the employee at the executive level has no effect on salary. Use $\alpha = .05$.

Pair	Salary (male)	Salary (female)
1	53,500	46,250
2	37,200	40,000
3	75,500	62,750
4	45,900	44,000
5	96,800	82,400
6	39,100	44,500
7	55,800	49,500
8	84,200	76,200

5. To determine if two tests are equally effective in evaluating job applicants for a certain position, the test questions are randomly intermixed and a combined test is given to each of 14 applicants. The answers to the two sets of test questions are then separated and the evaluative scores below were obtained. Test the hypothesis that both tests produce the same distributions. Use $\alpha = .01$.

Test 1	78	84	65	98	56	28	70	66	55	87	90	61	70	83
Test 2	74	81	73	98	60	13	58	74	59	88	93	66	88	90

6. A statistics professor is interested in whether there is a change in students' attitudes toward mathematics as a result of his course. He selects a random sample from his courses and administers a mathematics anxiety inventory test to them at the beginning of the course and again at its end. Use the results below to test the hypothesis than there is no change in student anxiety levels as a result of this course.

Subject	1	2	3	4	5	6	7	8	9	10	11	12
Pretest	54	36	85	75	45	22	42	85	57	65	33	80
Posttest	68	21	88	92	32	5	50	81	64	70	26	84

7. For a sample of $n = 35$ matched pairs, $T = 160$ was obtained. Use the normal distribution to decide if H_0 should be rejected in a two-tailed test with $\alpha = .05$. What is the p value of the test?

Part B

8. Compute the critical value $T_{.025}$ for each of the following choices of n using the normal distribution and compare it with the corresponding entry in Table X.
   a. $n = 30$.
   b. $n = 20$.
   c. $n = 10$.
   d. $n = 8$.
9. Use the normal distribution to obtain estimates of the p values for the tests of problem 1.
10. A shoe manufacturer wants to know if there is any difference in the lengths of the right and left feet of individuals. The measurements (in inches) of the lengths of the feet of 10 randomly selected subjects are given below. Do these results suggest a difference in the right and left feet? Use the Wilcoxon matched pair test with $\alpha = .05$.

Subject	1	2	3	4	5	6	7	8	9	10
Right Foot	11.2	10.1	8.7	7.0	6.8	12.3	10.6	10.0	9.2	8.2
Left Foot	10.9	10.3	8.5	7.0	6.6	12.1	10.7	10.1	9.0	8.0

11. Show that under the null hypothesis of identical distributions for X and Y, both T_+ and T_- have an expected value of $n(n + 1)/4$. Hint: If there are n ranks, what is the sum of these ranks?

12. When a large number of ties occur, an improved normal approximation for T is obtained by using

$$Z = \frac{T - \dfrac{n(n + 1)}{4}}{\sqrt{\dfrac{n(n + 1)(2n + 1)}{24} - \left[\dfrac{\Sigma c^3 - \Sigma c}{48}\right]}}$$

where c is the number of ties for a particular rank.

Rework problem 10 using this modified formula and compare with the results previously obtained.

13.7   THE SPEARMAN RANK CORRELATION COEFFICIENT

The Spearman rank correlation coefficient r_s, like the Pearson r (Chapter 11), is used for testing the hypothesis of no correlation between two variables X and Y. Unlike the test based on r, the test developed here makes no assumption about normality and may be used when only rankings (ordinal data) are available.

Given the sample $(x_1, y_1), (x_2, y_2), \ldots, (x_n, y_n)$, the first step in computing the coefficient r_s is to rank the x's and y's separately, ties being treated by assigning the mean of the ranks that would have been used had there been no ties. Next we compute the difference d_i of the ranks assigned to x_i and y_i. This is done for each pair and the sum of the squares of these difference d_i^2 is found. The coefficient r_s is then computed as shown below.

> ### DEFINITION 13.1
> The Spearman rank correlation coefficient is defined by
>
> $$r_s = 1 - \frac{6 \sum_{i=1}^{n} d_i^2}{n(n^2 - 1)}$$

Historically, the Spearman rank correlation coefficient was developed for the case of no ties by simply computing the Pearson coefficient r using the ranks rather than the actual observations. As a consequence, one can show that like r, r_s always has values between 1 and -1. A positive r_s indicates a tendency for both variables to increase or decrease together, while with a negative value one should increase while the other decreases. Unfortunately the size of r_s does not allow us to predict exactly how y changes with x. The coefficient r_s is best thought of as simply a measure of the strength of the correlation.

Example 13.10 Ten student teachers were ranked by their team of supervisors as to their aptitudes as potential teachers. They were also ranked by their grade point averages. Compute the Spearman coefficient r_s from the data in Table 13.9.

TABLE 13.9 APTITUDE AND GRADE POINT RANKS FOR 10 STUDENTS

Student	Aptitude rank (x)	Grade point rank (y)	d^2	
1	5	6	1	
2	8	3	25	
3	2	8	36	
4	1	2	1	
5	3	1	4	$\displaystyle\sum_{i=1}^{10} d_i^2 = 74$
6	4	4	0	
7	7	5	4	
8	6	7	1	
9	10	9	1	
10	9	10	1	

$$r_s = 1 - \frac{6 \displaystyle\sum_{i=1}^{10} d_i^2}{n(n^2 - 1)} = 1 - \frac{6 \cdot 74}{10 \cdot 99} = .55$$

This positive coefficient indicates that if a correlation exists, then the aptitude for teaching tends to improve or increase as the students' subject matter understanding increases. (We should be distressed otherwise.) ∎

Each sample of size n furnishes a value r_s that we can associate with a test statistic R_s whose distribution could be tabulated. However, for sample sizes even as small as 10, it has been found that the t distribution can be used. More specifically, in testing the null hypothesis of no correlation,

$$r_s \sqrt{\frac{n - 2}{1 - r_s^2}}$$

is the value of a variable having the t distribution with $v = n - 2$ degrees of freedom. We shall not cover the case of $n < 10$ except to provide the critical values (Table XI of the Appendix).

PROCEDURE FOR PERFORMING SPEARMAN'S RANK CORRELATION TEST

H_0: There is no correlation between X and Y.
H_1: A correlation exists between X and Y.

Given: The significance level α.

The sample $(x_1, y_1), (x_2, y_2), \ldots, (x_n, y_n)$.
1. Find the critical values $\pm t_{\alpha/2}$ using $\nu = n - 2$.
2. Rank $x_1, x_2, \ldots, x_n$ and $y_1, y_2, \ldots, y_n$ separately.
3. Compute the difference d_i of the ranks of x_i and y_i for each i.
4. Compute the coefficient

$$r_s = 1 - \frac{6 \sum_{i=1}^{n} d_i^2}{n(n^2 - 1)}$$

5. Compute the value of the test statistic

$$t = r_s \sqrt{\frac{n - 2}{1 - r_s^2}}$$

6. Compare the value of the test statistic with the critical values. Reject H_0 if $t > t_{\alpha/2}$ or $t < -t_{\alpha/2}$.

Example 13.11

Using the data of Example 13.10, test for a correlation between grade point average and teaching aptitude. The hypotheses are as follows:

H_0: There is no correlation between aptitude and grade point average.
H_1: A correlation exists between aptitude and grade point average.

For a 5% significance level test with $\nu = 10 - 2 = 8$, the critical values are $\pm t_{.025} = \pm 2.31$. Using $r_s = .55$, the t score is

$$t = r_s \sqrt{\frac{n - 2}{1 - r_s^2}} = .55 \sqrt{\frac{10 - 2}{1 - .55^2}} = 1.86$$

The null hypothesis is not rejected. ■

Example 13.12

Ten statistics texts were examined and ranked for acceptability by the instructors of a multisection course (a rank of 1 was assigned for the most acceptable). The texts were then stacked up and ranked by size (thickness), with a rank of 1 being assigned the smallest. The data are given in Table 13.10. Test the hypothesis that these two rankings are related.

Note that texts 8 and 9 were judged equally acceptable.

$$r_s = 1 - \frac{6(55.50)}{10 \cdot 99} = .66$$

$$t = .66 \sqrt{\frac{8}{1 - .66^2}} = 2.48$$

This clearly exceeds the critical value $t_{.025} = 2.31$. We reject the hypothesis of no correlation and conclude that the size of the text is related to its acceptability. (There's definitely a message here.)

TABLE 13.10   RANKS OF TEXTBOOKS BY ACCEPTABILITY AND SIZE

Text	Acceptability rank	Size rank	d^2	
1	6	8	4	
2	5	4	1	
3	2	6	16	
4	1	2	1	
5	10	7	9	$\sum_{i=1}^{10} d_i^2 = 55.50$
6	8	5	9	
7	9	9	0	
8	3.5	1	6.25	
9	3.5	3	0.25	
10	7	10	9	

EXERCISE SET 13.4

Part A

1. A sample of $n = 25$ pairs (x_i, y_i) produced a Spearman rank correlation coefficient of $r_s = .37$. Test for a correlation between X and Y using $\alpha = .05$.

2. Compute the rank correlation coefficient r_s for the ranks given in the table below and test the hypothesis of no correlation between X and Y.

x	y
1	2
5	3
9	6
7	8
4	5
6	7
2	1
10	4
3	9
8	10

3. A department chairman selected two committees to evaluate 10 new textbooks in elementary statistics. One committee was composed entirely of regular staff members, the second of part-time personnel. Their respective rankings of the ten texts are given at the right. Test for a correlation between the evaluations by staff and part-time personnel.

Text	Regular staff x	Part-time staff y
1	3	1
2	8	5
3	6	7
4	5	8
5	4	3
6	1	4
7	7	8
8	10	2
9	9	6
10	2	7

4. The entering SAT scores of 11 first-year students and their grade point averages after one year at college are given. Test for a correlation using $\alpha = .01$.

SAT score x	594	572	630	590	610	620	560	705	650	690	540
GPA y	2.7	2.9	3.0	2.5	2.1	2.8	3.0	3.4	3.3	4.0	2.1

5. A cosmetic company using door-to-door sales representatives ranked a sample of 20 sales personnel in terms of the size X of their assigned territories and in terms of their yearly sales Y in dollars. The sum of the squares of the differences of the ranked pairs was 2320. Using $\alpha = .05$, test the hypothesis of no correlation between sales volume and size of territory.

6. At a hospital, the director of nursing interviews applicants for licensed and vocational nursing positions and assigns them a score. The personnel director does likewise, only the score here is based on past employment history, references, and so forth. When the grades of 35 applicants were compared it was found that the sum of the squares of the differences of their corresponding ranks was 4010. Test for a correlation between these two evaluation procedures using $\alpha = .01$.

7. A mathematics instructor claims that he can rank students' preparation on the basis of a personal interview as well as the department's screening examination can. To verify this claim, 11 students are interviewed and ranked by the instructor and then ranked in terms of their examination grades as follows:

Student	Instructor's rank X	Examination rank Y
A	2	1
B	5	8
C	11	10
D	7	9
E	6	5
F	9	7
G	10	11
H	4	2
I	3	4
J	8	6
K	1	3

Test for a correlation between the instructor's rankings and placement examination grades.

Part B

8. The number of pages X and the ranking Y in terms of sales of a sample of five college algebra books is given below. Test for a correlation between size and success of a textbook in this area. Use $\alpha = .10$.

Number of pages X	Sales ranking
376	2
458	4
613	5
504	3
405	1

9. To test the reliability of a rating scale for parents of a learning-disabled child, 22 parents were interviewed twice (first and second assessments) over a 12 week period by two raters. Based on each of these interviews, scores were obtained that rated the families to be in five main dimensions: evaluation, permissiveness of autonomy, mutual affection, hostility, and pressuring. These were then combined to obtain a total parenting score (TPS). The results are presented below. (Source: "Assessing the Parents of the Learning Disabled Child: A Semistructured Interview Procedure," Leon Sloman and C. Webster, *Journal of Learning Disabilities*, vol. 11, no. 2, Feb. 1978.)

SPEARMAN'S RHO CORRELATIONS (ONE-TAILED) BETWEEN RATER 1 AND RATER 2 OVER THE TWO ASSESSMENTS

Dimension	Assessment 1		Assessment 2	
	r	p	r	p
Evaluation	+.88	<.001	+.83	<.001
Autonomy	+.74	<.001	+.51	<.007
Affection	+.68	<.001	+.83	<.001
Hostility	+.87	<.001	+.87	<.001
Pressuring	+.71	<.001	+.50	<.009
TPS	+.78	<.001	+.74	<.001

SPEARMAN'S RHO CORRELATIONS (ONE-TAILED) BETWEEN ASSESSMENT 1 AND ASSESSMENT 2 FOR EACH OF THE TWO RATERS

Dimension	Rater 1		Rater 2	
	r	p	r	p
Evaluation	+.63	<.001	+.66	<.001
Autonomy	+.44	<.020	+.12	NS
Affection	+.57	<.003	+.26	NS
Hostility	+.76	<.001	+.51	<.008
Pressuring	+.79	<.001	+.42	<.030
TPS	+.76	<.001	+.62	<.001

a. Was a significant correlation obtained between the raters' scores on all dimensions during the first assessment? During the second assessment?
b. Is there a strong correlation between the TPS scores of the two raters on the first assessment? On the second assessment?
c. In the first and second assessments, which dimensions consistently had the highest correlation?
d. Interpret the results in the second table.

10. As part of a study on the foraging habits of fish it was necessary to establish whether there was any relationship between the activity of a prey organism and its density in the aquarium. The scatter diagram below is for the two organisms used in the study. (Source: "Food Selection Versus Prey Availability," C. Magnhagen and A. Weiderholm, *Oecologia*, 1982, vol. 55.)

a. Does there appear to be a correlation between the density and the activity of either specie?

b. According to the authors, the Spearman rank coefficient $r_s = .84$ was obtained for *C. volutator*. Is this significant at $\alpha = .10$?

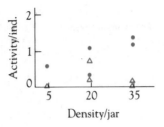

13.8

THE KRUSKAL-WALLIS ONE-WAY ANALYSIS OF VARIANCE TEST

It will be recalled that the one-way analysis of variance test developed in Chapter 12 is a procedure for testing the equality of several population means. This test relies on the F distribution, which in turn requires a normality assumption for each population.

The Kruskal-Wallis test that we now describe is a distribution-free test for the equality of several location parameters such as the mean. It uses only ranks and rank sums, as does the Mann-Whitney test, and for the case of $m = 2$ populations is equivalent to it.

The basic idea is to pool all the samples, rank the observations, and then compute a rank sum u_i for each sample. Using these sums we then compute the value h of a test statistic H by

$$h = \frac{12}{n(n+1)} \sum_{i=1}^{m} \left(\frac{u_i^2}{n_i}\right) - 3(n+1)$$

where

m = number of samples
n_i = size of the ith sample
$n = \sum_{i=1}^{m} n_i$ (the sum of the sample sizes)
u_i = rank sum of the ith sample

Under the null hypotheses that all m populations have the same distribution, and with all sample sizes at least 5, the distribution of H may be approximated by the χ^2 distribution having $\nu = m - 1$ degrees of freedom. For sample sizes below 5, tables of exact probabilities are available. We shall not, however, make use of these here.

The rationale for using H as a test statistic is mainly that when the assigned ranks are spread more or less uniformly over all the samples, the value of H is small. On the other hand, when some of the means are different the ranks tend to cluster, at least within the corresponding samples, and the value of H is large.

PROCEDURE FOR PERFORMING THE KRUSKAL-WALLIS ANOVA TEST

H_0: All m populations have identical distributions.
H_1: At least two of the populations do not have the same location.

Given: The significance level α.

m independent samples of size $n_1, n_2, \ldots, n_m$.

Assumptions: The variable under study is continuous. (This does not preclude ordinal rankings being used.)

All sample sizes are at least 5.

1. Using $\nu = m - 1$, find the critical value χ_α^2.
2. Combine the samples, rank the observations, and compute the rank sum u_i for each sample.
3. Compute the value of the test statistic H by

$$h = \frac{12}{n(n + 1)} \sum_{i=1}^{m} \left(\frac{u_i^2}{n_i}\right) - 3(n + 1)$$

where

$$n = \sum_{i=1}^{m} n_i$$

4. Compare the value of the test statistic with the critical value. Reject H_0 if $h > \chi_\alpha^2$.

Example 13.13

Random samples of graduates from the engineering, psychology, and sociology departments of a major university were asked to rate on a scale of 0 to 100 their satisfiaction with their degree. Using a 5% significance level and the data in Table 13.11, test the hypothesis that the degree of satisfaction is the same for all 3 departments.

TABLE 13.11 RANKS AND RATINGS OF DEGREES BY GRADUATES

Engineering majors		Psychology majors		Sociology majors	
Rating	Rank	Rating	Rank	Rating	Rank
84	18.5	93	20	76	15
81	17	84	18.5	72	12
75	14	80	16	61	9
65	11	73	13	60	8
52	3.5	64	10	58	7
50	1.5	57	6	52	3.5
50	1.5	55	5		
$n_1 = 7$		$n_2 = 7$		$n_3 = 6$	
$u_1 = 67$		$u_2 = 88.5$		$u_3 = 54.5$	
$\bar{x}_1 = 65.2$		$\bar{x}_2 = 72.3$		$\bar{x}_3 = 63.2$	

The ranks in the table were found by pooling all the ratings, exactly as in the Mann-Whitney test.

For $\alpha = .05$ and $\nu = 3 - 1 = 2$, the critical value is $\chi^2_{.05} = 5.99$. We reject the null hypothesis of identical distributions (or equal satisfaction) if $h > 5.99$. For h we find

$$h = \frac{12}{20(20 + 1)} \left(\frac{(67)^2}{7} + \frac{(88.5)^2}{7} + \frac{(54.5)^2}{6} \right) - 3(20 + 1) = 1.43$$

The null hypothesis is not rejected. ■

In the event that there are a great number of ties, the test statistic H can be adjusted by dividing by the number

$$c = 1 - \frac{\Sigma\,(t_i^3 - t_i)}{n^3 - n}$$

where t_i is the number of tied observations in the ith group of tied scores. The correction factor c is a number less than 1 but in most cases quite close to 1. As a result the adjusted test statistic

$$H' = \frac{H}{c}$$

is generally only slightly larger than H and may not be needed. In any event, if the null hypothesis is rejected with H, it will also be with H' and then it would not need to be computed.

Example 13.14

Consider again the ratings by majors of their university (Example 13.13). There were three groups of ties:

$$84, 84 \quad t_1 = 2$$
$$52, 52 \quad t_2 = 2$$
$$50, 50 \quad t_3 = 2$$

The correction factor is

$$c = 1 - \frac{(2^3 - 2) + (2^3 - 2) + (2^3 - 2)}{(20^3 - 20)}$$
$$= 1 - .0023 = .997$$

The adjusted value of H is

$$h' = \frac{h}{c} = \frac{1.43}{.9977} = 1.43 \text{ (to two decimal places)}$$

Only in the third decimal place does h differ from h'. ■

EXERCISE SET 13.5

Part A

1. Independent samples from three minority groups were asked to rate the services of a legal aid agency. The results are tabulated below. Test the hypothesis that the three groups are equally satisfied with the service of the agency.

Group I	Group II	Group III
38	63	50
76	90	61
87	81	61
71	86	65
77	74	70
88	81	57
56	72	74

2. Test the hypothesis that the following samples are from populations having the same distribution.

Sample I	Sample II	Sample III
145	123	198
155	176	138
163	156	144
130	170	201
140	150	143
159	140	184

3. The daily cost of food per patient in samples of hospitals from the eastern, central, and western United States are given below. Test whether the distribution of daily food costs is the same for these three regions.

Eastern	Central	Western
12.20	9.40	11.10
10.30	6.40	8.80
9.90	8.60	10.00
11.30	10.60	10.40
13.40	9.70	9.30
9.20	10.30	11.10

4. Samples of autos from three domestic and two foreign producers were carefully examined and ranked for quality of construction. Using the results below, test the hypothesis that the quality of construction is the same for all five producers.

Domestic I ranks	Domestic II ranks	Domestic III ranks	Foreign I ranks	Foreign II ranks
1	5	2	6	9
3.5	16	3.5	12	13
7	17	8	18	14
10	23	11	22	19
15	28	21	24	20
26	29	25	30	27

5. In each of four chambers with the same initial humidity level is placed a different make of dehumidifier (all equally rated). After the dehumidifiers operate for 27 hours the humidity levels are recorded. This experiment is then replicated four times (using different humidifiers) to produce the results below. Test for a difference in the performances of these four brands.

Brand			
1	2	3	4
23.2	21.4	20.3	19.0
14.7	22.1	18.2	15.3
17.4	20.6	19.9	16.2
20.0	19.4	20.5	18.3
22.4	21.3	17.9	17.2

6. Samples of fruit flies are placed in enclosed areas, each of which contains a trap with a different type of bait. The percentage of flies attracted to each trap is recorded. The experiment is then replicated six times to produce the results below. Test for a difference in the effectiveness of the three baits.

Bait		
1	2	3
63.2	60.5	69.6
47.4	78.3	64.5
59.7	65.5	67.2
62.0	61.8	68.4
55.4	52.3	62.9
80.3	61.3	66.2

Part B

7. The sugar content is measured in randomly selected samples of each of four brands of a certain type of breakfast cereal. The measurements (grams of sugar per pound of cereal) are given below. State and test a hypothesis.

Cereal I	Cereal II	Cereal III	Cereal IV
85.4	86.7	85.1	85.6
86.5	87.1	86.9	85.6
86.5	87.2	88.0	85.9
86.8	87.4	88 .0	86.1
87.0	87.4	88.3	86.2
87.1	87.5	88.4	86.4
	87.8	88.6	86.4
	88.1	88.7	86.6
		88.9	86.9
			87.9

8. Random samples of purchasers of subcompact, compact, and intermediate size autos are obtained. Using the ages of the purchasers given below, test for a difference in the median age of purchasers of these three categories of autos.

Sample 1: subcompact	Sample 2: compact	Sample 3: intermediate
30	28	26
32	33	31
32	35	37
33	37	45
34	38	47
34	38	48
36	43	49
42	45	50
43	45	
	46	

9. Six types of antiseptics were tested for effectiveness with random samples of size $n = 5$ for each. The rank sums were 40, 130, 110, 89, 26, and 70. Test for a difference in their effectiveness.

10. Suppose that in a Kruskal-Wallis ANOVA test there are three treatment levels and that the sample sizes corresponding to these are $n_1 = 5$, $n_2 = 8$, and $n_3 = 10$.

    a. What is the smallest rank sum for any treatment?

    b. What is the largest possible rank sum for any treatment?

11. Repeat problem 10 for the case of k treatments where each sample is of size n. You might begin by trying three samples of size 4 and then generalize from there.

13.9   THE FRIEDMAN TWO-WAY ANALYSIS OF VARIANCE TEST

Perhaps the most common application of the two-way analysis of variance involves a test for a treatment effect of the column factor that uses highly different blocks of matched subjects as the treatment levels of the row factor. In this so-called *randomized complete block design* the goals are to select the blocks so as to amplify any differences that might occur as a result of the column treatment levels and to select the subjects so as to minimize the variation between subjects within each block. The Friedman test is used with this design to test the null hypothesis of no treatment effect.

Following the development in Chapter 12, we shall assume that there are r blocks and c treatments and that the observations are arranged in a ANOVA table with the columns representing the treatments and the rows the blocks. The blocks are assumed to have no interaction and consist of one experimental subject for each treatment. Thus there will be one observation per cell.

The test statistic for the Friedman test has the chi-square distribution for all but extremely small choices of r and c. Accordingly it is denoted by χ_f^2. The first step in computing its value from the table is to rank all the observations in each row (block) from smallest to largest, assigning any ties the mean of the ranks that would have been used. Next we obtain the rank sums $u_1, u_2, \ldots, u_c$ of the columns (treatments). Finally we compute

$$\chi_f^2 = \frac{12}{rc(c + 1)} \sum_{i=1}^{c} u_i^2 - 3r(c + 1)$$

As a result of the squares $u_1^2, u_2^2, \ldots, u_c^2$ that enter into the formula, it is easy to see that χ_f^2 will be large if some of the rank sums exceed what might be expected. Also, under the null hypothesis of no treatment effect, all the rank sums should be nearly equal, since we then expect the ranks within a block to be more or less randomly assigned to the various columns. If, however, there is an effect, there will be a tendency for similar ranks to cluster within certain columns, which will produce some abnormally high (and also low) rank sums.

PROCEDURE FOR PERFORMING THE FRIEDMAN ANOVA TEST

H_0: All treatment populations are identical.

H_1: All treatment effects are not the same.

Given: The significance level α.

An $r \times c$ two-way ANOVA table with one entry per cell.

Assumptions: The r samples of blocks forming the rows are independent.

The number of treatments (columns) c is at least 5;

There is no interaction effect.

The variable of interest is continuous (ranks may be used).

1. Using $v = c - 1$ find the critical value χ_α^2.
2. Rank the observations within each row or block.
3. Compute the sum of the ranks $u_1, u_2, \ldots, u_c$ within each column.
4. Compute the value of the test statistic

$$\chi_f^2 = \frac{12}{rc(c+1)} \sum_{i=1}^{c} u_i^2 - 3r(c+1)$$

5. Compare the value of the test statistic with the critical value. Reject H_0 if $\chi_f^2 > \chi_\alpha^2$.

In his original article in 1937 in the *Journal of the American Statistical Association*, Friedman gave the distributions of χ_f^2 for $c = 3$, $c = 4$, and those values of r for which the chi-square approximation cannot be used. From these discrete distributions we have the following brief table of critical values (Table 13.12). These are all approximately 5% critical values of χ_f^2 with the exception of the $\mu = 2$, $c = 3$ entry. For it, $\alpha = .17 = 17\%$.

TABLE 13.12 SELECTED CRITICAL VALUES OF χ_f^2 FOR THE FRIEDMAN TEST

				r				
c	2	3	4	5	6	7	8	9
3	4.000	6.000	6.500	6.400	6.330	6.000	6.250	6.222
4	6.000	7.000	7.500					

For larger values of c and r, use the chi-square distribution with $v = c - 1$ df.

Example 13.15   Infants at two weeks of age are matched in blocks of three as to size, sex, bone structure, and the like. One child from each block is then placed on a particular feeding regimen. The weight gains in ounces over a 10 day period are summarized in Table 13.13, along with the ranks used in the Friedman test. Use these data to test the hypothesis that equal weight gains are produced by these three plans.

TABLE 13.13   ANALYSIS OF VARIANCE TABLE FOR RANKED WEIGHT GAINS OF INFANTS

Block	Regimen I		Regimen II		Regimen III	
	Weight gain	*Rank*	*Weight gain*	*Rank*	*Weight gain*	*Rank*
1	11.4	2	10.8	1	12.7	3
2	13.2	1	14.1	2	15.9	3
3	9.0	3	8.3	1	8.9	2
4	11.0	2	10.0	1	11.1	3
5	10.8	1	12.1	3	11.9	2
6	12.4	3	10.9	1	11.6	2
7	12.0	3	11.1	1	11.2	2
8	10.2	2	10.0	1	10.3	3
9	9.6	3	9.1	1	9.5	2
10	11.0	2	10.9	1	12.0	3
		$u_1 = 22$		$u_2 = 13$		$u_3 = 25$

Using $r = 10$, $c = 3$, and the rank sums in the table we find

$$\chi_f^2 = \frac{12}{10 \cdot 3 \cdot (3 + 1)} (22^2 + 13^2 + 25^2) - 3 \cdot 10(3 + 1) = 7.8$$

With $\alpha = .05$ and $\nu = 3 - 1 = 2$, the critical value is $\chi_{.05}^2 = 5.99$. The null hypothesis is rejected and we can conclude that there is a difference in the weight gains of infants on the different regimens. ∎

If there are a great many ties in the ranks (and this is rare), we can adjust the test statistic by dividing by

$$c = 1 - \frac{\Sigma (t^3 - t)}{rc(c^2 - 1)}$$

where t is the number of ties at any rank within a block and the summation extends over all ties in a block and over all blocks in the table.

EXERCISE SET 13.6

Part A

1. Matched blocks of three rats that have been bred to having varying degrees of susceptibility to liver cancer are placed on a diet highly laced with a cancer-causing agent. The treatment levels consist of different dosages of a chemotherapeutic drug. The number of days until death of the rats are given below. Test for an effect of the therapeutic drug.

	Dosage level		
Block	High	Medium	Low
1	197	164	172
2	239	186	145
3	412	154	286
4	275	243	201
5	314	205	214
6	536	181	152
7	102	176	150

2. Matched blocks of graduates from different departments of four universities are compared as to the starting salaries on their first job. Test for a difference in the mean starting salaries of students from these four universities.

	University			
Department	I	II	III	IV
Engineering	26500	27800	26000	25400
Mathematics	24800	26300	27000	26400
English	18900	16400	13800	21000
History	16000	13500	14200	17200
Kinesiology	19200	16400	14000	18500

3. Blocks of adults who have been matched by age, occupation, and education are used to test for a difference in the clarity of the instructions accompanying the short form of the federal income tax return. The hypothetical data below give the time in minutes needed to complete the form. Test for a difference due to the form of the instructions.

Block	Instruction set 1	Instruction set 2	Instruction set 3
1	66.9	60.2	54.1
2	84.3	78.4	69.3
3	99.0	85.4	86.1
4	102.4	112.3	101.5

4. Dust mites are present in nearly all homes. The question of whether repeated vacuuming on a regular basis will reduce their number is studied by vacuuming the master bedroom of five randomly selected homes over a six week period and microscopically counting the number of mites on the filter after each vacuuming. How might one rationalize that the Friedman ANOVA test can be applied to this discrete variable? Carry out the test using $\alpha = .05$.

	Vacuuming					
Room	First	Second	Third	Fourth	Fifth	Sixth
1	2180	2040	2310	2097	2035	2130
2	1540	1380	1478	1630	1510	1297
3	3064	3310	3100	3043	3512	3017
4	1945	2013	1735	1803	1840	1882
5	2239	2200	2345	1945	2180	2213

5. Entering college students are matched in groups of three on the basis of SAT scores, composite grade point averages in the sciences, rankings of their high schools, intended majors and so on. Each student of a matched triple was then randomly assigned to one of three calculus classes taught by different instructors using the same text and course outline. The students' grades on the common final are given below. Test for an effect of the class assignment (or possibly the instructor).

Block	Class 1	Class 2	Class 3
1	123	135	118
2	145	141	155
3	125	130	125
4	150	160	170
5	120	115	130
6	143	159	148

Part B

6. A study was conducted in a large metropolitan area of the mean price of three brands of jeans. Eight clothing stores, each handling the three brands, were selected and the price of each brand was determined. Use Friedman's test to test the null hypothesis that the mean selling price is the same for each brand.

Store	Brand A	Brand B	Brand C
1	18.95	19.95	19.95
2	15.00	15.00	13.95
3	17.95	19.95	18.95
4	12.50	15.00	15.00
5	21.00	26.00	21.00
6	18.95	22.95	20.95
7	15.00	18.00	18.00
8	13.95	16.95	16.99

7. Complex but often short computer programs that involve repeated loops, graphic displays, or file copying are often used as benchmarks for testing the speed of computers. The running times of eight of these benchmark programs on three popular makes of microcomputers are given below. Is there a significant difference in the median operating speeds of these three computers?

Benchmark	Make 1	Make 2	Make 3
1	19.6	11.2	7.8
2	5.4	3.9	3.3
3	25.9	17.2	10.3
4	65.8	74.7	103.5
5	55.4	50.3	37.9
6	45.8	33.9	30.0
7	65.0	25.8	15.8
8	37.2	32.0	33.5

8. Three appraisers are employed by a real estate financing agency to determine the market price of property for which financing is being arranged. Their appraisals on eight randomly selected pieces of property are given below. Use Friedman's test to determine whether these appraisers obtain significantly different results.

Property	Appraiser I	Appraiser II	Appraiser III
1	58,900	62,800	63,400
2	60,300	61,900	62,000
3	76,500	81,200	77,200
4	81,400	84,300	81,600
5	90,800	91,300	83,200
6	123,500	128,000	125,000
7	225,000	276,300	245,000
8	260,000	265,500	235,000

Library

Project

The following articles each employ at least one nonparametric test from this chapter. Select an article, locate it, and report on it.

a. Assessing Learning Capabilities in Sheltered Workshop Clients;" P. Freedman, *Human Learning*, vol. 2, no. 3, July 1983.

b. "The Effect of Swimming Training on Hormone Levels in Girls;" C. Carli et al., *Journal of Sports Medicine*, vol. 23, no. 1, March 1983.

c. "The Role of Inference in Children's Comprehension of Pronouns;" T. Wykes, *Journal of Experimental Child Psychology*, vol. 35, no. 1, 1983.

d. "A Multiclinic Double Blind Comparison of Timolo and Hydrochlorothiazide;" A. Leon and D. Hunninghake, *Journal of Clinical Pharmacology*, vol. 23, no. 1, Jan. 1983.

e. "Does the Surface Film of Lakes Provide a Source of Food for Animals Living in Lake Outlets?" R. Wooton, *Limnology and Oceanography*, vol. 27, no. 5, Sept. 1982.

f. "Stimuli for Male Mouse Ultrasonic Courtship Vocalizations;" J. Nyby and C. Wysocki, *Journal of Comparative and Physiological Psychology*, vol. 95, no. 4, Aug. 1981 (Friedman ANOVA).

g. "Pavlovian Conditioning in Tolerance to Amphetamine-Induced Anorexia;" C. Poulos et al., *Journal of Comparative and Physiological Psychology*, vol. 95, no. 5, Sept. 1981.

h. "Assessing the Parents of the Learning-Disabled Child;" L. Sloman et al., *Journal of Learning Disabilities*, vol. 11, no. 2, Feb. 1978.

i. "Attitudes Toward Fathering and Father-Child Activity;" J. Bigner, *Home Economics Research Journal*, vol. 6, no. 2, Dec. 1977.

j. "The Validity of the Spiral Aftereffect as a Clinical Tool for Diagnosis of Organic Brain Pathology;" E. Philbrick, *Journal of Consulting Psychology*, vol. 23, no. 1, 1959.

13.10   THE RUN TEST FOR RANDOMNESS

In studies where use is made of existing data, a frequent question of interest is whether a collection of data can be regarded as a random sample. The run test provides a means of testing whether this is the case.

We begin by examining whether a succession of two symbols, say S and F, can be regarded as having occurred in a random fashion. We then extend the test to a set of measurements. The basic idea is to look at successions of the same letter, which are called runs. Too many or too few runs are both taken to indicate a lack of randomness. The number of runs in a sample is denoted by r, the number of S's and F's by n_1 and n_2 respectively.

Example 13.16

Three sequences of $n_1 = 10$ S's and $n_2 = 11$ F's are given below. The runs are underlined.

$$\underline{F}\,\underline{S}\,\underline{F}\,\underline{S}\,\underline{F}\,\underline{S}\,\underline{F}\,\underline{S}\,\underline{F}\,\underline{S}\,\underline{F}\,\underline{S}\,\underline{F}\,\underline{S}\,\underline{F}\,\underline{S}\,\underline{F}\,\underline{S}\,\underline{F}\,\underline{S}\,\underline{F} \quad r = 21$$
$$\underline{F\,F\,F\,F\,F\,F\,F\,F\,F\,F}\,\underline{S\,S\,S\,S\,S\,S\,S\,S\,S\,S} \quad r = 2$$
$$\underline{F\,F}\,\underline{S\,S\,S}\,\underline{F}\,\underline{S\,S}\,\underline{F\,F\,F}\,\underline{S\,S\,S}\,\underline{F\,F}\,\underline{S\,S}\,\underline{F\,F\,F} \quad r = 9$$

In the first of these sequences we have the maximum number of runs, $r = 21$. Here there is an alternating pattern and a lack of randomness. In the second sequence we have the opposite extreme, with $r = 2$ runs and a clustering of symbols so extreme as to seemingly preclude randomness.

In the third sequence, $r = 9$; this number of runs would not appear to be too extreme for a random sequence. ∎

The number of runs R in a sample is a test statistic that we could use here. However, when the number of symbols n_1 and n_2 are both at least 10, the distribution of R is approximately normal. Specifically,

$$z = \frac{r - \left(\dfrac{2n_1 n_2}{n_1 + n_2} + 1\right)}{\sqrt{\dfrac{2n_1 n_2 (2n_1 n_2 - n_1 - n_2)}{(n_1 + n_2)^2 (n_1 + n_2 - 1)}}}$$

is the value of a variable having approximately the standard normal distribution. We shall phrase the test in terms of the statistic Z and not consider the case where n_1 and n_2 are small.

PROCEDURE FOR PERFORMING THE RUN TEST

H_0: The sequence of S's and F's is random.
H_1: The sequence is not random.

Given: The significance level α.
   The sequence consisting of n_1 S's and n_2 F's, where both n_1 and n_2 are at least 10.
1. Find the critical values $\pm z_{\alpha/2}$.
2. Find the number of runs r within the sequence.
3. Compute the value of the test statistic

$$z = \frac{r - \left(\dfrac{2n_1 n_2}{n_1 + n_2} + 1\right)}{\sqrt{\dfrac{2n_1 n_2(2n_1 n_2 - n_1 - n_2)}{(n_1 + n_2)^2(n_1 + n_2 - 1)}}}$$

4. Compare the value of the test statistic with the critical values. Reject H_0 if $z < -z_{\alpha/2}$ or $> z_{\alpha/2}$.

Example 13.17

Test whether the following sequence of F's and S's is random:

$$\underline{F\,F}\,\underline{S\,S\,S}\,\underline{F\,F\,F\,F}\,\underline{S\,S}\,\underline{F}\,\underline{S\,S}\,\underline{F\,F}\,\underline{S\,S}\,\underline{F}\,\underline{S\,S}$$

For a 5% two-tailed test the critical values are $\pm z_{.025} = \pm 1.96$. Using $r = 10$, $n_1 = 11$, and $n_2 = 10$,

$$z = \frac{10 - \left(\dfrac{2 \cdot 11 \cdot 10}{11 + 10} + 1\right)}{\sqrt{\dfrac{2 \cdot 11 \cdot 10(2 \cdot 11 \cdot 10 - 11 - 10)}{(11 + 10)^2(11 + 10 - 1)}}} = -.66$$

The null hypothesis is not rejected. Notice that the number of runs $r = 10$ is very close to the expected number

$$u_R = \left(\frac{2 \cdot 11 \cdot 10}{11 + 10} + 1\right) = 11.48 \quad\blacksquare$$

Now consider a sample of measurements that are arranged in order exactly as they were reported or observed. To test for randomness, we may classify each in terms of whether it is above (S) or below (F) the median and then apply the run test above to the total number of runs (values above and below the median).

Example 13.18

In an anthropological study of South American Indians, the following family unit sizes were reported. Can the data be treated as a random sample?

$$8, 14, 13, 12, 14, 8, 6, 7, 15, 16, 14, 7, 11, 10,$$
$$9, 9, 12, 13, 12, 14, 15, 9, 8, 9, 10, 11, 11$$

There are 27 observations. When they are arranged in order, the median is the 14th which is $\bar{x} = 11$. Any measurement below 11 is replaced with an F, any above with an S. We ignore the three 11s, which are at the median. The result is the following sequence with $r = 7$ runs:

$$\underline{F}\,\underline{S\,S\,S}\,\underline{F\,F\,F}\,\underline{S\,S}\,\underline{F\,F\,F\,F}\,\underline{S\,S\,S\,S\,S}\,\underline{F\,F\,F\,F}$$

With $\alpha = .05$, the critical values are $\pm z_{.025} = \pm 1.96$. Using $r = 7$, $n_1 = 12$, and $n_2 = 12$, the test value is

$$z = \frac{7 - \left[\dfrac{2 \cdot 12 \cdot 12}{12 + 12} + 1\right]}{\sqrt{\dfrac{2 \cdot 12 \cdot 12(2 \cdot 12 \cdot 12 - 12 - 12)}{(12 + 12)^2(12 + 12 - 1)}}} = -2.50$$

The null hypothesis is rejected. The evidence shows that the sample is not random. ■

EXERCISE SET 13.7

Part A

1. The following sequence of true-false answers was observed on an examination. Test for randomness.

    T T T T F F F F F T T T F F T T T T F T F T
    F F F T T T F F T T F F F F T T F F T F F

2. The following sequence of even (E) and odd (O) numbers were recorded for a roulette wheel. Test for randomness using $\alpha = .05$.

    E E E O O O E E O E O E O O O E O O
    E E O O E O E E O O E O E O O O E O

3. 51 students were asked to taste two brands of cola, A and B, and select their favorite. The following is their preferences in the order interviewed. Test for randomness.

    A B A B B B A A B B B B B A A B B B A A B A B A B B
    B B B A A A A A A A A B A B B A B B B A A A A B B

4. An employee of a data gathering company reported the following ages of a sample of 30 people she interviewed. Assuming these data are in the order in which they were obtained, test for randomness using runs above and below the median.

    25, 19, 27, 46, 52, 20, 23, 38, 43, 45, 21, 60, 18, 30, 50,
    18, 23, 29, 46, 49, 50, 55, 49, 21, 22, 24, 25, 29, 32, 36

5. Examination papers were kept in precisely the same order as turned in. Test for randomness if their grades are as follows:

    88, 74, 96, 15, 62, 77, 91, 100, 45, 72, 75, 69, 71, 65,
    89, 96, 23, 45, 41, 56, 71, 90, 54, 82, 79, 86, 31, 20

6. A regression formula for predicting demand for a product based on several economic indicators produced the following sequence of low (L) and high (H) predictions for 22 consecutive weeks. Test for randomness.

    L L L L L H H H L L L H H L L L L L H H H H

Part B

7. An examination has 100 true–false (T and F) questions. What is the expected number of runs if 50 T and 50 F answers are randomly assigned?

8. As part of its quality control on a critical casting, a foundry daily selects 10 of the castings and determines their mean weight. Suppose the means for 36 consecutive production days show a total of 12 means above and 18 means below the median. Using $\alpha = .05$ test for randomness in the production means if there are $r = 8$ runs in all.

9. For a random sequence of 3 S's and 2 F's,
   a. What is the minimum number of possible runs?
   b. What is the maximum number of possible runs?
   c. What is the expected number of runs?
10. One can show there are in all

$$\binom{n_1 + n_2}{n_1}$$

sequences consisting of n_1 S's and n_2 F's.
   a. How many sequences are there that have three S's and two F's?
   b. List these sequences and find the distribution of the number of runs X in a random sequence of three S's and two F's.

COMPUTER
PRINTOUT

The Minitab statistical package contains many of the nonparametric tests that we have discussed. The contingency table of the printout in Figure 13.1 is for the joint classification by age and sex of a random sample of 41 jurors who were accepted without contest by attorneys for plaintiffs and defendants in several civil trials. The male and female (column) categories are coded 0 and 1 respectively. Similarly the age (row) categories young, middle-aged, and old are coded 0, 1, and 2. Each data entry describes one juror. Following this table is the chi-square analysis.

FIGURE 13.1   MINITAB CONTINGENCY TABLE

```
MTB>READ AGE CATEGORIES INTO C1, SEX OF SUBJECTS INTO
C2
DATA>0   0
DATA>2   1
DATA>2   0
DATA>0   0
DATA>1   1
DATA>2   0
DATA>1   1
DATA>1   1
DATA>1   0
DATA>2   1
DATA>2   0
DATA>2   0
DATA>0   1
DATA>0   1
DATA>0   0
DATA>2   0
DATA>2   1
DATA>1   1
DATA>1   1
DATA>1   0
DATA>2   0
```

Continued.

FIGURE 13.1   MINITAB CONTINGENCY TABLE—cont.

```
DATA>2   1
DATA>1   0
DATA>1   1
DATA>2   0
DATA>2   0
DATA>1   0
DATA>0   1
DATA>0   0
DATA>0   1
DATA>2   0
DATA>2   0
DATA>2   1
DATA>1   1
DATA>1   0
DATA>1   1
DATA>0   1
DATA>0   1
DATA>1   1
DATA>2   0
DATA>1   0
DATA>END
MTB >CONTINGENCY TABLE AND CHI SQUARE ANALYSIS C1
DATA VS C2 DATA

EXPECTED FREQUENCIES ARE PRINTED BELOW OBSERVED
FREQUENCIES

    ROW CLASSIFICATION - C1
    COLUMN CLASSIFICATION - C2
        I     0   I     1   I TOTALS
-------I-------I-------I-------
   0  I     4  I     6  I     10
      I   5.1I   4.91
-------I-------I-------I-------
   1  I     7  I     9  I     16
      I   8.21   7.81
-------I-------I-------I-------
   2  I    10  I     5  I     15
      I   7.71   7.31
-------I-------I-------I-------
TOTALS I    21  I    20  I     41

TOTAL CHI SQUARE =

        0.25 +   0.26 +
        0.17 +   0.18 +
        0.70 +   0.73 +

             =   2.29

DEGREES OF FREEDOM = (3-1) × (2-2) = 2

NOTE 1 CELLS WITH EXPECTED FREQUENCIES LESS THAN 5
```

The Minitab printout for the Mann-Whitney test of the equality of the median fruit production of trees from two orchards is given

Continued.

in Figure 13.2. ETA1 and ETA2 represent the medians. The quantity $w = 90$ is the rank sum of the sample in column 1 (i.e., in the notation of the text, $w = w_1$).

FIGURE 13.2 MINITAB COMPUTER PRINTOUT: MANN-WHITNEY TEST

```
MTB > SET FRUIT YIELDS FOR TREES OF ORCHARD ONE INTO
C1
DATA>328, 340, 365, 382, 390, 390, 401, 412, 426
DATA>430, 441
DATA>END
MTB >SET FRUIT YIELDS FOR TREES OF ORCHARD TWO INTO
C2
DATA>290, 360, 410, 490, 503, 540, 555, 563, 570
DATA>END
MTB >MANN-WHITNEY TEST, USE DATA IN C1 AND C2
    C1      N =   11    MEDIAN =      390.00
    C2      N =    9    MEDIAN =      503.00
    A POINT ESTIMATE FOR ETA1-ETA2 IS     -113.0
    A 95.2 PERCENT C.I. FOR ETA1-ETA2 IS
(     -163.1,       2.1)

    W =       90.0
    TEST OF ETA1 = ETA2 VS ETA1 N.E. ETA2
    THE TEST IS SIGNIFICANT AT 0.0575

    CANNOT REJECT AT ALPHA = 0.05

MTB >
```

The Spearman rank correlation coefficient r_s is simply the Pearson correlation coefficient for the ranked observations. The ranks and the Pearson coefficient r are easily obtained with Minitab. The printout of the steps in obtaining the Spearman coefficient for the correlation between the reading and mathematics scores of a sample of sixth grade students is given in Figure 13.3.

FIGURE 13.3 MINITAB COMPUTER PRINTOUT: SPEARMAN RANK CORRELATION COEFFICIENT

```
MTB >
MTB >SET THE READING SCORES INTO C1
DATA>34,56,62,66,70,78,90,95,98,100,100
DATA>END
MTB >SET THE MATHEMATICS SCORES INTO COLUMN C2
DATA>43,48,65,59,70,79,84,82,90,85,88
DATA>END
MTB >RANK C1, PUT THE RANKS INTO C3
MTB >RANK C2, PUT THE RANKS INTO C4
MTB >PRINT C1,C3,C2,C4
```

Continued.

FIGURE 13.3   MINITAB COMPUTER PRINTOUT: SPEARMAN RANK CORRELATION COEFFICIENT—cont.

COLUMN	C1	C3	C2	C4
COUNT	11	11	11	11
ROW				
1	34.	1.0000	43.	1.
2	56.	2.0000	48.	2.
3	62.	3.0000	65.	4.
4	66.	4.0000	59.	3.
5	70.	5.0000	70.	5.
6	78.	6.0000	79.	6.
7	90.	7.0000	84.	8.
8	95.	8.0000	82.	7.
9	98.	9.0000	90.	11.
10	100.	10.5000	85.	9.
11	100.	10.5000	88.	10.

```
MTB >CORRELATION BETWEEN C3 AND C4
     CORRELATION OF        C3 AND C4        = 0.952
```

CHAPTER CHECKLIST

Vocabulary

Nonparametric test
Distribution-free test
Cross-classified
Contingency table
Expected frequencies
Marginal row totals
Marginal column totals

Homogeneity
Rank
Rank sum
Signed ranks
Rank correlation coefficient
Randomized complete block design
Runs

Techniques

Performing the chi-square test of independence
Performing the chi-square test of homogeneity
Performing the Mann-Whitney U test
Performing the Wilcoxon test
Computing the Spearman rank correlation coefficient and testing for a correlation
Performing the Kruskal-Wallis one-way ANOVA test
Performing the Friedman two-way ANOVA test
Performing the run test for randomness.

Notation

o_{ij} Observed frequency of the i, j cell of a contingency table
e_{ij} Expected frequency of the i, j cell of a contingency table
U The Mann-Whitney test statistic
T The Wilcoxon test statistic (not related to the t distribution)
r_s The Spearman rank correlation coefficient
H The Kruskal-Wallis test statistic
χ_f^2 The Friedman test statistic

CHAPTER TEST

1. Three chemotherapy treatments are evaluated on rats with malignant liver tumors. The tumors showed a definite reduction in size in 23 of the 50 rats receiving therapy 1, 34 of the 63 receiving therapy 2, and in 12 of the 20 receiving therapy 3. Test the hypothesis that these therapies are equally effective. Use $\alpha = .05$.

2. A student body president is interested in whether a student's class standing is related to the student's attitude toward a proposed campus child care center. A survey of 500 students produced the results below. At the 5% level of significance, test the hypothesis that the student attitude on this matter is independent of class standing.

Standing	In favor	Against	No opinion
Freshman	40	50	30
Sophomore	50	35	15
Junior	70	40	10
Senior	75	18	7
Graduate	40	12	8

3. A randomized block design was used with an experiment having four treatments and five blocks. The ranks of the measurements of each block are given below. Use the Friedman two-way analysis of variance test to test the hypothesis that the distributions of at least two of the treatment populations differ as to location. Use $\alpha = .05$.

Block	Treatment 1	Treatment 2	Treatment 3	Treatment 4
1	3	1	4	2
2	2	1	3	4
3	3	4	1	2
4	3	4	2	1
5	4	2	3	1

4. Four colors of automobile paint of the same manufacturer were tested for their resistance to fading. In the experiment, test areas were painted on the hoods of five automobiles with each of the four colors. After 10 months of exposure to road conditions and the elements, the amount of fading of each paint on each hood was measured. The results are given below. Can we conclude that there is a difference in the resistance to fading by these four colors?

Auto	Color			
	1	*2*	*3*	*4*
1	8.2	7.0	4.3	5.2
2	6.1	8.4	3.1	2.9
3	7.4	7.5	5.6	4.9
4	9.0	6.4	6.5	5.3
5	8.3	7.5	6.0	5.8

5. Sixteen applicants for a postal service examination are randomly assigned to one of two groups. The first group (A) was given the normal 60 minutes to complete the first part of the examination. The second group (B), however, was allowed only 45 minutes. The test scores for this part of the exam are given below. Using the Mann-Whitney test, test the hypothesis that reducing the time to 45 minutes does not change the average for this part of the examination.

Group	Scores							
A (60 minutes)	32	36	30	28	34	21	35	24
B (45 minutes)	38	30	20	24	28	33	31	27

6. The rankings of 10 restaurants by two television restaurant critics are given below. Test the hypothesis that a positive correlation exists between these two rankings. Use the Spearman rank correlation coefficient with $\alpha = .05$.

Restaurant	Critic 1	Critic 2
1	10	8
2	4	5
3	7	6
4	3	3
5	6	9
6	9	7
7	2	1
8	5	4
9	1	2
10	8	10

7. A firearms control group has prepared a film that they plan to use in a campaign to promote greater support for their position. To test the effectiveness, if any, of the film, 10 adults are selected at random and first given a test which measures their attitude toward firearms control. Next they show the subjects the film and then again give them the same attitude questionnaire. The results are shown below. Can we conclude that the film has an effect on attitudes toward firearms control? Use the Wilcoxon matched pair test with $\alpha = .05$.

Subject	1	2	3	4	5	6	7	8	9	10
Initial score	18	12	15	20	4	17	11	14	23	13
Final score	19	9	15	21	2	20	12	10	23	11

8. Independent random samples from three populations are given below. Use the Kruskal-Wallis test with $\alpha = .05$ and test the hypothesis that at least two of the populations do not have the same location.

Sample 1	Sample 2	Sample 3
25	70	40
20	60	70
35	45	25
65	50	30
70	30	55

9. The data below represents the number of employees who voluntarily left their employment with a large hospital during the first 30 months following its sale to a hospital conglomerate. Find the median, replace those observations below it with an F, those above it with an S, and perform a run test for randomness using $\alpha = .05$.

2, 0, 4, 3, 5, 5, 7, 6, 7, 10, 4, 3, 8, 9, 10,
6, 9, 12, 8, 7, 9, 11, 15, 10, 7, 10, 12, 7, 9, 14

10. Hospital patients are classified according to their age and number of prior admissions in the table below. Test the hypothesis that these two classifications are independent. Be careful: some frequencies are not as large as they should be.

Age	Admissions			
	0–1	2–3	4–5	>5
<40	8	6	4	3
40–60	25	8	5	4
60–70	9	29	18	12
>70	17	56	64	123

Appendix

TABLE I   RANDOM NUMBERS

	1	2	3	4	5	6	7	8	9
1	32942	95416	42339	59045	26693	49057	87496	20624	14819
2	07410	99859	83828	21409	29094	65114	36701	25762	12827
3	59981	68155	45673	76210	58219	45738	29550	24736	09574
4	46251	25437	69654	99716	11563	08803	86027	51867	12116
5	65558	51904	93123	27887	53138	21488	09095	78777	71240
6	99187	19258	86421	16401	19397	83297	40111	49326	81686
7	35641	00301	16096	34775	21562	97983	45040	19200	16383
8	14031	00936	81518	48440	02218	04756	19506	60695	88494
9	60677	15076	92554	26042	23472	69869	62877	19584	39576
10	66314	05212	67859	89356	20056	30648	87349	20389	53805
11	20416	87410	75646	64176	82752	63606	37011	57346	69512
12	28701	56992	70423	62415	40807	98086	58850	28968	45297
13	74579	33844	33426	07570	00728	07079	19322	56325	84819
14	62615	52342	82968	75540	80045	53069	20665	21282	07768
15	93945	06293	22879	08161	01442	75071	21427	94842	26210
16	75689	76131	96837	67450	44511	50424	82848	41975	71663
17	02921	16919	35424	93209	52133	87327	95897	65171	20376
18	14295	34969	14216	03191	61647	30296	66667	10101	63203
19	05303	91109	82403	40312	62191	67023	90073	83205	71344
20	57071	90357	12901	08899	91039	67251	28701	03846	94589
21	78471	57741	13599	84390	32146	00871	09354	22745	65806
22	89242	79337	59293	47481	07740	43345	25716	70020	54005
23	14955	59592	97035	80430	87220	06392	79028	57123	52872
24	42446	41880	37415	47472	04513	49494	08860	08038	43624
25	18534	22346	54556	17558	73689	14894	05030	19561	56517
26	39284	33737	42512	86411	23753	29690	26096	81361	93099
27	33922	37329	89911	55876	28379	81031	22058	21487	54613
28	78355	54013	50774	30666	61205	42574	47773	36027	27174
29	08845	99145	94316	88974	29828	97069	90327	61842	29604
30	01769	71825	55957	98271	02784	66731	40311	88495	18821
31	17639	38284	59478	90409	21997	56199	30068	82800	69692
32	05851	58653	99949	63505	40409	85551	90729	64938	52403
33	42396	40112	11469	03476	03328	84238	26570	51790	42122
34	13318	14192	98167	75631	74141	22369	36757	89117	54998
35	60571	54786	26281	01855	30706	66578	32019	65884	58485
36	09531	81853	59334	70929	03544	18510	89541	13555	21168
37	72865	16829	86542	00396	20363	13010	69645	49608	54738
38	56324	31093	77924	28622	83543	28912	15059	80192	83964
39	78192	21626	91399	07235	07104	73652	64425	85149	75409
40	64666	34767	97298	92708	01994	53188	78476	07804	62404
41	82201	75694	02808	65983	74373	66693	13094	74183	73020
42	15360	73776	40914	85190	54278	99054	62944	47351	89098
43	68142	67957	70896	37983	20487	95350	16371	03426	13895
44	19138	31200	30616	14639	44406	44236	57360	81644	94761
45	28155	03521	36415	78452	92359	81091	56513	88321	97910
46	87971	29031	51780	27376	81056	86155	55488	50590	74514
47	58147	68841	53625	02059	75223	16783	19272	61994	71090
48	18875	52809	70594	41649	32935	26430	82096	01605	65846
49	75109	56474	74111	31966	29969	70093	98901	84550	25769
50	35983	03742	76822	12073	59463	84420	15868	99505	11426

TABLE I *(CONTINUED)*

	1	2	3	4	5	6	7	8	9
51	12651	61646	11769	75109	86996	97669	25757	32535	07122
52	81769	74436	02630	72310	45049	18029	07469	42341	98173
53	36737	98863	77240	76251	00654	64688	09343	70278	67331
54	82861	54371	76610	94934	72748	44124	05610	53750	95938
55	21325	15732	24127	37431	09723	63529	73977	95218	96074
56	74146	47887	62463	23045	41490	07954	22597	60012	98866
57	90759	64410	54179	66075	61051	75385	51378	08360	95946
58	55683	98078	02238	91540	21219	17720	87817	41705	95785
59	79686	17969	76061	83748	55920	83612	41540	86492	06447
60	70333	00201	86201	69716	78185	62154	77930	67663	29529
61	14042	53536	07779	04157	41172	36473	42123	43929	50533
62	59911	08256	06596	48416	69770	68797	56080	14223	59199
63	62368	62623	62742	14891	39247	52242	98832	69533	91174
64	57529	97751	54976	48957	74599	08759	78494	52785	68526
65	15469	90574	78033	66885	13936	42117	71831	22961	94225
66	18625	23674	53850	32827	81647	80820	00420	63555	74489
67	74626	68394	88562	70745	23701	45630	65891	58220	35442
68	11119	16519	27384	90199	79210	76965	99546	30323	31664
69	41101	17336	48951	53674	17880	45260	08575	49321	36191
70	32123	91576	84221	78902	82010	30847	62329	63898	23268
71	26091	68409	69704	82267	14751	13151	93115	01437	56945
72	67680	79790	48462	59278	44185	29616	76531	19589	83139
73	15184	19260	14073	07026	25264	08388	27182	22557	61501
74	58010	45039	57181	10238	36874	28546	37444	80824	63981
75	56425	53996	86245	32623	78858	08143	60377	42925	42815
76	82630	84066	13592	60642	17904	99718	63432	88642	37858
77	14927	40909	23900	48761	44860	92467	31742	87142	03607
78	23740	22505	07489	85986	74420	21744	97711	36648	35620
79	32990	97446	03711	63824	07953	85965	87089	11687	92414
80	05310	24058	91946	78437	34365	82469	12430	84754	19354
81	21839	39937	27534	88913	49055	19218	47712	67677	51889
82	08833	42549	93981	94051	28382	83725	72643	64233	97252
83	58336	11139	47479	00931	91560	95372	97642	33856	54825
84	62032	91144	75478	47431	52726	30289	42411	91886	51818
85	45171	30557	53116	04118	58301	24375	65609	85810	18620
86	91611	62656	60128	35609	63698	78356	50682	22505	01692
87	55472	63819	86314	49174	93582	73604	78614	78849	23096
88	18573	09729	74091	53994	10970	86557	65661	41854	26037
89	60866	02955	90288	82136	83644	94455	06560	78029	98768
90	45043	55608	82767	60890	74646	79485	13619	98868	40857
91	17831	09737	79473	75945	28394	79334	70577	38048	03607
92	40137	03981	07585	18128	11178	32601	27994	05641	22600
93	77776	31343	14576	97706	16039	47517	43300	59080	80392
94	69605	44104	40103	95635	05635	81673	68657	09559	23510
95	19916	52934	26499	09821	87331	80993	61299	36979	73599
96	02606	58552	07678	56619	65325	30705	99582	53390	46357
97	65183	73160	87131	35530	47946	09854	18080	02321	05809
98	10740	98914	44916	11322	89717	88189	30143	52687	19420
99	98642	89822	71691	51573	83666	61642	46683	33761	47542
100	60139	25601	93663	25547	02654	94829	48672	28736	84994

TABLE II   CUMULATIVE BINOMIAL PROBABILITIES

N = 5

x	0.01	0.05	0.1	0.2	0.3	0.4	P 0.5	0.6	0.7	0.8	0.9	0.95	0.99	x
0	.9510	.7738	.5905	.3277	.1681	.0778	.0313	.0102	.0024	.0003	.0000	.0000	.0000	0
1	.9990	.9774	.9185	.7373	.5282	.3370	.1875	.0870	.0308	.0067	.0005	.0000	.0000	1
2	1.0000	.9988	.9914	.9421	.8369	.6826	.5000	.3174	.1631	.0579	.0086	.0012	.0000	2
3	1.0000	1.0000	.9995	.9933	.9692	.9130	.8125	.6630	.4718	.2627	.0815	.0226	.0010	3
4	1.0000	1.0000	1.0000	.9997	.9976	.9898	.9687	.9222	.8319	.6723	.4095	.2262	.0490	4

N = 6

x	0.01	0.05	0.1	0.2	0.3	0.4	P 0.5	0.6	0.7	0.8	0.9	0.95	0.99	x
0	.9415	.7351	.5314	.2621	.1176	.0467	.0156	.0041	.0007	.0001	.0000	.0000	.0000	0
1	.9985	.9672	.8857	.6554	.4202	.2333	.1094	.0410	.0109	.0016	.0001	.0000	.0000	1
2	1.0000	.9978	.9841	.9011	.7443	.5443	.3437	.1792	.0705	.0170	.0013	.0001	.0000	2
3	1.0000	.9999	.9987	.9830	.9295	.8208	.6562	.4557	.2557	.0989	.0158	.0022	.0000	3
4	1.0000	1.0000	.9999	.9984	.9891	.9590	.8906	.7667	.5798	.3446	.1143	.0328	.0015	4
5	1.0000	1.0000	1.0000	.9999	.9993	.9959	.9844	.9533	.8824	.7379	.4686	.2649	.0585	5

N = 7

x	0.01	0.05	0.1	0.2	0.3	0.4	P 0.5	0.6	0.7	0.8	0.9	0.95	0.99	x
0	.9321	.6983	.4783	.2097	.0824	.0280	.0078	.0016	.0002	.0000	.0000	.0000	.0000	0
1	.9980	.9556	.8503	.5767	.3294	.1586	.0625	.0188	.0038	.0004	.0000	.0000	.0000	1
2	1.0000	.9962	.9743	.8520	.6471	.4199	.2266	.0963	.0288	.0047	.0002	.0000	.0000	2
3	1.0000	.9998	.9973	.9667	.8740	.7102	.5000	.2898	.1260	.0333	.0027	.0002	.0000	3
4	1.0000	1.0000	.9998	.9963	.9712	.9037	.7734	.5801	.3529	.1480	.0257	.0038	.0000	4
5	1.0000	1.0000	1.0000	.9996	.9962	.9812	.9375	.8414	.6706	.4233	.1497	.0444	.0021	5
6	1.0000	1.0000	1.0000	1.0000	.9998	.9984	.9922	.9720	.9176	.7903	.5217	.3017	.0679	6

N = 8

x	0.01	0.05	0.1	0.2	0.3	0.4	P 0.5	0.6	0.7	0.8	0.9	0.95	0.99	x
0	.9227	.6634	.4305	.1678	.0576	.0168	.0039	.0007	.0001	.0000	.0000	.0000	.0000	0
1	.9973	.9428	.8131	.5033	.2553	.1064	.0352	.0085	.0013	.0001	.0000	.0000	.0000	1
2	.9999	.9942	.9619	.7969	.5518	.3154	.1445	.0498	.0113	.0012	.0000	.0000	.0000	2
3	1.0000	.9996	.9950	.9437	.8059	.5941	.3633	.1737	.0580	.0104	.0004	.0000	.0000	3
4	1.0000	1.0000	.9996	.9896	.9420	.8263	.6367	.4059	.1941	.0563	.0050	.0004	.0000	4
5	1.0000	1.0000	1.0000	.9988	.9887	.9502	.8555	.6846	.4482	.2031	.0381	.0058	.0001	5
6	1.0000	1.0000	1.0000	.9999	.9987	.9915	.9648	.8936	.7447	.4967	.1869	.0572	.0027	6
7	1.0000	1.0000	1.0000	1.0000	.9999	.9993	.9961	.9832	.9424	.8322	.5695	.3366	.0773	7

TABLE II *(CONTINUED)*

N = 9

x	0.01	0.05	0.1	0.2	0.3	0.4	P 0.5	0.6	0.7	0.8	0.9	0.95	0.99	x
0	.9135	.6302	.3874	.1342	.0404	.0101	.0020	.0003	.0000	.0000	.0000	.0000	.0000	0
1	.9966	.9288	.7748	.4362	.1960	.0705	.0195	.0038	.0004	.0000	.0000	.0000	.0000	1
2	.9999	.9916	.9470	.7382	.4628	.2318	.0898	.0250	.0043	.0003	.0000	.0000	.0000	2
3	1.0000	.9994	.9917	.9144	.7297	.4826	.2539	.0994	.0253	.0031	.0001	.0000	.0000	3
4	1.0000	1.0000	.9991	.9804	.9012	.7334	.5000	.2666	.0988	.0196	.0009	.0000	.0000	4
5	1.0000	1.0000	.9999	.9969	.9747	.9006	.7461	.5174	.2703	.0856	.0083	.0006	.0000	5
6	1.0000	1.0000	1.0000	.9997	.9957	.9750	.9102	.7682	.5372	.2618	.0530	.0084	.0001	6
7	1.0000	1.0000	1.0000	1.0000	.9996	.9962	.9805	.9295	.8040	.5638	.2252	.0712	.0034	7
8	1.0000	1.0000	1.0000	1.0000	1.0000	.9997	.9980	.9899	.9596	.8658	.6126	.3698	.0865	8

N = 10

x	0.01	0.05	0.1	0.2	0.3	0.4	P 0.5	0.6	0.7	0.8	0.9	0.95	0.99	x
0	.9044	.5987	.3487	.1074	.0282	.0060	.0010	.0001	.0000	.0000	.0000	.0000	.0000	0
1	.9957	.9139	.7361	.3758	.1493	.0464	.0107	.0017	.0001	.0000	.0000	.0000	.0000	1
2	.9999	.9885	.9298	.6778	.3828	.1673	.0547	.0123	.0016	.0001	.0000	.0000	.0000	2
3	1.0000	.9990	.9872	.8791	.6496	.3823	.1719	.0548	.0106	.0009	.0000	.0000	.0000	3
4	1.0000	.9999	.9984	.9672	.8497	.6331	.3770	.1662	.0473	.0064	.0001	.0000	.0000	4
5	1.0000	1.0000	.9999	.9936	.9527	.8338	.6230	.3669	.1503	.0328	.0016	.0001	.0000	5
6	1.0000	1.0000	1.0000	.9991	.9894	.9452	.8281	.6177	.3504	.1209	.0128	.0010	.0000	6
7	1.0000	1.0000	1.0000	.9999	.9984	.9877	.9453	.8327	.6172	.3222	.0702	.0115	.0001	7
8	1.0000	1.0000	1.0000	1.0000	.9999	.9983	.9893	.9536	.8507	.6242	.2639	.0861	.0043	8
9	1.0000	1.0000	1.0000	1.0000	1.0000	.9999	.9990	.9940	.9718	.8926	.6513	.4013	.0956	9

N = 15

x	0.01	0.05	0.1	0.2	0.3	0.4	P 0.5	0.6	0.7	0.8	0.9	0.95	0.99	x
0	.8601	.4633	.2059	.0352	.0047	.0005	.0000	.0000	.0000	.0000	.0000	.0000	.0000	0
1	.9904	.8290	.5490	.1671	.0353	.0052	.0005	.0000	.0000	.0000	.0000	.0000	.0000	1
2	.9996	.9638	.8159	.3980	.1268	.0271	.0037	.0003	.0000	.0000	.0000	.0000	.0000	2
3	1.0000	.9945	.9444	.6482	.2969	.0905	.0176	.0019	.0001	.0000	.0000	.0000	.0000	3
4	1.0000	.9994	.9873	.8358	.5155	.2173	.0592	.0093	.0007	.0000	.0000	.0000	.0000	4
5	1.0000	.9999	.9978	.9389	.7216	.4032	.1509	.0338	.0037	.0001	.0000	.0000	.0000	5
6	1.0000	1.0000	.9997	.9819	.8689	.6098	.3036	.0950	.0152	.0008	.0000	.0000	.0000	6
7	1.0000	1.0000	1.0000	.9958	.9500	.7869	.5000	.2131	.0500	.0042	.0000	.0000	.0000	7
8	1.0000	1.0000	1.0000	.9992	.9848	.9050	.6964	.3902	.1311	.0181	.0003	.0000	.0000	8
9	1.0000	1.0000	1.0000	.9999	.9963	.9662	.8491	.5968	.2784	.0611	.0022	.0001	.0000	9
10	1.0000	1.0000	1.0000	1.0000	.9993	.9907	.9408	.7827	.4845	.1642	.0127	.0006	.0000	10
11	1.0000	1.0000	1.0000	1.0000	.9999	.9981	.9824	.9095	.7031	.3518	.0556	.0055	.0000	11
12	1.0000	1.0000	1.0000	1.0000	1.0000	.9997	.9963	.9729	.8732	.6020	.1841	.0362	.0004	12
13	1.0000	1.0000	1.0000	1.0000	1.0000	1.0000	.9995	.9948	.9647	.8329	.4510	.1710	.0096	13
14	1.0000	1.0000	1.0000	1.0000	1.0000	1.0000	1.0000	.9995	.9953	.9648	.7941	.5367	.1399	14

TABLE II *(CONTINUED)*

N = 20

x	0.01	0.05	0.1	0.2	0.3	0.4	P 0.5	0.6	0.7	0.8	0.9	0.95	0.99	x
0	.8179	.3585	.1216	.0115	.0008	.0000	.0000	.0000	.0000	.0000	.0000	.0000	.0000	0
1	.9831	.7358	.3917	.0692	.0076	.0005	.0000	.0000	.0000	.0000	.0000	.0000	.0000	1
2	.9990	.9245	.6769	.2061	.0355	.0036	.0002	.0000	.0000	.0000	.0000	.0000	.0000	2
3	1.0000	.9841	.8670	.4114	.1071	.0160	.0013	.0000	.0000	.0000	.0000	.0000	.0000	3
4	1.0000	.9974	.9568	.6296	.2375	.0510	.0059	.0003	.0000	.0000	.0000	.0000	.0000	4
5	1.0000	.9997	.9887	.8042	.4164	.1256	.0207	.0016	.0000	.0000	.0000	.0000	.0000	5
6	1.0000	1.0000	.9976	.9133	.6080	.2500	.0577	.0065	.0003	.0000	.0000	.0000	.0000	6
7	1.0000	1.0000	.9996	.9679	.7723	.4159	.1316	.0210	.0013	.0000	.0000	.0000	.0000	7
8	1.0000	1.0000	.9999	.9900	.8867	.5956	.2517	.0565	.0051	.0001	.0000	.0000	.0000	8
9	1.0000	1.0000	1.0000	.9974	.9520	.7553	.4119	.1275	.0171	.0006	.0000	.0000	.0000	9
10	1.0000	1.0000	1.0000	.9994	.9829	.8725	.5881	.2447	.0480	.0026	.0000	.0000	.0000	10
11	1.0000	1.0000	1.0000	.9999	.9949	.9435	.7483	.4044	.1133	.0100	.0001	.0000	.0000	11
12	1.0000	1.0000	1.0000	1.0000	.9987	.9790	.8684	.5841	.2277	.0321	.0004	.0000	.0000	12
13	1.0000	1.0000	1.0000	1.0000	.9997	.9935	.9423	.7500	.3920	.0867	.0024	.0000	.0000	13
14	1.0000	1.0000	1.0000	1.0000	1.0000	.9984	.9793	.8744	.5836	.1958	.0113	.0003	.0000	14
15	1.0000	1.0000	1.0000	1.0000	1.0000	.9997	.9941	.9490	.7625	.3704	.0432	.0026	.0000	15
16	1.0000	1.0000	1.0000	1.0000	1.0000	1.0000	.9987	.9840	.8929	.5886	.1330	.0159	.0000	16
17	1.0000	1.0000	1.0000	1.0000	1.0000	1.0000	.9998	.9964	.9645	.7939	.3231	.0755	.0010	17
18	1.0000	1.0000	1.0000	1.0000	1.0000	1.0000	1.0000	.9995	.9924	.9308	.6083	.2642	.0169	18
19	1.0000	1.0000	1.0000	1.0000	1.0000	1.0000	1.0000	1.0000	.9992	.9885	.8784	.6415	.1821	19

N = 25

x	0.01	0.05	0.1	0.2	0.3	0.4	P 0.5	0.6	0.7	0.8	0.9	0.95	0.99	x
0	.7778	.2774	.0718	.0038	.0001	.0000	.0000	.0000	.0000	.0000	.0000	.0000	.0000	0
1	.9742	.6424	.2712	.0274	.0016	.0001	.0000	.0000	.0000	.0000	.0000	.0000	.0000	1
2	.9980	.8729	.5371	.0982	.0090	.0004	.0000	.0000	.0000	.0000	.0000	.0000	.0000	2
3	.9999	.9659	.7636	.2340	.0332	.0024	.0001	.0000	.0000	.0000	.0000	.0000	.0000	3
4	1.0000	.9928	.9020	.4207	.0905	.0095	.0005	.0000	.0000	.0000	.0000	.0000	.0000	4
5	1.0000	.9988	.9666	.6167	.1935	.0294	.0020	.0001	.0000	.0000	.0000	.0000	.0000	5
6	1.0000	.9998	.9905	.7800	.3407	.0736	.0073	.0003	.0000	.0000	.0000	.0000	.0000	6
7	1.0000	1.0000	.9977	.8909	.5118	.1536	.0216	.0012	.0000	.0000	.0000	.0000	.0000	7
8	1.0000	1.0000	.9995	.9532	.6769	.2735	.0539	.0043	.0001	.0000	.0000	.0000	.0000	8
9	1.0000	1.0000	.9999	.9827	.8106	.4246	.1148	.0132	.0005	.0000	.0000	.0000	.0000	9
10	1.0000	1.0000	1.0000	.9944	.9022	.5858	.2122	.0344	.0018	.0000	.0000	.0000	.0000	10
11	1.0000	1.0000	1.0000	.9985	.9558	.7323	.3450	.0778	.0060	.0001	.0000	.0000	.0000	11
12	1.0000	1.0000	1.0000	.9996	.9825	.8462	.5000	.1538	.0175	.0004	.0000	.0000	.0000	12
13	1.0000	1.0000	1.0000	.9999	.9940	.9222	.6550	.2677	.0442	.0015	.0000	.0000	.0000	13
14	1.0000	1.0000	1.0000	1.0000	.9982	.9656	.7878	.4142	.0978	.0056	.0000	.0000	.0000	14
15	1.0000	1.0000	1.0000	1.0000	.9995	.9868	.8852	.5754	.1894	.0173	.0001	.0000	.0000	15
16	1.0000	1.0000	1.0000	1.0000	.9999	.9957	.9461	.7265	.3231	.0468	.0005	.0000	.0000	16
17	1.0000	1.0000	1.0000	1.0000	1.0000	.9988	.9784	.8464	.4882	.1091	.0023	.0000	.0000	17
18	1.0000	1.0000	1.0000	1.0000	1.0000	.9997	.9927	.9264	.6593	.2200	.0095	.0002	.0000	18
19	1.0000	1.0000	1.0000	1.0000	1.0000	.9999	.9980	.9706	.8065	.3833	.0334	.0012	.0000	19
20	1.0000	1.0000	1.0000	1.0000	1.0000	1.0000	.9995	.9905	.9095	.5793	.0980	.0072	.0000	20
21	1.0000	1.0000	1.0000	1.0000	1.0000	1.0000	.9999	.9976	.9668	.7660	.2364	.0341	.0001	21
22	1.0000	1.0000	1.0000	1.0000	1.0000	1.0000	1.0000	.9996	.9910	.9018	.4629	.1271	.0020	22
23	1.0000	1.0000	1.0000	1.0000	1.0000	1.0000	1.0000	.9999	.9984	.9726	.7288	.3576	.0258	23
24	1.0000	1.0000	1.0000	1.0000	1.0000	1.0000	1.0000	1.0000	.9999	.9962	.9282	.7226	.2222	24

TABLE III AREAS OF THE STANDARD NORMAL DISTRIBUTION

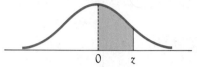

The entries in this table are the probabilities that a random variable having the standard normal distribution assumes a value between 0 and z; the probability is represented by the area under the curve shaded in the figure. Areas for negative values of z are obtained by symmetry.

				Second decimal place in z					
z	.00	.01	.02	.03	.04	.05	.06	.07	.08	.09
.0	.0000	.0040	.0080	.0120	.0160	.0199	.0239	.0279	.0319	.0359
.1	.0398	.0438	.0478	.0517	.0557	.0596	.0636	.0675	.0714	.0753
.2	.0793	.0832	.0871	.0910	.0948	.0987	.1026	.1064	.1103	.1141
.3	.1179	.1217	.1255	.1293	.1331	.1368	.1406	.1443	.1480	.1517
.4	.1554	.1591	.1628	.1664	.1700	.1736	.1772	.1808	.1844	.1879
.5	.1915	.1950	.1985	.2019	.2054	.2088	.2123	.2157	.2190	.2224
.6	.2257	.2291	.2324	.2357	.2389	.2422	.2454	.2486	.2517	.2549
.7	.2580	.2611	.2642	.2673	.2704	.2734	.2764	.2794	.2823	.2852
.8	.2881	.2910	.2939	.2967	.2995	.3023	.3051	.3078	.3106	.3133
.9	.3159	.3186	.3212	.3238	.3264	.3289	.3315	.3340	.3365	.3389
1.0	.3413	.3438	.3461	.3485	.3508	.3531	.3554	.3577	.3599	.3621
1.1	.3643	.3665	.3686	.3708	.3729	.3749	.3770	.3790	.3810	.3830
1.2	.3849	.3869	.3888	.3907	.3925	.3944	.3962	.3980	.3997	.4015
1.3	.4032	.4049	.4066	.4082	.4099	.4115	.4131	.4147	.4162	.4177
1.4	.4192	.4207	.4222	.4236	.4251	.4265	.4279	.4292	.4306	.4319
1.5	.4332	.4345	.4357	.4370	.4382	.4394	.4406	.4418	.4429	.4441
1.6	.4452	.4463	.4474	.4484	.4495	.4505	.4515	.4525	.4535	.4545
1.7	.4554	.4564	.4573	.4582	.4591	.4599	.4608	.4616	.4625	.4633
1.8	.4641	.4649	.4656	.4664	.4671	.4678	.4686	.4693	.4699	.4706
1.9	.4713	.4719	.4726	.4732	.4738	.4744	.4750	.4756	.4761	.4767
2.0	.4772	.4778	.4783	.4788	.4793	.4798	.4803	.4808	.4812	.4817
2.1	.4821	.4826	.4830	.4834	.4838	.4842	.4846	.4850	.4854	.4857
2.2	.4861	.4864	.4868	.4871	.4875	.4878	.4881	.4884	.4887	.4890
2.3	.4893	.4896	.4898	.4901	.4904	.4906	.4909	.4911	.4913	.4916
2.4	.4918	.4920	.4922	.4925	.4927	.4929	.4931	.4932	.4934	.4936
2.5	.4938	.4940	.4941	.4943	.4945	.4946	.4948	.4949	.4951	.4952
2.6	.4953	.4955	.4956	.4957	.4959	.4960	.4961	.4962	.4963	.4964
2.7	.4965	.4966	.4967	.4968	.4969	.4970	.4971	.4972	.4973	.4974
2.8	.4974	.4975	.4976	.4977	.4977	.4978	.4979	.4979	.4980	.4981
2.9	.4981	.4982	.4982	.4983	.4984	.4984	.4985	.4985	.4986	.4986
3.0	.4987	.4987	.4987	.4988	.4988	.4989	.4989	.4989	.4990	.4990
3.1	.4990	.4991	.4991	.4991	.4992	.4992	.4992	.4992	.4993	.4993
3.2	.4993	.4993	.4994	.4994	.4994	.4994	.4994	.4995	.4995	.4995
3.3	.4995	.4995	.4995	.4996	.4996	.4996	.4996	.4996	.4996	.4997
3.4	.4997	.4997	.4997	.4997	.4997	.4997	.4997	.4997	.4997	.4998
3.5	.4998									
4.0	.49997									
4.5	.499997									
5.0	.4999997									

C X^2

TABLE IV   CHI-SQUARE DISTRIBUTION

r d.f. c is the tabular value

$P(\chi^2 < c)$	$r = 1$	2	3	4	5	6	7	8	9	10	11	12	13	14	15
.001	.00	.00	.02	.09	.21	.38	.60	.86	1.15	1.48	1.83	2.21	2.62	3.04	3.48
.005	.00	.01	.07	.21	.41	.68	.99	1.34	1.73	2.16	2.60	3.07	3.57	4.07	4.60
.010	.00	.02	.12	.30	.55	.87	1.24	1.65	2.09	2.56	3.05	3.57	4.11	4.66	5.23
.05	.00	.10	.35	.71	1.15	1.64	2.17	2.73	3.33	3.94	4.57	5.23	5.89	6.57	7.26
.10	.02	.21	.58	1.06	1.61	2.20	2.83	3.49	4.17	4.87	5.58	6.30	7.04	7.79	8.55
.20	.06	.45	1.00	1.65	2.34	3.07	3.82	4.59	5.38	6.18	6.99	7.81	8.63	9.47	10.3
.30	.15	.71	1.42	2.19	3.00	3.83	4.67	5.53	6.39	7.27	8.15	9.03	9.93	10.8	11.7
.40	.28	1.02	1.87	2.75	3.66	4.57	5.49	6.42	7.36	8.30	9.24	10.2	11.1	12.1	13.0
.50	.46	1.39	2.37	3.36	4.35	5.35	6.35	7.34	8.34	9.34	10.3	11.3	12.3	13.3	14.3
.60	.71	1.83	2.95	4.04	5.13	6.21	7.28	8.35	9.41	10.5	11.5	12.6	13.6	14.7	15.7
.70	1.07	2.41	3.67	4.88	6.06	7.23	8.38	9.52	10.7	11.8	12.9	14.0	15.1	16.2	17.3
.80	1.64	3.22	4.64	5.99	7.29	8.56	9.80	11.0	12.2	13.4	14.6	15.8	17.0	18.2	19.3
.90	2.71	4.61	6.25	7.78	9.24	10.6	12.0	13.4	14.7	16.0	17.3	18.5	19.8	21.1	22.3
.95	3.84	5.99	7.81	9.49	11.1	12.6	14.1	15.5	16.9	18.3	19.7	21.0	22.4	23.7	25.0
.99	6.63	9.21	11.3	13.3	15.1	16.8	18.5	20.1	21.7	23.2	24.7	26.2	27.7	29.1	30.6
.995	7.88	10.6	12.8	14.9	16.7	18.5	20.3	22.0	23.6	25.2	26.8	28.3	29.8	31.3	32.8
.999	10.8	13.8	16.3	18.5	20.5	22.5	24.3	26.1	27.9	29.6	31.3	32.9	34.5	36.1	37.7

TABLE IV CHI-SQUARE DISTRIBUTION (CONTINUED)

r d.f.   c is the tabular value

$P(\chi^2 < c)$	r = 16	18	20	22	24	26	28	30	32	34	36	38	r^*
.001	3.94	4.90	5.92	6.98	8.08	9.22	10.4	11.6	12.8	14.1	15.3	16.6	$(3.09 - \sqrt{2r-1})^2/2$
.005	5.14	6.26	7.43	8.64	9.89	11.2	12.5	13.8	15.1	16.5	17.9	19.3	$(2.58 - \sqrt{2r-1})^2/2$
.01	5.81	7.01	8.26	9.54	10.9	12.2	13.6	15.0	16.4	17.8	19.2	20.7	$(2.33 - \sqrt{2r-1})^2/2$
.05	7.96	9.39	10.9	12.3	13.8	15.4	16.9	18.5	20.1	21.7	23.3	24.9	$(1.64 - \sqrt{2r-1})^2/2$
.10	9.31	10.9	12.4	14.0	15.7	17.3	18.9	20.6	22.3	24.0	25.6	27.3	$(1.28 - \sqrt{2r-1})^2/2$
.20	11.2	12.9	14.6	16.3	18.1	19.8	21.6	23.4	25.1	26.9	28.7	30.5	$(.84 - \sqrt{2r-1})^2/2$
.30	12.6	14.4	16.3	18.1	19.9	21.8	23.6	25.5	27.4	29.2	31.1	33.0	$(.52 - \sqrt{2r-1})^2/2$
.40	14.0	15.9	17.8	19.7	21.7	23.6	25.5	27.4	29.4	31.3	33.3	35.2	$(.25 - \sqrt{2r-1})^2/2$
.50	15.3	17.3	19.3	21.3	23.3	25.3	27.3	29.3	31.3	33.3	35.3	37.3	$(.00 - \sqrt{2r-1})^2/2$
.60	16.8	18.9	21.0	23.0	25.1	27.2	29.2	31.3	33.4	35.4	37.5	39.6	$(.25 + \sqrt{2r-1})^2/2$
.70	18.4	20.6	22.8	24.9	27.1	29.2	31.4	33.5	35.7	37.8	39.9	42.0	$(.52 + \sqrt{2r-1})^2/2$
.80	20.5	22.8	25.0	27.3	29.6	31.8	34.0	36.3	38.5	40.7	42.9	45.1	$(.84 + \sqrt{2r-1})^2/2$
.90	23.5	26.0	28.4	30.8	33.2	35.6	37.9	40.3	42.6	44.9	47.2	49.5	$(1.28 + \sqrt{2r-1})^2/2$
.95	26.3	28.9	31.4	33.9	36.4	38.9	41.3	43.8	46.2	48.6	51.0	53.4	$(1.64 + \sqrt{2r-1})^2/2$
.99	32.0	34.8	37.6	40.3	43.0	45.6	48.3	50.9	53.5	56.1	58.6	61.2	$(2.33 + \sqrt{2r-1})^2/2$
.995	34.3	37.2	40.0	42.8	45.6	48.3	51.0	53.7	56.3	59.0	61.6	64.2	$(2.58 + \sqrt{2r-1})^2/2$
.999	39.3	42.3	45.3	48.3	51.2	54.1	56.9	59.7	62.5	65.2	68.0	70.7	$(3.09 + \sqrt{2r-1})^2/2$

*Approximation to be used if r > 38.

Reprinted by permission of Holt, Rinehart and Winston, from Whitney, *Elements of Mathematical Statistics* (1959): 123–124.

TABLE V CRITICAL VALUES OF THE CHI-SQUARE DISTRIBUTION

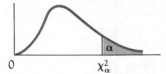

The entries in this table are the critical values for chi square for which the area to the right under the curve is equal to α.

df	Amount of α in right-hand tail									
	0.995	0.990	0.975	0.950	0.900	0.100	0.050	0.025	0.010	0.005
1	0.0000393	0.000157	0.000982	0.00393	0.0158	2.71	3.84	5.02	6.64	7.88
2	0.0100	0.0201	0.0506	0.103	0.211	4.61	5.99	7.38	9.21	10.6
3	0.0717	0.115	0.216	0.352	0.584	6.25	7.82	9.35	11.4	12.9
4	0.207	0.297	0.484	0.711	1.0636	7.78	9.50	11.1	13.3	14.9
5	0.412	0.554	0.831	1.15	1.61	9.24	11.1	12.8	15.1	16.8
6	0.676	0.872	1.24	1.64	2.20	10.6	12.6	14.5	16.8	18.6
7	0.990	1.24	1.69	2.17	2.83	12.0	14.1	16.0	18.5	20.3
8	1.34	1.65	2.18	2.73	3.49	13.4	15.5	17.5	20.1	22.0
9	1.73	2.09	2.70	3.33	4.17	14.7	17.0	19.0	21.7	23.6
10	2.16	2.56	3.25	3.94	4.87	16.0	18.3	20.5	23.2	25.2
11	2.60	3.05	3.82	4.58	5.58	17.2	19.7	21.9	24.7	26.8
12	3.07	3.57	4.40	5.23	6.30	18.6	21.0	23.3	26.2	28.3
13	3.57	4.11	5.01	5.90	7.04	19.8	22.4	24.7	27.7	29.8
14	4.07	4.66	5.63	6.57	7.79	21.1	23.7	26.1	29.1	31.3
15	4.60	5.23	6.26	7.26	8.55	22.3	25.0	27.5	30.6	32.8
16	5.14	5.81	6.91	7.96	9.31	23.5	26.3	28.9	32.0	34.3
17	5.70	6.41	7.56	8.67	10.1	24.8	27.6	30.2	33.4	35.7
18	6.26	7.01	8.23	9.39	10.9	26.0	28.9	31.5	34.8	37.2
19	6.84	7.63	8.91	10.1	11.7	27.2	30.1	32.9	36.2	38.6
20	7.43	8.26	9.59	10.9	12.4	28.4	31.4	34.2	37.6	40.0
21	8.03	8.90	10.3	11.6	13.2	29.6	32.7	35.5	39.0	41.4
22	8.64	9.54	11.0	12.3	14.0	30.8	33.9	36.8	40.3	42.8
23	9.26	10.2	11.0	13.1	14.9	32.0	35.2	38.1	41.6	44.2
24	9.89	10.9	12.4	13.9	15.7	33.2	36.4	39.4	43.0	45.6
25	10.5	11.5	13.1	14.6	16.5	34.4	37.7	40.7	44.3	46.9
26	11.2	12.2	13.8	15.4	17.3	35.6	38.9	41.9	45.6	48.3
27	11.8	12.9	14.6	16.2	18.1	36.7	40.1	43.2	47.0	49.7
28	12.5	13.6	15.3	16.9	18.9	37.9	41.3	44.5	48.3	51.0
29	13.1	14.3	16.1	17.7	19.8	39.1	42.6	45.7	49.6	52.3
30	13.8	15.0	16.8	18.5	20.6	40.3	43.8	47.0	50.9	53.7
40	20.7	22.2	24.4	26.5	29.1	51.8	55.8	59.3	63.7	66.8
50	28.0	29.7	32.4	34.8	37.7	63.2	67.5	71.4	76.2	79.5
60	35.5	37.5	40.5	43.2	46.5	74.4	79.1	83.3	88.4	92.0
70	43.3	45.4	48.8	51.8	55.3	85.5	90.5	95.0	100.0	104.0
80	51.2	53.5	57.2	60.4	64.3	96.6	102.0	107.0	112.0	116.0
90	59.2	61.8	65.7	69.1	73.3	108.0	113.0	118.0	124.0	128.0
100	67.3	70.1	74.2	77.9	82.4	114.0	124.0	130.0	136.0	140.0

Adapted from E. S. Pearson and H. O. Hartley, *Biometrika Tables for Statisticians*, vol. I (1962), pp. 130–131. Reprinted by permission of the Biometrika Trustees.

TABLE VI   CRITICAL VALUES OF STUDENT'S *t* DISTRIBUTION

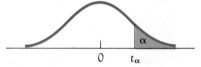

The entries in this table are the critical values for Student's *t* for an area of α in the right-hand tail. Critical values for the left-hand tail are found by symmetry.

Degrees of freedom	Amount of α in one-tail				
	.1	.05	.025	.01	.005
1	3.078	6.314	12.706	31.821	63.657
2	1.886	2.920	4.303	6.965	9.925
3	1.638	2.353	3.182	4.541	5.841
4	1.533	2.132	2.776	3.747	4.604
5	1.476	2.015	2.571	3.365	4.032
6	1.440	1.943	2.447	3.143	3.707
7	1.415	1.895	2.365	2.998	3.499
8	1.397	1.860	2.306	2.896	3.355
9	1.383	1.833	2.262	2.821	3.250
10	1.372	1.812	2.228	2.764	3.169
11	1.363	1.796	2.201	2.718	3.106
12	1.356	1.782	2.179	2.681	3.055
13	1.350	1.771	2.160	2.650	3.012
14	1.345	1.761	2.145	2.624	2.977
15	1.341	1.753	2.131	2.602	2.947
16	1.337	1.746	2.120	2.583	2.921
17	1.333	1.740	2.110	2.567	2.898
18	1.330	1.734	2.101	2.552	2.878
19	1.328	1.729	2.093	2.539	2.861
20	1.325	1.725	2.086	2.528	2.845
21	1.323	1.721	2.080	2.518	2.831
22	1.321	1.717	2.074	2.508	2.819
23	1.319	1.714	2.069	2.500	2.807
24	1.318	1.711	2.064	2.492	2.797
25	1.316	1.708	2.060	2.485	2.787
26	1.315	1.706	2.056	2.479	2.779
27	1.314	1.703	2.052	2.473	2.771
28	1.313	1.701	2.048	2.467	2.763
29	1.311	1.699	2.045	2.462	2.756
30	1.310	1.697	2.042	2.457	2.750
40	1.303	1.684	2.021	2.423	2.704
60	1.296	1.671	2.000	2.390	2.660
120	1.290	1.661	1.984	2.358	2.626
∞	1.282	1.645	1.960	2.326	2.576

Abridged from Table III of Fisher and Yates, *Statistical Tables for Biological, Agricultural, and Medical Research,* published by Longman Group Ltd., London (previously published by Oliver and Boyd Ltd., Edinburgh), and by permission of the authors and publishers.

TABLE VII   CRITICAL VALUES OF THE *F* DISTRIBUTION ($\alpha = 0.05$)

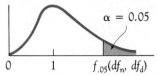

$\alpha = 0.05$

$f_{.05}(df_n, df_d)$

The entries in this table are critical values of F for which the area under the curve to the right is equal to 0.05.

		Degrees of freedom for numerator								
	1	*2*	*3*	*4*	*5*	*6*	*7*	*8*	*9*	*10*
1	161	200	216	225	230	234	237	239	241	242
2	18.5	19.0	19.2	19.2	19.3	19.3	19.4	19.4	19.4	19.4
3	10.1	9.55	9.28	9.12	9.01	8.94	8.89	8.85	8.81	8.79
4	7.71	6.94	6.59	6.39	6.26	6.16	6.09	6.04	6.00	5.96
5	6.61	5.79	5.41	5.19	5.05	4.95	4.88	4.82	4.77	4.74
6	5.99	5.14	4.76	4.53	4.39	4.28	4.21	4.15	4.10	4.06
7	5.59	4.74	4.35	4.12	3.97	3.87	3.79	3.73	3.68	3.64
8	5.32	4.46	4.07	3.84	3.69	3.58	3.50	3.44	3.39	3.35
9	5.12	4.26	3.86	3.63	3.48	3.37	3.29	3.23	3.18	3.14
10	4.96	4.10	3.71	3.48	3.33	3.22	3.14	3.07	3.02	2.98
11	4.84	3.98	3.59	3.36	3.20	3.09	3.01	2.95	2.90	2.85
12	4.75	3.89	3.49	3.26	3.11	3.00	2.91	2.85	2.80	2.75
13	4.67	3.81	3.41	3.18	3.03	2.92	2.83	2.77	2.71	2.67
14	4.60	3.74	3.34	3.11	2.96	2.85	2.76	2.70	2.65	2.60
15	4.54	3.68	3.29	3.06	2.90	2.79	2.71	2.64	2.59	2.54
16	4.49	3.63	3.24	3.01	2.85	2.74	2.66	2.59	2.54	2.49
17	4.45	3.59	3.20	2.96	2.81	2.70	2.61	2.55	2.49	2.45
18	4.41	3.55	3.16	2.93	2.77	2.66	2.58	2.51	2.46	2.41
19	4.38	3.52	3.13	2.90	2.74	2.63	2.54	2.48	2.42	2.38
20	4.35	3.49	3.10	2.87	2.71	2.60	2.51	2.45	2.39	2.35
21	4.32	3.47	3.07	2.84	2.68	2.57	2.49	2.42	2.37	2.32
22	4.30	3.44	3.05	2.82	2.66	2.55	2.46	2.40	2.34	2.30
23	4.28	3.42	3.03	2.80	2.64	2.53	2.44	2.37	2.32	2.27
24	4.26	3.40	3.01	2.78	2.62	2.51	2.42	2.36	2.30	2.25
25	4.24	3.39	2.99	2.76	2.60	2.49	2.40	2.34	2.28	2.24
30	4.17	3.32	2.92	2.69	2.53	2.42	2.33	2.27	2.21	2.16
40	4.08	3.23	2.84	2.61	2.45	2.34	2.25	2.18	2.12	2.08
60	4.00	3.15	2.76	2.53	2.37	2.25	2.17	2.10	2.04	1.99
120	3.92	3.07	2.68	2.45	2.29	2.18	2.09	2.02	1.96	1.91
∞	3.84	3.00	2.60	2.37	2.21	2.10	2.01	1.94	1.88	1.83

Degrees of freedom for denominator

TABLE VII α = .05 *(CONTINUED)*

		Degrees of freedom for numerator								
		12	15	20	24	30	40	60	120	∞
	1	244	246	248	249	250	251	252	253	254
	2	19.4	19.4	19.4	19.5	19.5	19.5	19.5	19.5	19.5
	3	8.74	8.70	8.66	8.64	8.62	8.59	8.57	8.55	8.53
	4	5.91	5.86	5.80	5.77	5.75	5.72	5.69	5.66	5.63
	5	4.68	4.62	4.56	4.53	4.50	4.46	4.43	4.40	4.37
	6	4.00	3.94	3.87	3.84	3.81	3.77	3.74	3.70	3.67
	7	3.57	3.51	3.44	3.41	3.38	3.34	3.30	3.27	3.23
	8	3.28	3.22	3.15	3.12	3.08	3.04	3.01	2.97	2.93
	9	3.07	3.01	2.94	2.90	2.86	2.83	2.79	2.75	2.71
	10	2.91	2.85	2.77	2.74	2.70	2.66	2.62	2.58	2.54
	11	2.79	2.72	2.65	2.61	2.57	2.53	2.49	2.45	2.40
	12	2.69	2.62	2.54	2.51	2.47	2.43	2.38	2.34	2.30
	13	2.60	2.53	2.46	2.42	2.38	2.34	2.30	2.25	2.21
Degrees of freedom for denominator	14	2.53	2.46	2.39	2.35	2.31	2.27	2.22	2.18	2.13
	15	2.48	2.40	2.33	2.29	2.25	2.20	2.16	2.11	2.07
	16	2.42	2.35	2.28	2.24	2.19	2.15	2.11	2.06	2.01
	17	2.38	2.31	2.23	2.19	2.15	2.10	2.06	2.01	1.96
	18	2.34	2.27	2.19	2.15	2.11	2.06	2.02	1.97	1.92
	19	2.31	2.23	2.16	2.11	2.07	2.03	1.98	1.93	1.88
	20	2.28	2.20	2.12	2.08	2.04	1.99	1.95	1.90	1.84
	21	2.25	2.18	2.10	2.05	2.01	1.96	1.92	1.87	1.81
	22	2.23	2.15	2.07	2.03	1.98	1.94	1.89	1.84	1.78
	23	2.20	2.13	2.05	2.01	1.96	1.91	1.86	1.81	1.76
	24	2.18	2.11	2.03	1.98	1.94	1.89	1.84	1.79	1.73
	25	2.16	2.09	2.01	1.96	1.92	1.87	1.82	1.77	1.71
	30	2.09	2.01	1.93	1.89	1.84	1.79	1.74	1.68	1.62
	40	2.00	1.92	1.84	1.79	1.74	1.69	1.64	1.58	1.51
	60	1.92	1.84	1.75	1.70	1.65	1.59	1.53	1.47	1.39
	120	1.83	1.75	1.66	1.61	1.55	1.50	1.43	1.35	1.25
	∞	1.75	1.67	1.57	1.52	1.46	1.39	1.32	1.22	1.00

TABLE VII   CRITICAL VALUES OF THE *F* DISTRIBUTION ($\alpha = 0.025$)

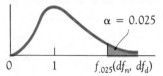

$\alpha = 0.025$

$f_{.025}(df_n, df_d)$

The entries in this table are critical values of F for which the area under the curve to the right is equal to 0.025.

		Degrees of freedom for numerator								
	1	*2*	*3*	*4*	*5*	*6*	*7*	*8*	*9*	*10*
1	648	800	864	900	922	937	948	957	963	969
2	38.5	39.0	39.2	39.2	39.3	39.3	39.4	39.4	39.4	39.4
3	17.4	16.0	15.4	15.1	14.9	14.7	14.6	14.5	14.5	14.4
4	12.2	10.6	9.98	9.60	9.36	9.20	9.07	8.98	8.90	8.84
5	10.0	8.43	7.76	7.39	7.15	6.98	6.85	6.76	6.68	6.62
6	8.81	7.26	6.60	6.23	5.99	5.82	5.70	5.60	5.52	5.46
7	8.07	6.54	5.89	5.52	5.29	5.12	4.99	4.90	4.82	4.76
8	7.57	6.06	5.42	5.05	4.82	4.65	4.53	4.43	4.36	4.30
9	7.21	5.71	5.08	4.72	4.48	4.32	4.20	4.10	4.03	3.96
10	6.94	5.46	4.83	4.47	4.24	4.07	3.95	3.85	3.78	3.72
11	6.72	5.26	4.63	4.28	4.04	3.88	3.76	3.66	3.59	3.53
12	6.55	5.10	4.47	4.12	3.89	3.73	3.61	3.51	3.44	3.37
13	6.41	4.97	4.35	4.00	3.77	6.60	3.48	3.39	3.31	3.25
14	6.30	4.86	4.24	3.89	3.66	3.50	3.38	3.28	3.21	3.15
15	6.20	4.77	4.15	3.80	3.58	3.41	3.29	3.20	3.12	3.06
16	6.12	4.69	4.08	3.73	3.50	3.34	3.22	3.12	3.05	2.99
17	6.04	4.62	4.01	3.66	3.44	3.28	3.16	3.06	2.98	2.92
18	5.98	4.56	3.95	3.61	3.38	3.22	3.10	3.01	2.93	2.87
19	5.92	4.51	3.90	3.56	3.33	3.17	3.05	2.96	2.88	2.82
20	5.87	4.46	3.86	3.51	3.29	3.13	3.01	2.91	2.84	2.77
21	5.83	4.42	3.82	3.48	3.25	3.09	2.97	2.87	2.80	2.73
22	5.79	4.38	3.78	3.44	3.22	3.05	2.93	2.84	2.76	2.70
23	5.75	4.35	3.75	3.41	3.18	3.02	2.90	2.81	2.73	2.67
24	5.72	4.32	3.72	3.38	3.15	2.99	2.87	2.78	2.70	2.64
25	5.69	4.29	3.69	3.35	3.13	2.97	2.85	2.75	2.68	2.61
30	5.57	4.18	3.59	3.25	3.03	2.87	2.75	2.65	2.57	2.51
40	5.42	4.05	3.46	3.13	2.90	2.74	2.62	2.53	2.45	2.39
60	5.29	3.93	3.34	3.01	2.79	2.63	2.51	2.41	2.33	2.27
120	5.15	3.80	3.23	2.89	2.67	2.52	2.39	2.30	2.22	2.16
∞	5.02	3.69	3.12	2.79	2.57	2.41	2.29	2.19	2.11	2.05

Degrees of freedom for denominator

TABLE VII   α = .025 *(CONTINUED)*

		Degrees of freedom for numerator								
		12	15	20	24	30	40	60	120	∞
Degrees of freedom for denominator	1	977	985	993	997	1,001	1,006	1,010	1,014	1,018
	2	39.4	39.4	39.4	39.5	39.5	39.5	39.5	39.5	39.5
	3	14.3	14.3	14.2	14.1	14.1	14.0	14.0	13.9	13.9
	4	8.75	8.66	8.56	8.51	8.46	8.41	8.36	8.31	8.26
	5	6.52	6.43	6.33	6.29	6.23	6.18	6.12	6.07	6.02
	6	5.37	5.27	5.17	5.12	5.07	5.01	4.96	4.90	4.85
	7	4.67	4.57	4.47	4.42	4.36	4.31	4.25	4.20	4.14
	8	4.20	4.10	4.00	3.95	3.89	3.84	3.78	3.73	3.67
	9	3.87	3.77	3.67	3.61	3.56	3.51	3.45	3.39	3.33
	10	3.62	3.52	3.42	3.37	3.31	3.26	3.20	3.14	3.08
	11	3.43	3.33	3.23	3.17	3.12	3.06	3.00	2.94	2.88
	12	3.28	3.18	3.07	3.02	2.96	2.91	2.85	2.79	2.72
	13	3.15	3.05	2.95	2.89	2.84	2.78	2.72	2.66	2.60
	14	3.05	2.95	2.84	2.79	2.73	2.67	2.61	2.55	2.49
	15	2.96	2.86	2.76	2.70	2.64	2.59	2.52	2.46	2.40
	16	2.89	2.79	2.68	2.63	2.57	2.51	2.45	2.38	2.32
	17	2.82	2.72	2.62	2.56	2.50	2.44	2.38	2.32	2.25
	18	2.77	2.67	2.56	2.50	2.44	2.38	2.32	2.26	2.19
	19	2.72	2.62	2.51	2.45	2.39	2.33	2.27	2.20	2.13
	20	2.68	2.57	2.46	2.41	2.35	2.29	2.22	2.16	2.09
	21	2.64	2.53	2.42	2.37	2.31	2.25	2.18	2.11	2.04
	22	2.60	2.50	2.39	2.33	2.27	2.21	2.14	2.08	2.00
	23	2.57	2.47	2.36	2.30	2.24	2.18	2.11	2.04	1.97
	24	2.54	2.44	2.33	2.27	2.21	2.15	2.08	2.01	1.94
	25	2.51	2.41	2.30	2.24	2.18	2.12	2.05	1.98	1.91
	30	2.41	2.31	2.20	2.14	2.07	2.01	1.94	1.87	1.79
	40	2.29	2.18	2.07	2.01	1.94	1.88	1.80	1.72	1.64
	60	2.17	2.06	1.94	1.88	1.82	1.74	1.67	1.57	1.49
	120	2.05	1.95	1.82	1.76	1.69	1.61	1.53	1.43	1.31
	∞	1.94	1.83	1.71	1.64	1.57	1.48	1.39	1.27	1.00

TABLE VII   CRITICAL VALUES OF THE *F* DISTRIBUTION (α = 0.01)

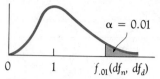

$\alpha = 0.01$

$f_{.01}(df_n, df_d)$

The entries in the table are critical values of F for which the area under the curve to the right is equal to 0.01.

		Degrees of freedom for numerator								
	1	2	3	4	5	6	7	8	9	10
1	4,052	5,000	5,403	5,625	5,764	5,859	5,928	5,982	6,023	6,056
2	98.5	99.0	99.2	99.2	99.3	99.3	99.4	99.4	99.4	99.4
3	34.1	30.8	29.5	28.7	28.2	27.9	27.7	27.5	27.3	27.2
4	21.2	18.0	16.7	16.0	15.5	15.2	15.0	14.8	14.7	14.5
5	16.3	13.3	12.1	11.4	11.0	10.7	10.5	10.3	10.2	10.1
6	13.7	10.9	9.78	9.15	8.75	8.47	8.26	8.10	7.98	7.87
7	12.2	9.55	8.45	7.85	7.46	7.19	6.99	6.84	6.72	6.62
8	11.3	8.65	7.59	7.01	6.63	6.37	6.18	6.03	5.91	5.81
9	10.6	8.02	6.99	6.42	6.06	5.80	5.61	5.47	5.35	5.26
10	10.0	7.56	6.55	5.99	5.64	5.39	5.20	5.06	4.94	4.85
11	9.65	7.21	6.22	5.67	5.32	5.07	4.89	4.74	4.63	4.54
12	9.33	6.93	5.95	5.41	5.06	4.82	4.64	4.50	4.39	4.30
13	9.07	6.70	5.74	5.21	4.86	4.62	4.44	4.30	4.19	4.10
14	8.86	6.51	5.56	5.04	4.70	4.46	4.28	4.14	4.03	3.94
15	8.68	6.36	5.42	4.89	4.56	4.32	4.14	4.00	3.89	3.80
16	8.53	6.23	5.29	4.77	4.44	4.20	4.03	3.89	3.78	3.69
17	8.40	6.11	5.19	4.67	4.34	4.10	3.93	3.79	3.68	3.59
18	8.29	6.01	5.09	4.58	4.25	4.01	3.84	3.71	3.60	3.51
19	8.19	5.93	5.01	4.50	4.17	3.94	3.77	3.63	3.52	3.43
20	8.10	5.85	4.94	4.43	4.10	3.87	3.70	3.56	3.46	3.37
21	8.02	5.78	4.87	4.37	4.04	3.81	3.64	3.51	3.40	3.31
22	7.95	5.72	4.82	4.31	3.99	3.76	3.59	3.45	3.35	3.26
23	7.88	5.66	4.76	4.26	3.94	3.71	3.54	3.41	3.30	3.21
24	7.82	5.61	4.72	4.22	3.90	3.67	3.50	3.36	3.26	3.17
25	7.77	5.57	4.68	4.18	3.86	3.63	3.46	3.32	3.22	3.13
30	7.56	5.39	4.51	4.02	3.70	3.47	3.30	3.17	3.07	2.98
40	7.31	5.18	4.31	3.83	3.51	3.29	3.12	2.99	2.89	2.80
60	7.08	4.98	4.13	3.65	3.34	3.12	2.95	2.82	2.72	2.63
120	6.85	4.79	3.95	3.48	3.17	2.96	2.79	2.66	2.56	2.47
∞	6.63	4.61	3.78	3.32	3.02	2.80	2.64	2.51	2.41	2.32

Degrees of freedom for denominator

TABLE VII α = .01 *(CONTINUED)*

		\multicolumn{9}{c}{Degrees of freedom for numerator}
		12	15	20	24	30	40	60	120	∞
	1	6,106	6,157	6,209	6,235	6,261	6,287	6,313	6,339	6,366
	2	99.4	99.4	99.4	99.5	99.5	99.5	99.5	99.5	99.5
	3	27.1	26.9	26.7	26.6	26.5	26.4	26.3	26.2	26.1
	4	14.4	14.2	14.0	13.9	13.8	13.7	13.7	13.6	13.5
	5	9.89	9.72	9.55	9.47	9.38	9.29	9.20	9.11	9.02
	6	7.72	7.56	7.40	7.31	7.23	7.14	7.06	6.97	6.88
	7	6.47	6.31	6.16	6.07	5.99	5.91	5.82	5.74	5.65
	8	5.67	5.52	5.36	5.28	5.20	5.12	5.03	4.95	4.86
	9	5.11	4.96	4.81	4.73	4.65	4.57	4.48	4.40	4.31
	10	4.71	4.56	4.41	4.33	4.25	4.17	4.08	4.00	3.91
	11	4.40	4.25	4.10	4.02	3.94	3.86	3.78	3.69	3.60
	12	4.16	4.01	3.86	3.78	3.70	3.62	3.54	3.45	3.36
	13	3.96	3.82	3.66	3.59	3.51	3.43	3.34	3.25	3.17
	14	3.80	3.66	3.51	3.43	3.35	3.27	3.18	3.09	3.00
	15	3.67	3.52	3.37	3.29	3.21	3.13	3.05	2.96	2.87
	16	3.55	3.41	3.26	3.18	3.10	3.02	2.93	2.84	2.75
	17	3.46	3.31	3.16	3.08	3.00	2.92	2.83	2.75	2.65
	18	3.37	3.23	3.08	3.00	2.92	2.84	2.75	2.66	2.57
	19	3.30	3.15	3.00	2.92	2.84	2.76	2.67	2.58	2.49
	20	3.23	3.09	2.94	2.86	2.78	2.69	2.61	2.52	2.42
	21	3.17	3.03	2.88	2.80	2.72	2.64	2.55	2.46	2.36
	22	3.12	2.98	2.83	2.75	2.67	2.58	2.50	2.40	2.31
	23	3.07	2.93	2.78	2.70	2.62	2.54	2.45	2.35	2.26
	24	3.03	2.89	2.74	2.66	2.58	2.49	2.40	2.31	2.21
	25	2.99	2.85	2.70	2.62	2.53	2.45	2.36	2.27	2.17
	30	2.84	2.70	2.55	2.47	2.39	2.30	2.21	2.11	2.01
	40	2.66	2.52	2.37	2.29	2.20	2.11	2.02	1.92	1.80
	60	2.50	2.35	2.20	2.12	2.03	1.94	1.84	1.73	1.60
	120	2.34	2.19	2.03	1.95	1.86	1.76	1.66	1.53	1.38
	∞	2.18	2.04	1.88	1.79	1.70	1.59	1.47	1.32	1.00

Degrees of freedom for denominator (row label, left side)

TABLE VIII   CRITICAL VALUES OF *r* FOR TESTING *p* = 0

For a two-sided test, α is twice the value listed at the heading of a column of critical *r* values, hence for α = .05 choose the .025 column.

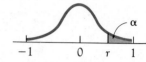

α n	.05	.025	.005
5	.805	.878	.959
6	.729	.811	.917
7	.669	.754	.875
8	.621	.707	.834
9	.582	.666	.798
10	.549	.632	.765
11	.521	.602	.735
12	.497	.576	.708
13	.476	.553	.684
14	.457	.532	.661
15	.441	.514	.641
16	.426	.497	.623

α n	.05	.025	.005
17	.412	.482	.606
18	.400	.468	.590
19	.389	.456	.575
20	.378	.444	.561
25	.337	.396	.505
30	.306	.361	.463
35	.283	.334	.430
40	.264	.312	.402
50	.235	.279	.361
60	.214	.254	.330
80	.185	.220	.286
100	.165	.196	.156

Reprinted by permission of John Wiley & Sons, Inc., from Hoel, *Elementary Statistics*, (1971):289.

TABLE IX CRITICAL VALUES OF *U* IN THE MANN-WHITNEY TEST

A. The entries are the critical values of U for a one-tailed test at 0.025 or for a two-tailed test at 0.05.

n_2 \ n_1	1	2	3	4	5	6	7	8	9	10	11	12	13	14	15	16	17	18	19	20
1																				
2								0	0	0	0	1	1	1	1	1	2	2	2	2
3				0	1	1	2	2	3	3	4	4	5	5	6	6	7	7	8
4			0	1	2	3	4	4	5	6	7	8	9	10	11	11	12	13	13
5		0	1	2	3	5	6	7	8	9	11	12	13	14	15	17	18	19	20
6		1	2	3	5	6	8	10	11	13	14	16	17	19	21	22	24	25	27
7		1	3	5	6	8	10	12	14	16	18	20	22	24	26	28	30	32	34
8	0	2	4	6	8	10	13	15	17	19	22	24	26	29	31	34	36	38	41
9	0	2	4	7	10	12	15	17	20	23	26	28	31	34	37	39	42	45	48
10	0	3	5	8	11	14	17	20	23	26	29	33	36	39	42	45	48	52	55
11	0	3	6	9	13	16	19	23	26	30	33	37	40	44	47	51	55	58	62
12	1	4	7	11	14	18	22	26	29	33	37	41	45	49	53	57	61	65	69
13	1	4	8	12	16	20	24	28	33	37	41	45	50	54	59	63	67	72	76
14	1	5	9	13	17	22	26	31	36	40	45	50	55	59	64	67	74	78	83
15	1	5	10	14	19	24	29	34	39	44	49	54	59	64	70	75	80	85	90
16	1	6	11	15	21	26	31	37	42	47	53	59	64	70	75	81	86	92	98
17	2	6	11	17	22	28	34	39	45	51	57	63	67	75	81	87	93	99	105
18	2	7	12	18	24	30	36	42	48	55	61	67	74	80	86	93	99	106	112
19	2	7	13	19	25	32	38	45	52	58	65	72	78	85	92	99	106	113	119
20	2	8	13	20	27	34	41	48	55	62	69	76	83	90	98	105	112	119	127

B. The entries are the critical values of U for a one-tailed test at 0.05 or for a two-tailed test at 0.10.

n_2 \ n_1	1	2	3	4	5	6	7	8	9	10	11	12	13	14	15	16	17	18	19	20
1																			0	0
2				0	0	0	1	1	1	1	2	2	2	3	3	3	4	4	4
3		0	0	1	2	2	3	3	4	5	5	6	7	7	8	9	9	10	11
4		0	1	2	3	4	5	6	7	8	9	10	11	12	14	15	16	17	18
5	0	1	2	4	5	6	8	9	11	12	13	15	16	18	19	20	22	23	25
6	0	2	3	5	7	8	10	12	14	16	17	19	21	23	25	26	28	30	32
7	0	2	4	6	8	11	13	15	17	19	21	24	26	28	30	33	35	37	39
8	1	3	5	8	10	13	15	18	20	23	26	28	31	33	36	39	41	44	47
9	1	3	6	9	12	15	18	21	24	27	30	33	36	39	42	45	48	51	54
10	1	4	7	11	14	17	20	24	27	31	34	37	41	44	48	51	55	58	62
11	1	5	8	12	16	19	23	27	31	34	38	42	46	50	54	57	61	65	69
12	2	5	9	13	17	21	26	30	34	38	42	47	51	55	60	64	68	72	77
13	2	6	10	15	19	24	28	33	37	42	47	51	56	61	65	70	75	80	84
14	2	7	11	16	21	26	31	36	41	46	51	56	61	66	71	77	82	87	92
15	3	7	12	18	23	28	33	39	44	50	55	61	66	72	77	83	88	94	100
16	3	8	14	19	25	30	36	42	48	54	60	65	71	77	83	89	95	101	107
17	3	9	15	20	26	33	39	45	51	57	64	70	77	83	89	96	102	109	115
18	4	9	16	22	28	35	41	48	55	61	68	75	82	88	95	102	109	116	123
19	0	4	10	17	23	30	37	44	51	58	65	72	80	87	94	101	109	116	123	130
20	0	4	11	18	25	32	39	47	54	62	69	77	84	92	100	107	115	123	130	138

Reproduced from the *Bulletin of the Institute of Educational Research at Indiana University*, vol. 1, no. 2.

TABLE X   CRITICAL VALUES OF T
FOR THE WILCOXON SIGNED-RANK TEST

| | α | | | |
| | .005 (one tail) .01 (two tails) | .01 (one tail) .02 (two tails) | .025 (one tail) .05 (two tails) | .05 (one tail) .10 (two tails) |
n				
5				1
6			1	2
7		0	2	4
8	0	2	4	6
9	2	3	6	8
10	3	5	8	11
11	5	7	11	14
12	7	10	14	17
13	10	13	17	21
14	13	16	21	26
15	16	20	25	30
16	19	24	30	36
17	23	28	35	41
18	28	33	40	47
19	32	38	46	54
20	37	43	52	60
21	43	49	59	68
22	49	56	66	75
23	55	62	73	83
24	61	69	81	92
25	68	77	90	101
26	76	85	98	110
27	84	93	107	120
28	92	102	117	130
29	100	111	127	141
30	109	120	137	152

Reject the null hypothesis if the test statistic T is less than or equal to the critical value found in this table. Fail to reject the null hypothesis if the test statistic T is greater than the critical value voune in this table.

"Table of the Critical Values of T for the Wilcoxon Signed Rank Test" From *Some Rapid Approximate Statistical Procedures*, Copyright © 1949, 1964, Lederle Laboratories Division of American Cyanamid Company, All Rights Reserved, and Reprinted With Permission.

TABLE XI   CRITICAL VALUES OF SPEARMAN'S RANK CORRELATION COEFFICIENT r_s

n	$\alpha = 0.10$	$\alpha = 0.05$	$\alpha = 0.02$	$\alpha = 0.01$
5	.900	—	—	—
6	.829	.886	.943	—
7	.714	.786	.893	—
8	.643	.738	.833	.881
9	.600	.683	.783	.833
10	.564	.648	.745	.794
11	.523	.623	.736	.818
12	.497	.591	.703	.780
13	.475	.566	.673	.745
14	.457	.545	.646	.716
15	.441	.525	.623	.689
16	.425	.507	.601	.666
17	.412	.490	.582	.645
18	.399	.476	.564	.625
19	.388	.462	.549	.608
20	.377	.450	.534	.591
21	.368	.438	.521	.576
22	.359	.428	.508	.562
23	.351	.418	.496	.549
24	.343	.409	.485	.537
25	.336	.400	.475	.526
26	.329	.392	.465	.515
27	.323	.385	.456	.505
28	.317	.377	.448	.496
29	.311	.370	.440	.487
30	.305	.364	.432	.478

For $n > 30$ use $r_s = \pm z/\sqrt{n-1}$, where z corresponds to the level of significance. For example, if $\alpha = 0.05$, then $z = 1.96$.

From E. G. Olds, "Distribution of sums of squares of rank differences to small numbers of individuals," *Annals of Statistics,* 9 (1938): 133–148, and Olds, E. G., with amendment in 20, 1949, 117–118 of *Annals of Statistics.*

Answers to Selected Questions

Chapter One

Set 1.1

1) a) Discrete
   b) Continuous
   c) Discrete
   d) Continuous
3) a) All graduating seniors or their responses
   b) The 200 graduating seniors or their responses
   c) Estimation
5) Observational
7) a) 90
   b) 280
   c) 70
   d) The average of the four grades
   e) $\dfrac{x_2 + x_3 + x_4}{3}$
9) a) Coronary prone adults or their levels of aspiration
   b) The 100 coronary prone adults or their levels of aspiration
   c) Coronary prone adults have higher levels of aspiration than do adults in general

Set 1.2

1) No. Houses toward the opposite end of the street are excluded from the sample.
3) 99, 112, 136, 193, 260, 329, 339, 373, 375, 392
5) Neither the patients nor the evaluators would know which patients received the new medication and which received the regular medication.
7) When relatively new on the scene, a drug is often more highly regarded by both patients and doctors than is justified. The result is an inflated success rate for the drug initially.
9) Select three families at random from each of the 8 floors.

11) 6, 10, 11, 15, and 19

13) 1) Non-listeners are excluded from
    participating.

  2) Only seriously motivated listeners will take
    the time and effort to respond.

  3) Special interest groups may marshall their
    forces to call in and bias the results.

Chapter Test

1) a) Continuous
  b) Discrete

2) Discrete: The Number of Chairs in the Room;
    The Number of Fire Sprinklers in the
    Room
  Continuous: The Blackboard Area; The Age of
    the Floor Covering

3) a) What is the proportion of students who
    drink coffee?
  b) Is it true that coffee is an aphrodesiac?
  c) Is there a relationship between elevated
    pulse rates and the amount of coffee
    consumed?

4) a) Military Personnel (Soldiers)
  b) The 56 marines selected.
  c) No
  d) Test of the hypothesis that "any soldier can
    use the weapon after a brief introduction."

5) a) 15
  b) 0

6) 9, 11, 12, 16, 25, 35, and 46

7) A positive response to a worthless medication
  or therapy.

8) Randomly assign the subjects to take either the
  placebo or the medication. Neither the subjects
  nor the evaluators should know how the
  assignments were actually made.

9) First renumber the houses on the north side as
  0, 1, 2, . . ., 9 and use the table of random
  digits to select three of these. Next renumber
  the houses on the south side as 0, 1, 2, . . ., 8
  and use the table of random digits to select 2 of
  these. Many other approaches are possible.

Chapter Two

Set 2.1

1) a) .18 = 18%
  b) .60 = 60%

c)

Interval	Relative Frequency
12500–13500	.04
13500–14500	.08
14500–15500	.18
15500–16500	.06
16500–17500	.04
17500–18500	.28
18500–19500	.18
19500–20500	.06
20500–21500	.06
21500–22500	.02

d) 16

3) a)

Interval	Relative Frequency
2.55–3.55	.24
3.55–4.55	.32
4.55–5.55	.20
5.55–6.55	.12
6.55–7.55	.08
7.55–8.55	.04

b)

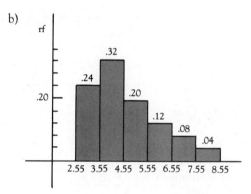

5) a)

Interval	Frequency	Relative Frequency
.005– .255	22	.379
.255– .505	12	.207
.505– .755	7	.121
.755–1.005	4	.069
1.005–1.255	3	.052
1.255–1.505	2	.034
1.505–1.755	3	.052
1.755–2.005	5	.086

b)

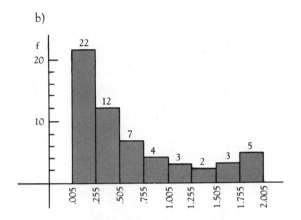

7)

Interval	Frequency
1950–2175	3
2175–2400	4
2400–2625	7
2625–2850	4
2850–3075	5
3075–3300	3

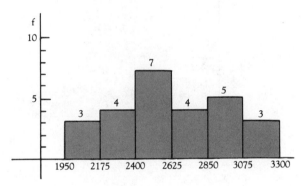

9)

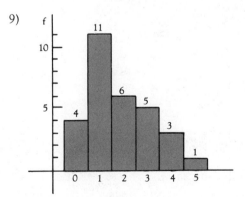

11)

8.0–8.1	
8.2–8.3	3
8.4–8.5	55555554
8.6–8.7	6677666
8.8–8.9	88999
9.0–9.1	00011
9.2–9.3	2333
9.4–9.5	445545454555

13)

Interval	Frequency
$-1.50 \le x < 1.25$	3
$1.25 \le x < 4.00$	2
$4.00 \le x < 6.75$	3
$6.75 \le x < 9.50$	5
$9.50 \le x < 12.25$	8
$12.25 \le x < 15.50$	4

There are many possibilities.

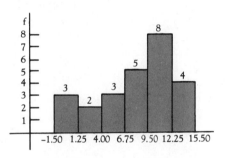

15)

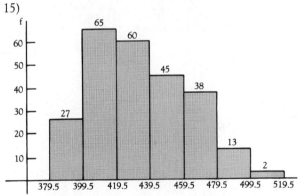

17) a) No. The number observed has increased but so has the number of laboratories.
b) Yes. However, see part (c).
c) Probably the increase in the number of observed quakes is due to improved techniques and new stations.

19) a) 30
    b) 36
    c) 4

    d)

Interval	Frequency
$1.05 < x \le 2.05$	8
$2.05 < x \le 3.05$	10
$3.05 < x \le 4.05$	12
$4.05 < x \le 5.05$	6
$5.05 < x \le 6.05$	2
$6.05 < x \le 7.05$	2

21) a) 45% for A and 70% for B
    b) Make B

    c)

Interval	Relative Frequency (A)	Relative Frequency (B)
18–19	.05	.05
19–20	.20	.10
20–21	.30	.15
21–22	.20	.20
22–23	.15	.30
23–24	.10	.20

    d)

Interval	Frequency	Relative Frequency
18–19	10	.050
19–20	26	.130
20–21	39	.195
21–22	40	.200
22–23	51	.255
23–24	34	.170

    e)

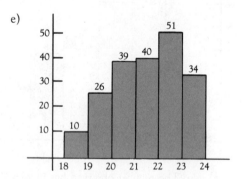

Set 2.2

1) a) 40–50
   b) 4%
   c) 4%

d)

Interval	Relative Frequency	Density
0–30	.12	.004
30–40	.30	.030
40–50	.40	.040
50–60	.10	.010
60–80	.08	.004

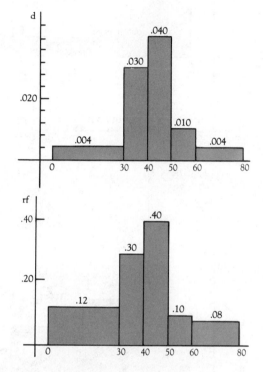

3)

Interval	Density
40–50	.0150
50–60	.0200
60–80	.0175
80–100	.0150

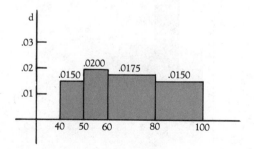

5) a) Bimodal
    b) Symmetric
    c) Skewed Right
    d) Skewed Left
    e) Skewed Right
7) .5
9) a) .025
    b) 2.5%
11) a) 72
    b) 192

Set 2.3

1)

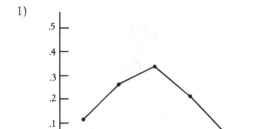

3)

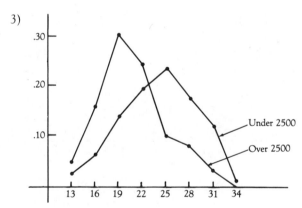

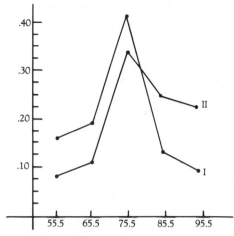

7) The lower level (0–1) contaminations have
    increased while the moderate level (1–2)
    contaminations have decreased.

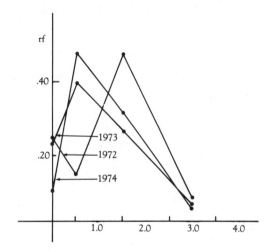

9) a) 15–70 (roughly)
    b) Female (Smokers)
    c) Figure 2: $\dfrac{26.5}{8.1} = 3.3$, $\dfrac{25.4}{16.0} = 1.6$

11) a) 16
    b) 5000
    c) About 1000 in each case.

5)

Interval	Section I		Section II	
	f	rf	f	rf
50.5–60.5	5	.161	2	.056
60.5–70.5	6	.194	5	.139
70.5–80.5	13	.419	12	.333
80.5–90.5	4	.129	9	.250
90.5–100.5	3	.097	8	.222

Chapter Test

1)

Interval	Frequency	Relative Frequency
181.5–187.5	3	.12
187.5–193.5	3	.12
193.5–199.5	5	.20
199.5–205.5	8	.32
205.5–211.5	3	.12
211.5–217.5	2	.08
217.5–223.5	1	.04

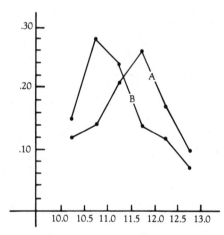

2) Those in environment "A" tend to require more time than do those in "B."

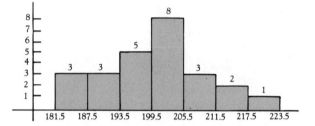

3) a) I.Q. scores of 10 year olds
   b) Assembly times for 10 speed bicycles by untrained adults (See Example of Text).
   c) Times until first failure of new engines
   d) Salaries of teachers at well established universities (Higher ranks tend to dominate).

4) a) 20

b)

x	f	rf
0	2	.10
1	3	.15
2	5	.25
3	5	.25
4	4	.20
5	1	.05

c) 50%

d)

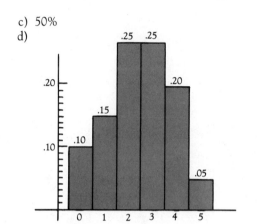

5) a)

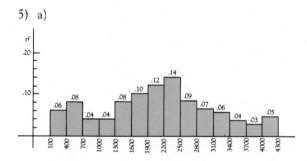

b)

Interval	Relative Frequency	Density
100–700	.14	.00023
700–1300	.08	.00013
1300–1900	.18	.00030
1900–2500	.26	.00043
2500–3100	.16	.00027
3100–3700	.10	.00017
3700–4300	.08	.00013

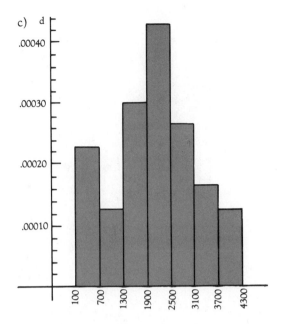

c) d)

b)

2	0022
2	5586
3	0004220
3	5567
4	302
4	5575
5	0002423
5	56895
6	0230
6	578
7	20
7	6

6) a) .16
   b) .12

7) 1

8) a)
0	75246
1	705984285634
2	40402233685657842
3	027435

   b)
0	24
0	756
1	04234
1	7598856
2	4040223342
2	6856578
3	0243
3	75

9) a)

2	50025862
3	55000674220
4	5302575
5	506895002423
6	5078230
7	620

10)
Interval	Relative Frequency
$5.00 \leq x < 5.20$	.000
$5.20 \leq x < 5.40$	.031
$5.40 \leq x < 5.60$	.141
$5.60 \leq x < 5.80$	.281
$5.80 \leq x < 6.00$	.422
$6.00 \leq x < 6.20$	.125

Chapter Three

Set 3.1

1) a) Mode = 2.4 and $\bar{x}$ = 3.8
   b) Mode = 3 and $\bar{x}$ = 3.5

3) Mode = 30, $\tilde{x}$ = 33.5, and $\bar{x}$ = 35

5) $\bar{x}$ = 2.2 and s = 1.6

7) $\bar{x}$ = 18.6 and s = 1.2

9) a) $\bar{x}$ = 8.6 and s = 1.9
   b) 70% and 100%

11) 1) Let the population be all scores on the 1985 SAT.
    2) Let the population be yearly salaries of all active players on the NFL roster.

13) a) 0
    b) 3

15) 1305 and 70
17) a) The first distribution
    b) The first distribution
19) 1365 million

21)

	First Graph	Second Graph	Third Graph
a)	$\tilde{x} = 5$	$\tilde{x} = 5$	$\tilde{x} = 5$
b)	$\overline{x} = 5$	$\overline{x} = 5.4$	$\overline{x} = 4.7$
c)	$\tilde{x} = 5$	$\tilde{x} = 5$	$\tilde{x} = 5$
	$\overline{x} = 5$	$\overline{x} = 5.25$	$\overline{x} = 4.875$

    d) For a symmetric distribution, $\overline{x} = \tilde{x}$
       For a skewed right distribution, $\overline{x} > \tilde{x}$
       For a skewed left distribution, $\overline{x} < \tilde{x}$

23) 0

Set 3.2

1) a) 132
   b) 82
   c) 126
   d) 107
3) a) $9 < x < 27$
   b) $6 < x < 30$
   c) $x > 36$
   d) $6 < x < 12$
5) a) 1.56
   b) $-.75$
   c) 0
   d) 211.8
   e) 117.4
   f) 199
7) a) 204
   b) 285
   c) 7.5 (7 or 8)
   d) 40.5 (40 or 41)
   e) 7.5 (7 or 8)
9) a) 57
   b) 1.5 (1 or 2)
   c) The distribution should be bell shaped.
11) a) $\overline{x} \approx 230$, $s \approx 40$ (using $6s \approx 350 - 110$)
    b) $\overline{x} \approx 60$, $s \approx 2.67$ (using $6s \approx 68 - 52$)
13) a) .5, 1, $-.3$, 0, -1
    b) No
15) a) $\overline{x} = 13.2$ and $s = 9.9$
    b) 10.5

17) a)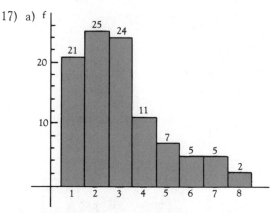
    b) Distribution is strongly skewed
    c) $\overline{x} = 3.0$ and $s = 1.8$
    d) 60%
19) a) 88.89%
    b) 55.56%
    c) 0%
21)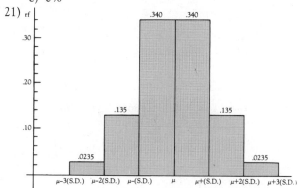

Set 3.3

1) a) 65
   b) 76.7
   c) 105.5
3) a) 17.34
   b) 20.15
   c) 16.40
   d) 18.65
5) 53.5
7) PR(0) = 17.5
   PR(1) = 42.5
   PR(2) = 62.5
   PR(3) = 80.0
   PR(4) = 92.5

9) No

11) The income tax you pay is at a low percentile for people in your income range.

13) 4.64

50% of all graduating seniors have ages between Q_1 and Q_3. These ages are within 4.64 years of one another.

Chapter Test

1) a) mode = 2, $\bar{x}$ = 3
   b) bimodal, $\bar{x}$ = 1.5

2) a) $\bar{x}$ = 5
   b) s = 1.947
   c) Sixteen observations are within 1.5 standard deviations of the mean

3) a) 952 (68%)
   b) 35 (2.5%)

4) a) 175
   b) 120.4
   c) 165.2
   d) 103.6

5) $\bar{x}$ = 2.8 and s = 1.3751 (1.4 rounded)

6) $\bar{x}$ = 3.3 and s = 1.4599 (1.5 rounded)

7) a) $x_1^2 + x_2^2 + x_3^2 + x_4^2$
   b) $(x_1 + x_2 + x_3 + x_4)^2$
   c) $x_1 f_1 + x_2 f_2 + x_3 f_3$
   d) $(x_1^2 f_1 + x_2^2 f_2 + x_3^2 f_3) - (x_1 f_1 + x_2 f_2 + x_3 f_3)^2$

8) 68

9) a) $\bar{x}$ = 0 and s = 10
   b) $\bar{x}$ = 0 and s = 1

10) a) 4.67
    b) 9.43
    c) 6.5
    d) 14

11) PR(6) = 13, PR(9) = 45, PR(10) = 74, and PR(12) = 92 (All rounded)

Chapter Four

Set 4.1

1) {(R,R), (R,D), (R,I), (D,D), (D,R), (D,I), (I,I), (I,D), (I,R)}

3) a) The probability that a randomly selected tire of the production will fail to pass inspection is .12
   b) The probability that a randomly selected person will find this new product superior to the present one is .58.

c) The probability is .30 that a randomly selected freshman will be on probation after one semester.

d) The probability is .05 that the mean of a randomly selected sample will be more than 2 standard deviations from the mean.

5) a) {(1,1), (1,2), (1,3), (1,4), (1,5), (1,6), (2,1), (2,2), (2,3), (2,4), (2,5), (2,6)}

   b)
Sum	Probability
2	$1/12$
3	$1/6$
4	$1/6$
5	$1/6$
6	$1/6$
7	$1/6$
8	$1/12$

   c) About 400

7) a) {(1,1), (1,2), (1,3), (1,4), (2,1), (2,2), (2,3), (2,4), (3,1), (3,2), (3,3), (3,4), (4,1), (4,2), (4,3), (4,4)}

   b) $\dfrac{3}{4}$

9) a) $\dfrac{1}{2}$

   b) 1

   c) Have the urn contain 1 white, 2 blue, 2 red, and 1 green balls. Let the event e_1 correspond to white, e_2 to blue, e_3 to red, and e_4 to green where the experiment is to draw one ball from the urn.

11) a) 216
    b) 27
    c) $\dfrac{1}{8}$

13) $\dfrac{1}{6}$

15) $\dfrac{1}{24}$

17) $\dfrac{7}{24}$

19) $\dfrac{3}{8}$

21) S = {1, 2, 3, 4, 5, 6, . . .} (The number of throws to obtain the 1st head)

23) a) 16
    b) 8

Set 4.2

1) a) No
   b) No
   c) Yes
   d) Yes
   e) Yes
3) a) No
   b) No
   c) Yes
   d) No
   e) No
5) a) Twenty or fewer smoke
   b) Three or more smoke
   c) Fewer than 5 smoke
   d) At least one smokes
   e) None smokes
7) a) The probability of a phrase being recognized when encountered a second time given that the phrase was previously read.
   b) The probability of a phrase being recognized when encountered a second time given that the phrase was previously heard.
   c) The probability that the phrase is not recognized when encountered a second time given that it was previously read.
   d) The probability that the phrase is recognized when encountered a second time given that it had not been previously read.
   e) No
9) a) .70
   b) .90
   c) .25
   d) .8235
   e) .9333
11) $\frac{1}{3}$
13) a) .02982
    b) .46718
    c) .05497
15) .92
17) 56%
19) a) .15
    b) .012
    c) .9811
21) a) A high probability by either the lonely or the not lonely group indicates the problem was rated in the top 5 of the problems by the respondents.
       A high probability by one of the groups and not by the other is desirable for the investigators since they have then identified a trait or tendency that could be used to differentiate between these two groups.
       A high probability for a problem by both groups might simply identify a common social problem
    b) Lonely: 3
       Not Lonely: 3 (3.15)
    c) There are 13 categories and each has probability $\frac{1}{2}$ of being ranked higher by the lonely than the non lonely group when the ranking is left strictly to chance. Thus p = $(1/2)(1/2). \ . \ . \ .(1/2) = (1/2)^{13}$
23) a) $(.25)^3 = .015625$
    b) $(.75)^3 = .421875$
    c) .421875
25) a) .30
    b) .40
    c) .90
    d) .5714
    e) .6667
    f) .4286
27) The probability of E and its complement are related by P(E) + P(notE) = 1. If the probability of the occurrence of either of these is affected by the occurrence of F, then so is the other.

Set 4.3

1) a) .15
   b) .40
2) .76
3) .24
4) .50
5) .50 (50%)

Chapter Test

1) a)   i) Experiment: A student is selected at random
          E: "The student selected is a freshman"
          F: "The student selected is a senior"
       ii) Experiment: Toss a pair of dice
          E: "The sum is even"
          F: "The sum is odd"
      iii) P(E or F) = P(E) + P(F)
   b)   i) Experiment: Toss a coin and a die
          E: "The coin shows a head"
          F: "The die shows an even number"
       ii) Experiment: Randomly select a man and a woman

E: "The age of the man exceeds 30"
F: "The age of the woman exceeds 30"
iii) $P(E \text{ and } F) = P(E) \cdot P(F)$

2) a) Observing the subjects who received the vaccine
   b) influenza contracted, influenza not contracted
   c) $p = \dfrac{1050}{1440} = .7291$

3) a) .65
   b) .35
   c) .90
   d) .25

4) (1,1), (1,2), (1,3), (1,4), (1,5), (2,1), (2,2), (2,3), (2,4), (2,5), (3,1), (3,2), (3,3), (3,4), (3,5), (4,1), (4,2), (4,3), (4,4), (4,5), (5,1), (5,2), (5,3), (5,4), (5,5)

5) a) .1
   b) .5
   c) .6

6) a) $\dfrac{5}{18}$
   b) $\dfrac{18}{36} = \dfrac{1}{2}$
   c) $\dfrac{4}{18} = \dfrac{2}{9}$

7) $\dfrac{24}{6497400} = .0000037$

8) a) $\dfrac{5}{18}$
   b) $\dfrac{5}{18}$
   c) $\dfrac{3}{8}$
   d) $\dfrac{5}{9}$

9) a) $\dfrac{400}{1000} = .4$
   b) $\dfrac{100}{1000} = .1$
   c) $\dfrac{300}{1000} = .3$
   d) $\dfrac{200}{260} = .7692$

10) a) $\dfrac{1}{10}$
    b) $\dfrac{3}{10}$
    c) $\dfrac{3}{5}$

Chapter Five

Set 5.1

1)

x	P(x)
0	⅛
1	⅜
2	⅜
3	⅛

3) Spin the wheel once and observe the color

Outcome	x	P(x)
White	0	¼
Red	1	⅛
Blue	2	⅜
Orange	3	¼

Range = {0, 1, 2, 3}

5) No

7) a) The probability that none of the 40 favor an increase.
   b) The probability that 8 or fewer (at most 8) favor an increase.
   c) The probability that between 20 and 30 (inclusively) favor an increase.
   d) The probability that more than 20 (at least 21) favor an increase.

9) a) $\dfrac{1}{12}$
   b) $\dfrac{5}{12}$
   c) $\dfrac{1}{12}$
   d) 0

11) a)

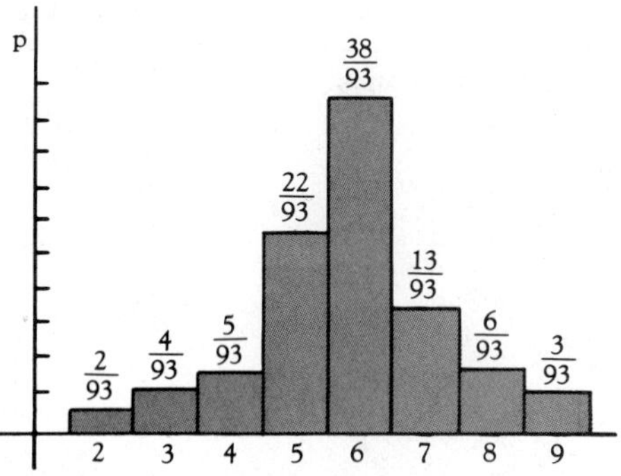

   b) $\dfrac{11}{93}$
   c) $\dfrac{22}{93}$

13)

x	P(x)
0	.941192
1	.057624
2	.001176
3	.000008

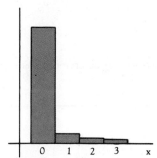

15) $P(X = x) = (1/2)^x$

$P(X \geq 3) = \dfrac{1}{4}$

17)

x	P(x)
1	$\frac{1}{10}$
2	$\frac{2}{5}$
3	$\frac{1}{5}$
4	$\frac{3}{10}$

Set 5.2

1) a) $\mu = 1.60$, $\sigma^2 = 1.04$, and $\sigma = 1.02$
   b) .6
   c) 1

3) a) $P(X = x) = \dfrac{1}{4}$

   b) $\mu = 1.5$, $\sigma^2 = 1.25$, and $\sigma = 1.12$

5) a) 2.95 minutes
   b) 5.53(6) clerks

7) .4775(48) cents

9) a)

x	P(x)
1	$\frac{1}{4}$
2	$\frac{1}{4}$
3	$\frac{1}{4}$
4	$\frac{1}{4}$

   b) 2.5
   c) The mean number of objects tried if the monkey repeats the experiment many, many times and does not learn from its experience.

11) -1.75
    You would lose around 175 dollars in 100 plays. Some friend.

13) a)

x	y = 3x	P(y)
0	0	$\frac{1}{6}$
2	6	$\frac{1}{2}$
3	9	$\frac{1}{3}$

   b) 6
   c) $E[3X] = 3E[X] = 3 \cdot 2 = 6$

15) $-.027$

17) Stock B has the highest expected value.

19) a)

x	P(x)
2	$\frac{1}{5}$
3	$\frac{2}{5}$
4	$\frac{2}{5}$

   $\mu = 3.2$

   b)

$\bar{x}$	$P(\bar{x})$
$\frac{5}{2}$	$\frac{2}{10}$
3	$\frac{3}{10}$
$\frac{7}{2}$	$\frac{4}{10}$
4	$\frac{1}{10}$

   c) 3.2
   d)

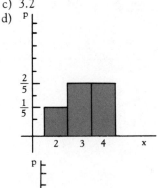

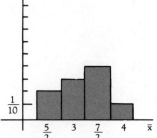

21) a) 24,400
    b) 23,200
    c) $E = 24400 - 100c$

d) Can the promoter stand a $12000 loss? If not, he should purchase the insurance.

23) a) 2
 b) $5.00
 c) $7.30
 d) $6.00

Set 5.3

1) a) .60
 b) .25
 c) 1.00
3) a) 0
 b) .5
 c) .5
 d) 1.0
 e) .175
5) a) .1
 b) 0.0
 c) .485
 d) .015
 e) 3
7) a) .15
 b) .75
 c) .55
9) $\mu \approx 85$ and $\sigma \approx 17.5$ (using $4\sigma \approx$ range) or $\sigma \approx 11.7$ (using $6\sigma \approx$ range)
11) The quantity p is the parameter.

Set 5.4

1) a)

x \ y	0	1	
0	¼	¼	½
1	¼	¼	½
	½	½	

 b)

x	P(x)
0	½
1	½

y	P(y)
0	½
1	½

3)

x	P(x)
1	.45
2	.30
3	.25

y	P(y)
0	.25
1	.10
2	.25
3	.25
4	.15

X and Y are not independent. For example, $P(X = 1, Y = 0) \neq P(X=1)P(Y=0)$

5) a)

x	P(x)
0	¼
1	½
2	¼

y	P(y)
1	⅙
2	⅙
3	⅙
4	⅙
5	⅙
6	⅙

 b)

x \ y	1	2	3	4	5	6	
0	1/24	1/24	1/24	1/24	1/24	1/24	¼
1	1/12	1/12	1/12	1/12	1/12	1/12	½
2	1/24	1/24	1/24	1/24	1/24	1/24	¼
	⅙	⅙	⅙	⅙	⅙	⅙	

7) a) The average 2 day catch is the sum of the first and second day averages.
 b) The average winning of a husband and wife team is the sum of the averages of the husbands' winnings and the wives' winnings.
9) a) 2(4/5) = 8/5
 b) 3(4/5) = 12/5
 c) n(4/5) = np where p is the probability of success on any trial.

Chapter Test

1) a) .2
 b) .50
 c) .70
 d) c = 3
2) a) $\frac{1}{2}$
 b) $\frac{1}{6}$
 c) c = 6
3) a)

x	P(x)
0	⅛
1	⅜
2	⅜
3	⅛

 b) $\mu = 1.5$ and $\sigma^2 = .75$

c)

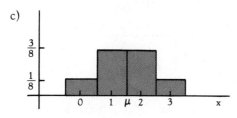

4) 5.2
5) 16.7 dozen
6) a) 6.20 (dollars)
   b) 620,000 (dollars)
7) a) $\dfrac{1}{9}$

b)

x	P(x)
1	1/3
2	19/36
3	5/36

c)

y	P(y)
0	1/4
1	14/45
2	79/180

d) $\dfrac{19}{36}$

8) X and Y are not independent. Note
   $P(X = 0, Y = 0) \neq P(X = 0) \cdot P(Y = 0)$

9) a)

x	P(x)
1	1/3
2	1/3
3	1/3

b) 2

10) a) $\dfrac{1}{2}$

b) $\dfrac{3}{4}$

c) $\dfrac{7}{8}$

Chapter Six

Set 6.1

1) a) Not binomial
   b) Binomial

c) Binomial
d) Not binomial. May be approximated as such.

3)

x	P(x)
0	.1296
1	.3456
2	.3456
3	.1536
4	.0256

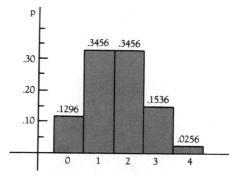

5) a) .1323
   b) .83692
   c) .83193
7) a) $\mu = 70$ and $\sigma = 4.58$
   b) $\mu = 30$ and $\sigma = 3.87$
   c) $\mu = 100$ and $\sigma = 8.66$
9) The average number of defective needles per shipment is 12. We need to assume that the defective needles occur in a random fashion during the production run.
11) a) .6098
    b) .9729
    c) .6964
13) .9095
15) $P(X \leq 16) = .0468 = 4.68\%$. About 4.68% of the time.
17) a) .0362(3.62%)
    b) .1841(18.41%)
    c) .6020(60.20%)
    This is not a very effective plan.
19) $P(X \geq 16) = .1148$. If he cannot tell forward from backward, there is only a 11.48% chance of his being 7 or more steps forward of where he started after 25 steps. Apparently he can tell the difference.
21) $\dbinom{3}{2}\left(\dfrac{5}{16}\right)^2\left(\dfrac{11}{16}\right) = .2014$
23) a) 8, 9, 10, 11, 12, . . .

b) $P(X = x) = \binom{x-1}{2-1}\left(\frac{1}{6}\right)^2\left(\frac{5}{6}\right)^{x-2}$

$P(X = 3) = \dfrac{5}{108}$

25) $\dfrac{1}{6}$

Set 6.2

1) a) .3925
   b) .1359
   c) .8384
   d) .0096
   e) .9332
   f) .0465
3) a) .4236
   b) .0139
   c) .1612
   d) .8904
   e) .0082
5) a) 2.33 (Nearest entry to .4900)
   b) .41 (Nearest entry to .1600)
   c) 1.64 (Nearest entry to .4500)
   d) 1.28
   f) 0.00
7) a) .8664
   b) .1401
   c) .9812
   d) 16.2
   e) 146.52
9) $P(X < 8000) = .0475$
11) $P(X < 80) = .0162$. About 1.62% will have to be replaced.
13) 103.92 (104 minutes would, no doubt, be used)
15) 24.10% not acceptable
17) 38.1 years
19) Every 27.48 months
21) $P(X \ge 300) = .0166$. About 1.66%.
23) $Y = 5X - 125$
25) $P(X > 2350) + P(X < 2050) = .0602$. About 6.02% fail to meet specifications.
27) $P(X \le 6.0) = .9772$

Set 6.3

1) a) $\mu = 520$
   $\sigma = 13.5$

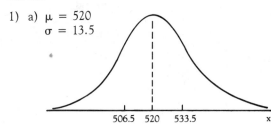

b) $\mu = 100$
   $\sigma = 7.07$

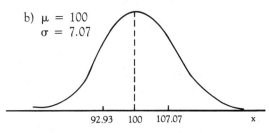

c) $\mu = 100$
   $\sigma = 8.66$

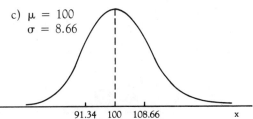

3) a) .6630
   b) .3409
   c) .0793
5) a) 38% or more of the sample of 200 or 38% or more of the 200 trials correspond to success.
   b) .1894
7) $P(X \le 260) = .0166$ (Normal Approximation)
9) $P(\hat{P} \ge .58) = .8186$
11) $P(\hat{P} < .1) = .1446$. The probability of having enough seats is .8554
13) $P(\hat{P} < .15) = .1056$
15) 70 correct answers are needed. $P(X \ge 70) = .0409$.

Set 6.4

1) a) .90
   b) .05
3) a) Table IV: $P(\chi^2 > 14.9) = .005$. Table V: $\chi^2_{.005} = 14.9$
   b) Table IV: $P(\chi^2 > 18.3) = .05$. Table V: $\chi^2_{.05} = 18.3$
5) $\chi^2 = 3.20$ and this is small in comparison to $\chi^2_{.05} = 11.1$ Do not reject the die as being fair.
7) $\chi^2 = 7.733 > \chi^2_{.10}$. The model should be rejected.
9) $\chi^2 = 2.123$ is considerably smaller than $\chi^2_{.05} = 5.99$. Do not reject the hypothesis of equal frequencies.
11) $P(\chi^2 < .437) = .20$ (Nearest Entry)
13) $\chi^2 = 16 > \chi^2_{.05}$. Reject the hypothesis that boys and girls are equally likely.
15) Table entry $\chi^2_{.05} = 67.5$. The normal approximation yields $\chi^2_{.05} = 66.5$

Chapter Test

1) a, b, and d
2) a) $\mu = 1.2$ and $\sigma = .85$

b)

x	P(X = x)
0	.216
1	.432
2	.288
3	.064

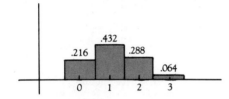

3) a) .2449
   b) .2483
4) .3504
5) .2061
6) a) $z_{.05} = 1.64$
   b) $z_{.01} = 2.33$
   c) $z_{.75} = -.67$
7) a) .0228
   b) 442.40
8) $\chi^2 = 1.8667$ and $\chi^2_{.05} = 5.99$ for $\nu = 2$ d.f.
   Do not conclude that any one procedure is preferred over the others.
9) $\chi^2 = 5.48$ and $\chi^2_{.05} = 5.99$ for $\nu = 2$ d.f.
   Do not conclude that a difference exists in the degree of job satisfaction for these three categories.
10) x = 119.2 (corresponds to $z_{.10}$)

Chapter Seven

Set 7.1

1) a) 8.4
   b) 0.3
   c) The standard error of the mean.
   d)

3) a) Normal, $\mu = 84$, and S.E. = 7.
      z = 1.0. $\bar{x} = 91.0$ is 1.0 standard errors above the mean.
   b) Normal, $\mu = 84$ and S.E. = 2
      z = -2.25. $\bar{x} = 79.5$ is 2.25 standard errors below the mean.
   c) Normal, $\mu = 84$, and S.E. = 0.7
      z = 3.57. $\bar{x} = 86.5$ is 3.57 standard errors above the mean.
   d) Normal, $\mu = 84$, and S.E. = 11.43
      z = 1.40. $\bar{x} = 100$ is 1.40 standard errors above the mean.
5) a) $P(\bar{X} > 132) = .0038$
   b) $P(126 < \bar{X} < 130) = .8164$
7) a) $P(X < 25000) = .3707$
   b) $P(\bar{X} < 25000) = .0968$.
   c) 37.07% of graduating electrical engineers have starting salaries below $25000. 9.68% of all samples of n = 16 graduating electrical engineers have a mean starting salary below $25000.
9) $P(\bar{X} > 205) = .0749$
11) $P(\bar{X} < 5.74) = .0015$
13) Large samples have a strong tendency to resemble the population. Their means should thus be close to the population mean.
   The variation in the sample means can be measured by the standard error. Since S.E. = $\dfrac{\sigma}{\sqrt{n}}$, this variation diminishes as the sample size n increases (The denominator becomes larger and the fractional value decreases.)
15) Continuous
17) The mean or expected value of the medians of all samples of any particular size n is equal to the population mean, that is,
$$E[\tilde{X}] = \mu$$
19) a) 100
    b) 100
    c) 30
    d) 3.0
    e) 3.0 (Same as d)

Set 7.2

1) The distribution of X is normal and the population standard deviation σ is being approximated by the standard deviation s of the sample.
3) a) $t_{.05} = 1.721$
   b) $t_{.10} = 1.337$
   c) $t_{.005} = 3.169$

5) $P(\overline{X} \geq 15.2)$ is about .01. Very unlikely.
7) When approximating the distribution of $\overline{X}$ by the t distribution there is more variability with which to contend as a result of the approximation $\sigma = s$
9) $P(-1.75 < t < 1.75) \approx .90$

Set 7.3

1) $\mu = .6$ and S.E. $= .0245$

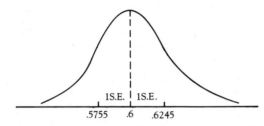

1S.E. | 1S.E.
.5755   .6   .6245

3) a) $P(.38 < \hat{P} < .42) = .6922$
   b) $P(\hat{P} > .44) = .0207$
5) $P(\hat{P} < .54) = .0170$
7) $P(\hat{P} > .50) = .0122$
9) $P(.64 < \hat{P} < .70) = .7330$
11) $P(\hat{P} < .55) = .0853$
13) $P(\hat{P} > .54) = .0082$
15) $P(\hat{P} > .265) = .2451$
17) 1776

Set 7.4

1) $P(S^2 > 30) \approx .20$
   (Nearest Entry: $P(\chi^2 > 19.56)$)
3) $P(\chi^2 > 24.89) \approx .05$
   (Nearest entry: $P(\chi^2 > 23.7) = .05$)
5) a) 3.14
   b) 3.16
7) $f > f_{.025}$. The probability of an increase at least as large as that observed is less than .025. This is a highly unlikely occurrence if indeed there has been no change in the variance of the emissions.
9) a) 3. Values exceeding, say, $f_{.05} = 9.01$
   b) 1.5. Values exceeding, say $f_{.05} = 4.39$
   c) 1.053. Values exceeding, say, $f_{.05} = 2.45$
11) $f_{.05}(10,\mu_2)$ decreases as the number of degrees of freedom μ_2 increases.
    $f_{.05}(\mu_1,10)$ decreases as μ_1 increases
    $f_{.05}(\mu_1,\mu_2)$ decreases to 1 as we move across and down the table, that is as both μ_1 and μ_2 get larger.

Since $f = \dfrac{S_1^2}{S_2^2}$, the probability is very high that this ratio is near 1 when both degrees of freedom (both sample sizes) are large.

13) a) $f_{.95}(4,6) = .1623$

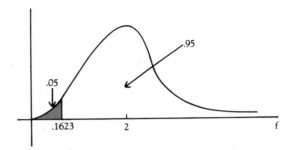

b) $f_{.99}(10,12) = .2133$

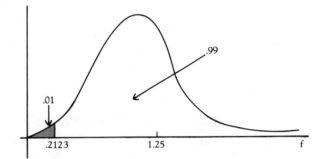

Set 7.5

1) Thirteen specialists underestimated the size of sphere number 1 while seven underestimated the size of sphere number 12.
2) May not be localized.
   May not be spherical
   May move around somewhat as it is measured.
3) Average bias $= -.3858$
   6.93%
4) Specialists 2, 3, 7, 8, 12, and 13 all the time
   All rounded to the nearest centimeter for the largest sphere.
5) 18.75% for the 25% criterion.
   7.81% for the 50% criterion.

Chapter Test

1) a) 1.753
   b) 3.07
   c) .2207

2) The distribution is approximately normal with
$\mu = 78$ and $\sigma = $ S.E. $ = 3$

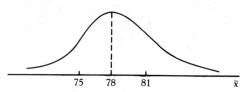

3) a) The distribution is approximately normal
with $\mu = .65$ and $\sigma = $ S.E. $ = .0337$

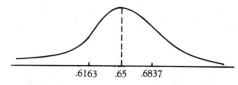

   b) .0015
4) .3830
5) .0000 (to 4 decimal places)
6) .0869
7) .0057
8) a) 34.46%
   b) .82%
9) .10 (approximately)
10) a) $f = 5.367$ and $f_{.05}(12,11) = 2.7876$. We
      conclude that there must have been an
      effect due to the treatment.
   b) The growths under both conditions should
      have normal distributions
11) $\mu = 1.25$

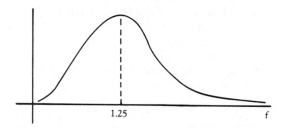

12)

$\bar{x}$	$P(\bar{x})$
1	1/9
2	2/9
2.5	2/9
3	1/9
3.5	2/9
4	1/9

Chapter Eight

Set 8.1

1) a) 95%
   b) 98%
   c) 90%
   d) 68.26%
3) a) 98%
   b) 90%
   c) 96%
5) a) $67.05 < \mu < 72.95$
   b) $66.47 < \mu < 73.53$
   c) $65.37 < \mu < 74.63$
   d) $67.70 < \mu < 72.30$
7) $36.84 < \mu < 40.16$
9) 98.76%
11) $n \geq 142$
13) $3.17 < \mu < 3.63$
   $317 and $363 are the lower and upper
   bounds on the expected cost of 100 castings.
15) $1828.27 < \mu < 1861.73$
17) White Group: $147.36 < \mu < 150.44$
   Black Group: $162.99 < \mu < 167.01$

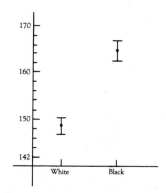

19)

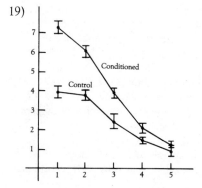

Set 8.2

1) $38.09 < \mu < 44.31$
3) $15.33 < \mu < 18.87$
5) $7.39 < \mu < 8.41$
7) $1186.5 < \mu < 1305.5$
9) The one constructed using $t_{.025}$
11) Before Group (I): $3.01 < \mu < 4.59$
    $0 - .5$ Group (II): $2.74 < \mu < 4.46$
    $.51 - 1.00$ Group (III): $1.74 < \mu < 3.26$
    $1.01 - 2.0$ Group (IV): $1.38 < \mu < 3.42$

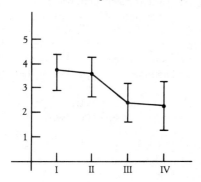

13) Caffeoyl Acid: $16.21 < \mu < 40.99$
    Coumaroyl Acid: $4.87 < \mu < 15.34$

Set 8.3

1) $.5266 < p < .6234$
3) $.2578 < p < .3722$
5) $.1817 < p < .2583$
    $21.8 < n < 31.0$
7) $.6799 < p < .7201$
9) $.6109 < p < .6891$
11) $n \geq 9604$
13) $n \geq 2305$
15) No Reminder Group (I): $.4566 < p < .6434$
    Letter Group (II): $.7615 < p < .9125$
    Telephone Group (III): $.7123 < p < .8877$

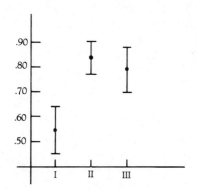

17) $.1616 < p < .1844$

Chapter Test

1) $80.83 < \mu < 83.17$
2) $15.02 < \mu < 21.58$
    The survival times should be normally distributed
3) $.2626 < p < .4174$
4) 96.34%
5) $107.62 < \mu < 168.38$
6) $n \geq 1068$
7) $n \geq 107$
8) $53,569,562 < x < 56,254,438$ (dollars)
9) 18–35 age group: $.4761 < p < .5739$
    35–55 age group: $.5347 < p < .6213$
    Over 55 age group: $.7226 < p < .7974$
    Combined: $.6026 < p < .6532$
10) $\bar{x} = 8.27$ and $s = .175$
    $7.98 < \mu < 8.56$

Chapter Nine

Set 9.1

1) a) H_0: $\mu = 57.5$ (The average speed is 57.5 m.p.h.)
        H_1: $\mu > 57.5$ (The average speed exceeds 57.5 m.p.h.)
        One tailed test to the right
    b) H_0: $p = .37$ (37% of the teenage population smoke.)
        H_1: $p > .37$ (More than 37% of the teenage population smoke.)
        One tailed test to the right
    c) H_0: $p = .67$ (67% of all students have never skiied.)
        H_1: $p \neq .67$ (The proportion of all students who have never skiied is not .67 (67%)
        Two tailed test
    d) H_0: $p = .67$ (The proportion of all licensed drivers who have never had a ticket nor been in an accident is .67 (67%))
        H_1: $p \neq .67$ (The proportion of all licensed drivers who have never had a ticket nor been in a major accident is not .67 (67%)
        Two tailed test
    e) H_0: $p = .635$ (63.5% of the present inventory of left over jeans are in small sizes.)
        H_1: $p < .635$ (Less than 63.5% of the present inventory of left over jeans is in small sizes.)
        One tailed test to the left

f) H_0: $\mu = 125$ (The mean grade of students which have a poor nutritional program is 125.)
H_1: $\mu < 125$ (The mean grade of students which have a poor nutritional program is below 125.)
One tailed test to the left

3) a) H_1: $\mu \neq 560$
b) H_1: $\mu < 560$
c) H_1: $\mu > 560$

5) H_0: $\mu = 45$ (The average age is 45 days)
H_1: $\mu > 45$ (The average age exceeds 45 days.)

7) Reject H_0 if $\bar{x} < 17.34$

9) a) H_0: $p = .60$
H_1: $p > .60$
Reject H_0 if $z > 3.0$ or equivalently if $\hat{p} > .6612$
b) H_0: $p = .60$
H_1: $p < .60$
Reject H_0 if $z < -3.0$ or equivalently if $\hat{p} < .5388$
c) H_0: $p = .60$
H_1: $p \neq .60$
Reject H_0 if $z > 3.0$ or $z < -3.0$.
Equivalently, reject H_0 if $\hat{p} > .6612$ or $\hat{p} < .5388$.

11) a) H_0: $\mu = 112$
H_1: $\mu > 112$
b) $Z > 2.54$
c) $P(Z > 2.54) = .0055$

13) a) $Z > 2.0$
$\bar{X} > 25250$
b) $Z < -2.0$
$\bar{X} < 23910$

15) 3.05 Reject H_0 and accept H_1.

17) Yes.

19) Reconsider the alternate hypothesis.

Set 9.2

1) a) H_0: The school should retain its certification
H_1: The school should lose its certification.
A type I error occurs if the school is decertified when in fact it should have retained its certification.
A type II error occurs if the school retains its certification when in fact it should have lost it.
From the schools point of view, the probability α of a type I error should be small. From the point of view of the students and the general public, the probability β of a type II error should be small.

b) H_0: The driver is not speeding.
H_1: The driver is speeding.
A type I error results if the driver is cited when, in fact, the driver is not speeding.
A type II error results if the driver is speeding and this is not detected
The probability α of a type I error should be small

c) H_0: The patient is healthy (no hypertension)
H_1: The patient has hypertension.
A type I error occurs if a healthy patient is diagnosed as having hypertension.
A type II error occurs if a patient with hypertension is diagnosed as healthy.
The probability β of a type II error should be small.

d) H_0: The instructor is effective
H_1: The instructor is ineffective
A type I error results if an effective teacher is judged ineffective.
A type II error results if an ineffective teacher is judged effective.
Ideally both α and β should be small to protect the good teacher on the one hand and the long suffering student on the other.

e) H_0: The area is normal (Not Depressed)
H_1: The area is economically depressed
A type I error results if a normal area is classified as depressed.
A type II error results if a depressed area is not so classified.
The probability β of a type II error should be small.

f) H_0: The child has normal intelligence
H_1: The child is mentally deficient.
A type I error results if a normal child is classified mentally deficient.
A type II error results if a mentally deficient child is not found to be so.
The probability α of a type I error should be small.

3) a) A type I error results.
b) No error.
c) No error.
d) A type II error results.

5) a) .0990
b) .0456
c) .0198

7) a) The critical value of Z is $z_{.05} = 1.64$
The critical value of $\bar{X}$ is $\bar{x} = 9485.52$
b) The critical value of Z is $z_{.01} = 2.33$
The critical value of $\bar{X}$ is $\bar{x} = 9588.82$

c) The critical value of Z is $z_{.10} = 1.28$
   The critical value of $\overline{X}$ is $\overline{x} = 9431.63$

9) a) H_0: $p = .67$
   H_1: $p \neq .67$
   b) The critical values of Z are $\pm z_{.005} = \pm 2.58$
   The critical values of $\hat{P}$ are .7050 and .6350

11) p value $= .0062$

13) a)

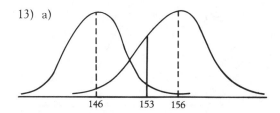

146   153   156

b) $\beta = .1587$
c) $\alpha = .1587$
   $\beta = .0099$

15) a) yes
    b) cannot tell. p value is needed.
    c) yes
    d) yes

17) a) .035
    b) .07

Set 9.3

1) a) $z = 2.29 \pm z_{.025} = \pm 1.96$. Reject H_0.
      p value $= .0110$
   b) $z = -2.92$, $-z_{.05} = -1.64$. Reject H_0.
      p value $= .0018$
   c) $z = 2.22$, $z_{.01} = 2.33$. Do not reject H_0.
      p value $= .0139$

3) H_0: $\mu = 16$, H_1: $\mu \neq 16$, $z = 5.0$, $\pm z_{.01} = \pm 2.33$. Reject H_0.
   p value $< .0001$

5) H_0: $\mu = 44.5$, H_1: $\mu > 44.5$, $z = .93$, $z_{.05} = 1.64$. Do not reject H_0

7) H_0: $\mu = 45$, H_1: $\mu > 45$, $z = 2.43$, $z_{.025} = 1.96$. Reject H_0

9) H_0: $\mu = 134$, H_1: $\mu < 134$, $z = -2.05$, $-z_{.01} = -2.33$. Do not reject H_0.

11) $n \geq 38$

13) H_0: $\mu = 45$, H_1: $\mu \neq 45$, $z = 1.59$, $\pm z_{.025} = \pm 1.96$. Do not reflect H_0.
    p value $= .0559$

15) p value $= .0096 \approx .01$. This is the smallest significance level for which H_0 can be rejected in this one tail test.

17) H_0: $\mu = 300$, H_1: $\mu < 300$, $z = -1.56$, $-z_{.001} = -3.08$. Do not reject H_0.
    p value $= .0594$

Set 9.4

1) $t = 1.81$, $t_{.01} = 2.718$. Do not reject H_0. p value $< .05$

3) $t = -3.07$, $\pm t_{.005} = \pm 2.861$. Reject H_0

5) H_1: $\mu > 80$, $t = 1.49$, $t_{.05} = 1.729$. Do not reject H_0. We cannot conclude that science students have a better self image.

7) H_1: $\mu > 29.2$, $t = 4.88$, $t_{.05} = 1.796$. Reject H_0 and conclude that the new engines have an emission level exceeding 29.2 ppm.

9) $\overline{x} = 39.92$, $s = 6.63$, $t = -1.71$, $-t_{.01} = -3.747$. Do not reject H_0

11) H_1: $\mu \neq 28$, $t = 2.56$, $\pm t_{.025} = \pm 2.306$. Reject H_0. Reject the claim of the report.

13) yes

Set 9.5

1) a) $z = .73$, $\pm z_{.025} = \pm 1.96$. Do not reject H_0. p value $= .2327$
   b) $z = 2.62$, $z_{.01} = 2.33$. Reject H_0. p value $= .0044$

3) H_0: $p = .4444$, H_1: $p > .4444$, $z = 1.47$, $z_{.01} = 2.33$. Do not reject H_0.
   We cannot conclude that the ratio is higher.

5) H_0: $p = .64$, H_1: $p < .64$, $z = -4.57$, $-z_{.01} = -2.33$. Reject H_0 and conclude that less than 64% of the population of the state are native born.

7) H_0: $p = .16$, H_1: $p > .16$, $z = 2.46$, $z_{.10} = 1.28$. Reject H_0 and conclude that the campaign increased the use of the product.

9) H_0: $p = \frac{1}{6}$, H_1: $p < \frac{1}{6}$, $z = -1.24$. $-z_{.01} = -2.33$. Do not reject H_0. We cannot conclude a bias against a 7 on the pair of dice.

11) H_0: $p = .5$, H_1: $p < .5$, $z = -3.16$, $-z_{.05} = -1.64$. Reject H_0 and conclude a bias toward the tail.

13) a) H_0: $p = .45$, H_1: $p > .45$, $z = 5.68$, $z_{.05} = 1.64$. Reject H_0 and conclude that more than 45% have a relapse free five year period.
    b) H_0: $p = .45$, H_1: $p > .45$, $z = 1.50$. Do not reject H_0.

Set 9.6

1) .35

3) Yes. P(H_0 rejected | H_0 false) + P(H_0 not rejected | H_0 false) = 1 and P(H_0 not rejected | H_0 false) = P(H_0 not rejected | H_1 true) = β

5) .9744

7) a) Decreases
   b) Increases

9) $n \geq 118$

11) H_0: $\mu = 39$, H_1: $\mu > 39$. The critical value of $\overline{X}$ is $\overline{x} = 41.73$. $P(\overline{X} > 41.73 \mid \mu = 45)$ $= .9750 =$ Power

13) a) .10
    b) .90
    c) .17
    d) .83
    e) .10
    f) .83

Chapter Test

1) a) A type I error results if the null hypothesis is rejected when in fact it is true. A type II error results if the null hypothesis is not rejected when in fact it is false, or equivalently, the alternate hypothesis is not accepted when, in fact, it is the alternate hypothesis which is true.

    b) The significance level of a test is the probability of a type I error by the test, that is,
$$\alpha = P(H_0 \text{ is rejected} \mid H_0 \text{ is true})$$

    c) The p value of a test is the probability of obtaining a value of the test statistic as extreme as actually found when in fact H_0 is true.

    d) Let β be the probability of a type II error. Then
$$\text{Power} = 1 - \beta$$
This is the probability of accepting H_1 when H_1 is true.

2) a) .04
    b) .96
    c) .85

3) a) yes
    b) Cannot be certain. We need to know the power of the test.
    c) yes

4) a) .08
    b) .04

5) H_0: $p = .75$, H_1: $p < .75$, $z = -1.49$, $-z_{.05} = -1.64$. Do not reject H_0
p value $= .0681$

6) H_0: $p = .60$, H_1: $p \neq .60$, $z = 1.56$, critical values: $\pm z_{.025} = \pm 1.96$. Do not reject H_0.
p value $= .0594$

7) H_0: $\mu = 55$, H_1: $\mu \neq 55$, $t = -1.1$, critical values: $\pm t_{.025} = \pm 2.131$ Do not reject claim.

8) H_0: $\mu = 16$, H_1: $\mu < 16$, $z = 3.16$, critical value: $-z_{.05} = -1.64$ Reject H_0 and conclude that the dissolved oxygen content is below 16 p.p.m.
p value $= .0008$

9) $\alpha = .0013 = .13\%$

10) $\beta = .0384$ and power $= .9616$

Chapter Ten

Set 10.1

1) a) $\mu = 0$, S.E. $= 4.74$

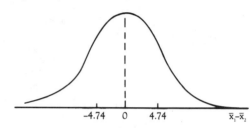

    b) $\mu = 4.2$, S.E. $= 1.51$

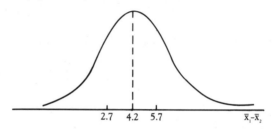

3) a) Approximately normal, $\mu = 0$, and S.E. $= 25.6$

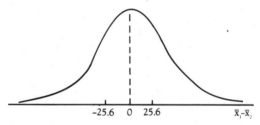

    b) .0594

5) $s_p^2 = 16.60$ and S.E. $= 3.11$

7) a) $(x_{11} - \overline{x}_1)^2 + (x_{12} - \overline{x}_1)^2 + (x_{13} - \overline{x}_1)^2$ = sum of the squares of the deviations of the observations of the first sample from their mean. This is simply $(3 - 1)s_1^2$ (assuming sample 1 has only 3 observations).

    b) $(x_{21} - \overline{x}_2)^2 + (x_{22} - \overline{x}_2)^2 + (x_{23} - \overline{x}_2)^2 + (x_{24} - \overline{x}_2)^2 = (4 - 1)s_2^2$ (assuming that sample 2 has only 4 observations).

    c) $\dfrac{(3 - 1)s_1^2 + (4 - 1)s_2^2}{3 + 4 - 2} = s_p^2$ (From parts (a) and (b))

9) a) Equipment, vehicles, road course, conditions, ... One design is to use the

same autos twice, with the same drivers, and the same road course. Simply change the fuel systems on the test autos.

b) Job classifications, duration of employment, geographic region of employment, sex of employee, . . .

c) Medium on which the test patches are applied, methods of application, types of exposure, ...

11) $s_p^2 = 60.31$, S.E. $= 3.33$, $P(\overline{X}_1 - \overline{X}_2 > 3.3)$ $+ P(\overline{X}_1 - \overline{X}_2 < -3.3) = .3222$

Set 10.2

1) a) $z = 1.71$, $\pm z_{.025} = \pm 1.96$. Do not reject H_0.

   b) $t = 1.37$, $\pm t_{.025} = \pm 2.086$. Do not reject H_0.

3) $-9.05 < \mu_1 - \mu_2 < 34.65$

5) $H_0: \mu_1 - \mu_2 = 0$, $H_1: \mu_1 - \mu_2 \neq 0$, $t = -1.71$, $\pm t_{.025} = \pm 2.306$ Do not reject H_0. p value $< .1$

7) $H_0: \mu_1 - \mu_2 = 0$, $H_1: \mu_1 - \mu_2 \neq 0$, $t = 2.26$, $\pm t_{.025} = \pm 2.021$. Reject H_0 and conclude that the reduction in blood pressure is not the same for men as for women.

9) $H_0: \mu_1 - \mu_2 = 0$, $H_1: \mu_1 - \mu_2 > 0$, $z = 3.16$, $z_{.025} = 1.96$. Reject H_0 and conclude an effect due to the distraction. This we interpret as a reduction in creative thinking due to the distractions.

11) $44.75 < \mu_1 - \mu_2 < 118.34$

13) a) Depression Scores: p value $= .0035$
    Obsession Scores: p value $= .1271$
    Somatization Scores: p value $= .0274$
    Anxiety Scores: p value $= .0495$

   b) Depression Scores: $.82 < \mu_1 - \mu_2 < 5.18$
   Somatization Scores: $-.04 < \mu_1 - \mu_2 < 3.44$
   Anxiety Scores: $-.20 < \mu_1 - \mu_2 < 2.40$

15) a) $H_0: \mu_1 = \mu_2$, $H_1: \mu_1 > \mu_2$, $t = 2.31$, $t_{.05} = 1.717$. Reject H_0 and conclude that the fasting plasma insulin level for normal men is less than that for hyperinsulinemic. Note the data is ± 1(S.E.).

   b) $H_0: \mu_1 = \mu_2$ (Mean heights are equal), $H_1: \mu_1 \neq \mu_2$ (Mean heights are not equal), $t = .68$, $\pm t_{.025} = 2.074$. Do not reject H_0.

Set 10.3

1) a) $\overline{d} = 5$, $s = 9.19$, $t = 1.33$, $\pm t_{.025} = \pm 2.571$. Do not reject H_0.

   b) $-4.65 < \mu_1 - \mu_2 < 14.65$

3) $H_0: \mu_1 = \mu_2$, $H_1: \mu_1 \neq \mu_2$, $\overline{d} = .14$, $s = .50$, $t = .62$, $\pm t_{.025} = \pm 2.776$ Do not reject H_0.

5) $H_0: \mu_1 = \mu_2$, $H_1: \mu_1 \neq \mu_2$, $\overline{d} = 55$, $s = 126.61$, $t = 1.06$, $\pm t_{.025} = \pm 2.571$. Do not reject H_0. We cannot conclude a difference in the means of the appraisals of these two services.

7) a) $H_0: \mu_A = \mu_B$, $H_1: \mu_B < \mu_A$, $t = 4.17$, $t_{.05} = 2.132$. Reject H_0 and conclude that system B has a lower mean operating cost than system A.

   b) $281.34 < \mu_1 - \mu_2 < 1398.66$

9) $H_0: \mu_1 = \mu_2$, $H_1: \mu_1 > \mu_2$, $\overline{d} = 1.7$, $s = 2.87$, $t = 2.65$, $t_{.05} = 1.729$. Reject H_0 and conclude a lower score due to the test form.

11) $H_0: \mu_1 = \mu_2$, $H_1: \mu_1 > \mu_2$, $\overline{d} = 56$, $s = 104.58$, $t = 1.69$, $t_{.05} = 1.833$. Do not reject H_0. We cannot conclude a reduction in mean usage due to the device.

Set 10.4

1) a) $\hat{p} = .3714$, $z = 1.67$, $\pm z_{.025} = \pm 1.96$. Do not reject H_0.

   b) $\hat{p} = .1920$, $z = 1.21$, $\pm z_{.005} = \pm 2.57$. Do not reject H_0.

3) $H_0: p_1 = p_2$, $H_1: p_1 > p_2$, $\hat{p} = .3897$, $z = 1.63$, $z_{.01} = 2.33$. Do not reject H_0.

5) $H_0: p_1 = p_2$, $H_1: p_1 > p_2$, $\hat{p} = .5775$, $z = 1.22$, $z_{.05} = 1.64$. Do not reject H_0.

7) $H_0: p_1 = p_2$, $H_1: p_1 \neq p_2$, $\hat{p} = .5557$, $z = 2.65$, $\pm z_{.025} = \pm 1.96$. Reject H_0 No.

9) $H_0: p_1 = p_2$, $H_1: p_1 > p_2$, $\hat{p} = .1625$, $z = 2.12$, $\pm z_{.005} = \pm 2.57$. Do not reject H_0. p value $= .0170$

11) $-.0475 < p_1 - p_2 < .0991$
    Since the interval includes 0 it is possible that p_1 and p_2 are equal.

13) Let $\hat{p}_0 = \hat{p}_1 = \hat{p}_2$ be the common value.
$$\hat{p} = \frac{n_1 \hat{p}_1 + n_2 \hat{p}_2}{n_1 + n_2} = \frac{n_1 \hat{p}_0 + n_2 \hat{p}_0}{n_1 + n_2} = \frac{(n_1 + n_2)\hat{p}_0}{n_1 + n_2} = \hat{p}_0$$

15) $n \geq 2049$.

Chapter Test

1) $z = 2.60$, critical values: $\pm z_{.025} = \pm 1.96$. reject H_0 and conclude that $\mu_1 \neq \mu_2$

2) $z = 1.86$, critical value: $z_{.05} = 1.64$ reject H_0 and conclude that $p_1 < p_2$

3) $t = 1.41$, critical values: $t_{.01} = 2.583$ do not reject H_0

4) $54.16 < \mu_2 - \mu_1 < 385.84$

5) $.0069 < p_2 - p_1 < .1131$

6) $H_0: \mu_1 = \mu_2$, $H_1: \mu_2 > \mu_1$, $\bar{d} = 7.2$, $s = 7.40$, $t = 2.18$, critical value: $t_{.05} = 2.132$
   Do not reject H_0. An improvement in the algebraic skills cannot be concluded.

7) $H_0: \mu_1 = \mu_2$, $H_1: \mu_1 \neq \mu_2$, $t = 3.01$, critical values: $\pm t_{.005} = \pm 3.250$
   do not reject H_0

8) $H_0: \mu_1 = \mu_2$, $H_1: \mu_1 < \mu_2$, $z = 2.18$, critical value: $z_{.05} = 1.64$. Reject H_0 and conclude that the salaries of area I are less than those of area II. p value $= .0146$

9) $H_0: p_1 = p_2$, $H_1: p_1 \neq p_2$, $z = 1.22$, critical values: $\pm z_{.025} = \pm 1.96$ Do not reject H_0. We cannot conclude that there has been a change of opinion.

10) $H_0: \mu_1 = \mu_2$, $H_1: \mu_1 \neq \mu_2$, $t = 1.57$, critical values: $\pm t_{.025} = \pm 2.120$. Do not reject H_0

Chapter Eleven

Set 11.1

1) a) Negative
   b) Positive
   c) Positive

3) a) $s_{XY} = 2.7$

   b) $s_{XY} = 0$

   c) $s_{XY} = 0$

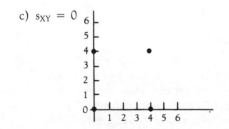

5) a)

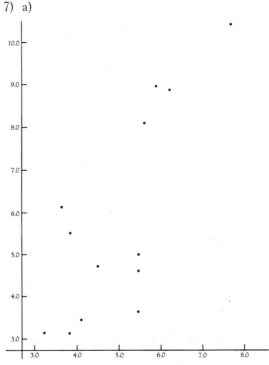

   b) $r = .9631$
   c) $H_0: \rho = 0$, $H_1: \rho > 0$, $r = .9631$, critical value: $r_{.05} = .621$
      reject H_0 and conclude that a positive correlation exists

7) a)

   b) $H_0: \rho = 0$, $H_1: \rho \neq 0$, $r = .7493$, critical values: $\pm r_{.025} = \pm .553$
      reject H_0 and conclude that a correlation exists between unemployment and interest rates

9) $r = 1$

11)

Quadrant	$(x - \bar{x})$	$(y - \bar{y})$	$(x - \bar{x})(y - \bar{y})$	$\dfrac{(x - \bar{x})(y - \bar{y})}{s_x s_y}$
I	+	+	+	+
II	−	+	−	−
III	−	−	+	+
IV	+	−	−	−

13) a) H_0: $\rho = 0$, H_1: $\rho > 0$, $r = .82$, critical value: $r_{.025} = .361$ ($n = 30$ used) reject H_0 and conclude that a positive correlation exists between the WISCR (verbal) and the SIT scores

b) H_0: $\rho = 0$, H_1: $\rho > 0$, $r = .50$, critical value $r_{.025} = .361$ ($n = 30$ used) reject H_0 and conclude that a positive correlation exists between WISCR (performance) and SIT scores

15) $n \leq 16$

Set 11.2

1) a) and c)

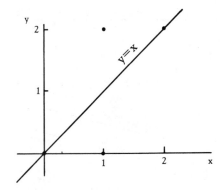

b) $y' = 0 + 1x = x$

c) $S_e = 1$

3) a) $y' = 104.24 + 2.36x$

b) $y' = 222.24$ when $x = 50$

c) $y' = 340.24$ when $x = 100$

d) $S_e = 6.31$

5) a)

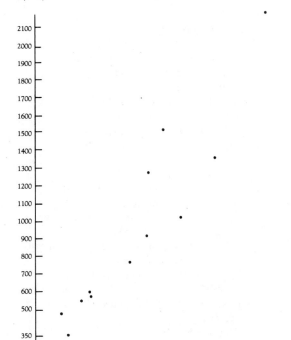

b) The number of employees tends to increase as the city size increases

c) $y' = -312.85 + 13.70x$

d) $S_e = 195.235$

e) $y' = 1194.15$ when $x = 110$

7) $y' = -41.39 + 4.79x$

9) a) $y' = 33.70 - .50x$

b) $y' = 14.7$ when $x = 38$

c) $S_e = .86$

11) $x' = 66.25 - 1.93y$

13) a) 4.25

b) 12.75

c) -17.00

Set 11.3

1) $.49 < \beta < 1.17$

3) $1.03 < \mu_{Y|68} < 3.93$

5) a) Total variation = 34.84
      Explained variation = 21.71
      Error variation = 13.08

   b) total variation = explained variation + error variation

   c) $r = .7898017$, $r^2 = .6237867$

      $$\frac{\text{explained variation}}{\text{total variation}} = .6231$$

      The small difference in these two quantities is due to rounding.

7) a) $y' = 55.06 - 78.57x$

   b) $S_e = 1.85$ (rounded)

   c) $-138.03 < \beta < -19.11$

      If the metering jet is changed by $\Delta x = 1$ unit, the mileage will decrease by an amount between 138.03 and 19.11 miles per gallon. More realistically, for a change of $\Delta x = .01$ in the metering jet, the predicted change is a decrease between 1.3803 and .1911 miles per gallon.

9) 99.95%

11) $z_Y = .9932 z_x$

13) error variation = total variation − explained variation

    = total variation − r^2(total variation)

    = $(1 - r^2)$ total variation ≈ 0 if r^2 is near 1

15) a) $t = .501$, critical value: $t_{.05} = 1.734$

       Do not reject H_0

    b) $t = -1.732$, critical values: $\pm t_{.005} = \pm 2.807$

       Do not reject H_0

17) Z is 1.84 times better than X

19) 91.48%

Set 11.4

1) a) $r_{12} = .51$ is the correlation coefficient between high school g.p.a.'s and S.A.T. scores

      $r_{1Y} = .45$ is the correlation coefficient between high school and college g.p.a.'s.

      $r_{2Y} = .97$ is the correlation coefficient coefficient between S.A.T. scores and college g.p.a.'s.

   b) $r_{12 \cdot Y} = -.2138 = -.21$ (rounded)

      $r_{2Y \cdot 1} = .9640 = .96$ (rounded)

   c) The correlation coefficient between high school and college g.p.a.'s when the S.A.T. score is held constant is $r = -.23$

      The correlation coefficient between S.A.T. scores and college g.p.a.'s is $r = .96$ when the high school g.p.a. is held constant

3) a) $6b_0 + 6b_1 + 8b_2 = 39$
      $6b_0 + 10b_1 + 5b_2 = 47$
      $8b_0 + 5b_1 + 22b_2 = 56$

   b) $b_0 = 2.206422$, $b_1 = 2.8256881$, $b_2 = 1.1009174$

      $y' = 2.21 + 2.83x_1 + 1.10x_2$

   c) $S_e = 1.22$

5) a) $y' = 84.7$ hours

   b) 6.5 (hours)

7) a) the age of the woman

   b) age, job stress, social stress, job security

9) a) graph: $y = 1.84 + .030x_1$
             $y = 7.84 + .030x_1$
             $y = 15.34 + .030x_1$

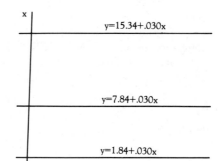

   b) graph: $y = 3.34 + .75x_2$
             $y = 12.34 + .75x_2$
             $y = 18.34 + .75x_2$

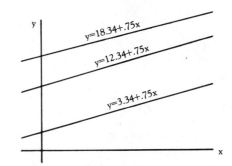

   c) When x_2 is held constant, b_1 is the slope of the resulting line

      When x_1 is held constant, b_2 is the slope of the resulting line

Chapter Test

1) a) positive for figure 2, negative for figure 1

   b) positive for figure 2, negative for figure 1

   c) positive for figure 2, negative for figure 1

2)
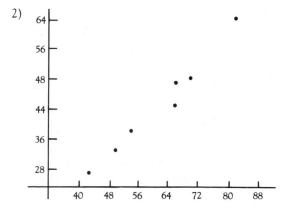

reject H_0 and conclude a positive correlation exists

3) a) $y' = -1.008 + .051x$
   b) $y' = 2.56$ when $x = 70$
   c) $S_e = .3261$

4) $H_0: \rho = 0$, $H_1: \rho > 0$, $r = .65$, $r_{.005} = .561$
   reject H_0 and conclude that a positive correlation exists between educational achievement and family income

5) $37.00 < \mu_{Y|40} < 56.40$

6) $.6382 < \beta_1 < 1.3618$

7) a) unchanged, $r = .9879$
   b) $z_Y = .9879 z_X$

8) $r^2 = .9443$
   About 94.43% of the variation in pulse rates is explained by the regression with dosage size.

9) a) 133
   b) $b_1 = 5$ predicts an increase of $\Delta y = 5$ pounds for each unit change $\Delta x_1 = 1$ inch in the height of the individual when age is held constant. $b_2 = -.4$ predicts a decrease of .4 pounds for each change $\Delta x_2 = 1$ year in the age of the individual when height is held constant

10) a)

	x_1	x_2	y
x_1	1.0000	.6349	.7314
x_2	.6349	1.0000	.9091
y	.7314	.9091	1.0000

   b) $r_{1Y \cdot 2} = .4789$
   when any effect due to the number of promotions is removed (or the number of promotions is held constant), the correlation coefficient between job satisfaction and years of employment is $r = .4789$

   c) $r_{2Y \cdot 1} = .8441$
   when any effect due to years of employment is removed (or the numbers of years of employment is held constant), the correlation coefficient between job satisfaction and number of promotions is .8441

11) z_{x_2} is 20.25 times as important as z_{x_1} in accounting for the total variation in z_y

Chapter Twelve

Set 12.1

1) a) square footage of the home, lot size, . . .
   b) I.Q.'s, high school g.p.a., . . .
   c) community size, mean income of residents of the community served by the store, . . .
   d) prevailing temperature, the amount of rain or snow during the week, . . .
   e) anxiety level, age, weight, . . .
   f) age, I.Q., diet, . . .

3) $SS = 180.25$, $SSTr = 61.58$, $SSE = 118.67$

5)

Source	SS	d.f.	MS	f
Package design	19763.33	2	9881.67	2.421
Error	48970.00	12	4080.83	
Total	68733.33	14		

$f = 2.421$, critical value: $f_{.05}(2,12) = 3.89$
Do not reject the null hypothesis of no effect on sales due to the package design

7) a) $k = 6$
   b) $n = 54$
   c) 9 per sample
   d) $f = 2.83$, critical value: $f_{.01}(5,48) = 3.51$
      (largest adjacent entry)
      Do not reject the null hypothesis

9)

Source	SS	d.f.	MS	f
Makes	7.25	2	3.625	3.50
Error	12.43	12	1.036	
Total	19.68	14		

$f = 3.50$, critical value: $f_{.025}(2,12) = 5.10$
Do not reject the null hypothesis

11)

Source	SS	d.f.	MS	f
Area	4890	2	2445	6.24
Error	4700	12	391.67	
Total	9590	14		

$f = 6.24$, critical value: $f_{.05}(2,12) = 3.89$
reject H_0 and conclude that a difference exists in the mean hospital rates for at least two of the areas

13) a) heating cost

b)
Source	SS	d.f.	MS	f
system	787.33	2	393.66	2.18
error	1801.75	10	180.18	
Total	2589.08	12		

$f = 2.18$, critical value: $f_{.05}(2,10) = 4.10$
do not reject H_0. An effect on the heating cost due to the type of heating system cannot be concluded. c) family size, house size, insulation, . . .

15)
Source	SS	d.f.	MS	f
design	74.32	2	37.16	4.19
error	213.04	24	8.88	
Total	287.36	26		

$f = 4.19$, critical value: $f_{.05}(2,24) = 3.40$
reject H_0 and conclude that at least two of the engine designs have different mean emission levels

17) a)
Source	SS	d.f.	MS	f
treatment	630	4	157.50	4.54
error	520	15	34.67	
Total	1150	19		

$f = 4.54$, critical value: $f_{.05}(4,15) = 3.06$
reject the null hypothesis
b) 5 treatment levels
c) sample size $= 4$ if all samples are of the same size

19)
Source	SS	d.f.	MS	f
arrangement	329.5	3	109.83	1.63
error	807.5	12	67.29	
Total	1137.0	15		

$f = 1.63$, critical value: $f_{.025}(3,12) = 4.47$
do not reject H_o. We cannot conclude an effect on the grades due to the arrangement of the problems on the exam form.

Set 12.2

1) a) 30
   b) 12.5
   c) 28.75
   d) 75
   e) 220
   f) 360

3)
Source	SS	d.f.	MS	f
Rows	533.33	1	533.33	2.17
Columns	1137.50	2	568.75	2.31
Interaction	4.17	2	2.09	.009
Error	1475.00	6	245.83	
Total	3150.00	11		

Rows: $f = 2.17$, critical value: $f_{.05}(1,6) = 5.99$
      do not reject $Hli0$
Columns: $f = 2.31$, critical value: $f_{.05}(2,6) = 5.14$
      do not reject H_0
Interaction: $f = .009$, critical value: $f_{.05}(2,6) = 5.14$
      do not reject H_0

5) a) $r = 5$
   b) $c = 3$
   c) $n = 2$
   d) Rows: $f = 3.93$, critical value: $f_{.05}(4,15) = 3.06$
      reject the null hypothesis of no row effect
      Columns: $f = 9.64$, critical value $f_{.05}(2,15) = 3.68$
      reject the null hypothesis of no column effect
      Interaction: $f = .71$, critical value: $f_{.05}(8,15) = 2.6408$
      do not reject the null hypothesis of no interaction effect

7)
Source	SS	d.f.	MS	f
company	77.44	2	38.71	3.70
price	489.65	2	244.83	23.41
interaction	186.11	4	46.53	4.49
error	94.16	9	10.46	
Total	847.36	17		

company: $f = 3.70$, critical value: $f_{.05}(2,9) = 4.26$
do not reject the null hypothesis of no effect on tube life due to the manufacturer of the tube
price: $f = 23.41$, $f_{.05}(2,9) = 4.26$

reject the null hypothesis of no effect on tube life due to the cost of the tube

interaction: $f = 4.45$, critical value: $f_{.05}(4,9) = 3.63$

reject the null hypothesis of no interaction effect due to the manufacturer and the cost of the tube

9) Rows: $f = 2.88$, critical value: $f_{.05}(1,8) = 5.32$
do not reject the null hypothesis of no row effect

Columns: $f = 3.08$, $f_{.05}(2,8) = 4.46$
do not reject the null hypothesis of no column effect

11)

Source	SS	d.f.	MS	f
rack	230104.17	1	230104.17	88.74
package	63508.33	2	31754.17	12.25
interaction	3008.33	2	1504.17	.58
error	46675.00	18	2593.06	
total	343295.83	23		

rack: $f = 88.74$, $f_{.05}(1,18) = 4.41$
reject H_0 and conclude an effect due to the display rack

design: $f = 12.25$, $f_{.05}(2,18) = 3.55$
reject H_0 and conclude an effect due to the package design

interaction: $f = .58$ (less than 1)
do not reject the null hypothesis of no interaction effect

Chapter Test

1) the common population variance when there is no treatment effect

2) The value of MSSTr depends on the variation between or among the various sample means, hence the *between group* name.
The value of MSSE depends on the variation of the observations within each sample about their respective means, hence the *within group* designation.

3) $SS = SSTr + SSE$
SS is called the total sum of squares
SSTr is called the treatment sum of squares
SSE is called the error sum of squares

4) $SS = SSR + SSC + SSRC + SSE$
SS is called the total sum of squares
SSR is called the treatment sum of squares for the factor R
SSC is called the treatment of squares for the factor C

SSRC is called the interaction treatment sum of squares
SSE is called the error sum of squares

$$SS = \sum_i \sum_j \sum_k (x_{ijk} - \bar{x})^2$$
$$SSR = \sum_i nc(\bar{x}_{i.} - \bar{x})^2$$
$$SSC = \sum_j nr(\bar{x}_{.j} - \bar{x})^2$$
$$SSE = \sum_i \sum_j \sum_k (x_{ijk} - \bar{x}_{ij})^2$$
$$SSRC = SS - SSR - SSC - SSE$$

5) a) $k = 5$
b) 4 observations per treatment group
c) $f = 7.12$, critical value: $f_{.05}(4,15) = 3.06$
reject H_0 and conclude that at least two of the population (treatment) means are different.

6)

Source	SS	d.f.	MS
Treatment	100	5	20
Error	180	36	5
Total	280	41	

7)

Source	SS	d.f.	MS	f
Airlines	509	3	169.67	16.42
Error	124	12	10.33	
Total	633	15		

Critical value: $f_{.05}(3,12) = 3.49$
Reject H_0 and conclude that a difference exists in the mean noise levels for at least two of the airlines.

8)

Source	SS	d.f.	MS	f
Variety	11.44	2	5.72	2.39
Diet	28.44	2	14.22	5.95
Interaction	13.56	4	3.39	1.42
Error	21.50	9	2.39	
Total	74.94	17		

Variety: $f_R = 2.39$, $f_{.05}(2,9) = 4.26$
do not reject the null hypothesis of no effect due to the variety of poultry used

Dietary Program: $f_C = 5.95$, $f_{.05}(2,9) = 4.26$
reject H_0 and conclude an effect due to the dietary program

Interaction: $f_{RC} = 1.42$, $f_{.05}(4,9) = 3.63$
do not reject the null hypothesis of no interaction effect

9)

Source	SS	d.f.	MS	f
Variety	11.44	2	5.72	2.12
Diet	28.44	2	14.22	5.27
Error	35.06	13	2.70	
Total	74.94	17		

Variety: $f_R = 2,12$, $f_{.05}(2,13) = 3.81$
do not reject H_0
Diet: $f_C = 5.27$, $f_{.05}(2,13) = 3.81$
reject H_0 and conclude an effect on weight gain
due to the diet

10) a) 161
   b) 20
   c) 7.83
   d) 7.17

Chapter Thirteen

Set 13.1

1) a) 75
   b) 40
   c) 26
   d) 8
   e) $e_{11} = 12.27$, $e_{12} = 13.87$, $e_{13} = 13.87$
      $e_{21} = 10.73$, $e_{22} = 12.13$, $e_{23} = 12.13$
   f) $\chi^2 = 5.82$, $\chi^2_{.05} = 5.99$. Do not reject the
     null hypothesis of independence.
3) $\chi^2 = 4.62$, $\chi^2_{.05} = 5.99$. Do not reject the null
  hypothesis of no relationship.
5) Because of the low cell frequencies in the last
  two rows, these two rows are combined to
  produce an "over 2" category.
  $\chi^2 = 45.03$, $\chi^2_{.05} = 12.59$. Reject the null
  hypothesis and conclude that a relationship
  exists between the number of tickets and the
  age of the driver.
7) $\chi^2 = 16.91$, $\chi^2_{.05} = 5.99$. Reject the null
  hypothesis of equal proportions for the three
  nationalities.
9) $\chi^2 = 5.71$, $\chi^2_{.05} = 7.81$. Do not reject the null
  hypothesis of equal rates for the four clinics.
11) Note that some guess work is needed in
   arriving at the observed frequencies due to the
   severe rounding used in reporting the results.
   a) Cell Frequencies Used: 304 112 52
                              287 226 38
     $\chi^2 = 34.58$, $\chi^2_{.05} = 5.99$. Reject the null
     hypothesis of homogeneity with respect to
     attitudes on home defense.

b) Black Sample
  Cell Frequencies Used: 123 30 15
                            179 84 39
  $\chi^2 = 9.17$, $\chi^2_{.05} = 5.99$. Reject the null
  hypothesis of no relationship between the
  sex and home defense orientation of
  blacks.

  White Sample
  Cell Frequencies Used: 128 108 15
                            160 120 24
  $\chi^2 = 1.19$, $\chi^2_{.05} = 5.99$. Do not reject the
  null hypothesis. We cannot conclude that
  a relationship exists between the sex and
  the home defense orientation of whites.

  Pooled Sample
  $\chi^2 = 3.56$, $\chi^2_{.05} = 5.99$. Do not reject the
  null hypothesis of no relationship between
  the sex and the home defense orientation
  of individuals.
c) Black Group
  Cell Frequencies Used: 45 17   9
                           115 31   9
                           129 59 33
                            14   7   3
  $\chi^2 = 12.41$, $\chi^2_{.05} = 12.59$
  Do not reject H_0 (the hypothesis of no
  relationship between region and
  orientation for blacks).

  White Group
  Cell Frequencies Used: 57 83   7
                             77 56 11
                         106 30   7
                         45 56 11
  $\chi^2 = 49.31$
  Reject H_0 and conclude that a relationship
  exists between orientation and region for
  whites.
d) The increasing sequence 27%, 41%, 49%,
  and 61% suggests that support for the
  police role increases with the size of the
  city of residence.

Set 13.2

1) a)

Observations	2	3	4	4	6	7	8	10	12
Ranks	1	2	3.5	3.5	5	6	7	8	9
Variable	y	x	x	y	x	y	x	y	y

b)

x	rank		y	rank
3	2		2	1
4	3.5		4	3.5
6	5		7	6
8	7		10	8
			12	9

$w_1 = 17.5$

$w_2 = 27.5$

c) Minimum Rank Sum = 10
Maximum Rank Sum = 35

3) $w_1 = 28$, $w_2 = 50$, $u_1 = 22$, $u_2 = 13$, $u = 13$, $u_{.025} = 5$. Do not reject the null hypothesis of identical distributions or equal mean room costs for the two areas.

5) $w_1 = 89.5$, $w_2 = 81.5$, $u_1 = 26.5$, $u_2 = 53.5$, $u = 26.5$, $u_{.025} = 17$. Do not reject the null hypothesis of no difference in the distributions of mastery scores.

7) $w_1 = 559.5$, $w_2 = 475.5$, $u_1 = 222.5$, $u_2 = 283.5$, $u = 222.5$, $\pm z_{.025} = \pm 1.96$, $z = -.28$. Do not reject the null hypothesis of equal mean expenditures for auto accessories by men and women.

9) $u_1 = 490$, $u_2 = 10$, $u = 10$, $z = -2.24$, $\pm z_{.005} = \pm 2.58$. Do not reject the null hypothesis of equal distributions for the attitude scores, that is, we cannot conclude a difference in attitudes of workers at the two factories using this measuring instrument.

11) a) $w_1 = 21.5$, $w_2 = 56.5$, $u_1 = 20.5$, $u_2 = 11.5$, $u = 11.5$.

b) yes. p value = .003 so result is significant for $\alpha = .003$ in a one tail test.

c) Reject H_0 and accept a difference in the mean latency times. p value = .006.

d) yes

Set 13.3

1) a) $T = 40$, $T_{.025} = 25$. Do not reject H_0.
b) $T = 26$, $T_{.005} = 7$. Do not reject H_0.

3) a) $T_+ = 10.5$, $T_- = 44.5$, $T = 10.5$, $T_{.025} = 8$. Do not reject H_0, that is, we cannot conclude a difference in the means of the dexterity scores due to the training.
b) $T_+ = 1$, $T_- = 27$, $T = 1$, $T_{.05} = 4$. Reject the null hypothesis of no change in the scores of the severely impaired

5) $T_+ = 30.5$, $T_- = 60.5$, $T = 30.5$, $T_{.005} = 10$. Do not reject H_0. We cannot conclude that scores from the two tests have different distributions.

7) $z = -2.54$, $\pm z_{.025} = \pm 1.96$. Reject H_0. p value = .0055

9) a) $z = -1.15$ p value = .1271
b) $z = -1.02$ p value = .1539

11) $T_+ + T_- = 1 + 2 + 3 + \ldots n = \dfrac{n(n + 1)}{2}$

Set 13.4

1) $r_s = .37$, $t = 1.91$, $\pm t_{.025} = \pm 2.069$. Do not reject the null hypothesis of no correlation.

3) $r_s = .200$, $t = .577$, $\pm t_{.025} = \pm 2.306$. Do not reject the null hypothesis of no correlation.

5) $r_s = -.74$, $t = -4.67$. Reject H_0 and accept the alternate hypothesis of a correlation between sales and the size of the assigned territory. The negative correlation might possibly be explained by the salesmen being unable to service their accounts adequately if there are too many or if they are spread out too much geographically.

7) $r_s = .85$, $t = 4.77$. Reject H_0 and conclude a correlation exists between the two rankings.

9) a) First Assessment: Yes (at $\alpha = .001$) since p value < .001 for all five dimensions. Second Assessment: Yes (at $\alpha = .01$) since p value < .01 for all 5 dimensions.
b) Yes. p value < .001 for both assessments
c) Hostility and Evaluation
d) The TPS scores by each rater were highly correlated for the two assessments. Both sets of rater's scores on the two assessments proved to be correlated ($\alpha = .05$) for evaluation, pressuring, and hostility.

Set 13.5

1) $u_1 = 82$, $u_2 = 102.5$, $u_3 = 46.5$, $h = 5.96$, $\chi^2_{.05} = 5.99$. Do not reject the null hypothesis. We cannot reject the hypothesis that the three populations are equally satisfied with the agency.

3) $u_1 = 73.5$, $u_2 = 39.5$, $u_3 = 58$, $h = 3.39$, $\chi^2_{.05} = 5.99$. Do not reject H_0. We cannot conclude a difference in the daily food costs for the three regions.

5) $u_1 = 57$, $u_2 = 76$, $u_3 = 51$, $u_4 = 26$, $h = 7.29$, $\chi^2_{.05} = 7.81$. Do not reject H_0. We cannot conclude that a difference exists in the performances of the three dehumidifiers.

7) H_0: All four brands have the same mean sugar content

H_1: At least one pair of the brands have different mean sugar contents.
$u_1 = 72.5$, $u_2 = 169.5$, $u_3 = 224.5$, $u_4 = 94.5$, $h = 15.22$, $\chi^2_{.05} = 7.81$. Reject H_0.

9) $h = 20.80$, $\chi^2_{.05} = 11.07$. Reject H_0 and conclude a difference in effectiveness for at least two of the antiseptics.

11) a) Smallest rank sum $= 1 + 2 + \ldots + n$

$= \dfrac{n(n + 1)}{2}$. This occurs when all the lowest ranks are in one sample.

b) Largest Rank Sum $= (k - 1)n + 1 + (k - 1)n + 2 \ldots + kn$

$= (k - 1)n^2 + \dfrac{n(n + 1)}{2}$

This occurs when the n highest ranks are all in one sample.

Set 13.6

1) $u_1 = 19$, $u_2 = 12$, $u_3 = 11$, $\chi^2_f = 5.43$, $\chi^2_{.05} = 5.99$. Do not reject H_0.

3) $u_1 = 11$, $u_2 = 8$, $u_3 = 5$, $\chi^2_f = 4.5$, Critical Value $= 6.5$. Do not reject H_0. We cannot conclude a difference in time to complete the form due to the instructions.

5) $u_1 = 9.5$, $u_2 = 13$, $u_3 = 13.5$, $\chi^2_f = 1.58$, Critical Value $= 6.3$. Do not reject H_0. We cannot conclude an effect due to class assignment.

7) $u_1 = 22$, $u_2 = 15$, $u_3 = 11$, $\chi^2_f = 7.75$, Critical value $= 6.3$. Reject H_0 and conclude that a difference exists in the (mean) operating speeds of these three computers.

Set 13.7

1) $r = 18$, $z = -1.23$, $\pm z_{.025} = \pm 1.96$. Do not reject H_0.

3) $r = 22$, $z = -1.25$, $\pm z_{.025} = \pm 1.96$. Do not reject H_0.

5) $\bar{x} = 71.5$, $r = 12$, $z = -1.16$, $\pm z_{.025} = \pm 1.96$. Do not reject the null hypothesis.

7) 51

9) a) Two
   b) Five
   c) 3.4

Chapter Test

1) $\chi^2 = 1.32$, $\chi^2_{.05} = 5.99$. Do not reject H_0.

2) $\chi^2 = 46.98$, $\chi^2_{.05} = 15.51$. Reject H_0 and conclude that attitude and class standing are related.

3) $\chi^2_f = 1.56$, $\chi^2_{.05} = 7.81$. Do not reject H_0.

4) $\chi^2_f = 10.92$, $\chi^2_{.05} = 7.81$. Reject H_0 and conclude a difference exists in the resistance to fading by (at least two of) these colors.

5) $u = 27$, $u_{.025} = 13$. Do not reject H_0. Note: A one tail test is more appropriate. For it, $u_{.05} = 15$. Still H_0 is rejected.

6) $t = 4.38$, $t_{.05} = 1.86$. Reject H_0 and conclude a positive correlation exists. (One tail test used)

7) $T = 12.5$, $T_{.025} = 4$. Do not reject H_0. We cannot conclude an effect on attitude due to the film.

8) $h = .665$, $\chi^2_{.05} = 5.99$ Do not reject H_0.

9) $z = -1.49$, $\pm z_{.025} = \pm 1.96$. Do not reject H_0.

10) Combine the 4–5 and > 5 categories. $\chi^2 = 107.4$. $\chi^2_{.05} = 12.6$. Reject H_0 and conclude that the age and admission classifications are not independent.

Bibliography

(Page number in color indicates the location in this text in which the work is cited.)

p. 303

Adams, S. A., Horner, J. K., and Vessey, M. P., "Delay in Treatment for Breast Cancer". Community Medicine, vol. 2, 1980. Used with permission of John Wright & Sons, Ltd.

p. 258

Adler, A., Stein, J. A., and Etraim, S. B., "Immunocompetence, Immunosuppression, and Human Breast Cancer", Cancer, vol. 45, No. 8, April 15, 1980. Used with permission of the authors and the American Cancer Society.

p. 85

"Administrative Compensation Surveys, 1967–68, 1975–76, and 1980–81". College and University Personnel Association (CUPA).

p. 3

"Alcohol Boosts Memory of Mice", Valley News, Van Nuys, Ca. 1979.

p. 3

Alkana, R. L., and Parker, E. S., Memory Facilitation By Post, Training Injection of Ethanol," Psychopharmacology, vol. 66, 1979. Abstracted with permission of the publisher and author.

p. 39

Allen, C., "The Southern California Cooperative Seismic Network", Earthquake Bulletin, vol. 11, No. 6, Nov.-Dec. 1979.

p. 38

"ATC Flow Management", Interavia, vol. 4, 1980.

p. 4

Baird W. P., "Sleep Apnea: A Non-Medical Presentation", The American Narcolepsy Association, July, 1976.

p. 290

Barlocher, Felix, "The Contribution of Fungal Enzymes to the Digestion of Leaves," Oecologia, vol. 52, Jan. 1982. Used with permission of the author and Springer-Verlag, Heidelberg.

p. 13

Benson, H., and Mccallie, D., "Angina Pectoris and the Placebo Effect," The New England Journal of Medicine, vol. 300, No. 25, June 21, 1979.

p. 506

Bhatnagar, J., "Educational Background at Entry and Drop Out from Part Time Undergraduate Studies", Journal of Experimental Education, vol. 47, No. 2, Winter 1978–1979, a publication of the Helen Dwight Reid Educational Foundation. Adapted with permission of the publisher.

p. 343

Bonadonna, G. and Valagussa, B. S., "Dose Response Effect of Adjuvant Chemotherapy on Breast Cancer", Adapted with permission of The New England Journal of Medicine, vol. 304, No. 1, Jan. 1981.

p. 416

Brand, C. J., and Keith, L. B., "Lynx Demography During a Snowshoe Hare Decline in Alberta", Journal of Wildlife Management, vol. 43, No. 4, 1979. Copyright 1979, The Wildlife Society. Adapted with permission.

p. 4

Carskdon, M. A., Dement, W. C., "Respiration During Sleep in the Aged Human", Journal of Gerontology, vol. 34, No. 4, pp 420–423, July 1981.

p. 370

Casper, R. C., Eckert, E. D., Halmi, K. A., Goldberg, S. C., and Davis, J. M., "Bulimia, Its Incidence and Clinical Importance in Patients with Anorexia Nervosa", Archives of General Psychiatry, vol. 37, pp 1034–1035, Sept. 1980. Used with permission. Copyright © 1980, American Medical Association.

p. 92

Chen, J. R., Francisco, R. B., and Miller, T., "Legionnaire's Disease: Nickel Levels", Science, vol. 196, No. 4292, May 20, 1977. Used with permission of the authors and the American Association for the Advancement of Science.

p. 515

Cheney, D. L. and Seytarth, R. M., "Vocal Recognition in Free Ranging Vervet Monkeys", Animal Behavior, vol. 28, No. 2, pp 362–367, 1980. Used with permission of authors and publisher.

p. 93

Clark, E., "Inflation and Unemployment: Are They Complements or Substitutes?", Agricultural Economic Research, vol. 32, No. 1, Jan. 1980, page 33. Used with permission.

p. 37

Cohen, H. and Oliver, J., "Look Whats Moving into Our Neighborhood", Astronomy magazine, April 1979. Adapted courtesy of Astronomy magazine.

p. 301

Dawber, T., "Risk Factors in Young Adults", Journal of American College Health, vol. 22, Dec. 1973, pp 84–90, a publication of the Helen Dwight Reid Educational Foundation. Adapted with permission.

p. 304

Delaney, J., "Routine Skull Films in Hospitalized Psychiatric Patients", The American Journal of Psychiatry, vol. 133, No. 1, Jan. 1976. Copyright 1976. Adapted by permission.

p. 4

Dement, W., Guilleminault, C., Eldridge, F.I., Tilkian, A., and Simons, F. S. "Sleep Apnea Syndromes Due to Upper Airway Obstruction," Archives of Internal Medicine, vol. 137, March 1977, pp 297–300. Abstracted with permission of author and publisher. Copyright 1977, American Medical Association.

p. 102

Evans, J. G., "Numbers Sometimes Lie, Cheat, and Steal", Psychology, vol. 16, No. 3, Fall 1979. Reprinted with permission of the editors of Psychology.

p. 507

Feagin, J. R., "Home Defense and the Police", American Behavioral Scientist, vol. 13, No. 5 and 6, Aug. 1970. Copyright © 1970, American Behavioral Scientist, Adapted by permission of Sage Publications.

p. 47

Fournier, D. V., Weber, E., Hoettken, W., Bauer, M., Kubli, F., and Barth, V., "Growth Rate of 147 Mammary Carcinomas", Cancer, vol. 45, April 1980. Adapted with permission of publisher and authors.

p. 160

Frerich, R. R., Beeman, B. L., and Coulson, A. H., "Los Angeles: Airport Noise and Mortality-Faulty Analysis and Public Policy", American Journal of Public Health, vol. 70, No. 4, April 1980. Adapted with permission by the authors and the American Public Health Association.

p. 81

Garman, E. T., "The Cognitive Consumer Education", The Journal of Consumer Affairs, vol. 13, No. 1. Used with permission of the American Council on Consumer Affairs.

p. 304

Gates, S. J., and Horner, J. K., "Lowering Appointment Failures in a Neighborhood Health Center", Medical Care, vol. 14, No. 3, 1976. Adapted with permission of the authors and the publisher, The American Public Health Association.

p. 326

Gibbons, J., and Pratt, J., "p Values: Interpretation and Methodology", American Statistician, vol. 19, 1975. Used with permission.

p. 93

Glisson, C. A., and Martin, P. Y., "Productivity and Efficiency", Academy of Management Journal, vol. 23, No. 1, March 1977. Used with permission of the author and the Academy of Management.

p. 423

Greenfield, H., and Maples, J., "Salting Food, A Function of Hole Size and Location of Shaker". Adapted with permission from Nature, vol. 301, p 332, Jan. 27, 1983. Copyright © 1983. Macmillian Journals, Limited.

p. 365 ·

Gregory, M. K., "Sex Bias in School Referrals", vol. 15, No. 1, 1977. Journal of School Pyschology. Used with permission.

p. 372

Halltrisch, J., Elkwood, K. C., Michaelis, U. E., Reiser, S., O'Dorisio, T. M., and Pratner, E. S., "Effects of Dietary Fructose on Plasma Glucose and Hormone Responses in Normal and Hyperinsulinemic Men", Journal of Nutrition, vol. 13, No. 9, pp 1819–1824, Sept. 1983. Used with permission.

p. 286

Hart, A., and Begon, M., "The Status of General Reproductive-Strategy Theories, Illustrated in Winkles", Oecologia, vol. 29, 1982. Used with permission of the publisher Springer-Verlag, Heidelberg.

p. 132

Horowitz, L. M., and French, R. S., "Interpersonal Problems of People Who Describe Themselves as Lonely", Journal of Consulting and Clinical Psychology, vol. 47, No. 4, 1979. Copyright © 1979, American Psychological Assn. Adapted by permission of the publisher and author.

p. 436

Joosens, J. V., Kesteloot, H., and Amery, A., "Salt intake and Mortality from Strokes". Reprinted by permission of the The New England Journal of Medicine, vol. 300, No. 24, June 14, 1979.

p. 38

Jung, A. F., "Price Variation for Refrigerators Among Retail Store Types and Cities", Journal of Consumer Affairs, vol. 13, No. 1, summer 1979. Used with permission of the publisher.

p. 56

Kanamori, H., "The Size of Earthquakes", Earthquake Information Bulletin, vol. 12, No. 1, Jan.-Feb. 1980.

p. 332

Kaufman, D. W., Slone, D., Rosenberg, L., Miettinen, U. S. and Shapiro, S., "Cigarette Smoking and Age at Natural Menopause", American Journal of Public Health, vol. 70, No. 4, 1980. Adapted with permission of the authors and publishers.

p. 71

King, D. R., "The Brutalization Effect: Execution Publicity and the Incidence of Homocide in South Carolina". Adapted with permission from Social Forces, vol. 57, No. 2, Dec. 1978. Copyright © 1978, The University of North Carolina Press.

p. 81

Klevay, L. M., Reck, S. J., and Barcome, D. F., "Evidence of Copper and Zinc Deficiencies", Journal of the American Medical Association, vol. 24, No. 118, pp 1916–1918, March 4, 1979. Used with permission.

p. 55

Laird, R., Perdrizet, S., Corriman, J., and Bioou, S., "Smoking and Chronic Respiratory Symptoms: Prevalence in Male and Female Smokers", American Journal of Public Health, vol. 70, No. 3, March 1980. Adapted with permission of the authors and The American Public Health Association.

p. 515

Lanza, F. L., Royer, G. L., and Nelson, R. S., "Endoscopic Evaluation of the Effects of Aspirin, Buffered Aspirin, and Enteric Coated Aspirin on the Gastric and Duodenal Mucasa". Adapted with permission from The New England Journal of Medicine, vol. 303, No. 3, July 1980.

p. 97

Lynds, G. B., Seyler, S. K., and Morgan, B. M., "The Relationship Between Elevated Blood Pressure and Obesity in Black Children", American Journal

608   Bibliography

of Public Health, vol. 70, No. 2, Feb. 1980. Used with permission of the author and the American Public Health Association.

p. 526

Magnhagen, C., and Wiederholm, A. M., "Food Selectivity versus Prey Availability: A Study Using the Marine Fish Pomatoschistus Microps", Oecologia, vol. 55, 1982. Used with permission of the author and the publisher, Springer-Verlag, Heidelberg.

p. 409

Martin, J., and Kidwell, J., "Intercorrelation of the Wechsler Intelligence Scale for Children—Revised, The Slossen Intelligence Test, and the National Education Development Test", Journal of Educational and Psychological Measurements, vol. 37, 1977. Used with permission of authors and publisher.

p. 294

Martin, L. T., and Albert, J. R., "Taste Aversion to Mother's Milk: The Age Related Role of Nursing in Acquisition and Expression of Learned Association", Journal of Comparative and Physiological Psychology, vol. 93, No. 3, 1979. Copyright © 1979 by The American Psychological Association. Adapted by permission of the authors and publisher.

p. 209

"MCAT Combined Results, April and October 1980", Permission to use granted by the Association of American Medical Colleges.

p. 230

Mentzer, T., "Response Biases in Multiple Choice Item Files", Educational and Psychological Measurements, vol. 42, 1982. Used with permission of the author and publisher.

p. 287

Moore, C., and Morelli, G. C. L., "Mother Rats Interact Differently with Male and Female Offsprings", Journal of Comparative and Physiological Psychology, vol. 93, No. 4, 1979. Copyright © by The American Psychological Association. Adapted with permission of publisher and authors.

p. 272

Mortel, C. G., and Hanley, J. A., "The Effect of Measuring Error on the Results of Therapeutic Trials in Advanced Cancer", Cancer, vol. 38, No. 1, July 1976. Adapted with permission of the authors and the American Cancer Society.

p. 101

Moutsoulas, M., and Peteri, S., "Impact Spacecraft Imagery and Comparative Morphology of Craters", The Moon and the Planets, vol. 21, No. 3, 1979. Used with permission. Copyright © 1979, D. Reidel Publishing Co., Dorarecht, Holland.

p. 295

Okamura, S. and Watanabe, M., "Determination of Phenolic Cinnamates in White Wines", Journal of Agricultural and Food Chemistry, vol. 45, No. 9, 1981. Copyright © 1981, American Chemical Society. Adapted with permission.

p. 284

Ossenbrugen, P. J., Pregent, G. P., and Meeker, L. D., "Offshore Wind Power Potential," Journal of the Energy Division, Preceedings of the American Society of Civil Engineers, vol. 105, No. EY1, Jan. 1979, pp 81–91. Adapted with permission of the American Society of Civil Engineers.

p. 135

Page, H. L., "Another Application of the Spiral Aftereffect in the Determination of Brain Damage", Journal of Consulting Psychology, vol. 21, 1957. Used with permission.

p. 135

Philbrick, Emily, "The Validity of the Spiral Aftereffect as a Clinical Tool for Diagnosis of Organic Brain Pathology", Journal of Consulting Psychology, vol. 23, No. 1, 1959. Used with permission.

p. 373

Pigge, F. L., Gibney, T. C., and Ginther, J. L., "School Teachers are Better Prepared to Teach Mathematics", Arithmetic Teacher, March 1979. Used with permission.

Prentice, S. A., "Lightning Risks", Search, Inc., vol. 9, No. 6, June 1978. Used with permission of Australian and New Zealand Association for the Advancement of Science (ANZAAS, Inc.).

p. 483

Prescott, T. E., Rausch, M. A., and Dewolfe, A. S., "Schrizophrenic and Aphasic Language, Discernable or Not", Journal of Consulting and Clinical Psychology, vol. 48, No. 1, Feb. 1980. Copyright © 1980, American Psychological Association. Adapted by permission of the authors and publisher.

p. 134

Price, A. C., and Deabler, H. L., "Diagnosis of Organicity by Means of the Spiral Aftereffect", vol. 19, 1955. Journal of Consulting Psychology. Used with permission.

"Report #25", NFLB Quarterly, 1979. Used with permission of the National Federation of Independent Business.

p. 429

Richter, E. D., Yatte, Y., and Gruener, N., "Air and Blood Lead Levels in a Battery Factory", vol. 20, 1979. Used with permission.

p. 481

Rosen, B. and Mericle, M. F., "Influence of Strong versus Weak Fair Employment Policies and Applicants Sex on Selection Decisions and Salary Recommendations", Journal of Applied Psychology, vol. 64, No. 4, 1979. Copyright 1979 by the American Psychological Association. Adapted with permission of the authors and publisher.

p. 15

Ryan, T., Ryan, B., Minitab, Inc., Statistics Department, 215 Pond Laboratory, The Pennsylvania State University, University Park, Pa. 16802. All minitab programs with the permission of Minitab, Inc.

p. 29

Saari, J. T., and Lahtela, J., "Job Enrichment, Cause of Increased Accidents?" Reprinted with permission from Industrial Engineering Magazine, vol. 10, No. 10, October 1978. Copyright Institute of Industrial Engineers, 25 Technology Park/Atlanta, Norcross, Ga 30092.

p. 248

"Selected Hospital Statistics for February 1981". Reprinted with permission from HOSPITAL, Published by American Hospital Publishing, Inc., copyright © June 1, 1981, vol. 55, No. 11.

p. 488

Simas, K. and McCarrey, M., "Impact of Recruiter Authoritarianism and Applicant Sex on Evaluation and Selection Decisions in a Recruitment Interview Study", Journal of Applied Psychology, vol. 64, No. 5, 1979. Copyright © 1979 by the American Psychological Association. Adapted by permission of the authors and publisher.

p. 526

Sloman, L. and Webster, C., "Assessing the Parents of the Learning Disabled Child: A Semi Structured Interview Procedure," Journal of Learning Disabilities, vol. 11, No. 2, Feb. 1978. Adapted with permission of Professional Press. Copyright © 1978, Professional Press, Inc.

p. 289

Smith, J. S., McCelland, C. L., Reid, F. L., and Baer, G. M., "Dual Role of the Immune Response in Street Rabiesvirus Infection of Mice", Infection and Immunity, vol. 35, No. 1, Jan. 1982. Adapted with permission of the authors and The American Society for Microbiology.

p. 467

Snyder, C. E., and Clair, M., "Effects of Expected and Obtained Grades on Teaching Evaluations and Attributes of Performance", Journal of Educational Psychology, vol. 68, No. 1, 1976. Copyright © 1976 by the American Psychological Association. Adapted with the permission of the authors and publisher.

p. 445

Stack, S., "Suicide, A Comparative Study", Social Forces, vol. 57, No. 2, Dec. 1978. Adapted with permission. Copyright © 1978. The University of North Caroline Press.

p. 248

Stevens, S. J., Cooper, P. E., and Thomas, L. E., "Age Norms for Templer's Death Anxiety Scale", Psychological Report, 1980, vol. 46, 205–206, Table 1. Used with permission of authors and publisher.

p. 134

Stilson, D., Probability and Statistics in Psychological Research and Theory, Holden Day, 1966.

p. 53

Storch, G., McFarland, L. M., Kelso, K., Heilman, C. J., and Caraway, C. T., "Viral Hepatitis Associated with Day-care Centers", Journal of the American Medical Association, vol. 242, No. 14, Oct. 5, 1979. Adapted with permission of publisher and authors. Copyright © 1979, American Medical Association.

p. 448

Sundstrom, E., Lounsbury, J. W. Schuller, C. R., Fowler, J. R., and Mattingly, I. J., "Community Attitudes Toward a Proposed Nuclear Power Generating Facility as a Function of Expected Outcomes", Journal of Community Psychology, vol. 5, 1977. Used with permission.

p. 54, p. 258

Szklo, M., Tonascia, J., Goldberg, R., and Kennedy, H. L., "Additional Data Favoring Use of Anticoagulant Therapy in Myocardial Infarctions, A Population Based Study", Journal of the American Medical Association, vol. 242, No. 12, Sept. 21, 1979. Adapted with permission of authors and publisher. Copyright © 1979, American Medical Association.

p. 402

Trotman, F., "Race, IQ, and the Middle Class", Journal of Educational Psychology, vol. 69, No. 3, June 1977. Copyright © 1977, American Psychological Association. Adapted with permission of the author and publisher.

p. 4

UPI, "Sleep Disorders Can Prove to be Fatal" Los Angeles Times, March 30, 1980.

p. 295

Vallner, J. J., Needham, T. E., Jun, H. W., Brown, W. J., Stewart, J. T., Kotzan, J. A., and Honigberg, I. L., "Plasma Levels of Clobazam After Three Oral Dosage Forms in Healthy Subjects", Journal of Clinical Pharmacology, July 1978. Used with permission of the publisher.

p. 214

Waser, N. M., "Comparison of Distances Flown by Different Visitors to Flowers of the Same Species", Oecologia, vol. 55, 1982. Used with permission of the author and Springer-Verlag, Heidelberg.

p. 292

Wettenhall, N., "Tall Girls: 15 Years of Management and Treatment", Journal of Pediatrics, vol. 86, No. 4, April 1975, pp 602–610. Adapted by permission of the publisher.

p. 360

Witcher, J. C., "Thermal Springs of Arizona", Arizona Bureau of Geology and Mineral Technology, Field Notes, vol. 11, No. 2, pp 1–4, 1981. Used with permission.

p. 57

Zagier, Don, "The First 50 Million Prime Numbers", The Mathematical Intelligeneer, vol. 0, 1977. Adapted with permission of Springer-Verlag, NY City.

SELECTED REFERENCES

Blum, J. R. and Rosenblatt, J. I., Probability and Statistics, (W. B. Saunders Co., Philadelphia, Pennsylvania)

Bowen, E. K. and Starr, M. K., Basic Statistics for Business and Economics, (McGraw Hill Co., New York, N.Y.)

Daniels, W. D., Applied Non Parametric Statistics, (Houghton Mifflin Co., Boston, Mass.)

Dwass, Meyer, Probability and Statistics, (W. A. Benjamin, Inc., New York, N.Y.)

Freedman, D., Pisani R., and Purves, R., Statistics, (Norton, Inc., New York, N.Y.)

Gilbert, Norma, Statistics, 2nd ed., (Saunders Publishing Co., Philadelphia, Pa.)

Hays, W. L., Statistics for Psychologists, (Holt, Rhinehart & Winston, New York, N.Y.)

Hoel, Paul G., Elementary Statistics, 3rd ed., (Wiley and Sons, Inc., New York, N.Y.)

Huntsberger, D. V., and Billingsley, P., Elements of Statistical Inference, 5th ed., (Allyn and Bacon, Boston, Massachusetts)

Johnson, R., Elementary Statistics, 3rd ed., (Duxbury Press. North Scituate, Mass.)

Kachigan, S. K., Multivariate Statistical Analysis, (Radius Press, New York, N.Y.)

Kennedy, J. B. and Neville, A. M., Basic Statistical Methods of Engineers and Scientists, 2nd ed., (I.E.P., New York, N.Y.)

Mendenhall, W., Introduction to Probability and Statistics, 4th ed., (Duxbury Press, North Scituate, Mass.)

Mood, A. M. and Graybill, F. A., Introduction to the Theory of Statistics, (McGraw Hill, New York, N.Y.)

Ryan, T. A., Joiner, B. L., and Ryan, B. F., Minitab Student Handbook, (Duxbury Press, Boston, Mass.)

Sincich, T., Statistics by Example, (Dellen Publishing Co., San Francisco, Ca.) Cumulative Binomial Tables. Reprinted by permission.

Steel, R. G. D. and Torrie, J. H., Principles and Procedures of Statistics, (McGraw-Hill Book Co., New York, N.Y.)

Walpole, R. L., and Myers, R. H., Probability and Statistics for Engineers and Scientists, 2nd ed., (Macmillan Publishing Co. Inc., New York, N.Y.)

Whitney, D. R., Elements of Mathematical Statistics, (Holt and Company, New York, N.Y.)

Index

AREAS OF THE STANDARD NORMAL DISTRIBUTION

The entries in this table are the probabilities that a random variable having the standard normal distribution assumes a value between 0 and z; the probability is represented by the area under the curve shaded in the figure. Areas for negative values of z are obtained by symmetry.

z	Second decimal place in z									
	.00	.01	.02	.03	.04	.05	.06	.07	.08	.09
.0	.0000	.0040	.0080	.0120	.0160	.0199	.0239	.0279	.0319	.0359
.1	.0398	.0438	.0478	.0517	.0557	.0596	.0636	.0675	.0714	.0753
.2	.0793	.0832	.0871	.0910	.0948	.0987	.1026	.1064	.1103	.1141
.3	.1179	.1217	.1255	.1293	.1331	.1368	.1406	.1443	.1480	.1517
.4	.1554	.1591	.1628	.1664	.1700	.1736	.1772	.1808	.1844	.1879
.5	.1915	.1950	.1985	.2019	.2054	.2088	.2123	.2157	.2190	.2224
.6	.2257	.2291	.2324	.2357	.2389	.2422	.2454	.2486	.2517	.2549
.7	.2580	.2611	.2642	.2673	.2704	.2734	.2764	.2794	.2823	.2852
.8	.2881	.2910	.2939	.2967	.2995	.3023	.3051	.3078	.3106	.3133
.9	.3159	.3186	.3212	.3238	.3264	.3289	.3315	.3340	.3365	.3389
1.0	.3413	.3438	.3461	.3485	.3508	.3531	.3554	.3577	.3599	.3621
1.1	.3643	.3665	.3686	.3708	.3729	.3749	.3770	.3790	.3810	.3830
1.2	.3849	.3869	.3888	.3907	.3925	.3944	.3962	.3980	.3997	.4015
1.3	.4032	.4049	.4066	.4082	.4099	.4115	.4131	.4147	.4162	.4177
1.4	.4192	.4207	.4222	.4236	.4251	.4265	.4279	.4292	.4306	.4319
1.5	.4332	.4345	.4357	.4370	.4382	.4394	.4406	.4418	.4429	.4441
1.6	.4452	.4463	.4474	.4484	.4495	.4505	.4515	.4525	.4535	.4545
1.7	.4554	.4564	.4573	.4582	.4591	.4599	.4608	.4616	.4625	.4633
1.8	.4641	.4649	.4656	.4664	.4671	.4678	.4686	.4693	.4699	.4706
1.9	.4713	.4719	.4726	.4732	.4738	.4744	.4750	.4756	.4761	.4767
2.0	.4772	.4778	.4783	.4788	.4793	.4798	.4803	.4808	.4812	.4817
2.1	.4821	.4826	.4830	.4834	.4838	.4842	.4846	.4850	.4854	.4857
2.2	.4861	.4864	.4868	.4871	.4875	.4878	.4881	.4884	.4887	.4890
2.3	.4893	.4896	.4898	.4901	.4904	.4906	.4909	.4911	.4913	.4916
2.4	.4918	.4920	.4922	.4925	.4927	.4929	.4931	.4932	.4934	.4936
2.5	.4938	.4940	.4941	.4943	.4945	.4946	.4948	.4949	.4951	.4952
2.6	.4953	.4955	.4956	.4957	.4959	.4960	.4961	.4962	.4963	.4964
2.7	.4965	.4966	.4967	.4968	.4969	.4970	.4971	.4972	.4973	.4974
2.8	.4974	.4975	.4976	.4977	.4977	.4978	.4979	.4979	.4980	.4981
2.9	.4981	.4982	.4982	.4983	.4984	.4984	.4985	.4985	.4986	.4986
3.0	.4987	.4987	.4987	.4988	.4988	.4989	.4989	.4989	.4990	.4990
3.1	.4990	.4991	.4991	.4991	.4992	.4992	.4992	.4992	.4993	.4993
3.2	.4993	.4993	.4994	.4994	.4994	.4994	.4994	.4995	.4995	.4995
3.3	.4995	.4995	.4995	.4996	.4996	.4996	.4996	.4996	.4996	.4997
3.4	.4997	.4997	.4997	.4997	.4997	.4997	.4997	.4997	.4997	.4998
3.5	.4998									
4.0	.49997									
4.5	.499997									
5.0	.4999997									